Next-Generation Electronic Textiles and Conductive Materials for Smart Wearables

Pranshu Saxena
https://orcid.org/0000-0002-0822-1532
Bennett University, India

Mandeep Singh
https://orcid.org/0000-0003-3278-4877
Bennett University, India

Sanjay Kumar Singh
University School of Automation and Robotics, Guru Gobind Singh Indraprastha University, East Delhi, India

Mamoon Rashid
https://orcid.org/0000-0002-8302-4571
Bahrain Polytechnic, Bahrain

IGI Global
Scientific Publishing
Publishing Tomorrow's Research Today

Vice President of Editorial | Melissa Wagner
Director of Acquisitions | Mikaela Felty
Director of Book Development | Jocelynn Hessler
Production Manager | Mike Brehm
Cover Design | Jose Rosado

Published in the United States of America by
IGI Global Scientific Publishing
701 East Chocolate Avenue
Hershey, PA, 17033, USA
Tel: 717-533-8845 | Fax: 717-533-7115
Website: https://www.igi-global.com E-mail: cust@igi-global.com

Library of Congress Cataloging-in-Publication Data

Names: Saxena, Pranshu, 1987- editor | Singh, Mandeep, 1989- editor | Singh, Sanjay, 1986- editor | Rashid, Mamoon editor | Kumar, Prabhat (Software engineer) contributor
Title: Next-generation electronic textiles and conductive materials for smart wearables / edited by: Pranshu Saxena, Mandeep Singh, Sanjay Kumar Singh, Mamoon Rashid, Prabhat Kumar.
Description: Hershey, PA : IGI Global Scientific Publishing, [2026] | Includes bibliographical references. | Summary: "This book aims to provide a comprehensive, interdisciplinary exploration of the latest advancements in electronic textiles and conductive materials, with a focus on their integration into smart wearable systems"-- Provided by publisher.
Identifiers: LCCN 2025028294 (print) | LCCN 2025028295 (ebook) | ISBN 9798337342870 hardcover | ISBN 9798337342887 paperback | ISBN 9798337342894 ebook
Subjects: LCSH: Electronic textiles--Materials | Wearable technology--Materials | Smart materials
Classification: LCC TS1770.E44 N498 2026 (print) | LCC TS1770.E44 (ebook)
LC record available at https://lccn.loc.gov/2025028294
LC ebook record available at https://lccn.loc.gov/2025028295

British Cataloguing in Publication Data
A Cataloguing in Publication record for this book is available from the British Library.

Table of Contents

Detailed Table of Contents

This chapter explores the exciting area of smart fabrics and wearable technology, where regular clothing is combined with electronic part. As electronics become smaller and more flexible, their integration into textiles opens up new possibilities in health monitoring, fitness tracking, communication, and even fashion. Smart fabrics create functional clothing that can sense and respond to changes in the environment or the wearer's body. Wearable technology is rapidly growing due to the increasing demand for real-time data and personalized user experiences. This chapter explores the basic concepts behind smart fabrics, including the materials used, the methods of embedding electronic components, and the types of applications that benefit from these innovations. It also highlights the role of conductive materials in enabling seamless integration without compromising the comfort and flexibility of the fabric. The chapter aims to make the reader familiar with the core ideas and spark interest in advanced developments in this growing domain.

Smart fabrics and wearable technology represent of form of textile engineering data science and electronics. Such innovation have led creation of smart fabric that can react communicate and sense, transformed many sectors like healthcare, fashion, defense, sports and many more. This chapter explodes the foundation and diverse application of smart and wearable textile into active, passive and ultra-smart variants of fabrication methods weaving to printing and 3D knitting (Stoppa & Chiolerio, 2014;

The integration of nanomaterials and conductive polymers is transforming electronic textiles (e-textiles) by combining textile attributes like softness and comfort with advanced functionalities such as physiological monitoring, interactive interfaces and energy storage. Nanomaterials like carbon nanotubes (CNTs) and graphene enhance electrical conductivity and mechanical robustness, allowing for the development of flexible textile sensors, batteries, and heating textiles. Conductive polymers, such as polyaniline and polythiophene, offer durability and breathability without the oxidation issues of metals, enabling specific applications like physiological monitoring. However, challenges remain, including stability under mechanical stress, long-term functionality and ecological concerns related to nanomaterials. Addressing cost-effectiveness and sustainability through recyclable designs is also essential. Ultimately, the goal is to create smart fabrics that seamlessly integrate advanced electronic capabilities while enhancing health monitoring and communication functions.

Imagine wearing the clothes that don't just provide you comfort but also keep a supervising eye on your health and the environment around you. Smart textiles fabric embedded with sensors is making it all a reality. The latest-gen apparel can monitor vital parameters, sense changes in the environment, and collect real-time data without administering any invasive therapies. This chapter explores how materials science, flexible electronics, and biomedical innovation come together to create these smart fabrics. We will walk you through some of the different types of sensors from those that monitor heart rate and levels of hydration to those that detect pollutants and UV radiation. We also examine how sensors are incorporated into clothing using methods like conductive yarns and printed electronics. with ongoing research and breakneck advancements, smart textiles are about to transform the way we care for our wellbeing and our environment stylishly and comfortably.

Smart wearables and e-textiles are transforming personal technology by combining materials science, embedded computing, and wireless communication. At their core, embedded systems and microcontrollers (MCUs) enable sensing, data processing, connectivity, and user interaction within strict limits of size, power, and comfort. This work reviews the fundamentals of embedded design, including processors, memory, sensors, actuators, and architectures, while addressing challenges such as miniaturization, textile integration, durability, and data security. Popular MCU platforms are analyzed for energy efficiency, processing power, and connectivity. Power management strategies, including low-power modes and energy harvesting, are emphasized for extending battery life. The chapter also highlights emerging directions such as on-device machine learning, hybrid architectures, and flexible electronics. By integrating research and practical insights, it underscores the central role of embedded systems in enabling next-generation, intelligent, and user-centric wearable technologies.

Smart fabrics integrate sensors, compute, and radios into deformable, washable textiles. This chapter is a practical guide to end-to-end data links in e-textiles, covering wired buses (I^2C/SPI/UART/CAN-FD), short-range wireless (BLE 5.x, IEEE 802.15.4/Thread), body-centric options (UWB for ranging, capacitive human-body links), and backhaul (LoRaWAN, NB-IoT/LTE-M). We map core requirements—range, latency/jitter, reliability, energy per bit, security/privacy, comfort, and washability—to protocol and topology choices. RF topics include textile/embroidered antennas, fabric transmission lines, impedance control, and

SAR. A compact link-budget method models on-body loss and duty cycling for early feasibility and battery-life estimates. We define a reproducible test regimen (throughput, latency, packet-error rate, energy profiling, and S-parameter stability under bend/stretch/laundering). Four garment-level case studies illustrate sensor-to-cloud integration trade-offs. The chapter concludes with checklists for resilient interconnects, secure OTA updates, and on-garment TinyML.

Monika Kumari, Poornima University, Jaipur, India
Rajat Kumawat, Vivekananda Global University, Jaipur, India & Sunstone Education Technology Pvt. Ltd, India
Ritam Dutta, Poornima Institute of Engineering and Technology, Jaipur, India
Sheena Tahira Khan, Poornima University, Jaipur, India
Nikhil Kumar Goyal, Poornima University, Jaipur, India
Mohammed Firdos Alam Sheikh, Poornima University, Jaipur, India

This chapter explores the convergence of wireless communication technologies and the Internet of Things (IoT) in the development of smart clothing—intelligent garments embedded with sensors, actuators, and microelectronics for data sensing, processing, and transmission. It presents a comprehensive technological framework, covering wearable electronics, body area networks, and key communication protocols such as Bluetooth Low Energy (BLE), Wi-Fi, Zigbee, and LoRa. The chapter also examines the design and integration of textile-based sensors, flexible antennas, and energy-efficient microcontrollers, addressing critical challenges in comfort, durability, and washability. Reliability, latency, and data security issues are analyzed in the context of applications in healthcare, sports, industry, and defense. Current research trends including energy harvesting, AI-based data processing, and 6G-ready textiles are discussed alongside real-world case studies.

Pawan Kumar, Ajay Kumar Garg Engineering College, Ghaziabad, India
Mandeep Singh, Bennett University, Greater Noida, India
Juli Yadav, Ajay Kumar Garg Engineering College, Ghaziabad, India
Shalbha Chaudhary, R.D. Engineering College, Ghaziabad, India
Pranshu Saxena, Bennett University, Greater Noida, India

Due to the black-box nature of AImodels, they face significant critical challenges, particularly in safety domains such as healthcare, banking, and autonomous

The study presents a novel concept for Smart Fabrication and Autonomous Manufacturing Work-flows based on the convergence of Generative AI models and the Digital Twin paradigm. The aim is the prediction and optimization of the most significant machining parameters such as surface roughness, tool life, and energy consumption using advanced AI models and simulation of real-time machining with a responsive digital twin. 50 work-pieces' tool wear, surface finish, and energy consumption data were obtained. Four generative AI models VAE, GAN, Reinforcement Learning (RL), and CAM Code Transformer were trained and evaluated with respect to the RMSE, MAE, R^2, and MAPE criteria. The RL model exhibited the highest accuracy and served as the digital twin input engine. The twin simulated machining operations dynamically with rein-forcement learning and allowed for proactive adjustment for performance enhancement. Compara-tive analysis revealed high correlation of predicted and measured values. Optimization using AI and digital twin improved tool life and efficiency

Preface

INTRODUCTION

The rapid convergence of materials science, electronics, and wearable technology has catalyzed the emergence of electronic textiles (e-textiles) and conductive materials as foundational elements in the next generation of smart wearables. These innovative fabrics extend far beyond traditional clothing, integrating sensing, communication, and energy management capabilities that transform how we engage with the digital world. As demand continues to grow for flexible, durable, and comfortable wearable systems capable of monitoring physiological signals, supporting remote healthcare, and enhancing human–machine interaction, e-textiles are poised to redefine both consumer applications and industrial solutions.

This edited volume brings together leading interdisciplinary research at the intersection of smart materials, nanoengineering, textile manufacturing, and biomedical integration. It highlights the latest advances in conductive polymers, graphene-infused fabrics, flexible printed electronics, and fiber-based sensors—while also addressing the critical challenges that accompany these innovations, including washability, signal fidelity, power efficiency, and data security.

The overarching objective of this book is to provide a comprehensive, cross-disciplinary exploration of the latest developments in electronic textiles and conductive materials, emphasizing their integration into intelligent wearable systems. By uniting cutting-edge research, real-world applications, and forward-looking perspectives, the book seeks to bridge the gap between material innovation and functional wearable technologies. It aims to serve as a valuable resource for researchers, engineers, and practitioners by addressing current challenges, showcasing scalable fabrication methods, and identifying emerging trends in healthcare, human performance monitoring, and ambient intelligence. Ultimately, this volume aspires to advance the field by inspiring novel research directions and fostering cross-sector collaboration.

This book is intended primarily for researchers, engineers, and graduate students working in electronic textiles, wearable technology, materials science, biomedical engineering, and flexible electronics. It will also benefit professionals in healthcare, sports technology, defense, and fashion-tech industries who are engaged in the design, development, and deployment of smart wearable systems. Product developers, innovators, and academic instructors seeking in-depth and up-to-date insights into next-generation conductive materials and their practical applications will likewise find this collection highly valuable.

By offering both foundational knowledge and emerging research perspectives, *Next-Generation Electronic Textiles and Conductive Materials for Smart Wearables* aims to provide a timely, authoritative reference that captures the momentum and promise of this transformative field. We hope this volume not only informs but also inspires continued exploration and innovation at the frontiers of wearable technology.

ORGANIZATION OF THE BOOK

Chapter 1: Introduction to Smart Fabrics and Wearable Technology

The opening chapter introduces readers to the foundational concepts of smart fabrics and wearable technology, where traditional textiles merge with miniaturized electronic components to form intelligent, responsive materials. The editors highlight how advancements in flexible electronics have enabled garments capable of sensing physiological and environmental parameters while maintaining comfort and wearability. The chapter offers a grounding in essential materials, fabrication techniques, and functional applications—from healthcare monitoring to communication and fashion—emphasizing the central role of conductive materials in enabling seamless, flexible integration. Serving as both an introduction and an inspiration, this chapter lays the conceptual groundwork for the interdisciplinary themes developed throughout the volume.

Chapter 2: Exploring Smart Textiles and Wearable Tech: An Introductory Overview

Expanding on the introduction, this chapter delves into the multidisciplinary fusion of textile engineering, data science, and electronics that underpins smart textile development. The editors highlight how innovations in weaving, printing, and 3D knitting have produced active, passive, and ultra-smart textiles capable of sensing, reacting, and communicating with users. Addressing durability, comfort, power management, and ethical design, the chapter underscores the social and environmental implications of wearable technology. Looking toward the future, it explores AI integration, personalization, and sustainable materials as defining directions for next-generation smart wearables. This comprehensive overview establishes the intellectual foundation for understanding both the functional and expressive potential of smart textiles.

Chapter 3: Foundations and Emerging Trends in Electronic Textiles: Bridging Conductive Materials with Smart Wearable Applications

This chapter provides a technical and strategic bridge between material science and practical applications in e-textiles. It classifies key conductive materials—including metallic fibers, conductive polymers, and carbon-based nanomaterials—and compares their integration through weaving, knitting, printing, and coating techniques. With careful attention to mechanical robustness, washability, and signal fidelity, the authors review performance standards and field deployments across healthcare, sports, and defense. The editors emphasize enduring challenges such as durability, power autonomy, and data privacy, while presenting forward-looking strategies in self-healing materials, hybrid energy systems, and intelligent textile networks. The chapter serves as a roadmap for sustainable, scalable, and user-centered wearable design.

Chapter 4: Nanomaterials and Conductive Polymers in E-Textiles: Toward Smart and Sustainable Wearables

Focusing on the transformative role of nanomaterials and conductive polymers, this chapter explores how advances in carbon nanotubes, graphene, and conductive polymers have redefined e-textile performance. The editors emphasize the importance of achieving balance between conductivity, flexibility, and user comfort, enabling fabrics to monitor vital signs, store energy, or even provide localized heating. Discussions of environmental impact and recyclability position sustainability as a guiding principle for the next generation of textile electronics. By integrating scientific depth with real-world context, this chapter highlights the material innovations driving smart and eco-conscious wearable systems.

Chapter 5: Smart Sensors Embedded in Textiles: Health and Environmental Monitoring

This chapter vividly illustrates how textiles can evolve into intelligent systems capable of monitoring both the human body and the surrounding environment. It explores sensor technologies that detect heart rate, hydration levels, air quality, and UV exposure—demonstrating how these data streams contribute to personalized healthcare and ecological awareness. Through examples of conductive yarns, printed sensors, and flexible circuits, the authors show how engineering ingenuity meets daily usability. The editors emphasize how these innovations promise to revolutionize preventive healthcare and environmental stewardship through stylish, comfortable, and non-invasive design.

Chapter 6: Embedded Systems and Microcontrollers in Wearables

At the heart of every smart wearable lies an embedded system—a microcontroller coordinating sensing, data processing, and communication. This chapter reviews the architecture and operation of embedded platforms, discussing processor selection, memory management, and connectivity optimization. The editors highlight energy efficiency and miniaturization as ongoing challenges and explore innovations in energy harvesting, low-power design, and on-device AI. This comprehensive overview reveals how embedded systems form the "nervous system" of modern wearables, translating raw data into actionable intelligence while maintaining user comfort and autonomy.

Chapter 7: Data Transmission and Wireless Communication in Smart Fabrics

The seventh chapter provides a practical exploration of data communication within deformable, washable textiles. It examines both wired and wireless transmission methods—from I^2C and SPI to BLE, UWB, and LoRaWAN—mapping key performance parameters such as range, latency, and energy efficiency. Textile antenna design, signal integrity, and electromagnetic safety are analyzed alongside real-world case studies. The editors highlight reproducible testing and reliability benchmarks that ensure robust communication in flexible garments. This chapter serves as a guide to building secure, efficient, and responsive networks within fabric-based systems.

Chapter 8: Wireless Communication and IoT Integration in Smart Clothing

This chapter extends the discussion of connectivity into the realm of the Internet of Things (IoT), presenting a holistic view of smart clothing as part of an interconnected ecosystem. Covering BLE, Wi-Fi, Zigbee, and LoRa protocols, it discusses how textile-integrated antennas and sensors support seamless data flow in healthcare, sports, and defense applications. Energy-efficient microcontrollers and flexible architectures are emphasized as enablers of comfort and durability. The editors draw attention to pressing concerns in reliability, latency, and data security while exploring future trajectories such as AI-driven analytics, energy harvesting, and 6G-ready smart textiles.

Chapter 9: Explainable Artificial Intelligence for Trustworthy Decision-Making in Smart Wearable Textiles

As AI increasingly powers wearable technologies, the need for transparency and accountability becomes paramount. This chapter introduces Explainable Artificial Intelligence (XAI) as a means to ensure fairness, interpretability, and trust in data-driven decision-making. Focusing on healthcare wearables, it explores how XAI methods enhance diagnostic reliability and patient confidence by translating algorithmic outcomes into comprehensible insights. The editors underscore XAI's role in enabling safer, more ethical applications of AI in real-time medical monitoring and decision support.

Chapter 10: Uncertainty-Resilient Signal Processing in Smart Textiles Using Fuzzy Logic

Recognizing the inherent variability of textile-based sensors, this chapter proposes fuzzy logic as a robust framework for managing uncertainty in wearable signal processing. The discussion spans fuzzy sets, soft computing, and fuzzy inference systems applied to noise reduction and adaptive classification. The editors highlight how fuzzy systems enhance the resilience of wearable data interpretation, ensuring accuracy despite movement, humidity, or environmental noise. The chapter concludes with future perspectives on merging fuzzy logic with machine learning and quantum computing to achieve next-generation wearable intelligence.

Chapter 11: Review of Smart Wearables Sensor-Based Techniques and Machine Learning for Change of Direction Detection in Sports

Focusing on athletic performance, this chapter reviews machine learning techniques for detecting changes of direction (COD) in sports through wearable sensors. The editors emphasize how such detection can inform injury prevention, training optimization, and performance analytics. Despite significant progress, challenges persist in sensor standardization and dataset availability. The chapter concludes by proposing an AI-based framework for automated COD detection, offering a pathway toward data-driven sports performance monitoring and assessment.

Chapter 12: Deep Learning-Based Defect Detection and Quality Assurance in Advanced 3D Printing and Microfabrication Processes

This chapter bridges additive manufacturing and AI, showcasing how deep learning architectures such as YOLOv8, ResNet-50, and U-Net enhance defect detection in 3D printing. The editors highlight YOLOv8's superiority for real-time monitoring and fault mitigation, emphasizing its implications for manufacturing reliability and efficiency. By integrating image analysis with intelligent control, this work illustrates how smart fabrication processes mirror the adaptivity and precision of intelligent textiles.

Chapter 13: Integrating Generative AI and Digital Twin Technologies for Smart Fabrication and Autonomous Manufacturing Workflows

Here, the focus shifts to the fusion of Generative AI and Digital Twin technologies for autonomous manufacturing. The chapter presents an innovative framework for predicting machining parameters such as tool wear, surface roughness, and energy use through reinforcement learning and real-time simulation. The editors highlight the power of AI-driven optimization in creating closed-loop systems capable of self-adjustment, reflecting the same adaptive intelligence that defines next-generation e-textiles.

Chapter 14: AI-Driven Adoption and Management Strategies for Advanced Smart Fabrication in Global Manufacturing Enterprises

This chapter outlines a comprehensive, AI-driven management framework for industrial-scale smart fabrication. By combining deep learning, reinforcement learning, and sensor-based predictive analytics, the approach enables real-time fault detection, maintenance prediction, and process optimization in CNC systems. The editors emphasize the strategic importance of integrating AI models for efficiency, safety, and sustainability—principles that resonate across the broader smart materials ecosystem presented in this volume.

Chapter 15: Applications of E-Textiles in Healthcare, Sports, and Defense

Chapter 16 synthesizes the real-world impact of e-textiles across critical domains. The editors highlight how smart garments enable continuous, non-invasive health monitoring, athletic optimization, and enhanced situational awareness in defense applications. Through case studies, the chapter illustrates how comfort, adaptability, and intelligence converge in practical, life-improving technologies. It also underscores remaining challenges in power management and scalability, setting the stage for future research and commercialization.

Chapter 16: Sports and Fitness Applications of Wearable Technology

The concluding chapter explores how wearable technology is reshaping sports science and personal fitness. From data-driven coaching to performance tracking and injury prevention, wearables now serve as integral tools for athletes at all levels. The editors discuss advances in sensor design, analytics, and augmented reality that are redefining the user experience. Concluding the volume, this chapter reinforces the overarching theme of human-centered innovation—illustrating how intelligent textiles and wearable systems are ushering in a new era of personalized, connected, and adaptive living.

CONCLUSION

As we conclude this volume, *Next-Generation Electronic Textiles and Conductive Materials for Smart Wearables*, we reflect on the remarkable convergence of disciplines that has made this field one of the most dynamic frontiers in science and engineering today. The chapters within this book collectively illuminate a landscape where materials science, electronics, data analytics, and human-centered design intersect to create textiles that are not merely worn, but lived with—interfaces that sense, think, and respond to the world and the body in real time.

From the foundations of conductive fibers and nanomaterials to advances in artificial intelligence, embedded systems, and sustainable fabrication, this collection presents an integrated vision of how wearable technologies are evolving toward greater functionality, adaptability, and intelligence. What emerges is a compelling narrative of transformation: fabrics becoming networks; garments becoming healthcare systems; and the human body becoming an active node in a digitally connected ecosystem.

Equally significant are the challenges that lie ahead. Durability, washability, energy management, data security, and ethical design must continue to shape our collective research agenda. The editors believe that addressing these challenges requires not only technical innovation but also interdisciplinary collaboration—among engineers, materials scientists, designers, clinicians, and policymakers alike. The success of next-generation smart wearables will depend as much on human insight and responsibility as on technological sophistication.

Looking forward, the promise of e-textiles and conductive materials extends well beyond the laboratory or marketplace. These technologies have the potential to redefine our relationship with clothing, the environment, and ourselves—enabling a more responsive, sustainable, and health-conscious society. As editors, we hope this book serves as both a reference and a catalyst: a foundation for continued discovery, innovation, and dialogue across disciplines.

We extend our deepest gratitude to the contributing authors, whose expertise and vision have made this work possible, and to the readers, who will carry these ideas into future explorations. The journey of smart wearables is far from complete; rather, it is only beginning. The next generation of electronic textiles will not only connect devices—but connect people, knowledge, and purpose in ways that continue to evolve the very fabric of modern life.

Chapter 1

Introduction to Smart Fabrics and Wearable Technology

Isha Gupta
https://orcid.org/0009-0000-3924-7818
Vivekananda Institute of Professional Studies - Technical Campus, India

Deepika Rawat
https://orcid.org/0009-0002-4852-6610
Raj Kumar Goel Institute of Technology, India

Chaitali Bhowmik
https://orcid.org/0000-0001-9591-9818
Vivekananda Institute of Professional Studies - Technical Campus, India

Monika Bansal
https://orcid.org/0000-0003-0178-4727
Vivekananda Institute of Professional Studies - Technical Campus, India

Nishi Jain
https://orcid.org/0000-0002-0311-3866
Vivekananda Institute of Professional Studies - Technical Campus, India

ABSTRACT

This chapter explores the exciting area of smart fabrics and wearable technology, where regular clothing is combined with electronic part. As electronics become smaller and more flexible, their integration into textiles opens up new possibilities in health monitoring, fitness tracking, communication, and even fashion. Smart fabrics create functional clothing that can sense and respond to changes in the environment

DOI: 10.4018/979-8-3373-4287-0.ch001

or the wearer's body. Wearable technology is rapidly growing due to the increasing demand for real-time data and personalized user experiences. This chapter explores the basic concepts behind smart fabrics, including the materials used, the methods of embedding electronic components, and the types of applications that benefit from these innovations. It also highlights the role of conductive materials in enabling seamless integration without compromising the comfort and flexibility of the fabric. The chapter aims to make the reader familiar with the core ideas and spark interest in advanced developments in this growing domain.

1. INTRODUCTION

This chapter's unique contribution lies in its integrative lens: rather than isolating materials, electronics, or applications, it connects technical, ergonomic, and deployment realities into a coherent roadmap for smart fabrics and wearables. We foreground persistent gaps—durability under washing and strain, power management under real usage, and long-term comfort and safety, and translate them into design principles that bridge lab prototypes and field-ready systems.

Smart fabrics, or electronic textiles (e-textiles), are textile substrates that embed digital functionality, sensing, actuation, processing, and communication, while retaining softness, flexibility, and drape (Stoppa & Chiolerio, 2014). Functionally, they span a progression from passive (sense-only) to active (sense-and-respond) to ultra-smart systems that incorporate on-fabric intelligence and can adapt or learn in context (Xiang et al., 2024). This capability stack depends on how electronics are integrated into cloth: knitting, printing, coating, and weaving enable conductive pathways and device attachment without sacrificing wearability. Conductive materials such as carbon nanotubes and graphene, conductive polymers like PEDOT:PSS, and metalized yarns (e.g., silver-coated fibers) provide electrical performance while conforming to the body and surviving cyclic deformation.

Wearable systems built on these fabrics typically comprise sensors (physiological and biomechanical), actuators (thermal, haptic, visual), low-power microcontrollers for on-garment processing, and short-range communication (e.g., Bluetooth, NFC) for phone- or hub-tethered services, alongside power sources that range from micro-batteries to energy-harvesting units (Ali & Islam, 2024). Real-world viability, however, hinges on end-to-end engineering: matching link budgets and latency to workload demands; packaging interconnects and antennas to survive bending and laundering; allocating computation between edge and cloud for privacy and responsiveness; and designing power budgets that sustain multi-day use without bulky form factors. By synthesizing advances in materials, circuits, wireless links, and system software with human-factors constraints, this chapter outlines practical pathways to robust, comfortable, and secure smart-fabric deployments at scale.

Figure 1. Overview of Smart Fabrics and Wearable Systems, including fabric classifications, technological components, and key applications.

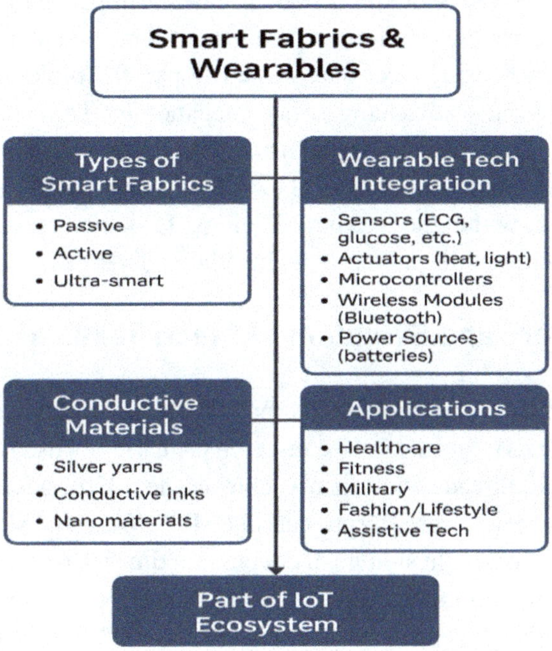

1.1 Importance and Relevance in Today's World

The demand for connected, individualized, and health-centric technology has amplified the significance of smart fabrics in today's world. Their relevance spans multiple industries. In healthcare, wearable technology enables remote and continuous monitoring of vital signs such as respiration, glucose levels, and electrocardiograms (Le et al., 2022)This improves medical responsiveness and patient care outcomes.

In the fitness industry, smart apparel assists in tracking physical performance and biomechanics, enabling users to optimize their training routines. Industrial safety also benefits from these innovations. Workers can be outfitted with garments that detect radiation, gas leaks, or signs of fatigue, thereby enhancing workplace safety procedures.

Military applications make use of smart fabrics that incorporate sensors and communication systems to monitor soldier health and provide real-time situational awareness, greatly enhancing decision-making and safety in the field (Hossain & Bradford, 2020). In the fashion and lifestyle sector, interactive clothing that responds to touch or light integrates technology with self-expression, pushing the boundaries of modern apparel.

Sustainability also plays a critical role in the development of smart fabrics. The use of recyclable materials helps reduce environmental impact. Moreover, integrating electronic capabilities into clothing reduces the reliance on standalone electronic devices, thereby decreasing electronic waste (Stoppa & Chiolerio, 2014).

Smart fabrics are also advancing assistive technology. They enable features such as GPS tracking and fall detection, which are particularly useful for older adults and individuals with disabilities (Xiang et al., 2024). In addition, these textiles function as connected nodes within the Internet of Things (IoT), facilitating integration with digital systems across cities, homes, and healthcare facilities.

1.2 Brief History and Evolution of Smart Textiles

The development of smart textiles has paralleled the evolution of computing and flexible electronics. Before the 1990s, research primarily focused on basic elements such as conductive threads and sensors. During the 1990s, advances in wearable medical sensing were pioneered by projects like MIT's Wearable Motherboard, which was designed to monitor physiological data in real time (Park & Jayaraman, 2003).

In the 2000s, the miniaturization of electronics led to the emergence of commercial smart clothing products, such as heated jackets, bringing the technology to a wider market. The 2010s witnessed a proliferation of stretchable electronics and consumer fitness trackers. Projects like Google Jacquard further demonstrated the viability of fusing fashion with functional electronics.

As we move into the 2020s, smart wearables enhanced with artificial intelligence are beginning to provide adaptive, real-time feedback and predictive insights based on user data (Yin & Reddy, 2023). These developments mark a shift toward truly personalized and interactive garments. Current research continues to push the boundaries of what smart textiles can achieve. Key challenges being addressed include improving washability, enhancing durability, and ensuring power efficiency. The goal is to make these technologies more robust and suitable for everyday use.

2. FUNDAMENTALS OF SMART FABRICS

2.1 Structure of a smart fabric system

Smart fabrics, also known as electronic textiles or e-textiles, represent a convergence of traditional textile technology with modern electronics. These fabrics are designed to sense, process, and transmit data while retaining the comfort and flexibility of ordinary clothing. The integration of sensors, microelectronics, and wireless communication within the fabric enables real-time monitoring of physiological and environmental parameters. Fig 2.1 illustrate the layered structure of a smart fabric system and explain how embedded electronics interact with the fabric to collect and transmit data.

A) **Fabric Layer:** The fabric layer serves as the foundation of the smart fabric system. It is the base textile material into which electronic components are integrated. The fabric may be woven, knitted, or coated with conductive fibers, which act as electrical pathways for the transmission of signals. Importantly, this layer is designed to preserve the comfort, flexibility, and wearability of traditional textiles while simultaneously functioning as a substrate for sensors and electronic circuits.

Figure 2. Structure of a smart fabric system and the interaction of embedded electronics with the fabric

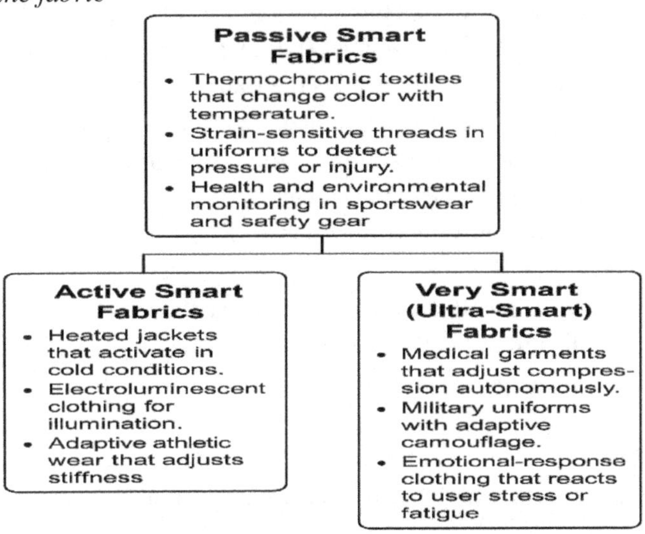

B) **Sensors:** At the core of the system are miniaturized sensors, which are either embedded within or attached to the fabric. These sensors continuously monitor various parameters such as temperature, pressure, motion, heart rate (ECG), and humidity. For example, a temperature sensor integrated into a smart shirt can detect changes in body heat, enabling real-time health monitoring. The ability to embed multiple sensor types allows smart fabrics to serve diverse applications in healthcare, sports, military, and environmental monitoring.

C) **Microprocessor (Embedded Electronics):** The data collected by sensors is directed to a microprocessor or microcontroller, which is typically mounted on a small, flexible circuit board integrated into the fabric. This processing unit plays a vital role in handling and analyzing raw sensor signals.

Its key functions include:

- **Data conversion**: Transforming analog signals into digital form.
- **Filtering and preprocessing**: Removing noise and preparing data for analysis.
- **Algorithm execution**: Running real-time programs, such as detecting irregular heartbeats or unusual motion patterns.

Thus, the microprocessor acts as the intelligence hub of the smart fabric system.

D) **Data Transmission Module:** Once processed, the information must be communicated to external devices for storage, visualization, or further analysis. This is achieved through wireless communication technologies such as Bluetooth, Wi-Fi, or Near Field Communication (NFC). In the system diagram (Figure 1), the wireless signal symbol represents the transmission of data from the fabric to an external device such as a computer, smartphone, or cloud server.

E) **Output and Analysis:** The final stage of the system involves output and analysis of the transmitted data. The data can be:

- Displayed in real-time on connected devices for immediate feedback to the user.
- Analyzed using advanced software or artificial intelligence (AI) for detecting health anomalies, performance tracking, or sending automated alerts in emergency conditions.

This capability enables smart fabrics to act not only as data collectors but also as decision-support tools in healthcare, sports science, defense, and everyday wellness.

2.2 Types of Smart Fabrics

Smart fabrics, or electronic textiles (e-textiles), integrate electronic components directly into fabrics, enhancing their functional range and interactivity. These advanced textiles fall into three main categories: passive, active, and very smart (or ultra-smart) fabrics.

Passive smart fabrics are capable of detecting environmental or physiological changes, but they do not have the ability to respond to those changes. A typical example includes thermochromic textiles that change colour in response to temperature variations, offering both aesthetic and functional benefits. Another example is adding strain-sensitive threads to uniforms to identify pressure areas or possible injuries. Furthermore, without requiring user engagement, these textiles are utilized in safety equipment and athletics to track environmental or health metrics.

Both sensors and actuators allow them to react dynamically in addition to sensing changes are a part of active smart textiles. An example of this kind of fabric is heated coats, when exposed to chilly temperatures they turn on themselves (Cheng et al., 2025). Electroluminescent clothing which are designed for illumination and visibility is another practical application. Some athletic wear even incorporates mechanisms that adapt their stiffness in response to movement or strain, improving support and performance.

Very smart or ultra-smart fabrics represent the most sophisticated class. These combine sensing, actuation, and onboard data processing capabilities, often with embedded microcontrollers or AI-based systems (Liang & Deng, 2025). Examples include medical garments that can automatically adjust compression based on real-time physiological feedback, military uniforms equipped with adaptive camouflage systems, and clothing that alters its behaviour in response to the emotional state of the wearer, such as detecting and reacting to stress or fatigue.

Figure 3. Classification of Smart Fabrics

Here is a comparative table summarizing the main types of smart fabrics, passive, active, and ultra-smart (very smart) fabrics, along with their sensing capabilities, power requirements, durability, and application domains for increased technical clarity.

Table 1. Technical Comparison of Smart Fabric Types and Their Functional Performance

Type of Smart Fabric	Sensing Capabilities	Power Requirements	Durability/ Washability	Application Domains
Passive	Detects environmental/ physiological changes (e.g., temperature, pressure)	None or minimal (no actuation)	High; mainly textile preservation needed	Sports, safety, basic medical monitoring
Active	Sensing with dynamic feedback (heat, light, mechanical response)	Moderate (requires power for actuation)	Moderate; encapsulation often required	Therapeutic garments, illumination for visibility, adaptive athletic wear
Ultra-Smart (Very Smart)	Sensing + data processing + actuation (AI, onboard CPU; adaptive, learning, prediction)	High (onboard processing, wireless communication)	Challenging; advanced washability, flexible encapsulation	Medical, military (adaptive camouflage, health monitoring), assistive technologies

2.3 Key Features of Smart Fabrics

The performance of smart fabrics relies on several key features that influence usability and functionality. Flexibility is essential, allowing textiles to move and stretch without breaking. This is especially important in applications like compression garments embedded with stretchable sensors that must conform to the human body during movement.

Sensitivity is another crucial characteristic, enabling fabrics to detect subtle physiological or environmental changes. An example is sweat-sensing textiles that monitor hydration levels, contributing to health monitoring applications. Conductivity must also be maintained under mechanical stress to ensure uninterrupted electrical performance. Silver-coated threads are often used in biosignal transmission systems due to their stable conductive properties.

Breathability and comfort are vital for long-term wear. Heat regulation and moisture management are supported through the use of porous inks on mesh textiles, making them more suitable for everyday or sports use. Durability is also essential, as smart textiles must endure repeated washing, exposure to sweat and UV light, and mechanical wear. Innovations in this area have led to e-textiles that continue functioning after more than fifty washing cycles (Hossain & Bradford, 2020).

Lightweight design contributes to wearability. Thin and flexible components, such as slim batteries or printed circuits, are often used in health-monitoring garments to ensure minimal impact on comfort. Lastly, user-friendly integration plays a major role. Wireless connectivity and intuitive operation make the user experience more seamless. For instance, garments that automatically sync with mobile apps enhance accessibility and real-time monitoring.

Figure 4. Key Functional Features of Smart Fabrics

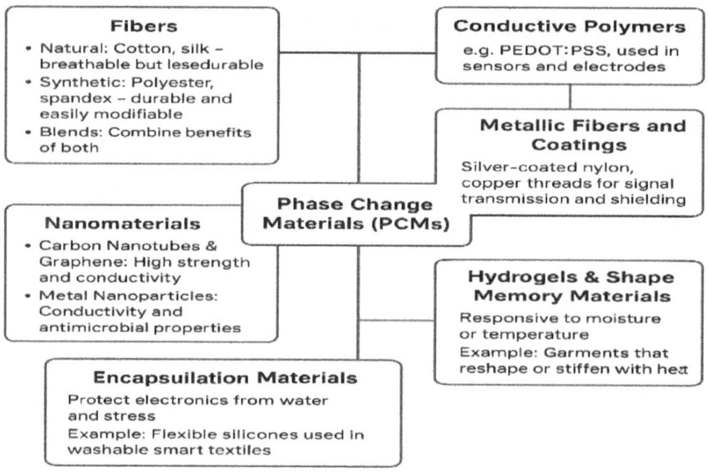

2.4 Materials Used in Smart Fabrics

The choice of materials is fundamental to the success of smart fabric applications. Natural fibers such as cotton and silk offer excellent breathability but generally lack the durability required for robust smart textile systems. Synthetic fibers like polyester and spandex, on the other hand, are more durable and offer better compatibility with conductive treatments. Blended fibers combine the strengths of both types, improving functionality and comfort (Xiang et al., 2024).

Conductive polymers, such as PEDOT:PSS, are commonly used for sensors and electrodes due to their flexibility and electrical performance. Metallic fibers and coatings play a significant role in signal transmission and shielding. Silver-coated nylon and copper threads are frequently employed for their high conductivity and integration potential with fabric substrates.

Smart textile design takes on new dimensions because of nanomaterials. High-performance applications benefit from the mechanical strength and electrical conductivity that carbon nanotubes (CNTs) and graphene provide (Keawploy et al., 2020).

In addition to enhancing conductivity by providing antibacterial qualities which is crucial for hygiene in wearable health monitoring devices by metal nanoparticles.

Phase change materials (PCMs) are used to control temperatures. These materials store and release heat, which is an essential component of outdoor and performance apparel, to allow clothing to maintain comfortable temperatures in a range of circumstances. Shape memory materials and hydrogels are affected by moisture and temperature changes. These materials are used in clothing that can stiffen or change shape based on the surroundings.

Sensitive electronic components must be protected from moisture, mechanical stress, and washing using encapsulation materials. Flexible silicones are frequently employed as encapsulants in order to retain fabric flexibility and guarantee long-lasting durability.

Figure 5. Smart Fabric Material Categories

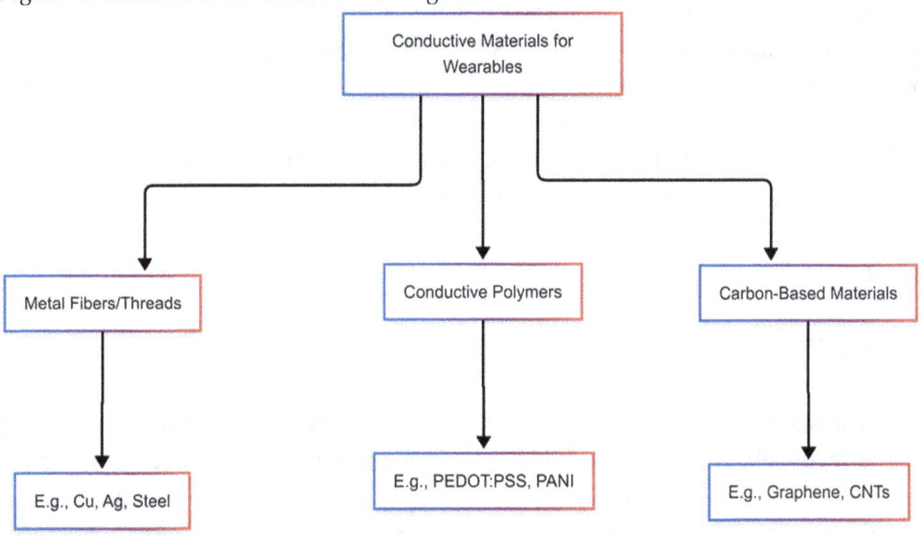

3. COMPONENTS AND TECHNOLOGIES

Smart textiles are changing the way we connect with clothing and environments made of fabric by incorporating technology that allows for sensing, acting, communicating, and managing energy. Previously passive materials can be transformed into interactive systems with real-time data gathering and response capabilities by integrating sensors, actuators, power sources, and communication modules.

3.1 Sensors in Smart Fabrics

Because they enable the monitoring of external factors as well as physiological signs, sensors are the cornerstone of smart textiles. By weaving, printing, or embedding these sensors into textiles, they can be seamlessly incorporated with little to no negative effect on user comfort (Le et al., 2022). Smart fabrics employ a variety of sensor kinds. Clothing can sense body or environmental temperature thanks to temperature sensors that work with thermistors, thermocouples, or capacitive changes. Fever-detection clothing and climate-responsive clothes benefit greatly from these features. Based on piezoresistive or capacitive systems, mechanical stress is measured by pressure and strain sensors. Applications such as smart insoles, posture-tracking chairs, and rehabilitative clothing depend on these sensors (Cheng et al., 2025).

For fall detection, physical rehabilitation, and gesture-based control systems motion sensors are very essential. They include accelerometers, gyroscopes, and inertial measurement units to track user movement. (IMUs). On the other hand, through the analysis of sweat composition, skin impedance, or other bio-signals, biosensors are used to monitor biological states like stress, glucose levels, and hydration. Applications related to health and fitness benefit greatly from these sensors (Xiang et al., 2024). Sensor systems in smart textiles must overcome a number of obstacles, such as guaranteeing washability, preserving low power consumption, offering materials that are compatible with skin, and obtaining steady signal performance when moving.

3.2 Actuators and Their Role

As actuators are integrated into textiles active features like warmth, lighting, tactile feedback, and chemical release are made possible. Smart textiles are given responsive element by transforming electrical energy into mechanical, thermal, optical, or chemical reactions by actuators. Thermal actuators are frequently used by therapeutic clothing and cold-weather equipment, which use conductive fibres to produce localized heating (Dolez, 2021). Shape-memory materials change shape in reaction to heat or moisture are especially helpful in compression clothing and dynamic orthotic devices.

To enhance their safety features and appeal lighting components like organic light-emitting diodes (OLEDs) or light-emitting diodes (LEDs) can be incorporated into textiles. Haptic actuators provide tactile feedback mechanisms that are helpful in wearable navigation interfaces and alert systems. Chemical actuators are also useful for personal care and medical applications since they allow the regulated release of medications or perfumes through the fabric. High energy needs, safe and effective heat regulation, attaining quick actuation response, and preserving the fabric's natural flexibility are some of the engineering issues associated with integrating actuators into textiles.

3.3 Power Sources

One of the biggest issues in the industry is powering smart materials while maintaining their wearability and comfort (Ali & Islam, 2024). Batteries, such as flexible and flat lithium-based batteries or textile-integrated energy storage devices, are commonly used in conventional power systems. Despite their effectiveness, these power sources need to be charged frequently and could make the garment bulkier (Hossain & Bradford, 2020). Researchers are looking into energy harvesting technology as a way to get around these restrictions. For example, body heat can be transformed into power using thermoelectric systems (Singh & Malik, 2024). In a similar vein, piezoelectric and triboelectric systems use pressure and body movement to produce energy (Dolez, 2021).

Another sustainable energy source for wearable electronics is provided by solar-powered textiles that incorporate photovoltaic cells. More creatively, methods for harvesting bioenergy have been created to turn body fluids like perspiration into useful electricity (Dolez, 2021). To increase operating time, several systems include energy harvesting and battery storage; these hybrid systems show promise for long-term use cases (Xiang et al., 2024). Minimizing component weight and size, guaranteeing safety for extended skin contact, sustaining steady power production, and accomplishing seamless textile integration are all crucial factors in the design of smart textile power systems.

3.4 Communication Modules

For smart textiles to function as interactive systems, they must be capable of wireless data transmission. Communication modules allow smart garments to interface with external devices and cloud services, enhancing their usability and connectivity. Several wireless technologies are commonly employed. Bluetooth Low Energy (BLE) is ideal for short-range communications, such as linking to smartphones, while consuming minimal power. Near-field communication (NFC) enables quick interactions for applications like authentication and access control, where a simple tap is sufficient.

Wi-Fi modules are used for more complex data exchanges, such as remote healthcare monitoring and industrial applications, providing direct cloud access. For broader communication capabilities, cellular technologies such as 4G and 5G support wide-area, untethered data transmission, making them ideal for real-time applications in healthcare and logistics. Other communication systems include mesh networks, LoRa, and Zigbee, which are particularly suited for Internet of Things (IoT) configurations in smart textile ecosystems. Each of these technologies requires careful consideration of trade-offs between power consumption, transmission range, latency, bandwidth, and data privacy.

4. CONDUCTIVE MATERIALS FOR WEARABLE INTEGRATION

The core function of smart textiles lies in their ability to sense, transmit, and sometimes store electrical signals while maintaining textile properties such as flexibility, comfort, and breathability. Conductive materials form the fundamental building blocks of this functionality, enabling the electrical pathways required for smart wearable operation. To impart conductivity, non-textile metallic filament wires made from silver, stainless steel, nickel, aluminum and copper can be inserted into the textile structure.

Fig 4.1. illustrate three principal categories of conductive materials utilized in wearable integration: Metallic fibers and threads, Conductive polymers and Carbon-based materials such as graphene and carbon nanotubes (CNTs). Their mechanisms, properties, integration techniques, and current research progress are explored to provide a foundation for future development in the field.

Figure 6. Principal Categories of Conductive Materials

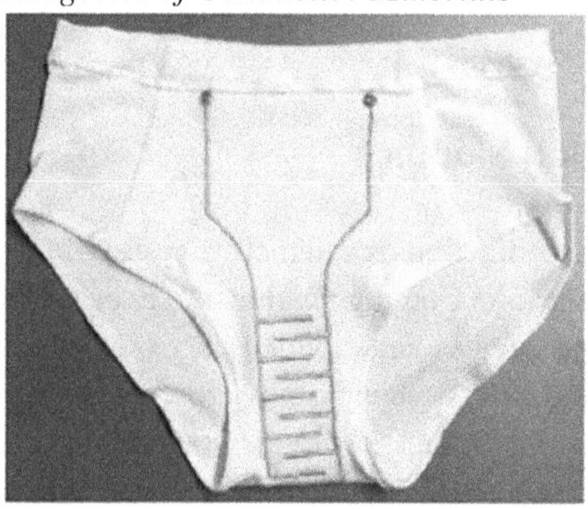

4.1 Metallic Fibers and Threads

Metallic conductors, notably silver, copper, stainless steel, and gold, offer high electrical conductivity and are the most established conductive materials in smart textiles. They are typically engineered into metallic microfibers, yarns, or coated threads, which are then woven, knitted, or embroidered into fabrics.

Silver-plated nylon and copper-core threads are widely used due to their balanced conductivity, cost, and ease of textile processing. For example, silver-coated threads embedded in compression garments enable real-time ECG monitoring and bio-impedance sensing.

4.1.2 Integration Techniques

Embroidery & Stitching: An embroidery, needles are used to apply conductive filaments or yarns onto textile fabrics and other materials.It gives the flexibility to design and embroider traces of required shapes or contours on a plane. Embroidering has been also used to develop a textile-based sensor and antenna.

Metal Coating: It is a process of adding a layer of metal components on the surface of textile materials. Before plating, the substrate must be cleaned to remove impurities, which offer assistance for effective attachment of the metal particles to the surface of the substrate.

Weaving/Knit Structures: Knitting is a continuous and efficient fabric manufacturing process. Apart from creating the textile as a substrate, knitting allows the inclusion of active elements and conductive yarns during the fabrication process, making them integral to the textile structure.

4.2 Conductive Polymers

Electrical connecting devices such as wires, antennas, Bluetooth chips, USB connections, LED communicators, and diode actuators are typically made of conductive polymers. Naturally conductive polymers (ICPs), such as polyaniline (PANI), polypyrrole (PPy), polytiophene (PT), and poly (perinapthalene) (Pna), have several applications. ICP was first used in science in 1977 because of its stronger tight-paced structure and better oxygen resistance, but its primary drawback is that it is difficult to process. In later years, the use of mixed solution-dipped ICP with colloidal dispersions that are sensitive to electric discharge, such as polyisoprene, polystyrene, or polybutadiene mixtures. Low defect content and consistent optical properties are characteristics of the stability of Ar-doped nanoparticles (SWNCT) of polyacetylene blends. The Swiss-Shield Company, located in Flums, Switzerland, specializes in producing it. CNTs can be spun using the dry spinning method in an acidic solution. A simple way for creating conductive polymer based on carbon nanotubes is vertical alignment. As seen in Table 2, intrinsically conductive materials, including metallic fibers derived from ferrous alloys, nickel, stainless steel, titanium, aluminum, and copper, offer excellent electrical properties for the development of wearable electronic devices. Conductive materials are superior for optical microscopy once these metals have dried.

Table 2. Electrical properties of various nanofilament conducting polymers

Metal type	Conductivity (S m/ mm^2)	Resistivity ($\Omega mm^2/m$)	Thermal coefficient resistance (10^{-6} K^{-1})		
			Max	Min	Type
Cu/Ag	58.5	0.0171	4300	3900	4100
Bronze	7.5	0.1333	700	600	650
Ag 996	62.5	0.0160	4100	3800	3950
AgCu	57.5	0.0174	4100	3800	3950
Ms/Ag	16.0	0.0625	1600	1400	1500
Steel 304	1.4	0.7300	-	-	1020
Ms*70	16.0	0.0625	1600	1400	1500
Steel 316L	1.3	0.7500	-	-	1020

4.2.1 Deposition and Integration Methods

Dip-Coating: In this technique, textile materials such as fiber, yarn or fabric are immersed for a certain duration of time in a bath contain conductive dispersion.

In-Situ Polymerization: Intrinsically conductive polymers are widely used in the development of electroconductive textiles. Traditional organic polymers are electrical insulators or semiconductors, so the discovery of conductive polymers in 1970s, opened a new opportunity to produce electroconductive textile.

Screen Printing: This process involves printing a viscous conductive paste into a screened stencil on a substrate, which is then cured with that depends on the conductive compound properties and the type of substrate used. Among all methods of printing, it is one of the most recommended because it simplifies fabrications.

4.2.2 Performance Considerations

Conductive polymers offer good mechanical flexibility, but suffer from lower conductivity and environmental degradation (moisture, UV). Research is ongoing into cross-linked or composite forms (e.g., PEDOT:PSS with polyurethane) to improve longevity in wearable applications .

4.3 Carbon-Based Materials

Graphene and CNTs represent a new generation of lightweight, high-performance conductive materials with exceptional electrical, thermal, and mechanical properties. Their nano-scale morphology allows for integration into fabrics without compromising softness or flexibility. Graphene offers high carrier mobility and transparency—ideal for sensors and touch interfaces.CNTs, especially multi-walled types, exhibit tunable conductivity and excellent strain sensitivity.

4.3.1 Integration with Smart Fabrics

Integration methods include:

Spray Coating: A spray of conductive particles or droplets are deposited to a textile substrate. A screen mesh or frame can be used to apply the conductive components at required locations as shown in fig 4.4. This technique has also extensively investigated in textile applications. For

Figure 7. Layer-by-layer spray-coated conductive polyester/cotton

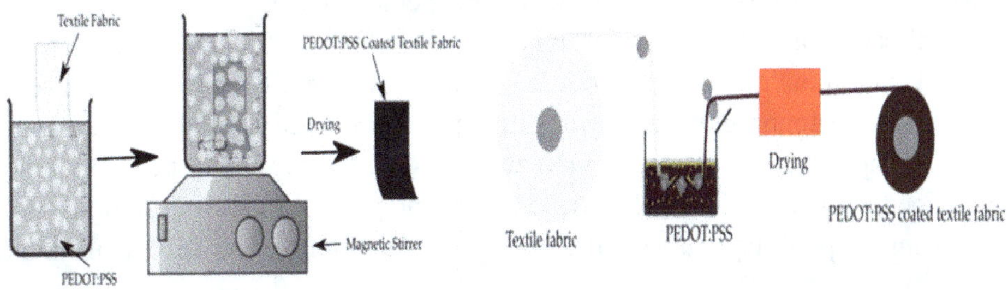

Electrospinning: In electrospinning, smart textiles can be also produced by adding sensor materials to a polymer spinning solution in nanotechnology level and microencapsulation and/or electrospinning technology. A process to encapsulate tiny particles or droplets into wall materials is quickly becoming a well-used technology for use in smart textiles.

Inkjet Printing: Ink jet printing is the most popular direct-write deposition tool that rapidly transfers to electronics fabrication. In inkjet printing images and structures are constructed in a droplet-by-droplet manner. The user has control over the jetting parameters which could be altered alongside the ink.

5. APPLICATIONS OF SMART FABRICS AND WEARABLES

Smart fabrics, integrated into wearable technology, are transforming industries through their capacity for real-time data collection, user interaction, and environmental responsiveness. These textiles have moved from niche innovations to everyday, professional, and industrial applications, demonstrating their adaptability and potential impact across multiple domains.

5.1 Healthcare and Medical Monitoring

Because smart textiles can continually and non-invasively monitor patient health, they are quickly becoming essential in the healthcare industry. These textiles can monitor biochemical markers present in biological fluids as well as physiological indicators including heart rate, temperature, and respiration. Electrocardiogram (ECG) or photoplethysmogram (PPG) sensors implanted in clothing can track heart activity and assist in the early detection of arrhythmias or cardiac stress (Le et al., 2022). Wearable textiles can assess important indicators like glucose levels, oxygen saturation, and hydration status for those managing chronic diseases like diabetes, heart problems, or respiratory disorders (Xiang et al., 2024).

By directing physical therapy exercises and monitoring patient progress, sensor-embedded clothing aids in rehabilitation during post-operative or injury recovery. These devices support eldercare by tracking mobility, identifying falls, and guaranteeing prompt emergency responses (Keawploy et al., 2020). When taken as a whole, smart medical textiles provide significant advantages by improving remote care, decreasing hospital stays, and facilitating highly individualized health management.

5.2 Sports and Fitness Tracking

In sports and fitness real-time feedback on physical performance is very critical, which is where smart fabrics have gained a lot of traction. In order to maximize performance and reduce the risk of injury, these clothing items and accessories enable users to monitor heart rate, muscular activity, gait, and tiredness levels throughout training sessions (Xiang et al., 2024). Athletes can improve technique and efficiency by using smart insoles and advanced compression clothing, which offer detailed insights on posture and motion (Le et al., 2022).

To avoid dehydration and performance declines, wearable biosensors integrated into textiles track the electrolyte balance, hydration levels, and sweat composition (Yin & Reddy, 2023). Additionally, these smart clothes can notify users of any injuries or overtraining by monitoring the strain and stress on muscles and joints, enabling prompt recovery measures (Le et al., 2022). The information that is produced enables fitness enthusiasts and professional athletes to improve their performance, exercise more intelligently, and recuperate more quickly.

5.3 Military and Defence Uses

In the military and defence industries, smart textiles are also becoming more popular because they improve situational awareness, safety, and operational efficiency. During high-intensity missions or harsh weather circumstances, soldiers can be monitored for fatigue, injuries, or heat stress by wearing uniforms equipped with sensors. By changing their colour or texture to blend in with their environment, certain adaptable fabrics are made to act as smart camouflage.

In terms of energy, military uniforms are being developed with integrated solar panels or energy-harvesting fibres, reducing dependence on external power sources in the field (Ali & Islam, 2024). Communication modules woven into the fabric enable discreet, hands-free data exchange that can be vital during tactical operations (Keawploy et al., 2020). Smart body armor equipped with pressure or impact sensors can detect bullet strikes or blunt force trauma, enhancing survivability. These innovations collectively provide a strategic edge, improving logistics, mission effectiveness, and soldier safety.

5.4 Fashion and Entertainment

Smart textiles are bringing in a new era of digital expression and interaction in the fashion and entertainment industries. These textiles combine style and utility to create clothing that reacts dynamically to touch, sound, or emotional states. Certain interactive garments combine art and technology by changing color or lighting up in response to outside stimuli.

Clothing can now show dynamic patterns, messages, or images thanks to LED-embedded fabrics, transforming them into wearable screens. According to (Xiang et al., 2024), clothes that incorporate motion or biometrically induced graphics improve audience immersion and storytelling in the performing arts. Additionally, smart clothing that incorporates haptic feedback and motion tracking has expanded its applicability to virtual and augmented reality experiences, producing more engaging AR/VR settings. Through digital fashion, these advances provide new avenues for creative participation and self-expression.

5.5 Industrial and Workplace Safety

Additionally, smart textiles are revolutionizing dangerous industries like construction, mining, and manufacturing. To improve safety and keep an eye on environmental hazards, these industries are progressively implementing smart clothing. Sensor-equipped wearable devices can identify exposure to radiation, dangerous substances, or extremely high or low temperatures, enabling prompt protective measures.

Additionally, posture and movement tracking through ergonomic feedback systems can help workers reduce strain and avoid musculoskeletal injuries. Smart clothing with visual or haptic alerts warns of nearby machinery, potential falls, or other threats. Compliance monitoring is also improved, as smart garments can verify whether workers are following safety protocols, thereby supporting accountability and reducing incidents. The cumulative effect is a safer, more efficient, and more regulated workplace.

6. CHALLENGES AND LIMITATIONS

Smart fabrics and wearable technology represent a convergence of textiles and electronics, offering innovative solutions across healthcare, sports, fashion, and more. However, several challenges and limitations hinder their widespread adoption and functionality. These include durability and washability, energy management, user comfort, calibration, manufacturing costs, interoperability, and data privacy.

6.1 Durability and Washability

Durability and washability are among the most significant hurdles in integrating electronics into textiles. Traditional garments undergo regular wear, mechanical stress, and laundering. Embedding electronics without compromising textile integrity or function poses a complex engineering challenge.

6.1.1 Key Issues

- Electronic components like conductive threads, sensors, and microchips are often sensitive to moisture, friction, and heat. These electronic parts are delicate and can be easily damaged when exposed to water, rubbing, or heat. Since clothes are worn daily and washed regularly, these conditions are common. Protecting the electronics from such stress is a big challenge.
- Standard washing methods can degrade or destroy embedded components, limiting product lifespan. Washing machines use water and mechanical force that can break or weaken the electronics inside fabrics. This makes smart clothing less durable and shortens its useful life. Most standard laundry processes are not designed for such embedded systems.

- Encapsulation techniques—such as polymer coatings—can improve water resistance but may reduce breathability or flexibility. Coating electronics with waterproof materials can help protect them during washing. However, this can make the fabric less comfortable, stiffer, or less breathable. There is a trade-off between protection and wearer comfort.

6.1.2 Recent Research

A 2024 study in Smart Materials and Structures titled "Flexible Washable Sensors for Smart Clothing: Challenges and Design Guidelines" highlights how substrate geometry and multilayer encapsulation improve washability without compromising sensor functionality. further propose ISO-based durability testing protocols specific to smart textiles, providing a foundation for standard evaluation methods.

Figure 8. Architecture of the proposed smart clothing system.

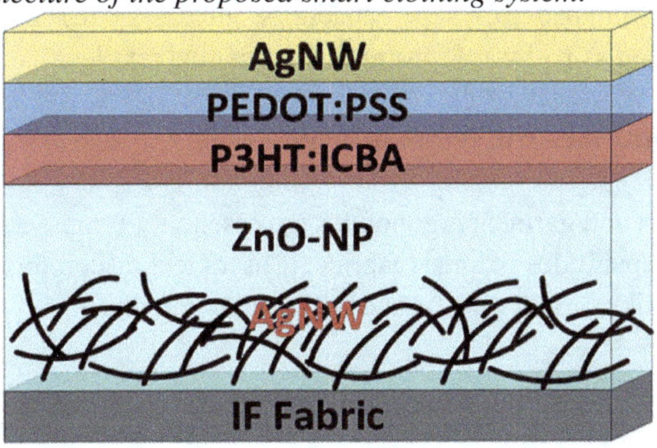

6.2 Power and Energy Management

Smart wearables rely on continuous power to collect data, perform computations, and maintain connectivity. Powering these systems effectively while maintaining wearability is a critical bottleneck.

6.2.1 Key Issues

- Traditional lithium-ion batteries are bulky, rigid, and require frequent charging. These batteries are not designed for flexible clothing. They are heavy and can make wearables uncomfortable. Also, the need for regular charging makes them inconvenient for daily use.

- On-body energy harvesting is still nascent and limited in power output. Some smart wearables try to generate power from the body, like from heat or movement. However, the energy they produce is very small. This is not enough to power most sensors or data-processing tasks for long periods.
- Trade-offs exist between energy capacity, recharge cycles, flexibility, and garment aesthetics. A bigger battery lasts longer but may make the clothing bulky. More flexible power sources may not store enough energy. Designers have to balance power needs with looks and comfort.

6.2.3 Emerging Solutions

Introduction to wearable thermoelectric materials that leverage body heat for energy harvesting, though they currently generate low voltage. Hybrid approaches combining thin-film batteries, supercapacitors, and energy-harvesting modules are being investigated as pathways toward self-sustaining textiles.

6.3 User Comfort and Wearability

User comfort and garment ergonomics are essential for mass adoption. Despite the advanced capabilities of smart fabrics, if users feel discomfort or perceive the wearables as bulky or irritating, usage will decline rapidly.

6.3.1 Key Challenges

- **Flexibility**: Rigid electronics may hinder natural fabric stretch and restrict body movement.
- **Skin Compatibility**: Poor ventilation and constant skin contact can cause irritation, sweating, or dermatitis.
- **Weight Distribution**: Heavy or unevenly integrated components can alter wearer biomechanics, leading to discomfort over time.

6.3.2 Research Perspective

A comprehensive review titled "Recent Advances in Smart Fabric-Type Wearable Electronics Toward Comfortable Wearing" published in Advanced Functional Materials, emphasizes that future smart textiles must integrate electronics at the yarn or fiber level. The review showcases examples of nanomaterial-infused threads that maintain garment softness while ensuring electrical conductivity (Xiang et al., 2024).

To reduce skin irritation, researchers at MIT Media Lab are exploring hydrogel interfaces between the skin and sensors to create breathable, skin-safe wearables.

6.4 Calibration and Reliability

Accurate sensing is central to smart wearables, yet calibration remains a persistent issue.

- **Sensor Drift**: Variations in placement, fabric stretch, and sweat can lead to measurement errors.
- **Dynamic Deformation**: Unlike rigid devices, textiles are constantly stretched, bent, or compressed, making consistent calibration difficult.

Recent studies suggest machine-learning–based auto-calibration algorithms and **self-correcting sensors** as potential solutions, but robust real-world validation is still limited.

6.5 Manufacturing Costs and Scalability

The fabrication of smart textiles often involves complex, multi-step processes such as conductive fiber weaving, screen printing of nanomaterials, or manual module attachment. These techniques increase production costs and reduce scalability. Small-batch prototypes can demonstrate high performance, but translating them into mass-market, affordable products requires automated, standardized manufacturing pipelines. The lack of cost-effective production remains one of the most significant limitations to commercialization (Singh et al., 2024).

6.6 Interoperability and Standards

Smart fabrics frequently use proprietary hardware and communication protocols, creating compatibility issues across devices and ecosystems. Without standardized frameworks for data formatting, wireless communication, and cloud integration, interoperability is limited.

Recent industry initiatives have begun working toward IoT standardization for wearable data, but the field remains fragmented. This lack of common standards hinders integration with healthcare platforms, sports analytics systems, and consumer devices (Saxena et al., 2024).

6.7 Data Privacy and Security

Smart fabrics collect physiological and behavioural data that are personal and sensitive. Ensuring ethical and secure data management is critical to user trust and regulatory compliance (Singh et al., 2021).

6.7.1 Key Risks

- Unauthorized use of biometric information. Smart clothing continuously collects personal information about a person's physical activity and health. Malicious actors or hackers could access this data if appropriate security measures are not in place. Such violations may result in abuse, such as identity theft, stalking, or discrimination in insurance. Because users frequently wear these devices all day, unauthorized access is a major risk due to the volume and sensitivity of data collected.
- Inadequate transparency in the gathering and storing of data. A lot of wearable technology passively gathers and sends data, frequently without the user's knowledge. Users might not understand what information is being gathered, where it is kept, who can access it, or how long it will be kept on file. Particularly when the data is shared with third-party apps or services, this lack of transparency erodes consumer trust and could result in problems with informed consent.
- Low-power devices have inadequate encryption techniques. Wearable devices frequently have limited processing and storage capacity due to energy constraints. Because of this restriction, it is challenging to use robust encryption methods without compromising device performance. Because of this, information sent between the wearable and its accompanying app or cloud server could be intercepted. Insecure data channels can be used to retrieve stored personal information or for real-time surveillance.

6.7.2 Legal and Technical Safeguards

Wearable technology is subject to legal frameworks like the California Consumer Privacy Act (CCPA) and the General Data Protection Regulation (GDPR) of the European Union. The restricted user interfaces and passive data collecting characteristic of smart clothing, however, make compliance challenging.

In their groundbreaking study "GaitCrypt: Continuous Biometric Encryption for Wearable Devices," present gait-based encryption models that reduce the requirement for human authentication by using each user's walking pattern as a dynamic key for data security (Su et al., 2023).

7 ACTIONABLE RESEARCH DIRECTION IN SMART FABRICS AND WEARABLE TECHNOLOGY

Recent developments in smart fabrics have started to change how wearable technology works and interacts with daily living. Improvements in materials, electronics, and system integration are expanding the possibilities of smart fabrics in a variety of fields, including as healthcare, fitness, fashion, and military applications.

Elastic and flexible electronics, self-healing and adaptable fabrics, and integration with AI and the Internet of Things (IoT) are the primary subjects of this chapter.

7.1 Flexible and Stretchable Electronics

Electronics that are flexible and stretchable are essential to the future of wearable technology. Unlike traditional inflexible electronics, these devices can bend, stretch, and fit the curves of the body without impairing their functionality. Materials such as graphene, silver nanowires, and conductive polymers are being used to create circuits that maintain conductivity under stress. Additionally, elastic screens and batteries are being developed to enable entirely flexible electronics.

Improved comfort and seamless design are made possible by the direct integration of these technologies into fabrics. Applications include smart clothing that can detect movement, temperature, and other physiological indicators, wearable ECG monitors, and electronic skins. Table 3 shows the comparison of different types of flexible and stretchable materials.

Table 3. Comparison of Materials for Flexible and Stretchable Electronics

Material	Properties	Applications
Graphene	High conductivity, flexibility	Sensors, e-textiles, electrodes
Silver Nanowires	Excellent conductivity, stretchable	Wearable circuits, conductive fabrics
Conductive Polymers	Tunable conductivity, flexibility	Smart clothing, biomedical sensors

7.2 Self-Healing and Adaptive Textiles

Self-healing textiles represent a significant advancement in material science and are inspired by the ability of biological systems to regenerate. Thermally responsive materials, dynamic covalent chemistry, or microcapsules containing healing agents can be used to help these textiles self-heal tears and abrasions.

On the other hand, adaptive textiles can change their properties in response to external inputs. For instance, textiles that change color in reaction to light or moisture or that adjust to temperature by altering their insulating properties are now being developed. These materials enhance the user experience by being more comfortable, practical, and long-lasting.

7.3 AI-Integrated Fabric With IoT Systems

Wearable technology is becoming smarter, more connected, and more self-sufficient as a result of the combination of smart textiles, AI, and IoT. In order to identify trends, identify illnesses, and forecast outcomes, artificial intelligence (AI) systems analyze enormous volumes of sensor data. Gait tracking, for instance, and Machine learning enables the early diagnosis of neurological diseases.

Through Bluetooth, Wi-Fi, and NFC, wearables can interact in real time with other gadgets and cloud platforms thanks to Internet of Things connectivity. This makes it possible to use apps like real-time fitness tracking, remote health monitoring, and connection with smart home settings (Sharma et al., 2022). The development of customized, flexible systems that can adjust to consumers' needs is being made possible by the convergence of AI, IoT, and smart textiles (Jain et al., 2024).

7.4 Biodegradable Conductive Material

Biodegradable conductive materials combine electrical conductivity with biodegradability which decompose naturally by biological process. These types of fabric are very useful in transient electronics, biomedical implants, wearable sensors, environmental devices, and green energy systems, since they perform their electronic function for a limited time and then safely degrade without causing long-term waste or pollution. Conductive polymers, carbon-based materials, and metallic nanoparticles are some key components of biodegradable materials. Different mechanisms like Hydrolysis (PLA, PCL), Enzymatic degradation (cellulose, chitosan, proteins like silk/gelatin), Metal corrosion (Mg, Zn, Fe), Oxidation & microbial degradation (carbon-based biomaterials) can be used for biodegradability.

7.5 Energy Harvesting Textiles

Energy harvesting textiles are smart fabrics that can generate and store energy from the environment while remaining lightweight, flexible, and wearable. Different mechanisms are used for harvesting: triboelectric textiles generate electricity from friction and body motion; piezoelectric textiles convert mechanical strain into electricity; thermoelectric textiles use body heat via Seebeck effect; and photovoltaic textiles convert light into electricity through organic solar cells, dye-sensitized solar cells. Hybrid designs combine multiple mechanisms (e.g., solar + motion) to improve efficiency. Key materials include conductive yarns (silver-coated nylon, stainless steel, CNT fibers, graphene-coated cotton), polymer coatings (PVDF, PDMS, PEDOT:PSS), and nanomaterials (graphene, $BaTiO_3$, ZnO, perovskites). Applications range from wearable electronics, healthcare monitoring, and military uniforms to sustainable fashion and IoT-enabled textiles. Despite their promise, challenges remain in terms of durability (washing resistance), scalability, comfort, and seamless energy storage integration with supercapacitors or micro-batteries (Singh & Malik, 2025).

Future developments in wearable technology and smart fabrics will most likely focus on durability, multifunctionality, and shrinkage. To lessen reliance on outside sources, research is being done to develop fabrics that can absorb energy from movement or the surroundings. Power sources (Stoppa & Chiolerio, 2014). Furthermore, when these technologies are used more often, ethical considerations around privacy and data security will become increasingly significant.

8 CONCLUSION AND FUTURE SCOPE

The integration of smart fabrics and wearable technology represents a transformative advancement in both the textile and electronics industries. This chapter has explored the fundamentals, including material composition, embedded components, communication systems, and wide-ranging applications across healthcare, fitness, military, fashion, and industrial domains. Through real-time feedback, personalized monitoring, and improved functionality while maintaining comfort and flexibility smart fabrics have proven their potential to enhance user experience. Key challenges such as durability, washability, power management, user comfort, and data security remain critical barriers to mass adoption despite of impressive progress. However, these limitations are taken care by research advancements in flexible electronics, nanomaterials, energy harvesting, and bio-compatible encapsulation. A rich ground for innovation is still ensured by the multidisciplinary nature of this field including spanning materials science, electrical engineering, data analytics, and design.

Future smart fabrics will evolve into fully autonomous, multifunctional systems that merge energy harvesting, AI integration, and advanced materials. Progress in thermoelectric, piezoelectric, and bioenergy harvesting will enable long-term, self-powered operation, while AI-driven textiles will provide adaptive behavior and personalized feedback. Self-healing, stretchable, and biodegradable materials will enhance comfort and sustainability, and advanced communication protocols will position garments as vital nodes within the IoT ecosystem. Ensuring privacy-centric, user-controlled data architectures will remain essential for secure and ethical deployment.

REFERENCES

Ali, I., Islam, M. R., Yin, J., Eichhorn, S. J., Chen, J., Karim, N., & Afroj, S. (2024). Advances in smart photovoltaic textiles. *ACS Nano*, *18*(5), 3871–3915. DOI: 10.1021/acsnano.3c10033 PMID: 38261716

Cheng, X., Shen, D., Zheng, K., Wu, Z., Shi, L., & Hu, X. (2025). A wearable strain sensor for medical rehabilitation based on piezoresistive knitting textile. *Sensors and Actuators. A, Physical*, *387*, 116379. DOI: 10.1016/j.sna.2025.116379

Dolez, P. I. (2021). Energy harvesting Materials and structures for smart textile Applications: Recent progress and Path forward. *Sensors (Basel)*, *21*(18), 6297. DOI: 10.3390/s21186297 PMID: 34577509

Hossain, M. M., & Bradford, P. D. (2020). Durability of smart electronic textiles. In Elsevier eBooks (pp. 27–53). DOI: 10.1016/B978-0-12-820777-2.00003-0

Jain, P., Sharma, S., Arora, S., Shokeen, V., Aggarwal, P. K., & Singh, M. (2024). Edge computing-based design for IoT security. In Network Optimization in Intelligent Internet of Things Applications: Principles and Challenges (pp. 298–309). CRC Press. DOI: 10.1201/9781003405535-22

Keawploy, N., Venkatkarthick, R., Wangyao, P., & Qin, J. (2020). Screen printed textile electrodes using graphene and carbon nanotubes with silver for flexible supercapacitor applications. *Journal of Metals Materials and Minerals*, *30*(4), 39–44. DOI: 10.55713/jmmm.v30i4.892

Le, K., Narayana, H., Servati, A., Bahi, A., Soltanian, S., Servati, P., & Ko, F. (2022). Electronic textiles for electrocardiogram monitoring: A review on the structure–property and performance evaluation from fiber to fabric. *Textile Research Journal*, *93*(3–4), 878–910. DOI: 10.1177/00405175221108208

Liang, Z., Deng, Y., Shi, Z., Liao, X., Zong, H., Ren, L., Li, X., Zeng, X., Hu, P., Ke, W., Wu, B., Wang, K., Qian, J., Xu, W., & Chen, F. (2025). AI-driven design of powder-based nanomaterials for smart textiles: From data intelligence to system integration. *Advanced Powder Materials*, *5*(1), 100356. DOI: 10.1016/j.apmate.2025.100356

Park, S., & Jayaraman, S. (2003). Enhancing the quality of life through wearable technology. *IEEE Engineering in Medicine and Biology Magazine*, *22*(3), 41–48. DOI: 10.1109/MEMB.2003.1213625 PMID: 12845818

Saxena, P., Jain, P., Aggarwal, P. K., Singh, M., Goel, S., & Batra, M. (2024). Communication Requirements and Performance Metrics for Electric Vehicle Charging. In Optimized Energy Management Strategies for Electric Vehicles (1st ed., pp. 35–60). DOI: 10.4018/979-8-3693-6844-2.ch002

Sharma, A., Singh, M., Gupta, M., Sukhija, N., & Aggarwal, P. K. (2022). IoT and blockchain technology in 5G smart healthcare. In *Blockchain Applications for Healthcare Informatics: beyond 5G* (pp. 137–161). Elsevier., DOI: 10.1016/B978-0-323-90615-9.00004-9

Singh, M., & Malik, A. (2024). Multi-Hop Routing Protocol in SDN-Based Wireless Sensor Network. In Software-Defined Network Frameworks (1st ed., pp. 121–141). CRC Press. DOI: 10.1201/9781040018323-8

Singh, M., & Malik, A. (2025). Energy-Efficient Cluster Head Selection in Heterogeneous WSNs Using Aquila Optimizer and Fuzzy Logic (AO-FL). *International Journal of Communication Systems*, *38*(16), 1–22. DOI: 10.1002/dac.70272

Singh, M., Sharma, R., Singh, R., Malik, A., & Aggarwal, P. K. (2024). Advancements in renewable energy harvesting for EV charging infrastructure. In *Optimized Energy Management Strategies for Electric Vehicles* (pp. 119–152). IGI Global., DOI: 10.4018/979-8-3693-6844-2.ch005

Singh, M., Sukhija, N., Sharma, A., Gupta, M., & Aggarwal, P. K. (2021). Security and Privacy Requirements for IoMT-Based Smart Healthcare System: Challenges, Solutions, and Future Scope. In Big Data Analysis for Green Computing: Concepts and Applications. DOI: 10.1201/9781003032328-2

Stoppa, M., & Chiolerio, A. (2014). Wearable electronics and smart textiles: A critical review. *Sensors (Basel)*, *14*(7), 11957–11992. DOI: 10.3390/s140711957 PMID: 25004153

Su, Y., Li, Y., & Cao, Z. (2023). Gait-based privacy protection for smart wearable devices. *IEEE Internet of Things Journal*, *11*(2), 3497–3509. DOI: 10.1109/JIOT.2023.3296650

Xiang, H., Li, Y., Liao, Q., Xia, L., Wu, X., Zhou, H., Li, C., & Fan, X. (2024). Recent advances in smart fabric-type wearable electronics toward comfortable wearing. *Energies*, *17*(11), 2627. DOI: 10.3390/en17112627

Yin, J., Li, J., Reddy, V. S., Ji, D., Ramakrishna, S., & Xu, L. (2023). Flexible Textile-Based sweat sensors for wearable applications. *Biosensors (Basel)*, *13*(1), 127. DOI: 10.3390/bios13010127 PMID: 36671962

Chapter 2
Exploring Smart Textiles and Wearable Tech: An Introductory Overview

Ranu Burad
https://orcid.org/0000-0001-6876-6975
Poornima University, India

ABSTRACT

Smart fabrics and wearable technology represent of form of textile engineering data science and electronics. Such innovation have led creation of smart fabric that can react communicate and sense, transformed many sectors like healthcare, fashion, defense, sports and many more. This chapter explodes the foundation and diverse application of smart and wearable textile into active, passive and ultra-smart variants of fabrication methods weaving to printing and 3D knitting (Stoppa & Chiolerio, 2014; Cherenack & van Pieterson, 2012). This chapter highlights the issues and challenges linked to wearable technology like durability privacy, comfort, power management and cost, keeping in mind about environmental and ethical consideration (Zhao et al., 2021). It outline future direction, including AI integration personalized design and sustainable material. This comprehensive overview established understanding of both functional and expressive value and how smart and wearable textiles re-shape interaction with clothing (Castano & Flatau, 2014).

INTRODUCTION

In the era of rapid change & rapid growth, everything is collaborating with electronics. Textile is witnessed the convergence with technology & creating new domain as smart textile or smart wearable textile. Such smart textile is combined

DOI: 10.4018/979-8-3373-4287-0.ch002

with sensors, microprocessors, actuator, and communication module which make a garment interactive with humans & their environment with real time (Stoppa & Chiolerio, 2014). The integration of human with the clothing, leads to a support system as clothing helps to regulating & monitoring health, performance, helps in assist people with disabilities & it is a platform of connectivity & personal expression.

Smart wearable technology refers to e-devices that are worn by person, it is design to regulate movement & monitor health, physical symptoms, signals, communication with other systems, also give feedback (Heikenfeld et al., 2018). When such technology is embedded into textiles through threads, sensors that are flexible, printed circuits, it provides an unobtrusive user experience along with comfort. Unlike traditional wearable gadgets such as wristbands and clip-on monitors, smart fabrics maintain the look and feel of everyday clothing while adding layers of functionality, making them particularly appealing for applications requiring long-term comfort and discretion (Cherenack & van Pieterson, 2012).

The impetus behind the development of smart fabrics is multifaceted. In healthcare, there is a growing demand for non-invasive, continuous monitoring of vital signs to support remote patient care and early diagnosis of diseases (Jee et al., 2019). In sports and fitness, wearable garments help athletes and recreational users monitor their physical activity, biomechanics, and hydration in real time, contributing to performance optimization and injury prevention (Castano & Flatau, 2014). The military sector explores smart uniforms for soldiers that can monitor physiological stress, detect chemical exposure, and enhance communication during combat operations. Moreover, in the context of disability and aging, smart textiles offer promising solutions to improve independence and safety through features like fall detection, automated closures, and adaptive fits (Paterson et al., 2022).

Historically, smart wearable textile was first introduced by the military sector to help them in extreme situation in such situation, in such situation, smart clothing was them to locate the changes, but from last two decades such clothing has been redesigned for local market due to technological advancement like minimize electronics, wireless device in flexible material. Now there is end number of options available for consumer application like jackets, socks, shirts etc. to monitor various parts of body.

Figure 1. Taxonomy of smart textile types and their application areas

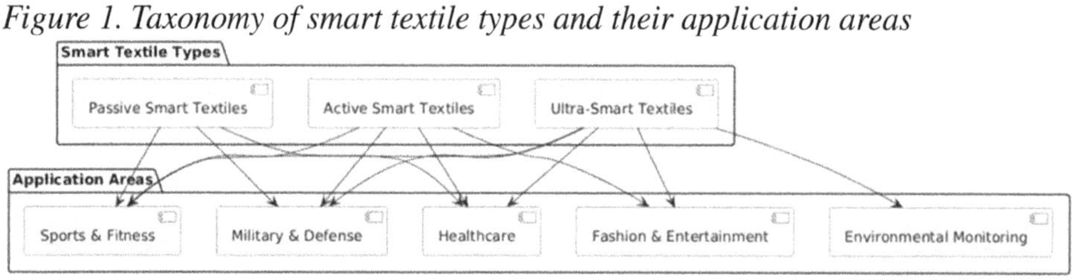

Figure 1 classified smart textiles into three broad categories namely, Passive Smart Textiles, Active Smart Textiles, and Ultra-Smart Textiles. Passive smart textiles are the traditional fabrics which may possess some straightforward minimally-augmented functions but only few or no active electronics. Active smart textiles are those that contain integrated sensor along with the actuators that enable it to react to figures or stimuli in the environment in real time. The ultra-smart type integrates embedded electrics, artificial intelligence, and connectivity to enable ultra-ruggedization with a full range of sensing possibilities, data processing and on-demand behaviors. The categories are given as relating to a variety of application fields that comprise medical care, sport and health, army and defense arm, style and entertainment, and environment surveillances. It is introduced as the diagram representing the grounded outcomes of being more technologically mature by expanding the scope of utilization, where ultra-smart textile is the most wholesome scope of disciplinary scopes, including the new emerging applications, like sensing the environment and the use of the fabric as attire. Through this visual strategy, readers should be able to comprehend how smart textile technologies were developed and applied as well as through very simple to highly integrated systems.

Despite of these technology advancement & availability of such clothing, this section of the clothing sector faced lot of challenges and issues like durability, compatibility, energy consumption and the most important concern is hinder the privacy of wearer because of regular monitoring along with this it also increases the environmental carbon footprint due to non-degradable sensors, fabric batteries etc. sustainability has become too critical concentration (Stoppa & Chiolecio, 2014)

As society has become more tech friendly, in the era of technology such clothing plays the central role. This study provides the structured information about smart textile and wearable technology. This study discusses about the key components, classification technique, application, manufacturing, issues, challenges, future scope and directions. Such technology aims to provide the broader understanding of how intelligent clothing is reshaping the future of clothing sector & wearer. Wearable technology & smart fabric together represent the Synergy of textile with digital innovation and electronics this convergence has given rise to garment that can reacts and adapt to human activity and environmental stimuli.

In spite of the high rate of development in material science and integration of IoT, the body of lactate concerning the issues and key supporting technologies in smart textiles have been lagging behind with current practices and recent material advancements, especially in the areas of user privacy, power autonomy, and regulatory issues in real-to-life applications. The fact that the wearables market totaling to 61.3 billion USD globally in 2022 and forecasted to reach 186.1 billion USD by 2030 (Khan et al., 2020) is experiencing an exponential growth makes the topic of scalable, comfortable, and secure devices increasingly urgent to research on.

EVOLUTION AND HISTORY OF SMART TEXTILES

The evolution of smart textiles has been marked by a steady convergence of materials science, electrical engineering, computing, and fashion design. From ancient times, textiles have served not only as a basic human need for protection but also as a medium of cultural identity, social expression, and innovation. The historical trajectory of smart textiles reflects the transition from purely functional garments to intelligent systems capable of sensing, reacting, and communicating.

Early Foundations: Functional and Technical Textiles

Before the urgency of smart & wearable textiles, functional textiles like water, fire resistance, thermal insulation & UV protection garment were considered. With the rising need of such garments, especially in military and industry such smart garments were designed. At the time world war II, to protect soldiers and pilots, durable & protective garments were developed (Mattila, 2006).

The main development was created on the basics of technical textile, driven by new ideas of synthetic fiber as polyester, nylon & Kevlar which was started in 20th century. These material were soft, light weighted, flexible and durable which makes ideal for protective clothing, industrial equipment & aerospace (Horrocks & Anand, 2000). This development in garment sector created a platform for electronic functionalities

The Birth of Wearable Electronics: 1980s–1990s

MIT wearable computing project was one of the earliest inventions launched in 1990s which were portable computing system embedded in garments (Mann, 1997). The idea of smart & wearable technology emerged in late 20s; however the garments were used to heavy in weight, bulky & less flexible. At this point of development the researchers started experimenting microelectronics into clothing.

The defense industry was the first to introduce such smart uniforms & electronic textiles for army person that could monitor vital physical signs, gives situational report at the time of battlefield, helps in communicate (Tao, 2005).

Integration of Electronics into Textiles: 2000s

The 2000s notices the drastic shifting toward miniaturization and commercialization of electronic elements, due to these shifting conductive yarns & fabrics were widely available, enabling the fixing of actuators, sensors & power supplies into garments along with comfort & flexibility (Stoppa & Chiolerio, 2014).

Some of the notable examples of commercial products are:

1. The "Smart Shirt" by Sensatex: This shirt can monitor vital signs, like heart rate, body temperature & respiration which was designed & developed with collaboration with U.S. military (Park & Jayaraman, 2003).
2. Levi's ICD+ Jacket: A denim jacket which can controls MP3 player & mobile communication which was designed & developed between the partnerships of Levi & Philips (Tao, 2005).

All such ideas & innovations led the transitions from concept to application, giving user comfort, durability & washability which lacks in early products.

Advances in Sensor Technology and Flexible Electronics: 2010s

Research & development saw a boom in the area of stretchable and flexible electronics in 2010s, which allows the fabrics for seamless integration of circuits. Such smart wearable garments are capable of wireless communication, haptic feedback & real time sensing.

Important contributions came from interdisciplinary collaborations between electrical engineers, textile scientists, and designers.

1. Medical & Healthcare: Garments which monitor EMG, ECG, and respiration (Heikenfeld et al., 2018).
2. Fitness & Sports: Outfits that tracked muscle activity, hydration & movement (Castano & Flatau, 2014).
3. Fashion & Entertainment: color-changing fabrics, responsive textiles for stage, LED-fixed dresses.

Moreover, smart wearable technologies like Google Glass, Fitbit, & Apple Watch, gain popularity which helps in normalizing such wearable technologies & fixes the consumer expectations for interactivity and data-driven functionality.

2020s and Beyond: Toward Sustainable, Intelligent, and Ubiquitous Textiles

2020s witnessed the drastic change with increased use of smart textile. This evolution is shaped by AI (Artificial Intelligence), IoT (Internet of things), connectivity

& sustainable designs. Smart garments have become smarter which are capable of sensing also reacting with learning their user behavior & adapt accordingly.

The aim of smart fabric in new generation is to be:

1. Energy Saving: using kinetic energy harvesting, & thermoelectric materials, solar fibers, for power (Zhou, Gao, & Wang, 2020).
2. eco-friendly & Biocompatible: through recyclable textile substrates & biodegradable sensors.
3. Mass-customizable,gives User based solutions for wellness, health or fashion through 3D knitting & digital fabrication (Cherenack & van Pieterson, 2012).

For PWDs (People with disabilities) such technology is game changer, as this help PWDs to assist in health monitoring, posture correction & mobility (Paterson et al., 2020).

Researchers continues to address the issues & key challenges of smart garments regarding flexibility, comfort, data privacy, cost effectiveness & washablity – smart wearable textile has become the important part of day to day life. The evolution of smart textiles reflects the user centric designs with the innovation of computation & clothing

Facilitating Electronics and Signal Processing

This transition into novel wearable systems using new smart material systems will require the seamless integration of both enabling electronics and smart signal processing. Integrating sensors, microcontrollers, and wireless communication modules into fabrics and turning them into active information systems opens up new possibilities in the use of textiles. Signal conditioning, noise filtering and sensor fusion algorithms are employed to insure proper, reliable performance in real world environments, especially in wearable systems that are subject to both movement and environmental changes. This intermediate level brings the innate functionality of these smart materials up to the relationship with their ultimate use environments, which leads to a discussion of the various real life applications of these materials in healthcare, defense, sports and many more.

CLASSIFICATION OF SMART TEXTILES

Smart wearable textiles also known as electronic or intelligent textiles, is a results of fusion of multidisciplinary science materials, textile engineering & electronics. They are divided onto basis of functionality, integration level & intended application.

To understand the developmental complexity, suitability & capabilities, for various real time applications, ranging from defense to healthcare, fashion, it is necessary to understand the classification of smart wearable textiles.

Classification Based on Functionality

The most widely accepted methods of classification of smart & wearable textile is depends on its functionality & capability- how such textiles adapts or reacts to thermal, mechanical, magnetic, electrical, or chemical inputs. Based on these external stimuli capabilities, it further divides into three parts active, passive & very smart textiles (Tao, 2005; De Rossi et al., 2003).

Passive Smart Textile

Passive textiles don't have any actuating capabilities, these are the simplest form of textile which are specially designed to sense environmental changes or conditions like radiation, pressure or temperature, but they can't react to these changes. The reliability of such textiles is dependent on their inbuilt sensors for data collection. For instance clothing which are sensitive to UV rays or temperature, color changing textiles, fabric with optical fibers which monitor light exposure (Cherenack & van Pieterson, 2012).

Active Smart Textiles

Active textiles have both actuating and sensing capabilities. These kinds of materials can react & sense to external stimuli & react accordingly. For example PCMs (Phase change materials) fixed in garment can release thermal energy according to body requirement. Some other examples are shape memory polymers or alloys that change their shape according to heat response or garments with fixed heating pads which can regulate the temperature when it detect cold weather (Stoppa & Chiolerio, 2014).

Very Smart Textiles

The smartest & intelligent textile which can sense, react, adapt also is capable of reasoning to external stimuli in autonomous & dynamic way. By using feedback & AI (artificial intelligence) control mechanisms, such textile are able to react & sense stimuli. For instance, garments for PWDs (people with disabilities) that can sense body posture and can change stress, stiffness, mobility & tension to help them to move freely, (Park & Jayaraman, 2003). Such garments are fixed with sensors, microcontrollers, memory devices & communication systems which make them computing platforms.

Classification Based on Integration Level

Smart textiles can also be divided on the level of integration between textile substrates & electronic components

Textile-Based Smart Systems

These smart systems are made up of conductive fibers where electronic functionality is embedded into textiles. Conductive fabric made from carbon nanotubes or silver coated yarns are knitted or woven to form sensors & circuits (Castano & Flatau, 2014). These textile based systems are best for wearable applications due to their stretchability, washability & flexibility.

Attached Smart Components

Attached components are electronic components which are attached to textiles like batteries, processors & sensors. Due to their attached electronic components these modules are semi rigid or rigid which reduce their wearability or comfort. Independent Components in such textile which can be upgraded or replaced makes them greater modular & provides ease of maintenance (Stoppa & Chiolerio, 2014).

Hybrid Smart Textiles

These textiles are the combination of electronic modules & textile elements. Such textiles are the example of compromise between functionality & wearability. For instance, a sport jersey with textile sensors and a data processing unit which records and transmits the performance chart to phone application.

Figure 2. System block diagram of IoT integration in smart textiles

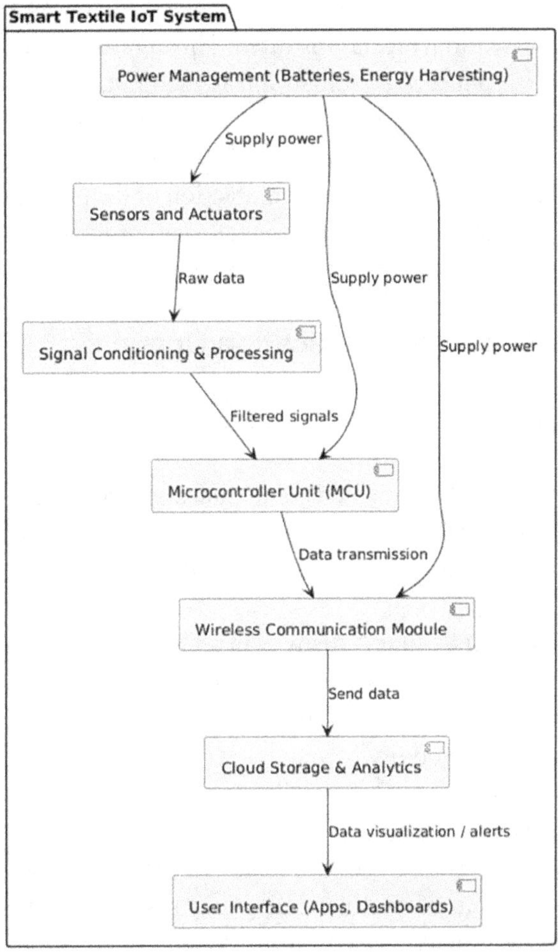

Figure 2 is a graphic demonstration of key elements of the data flow and structure of a wearable IoT-enabled smart textile system. It starts with impregnated sensors and actuators that assemble rough physiological or local surroundings data of the user or the atmosphere. Raw data can be processed through signal conditioning and processing in order to remove noise and highlight useful parameters and hand it over to a microcontroller unit (MCU), where it controls and operates the system. The processed data is then wirelessly relayed via communication modules e.g. Bluetooth, Wi-Fi, or NB-IoT to cloud-storage levels and analytics platforms by the MCU. There the information is captured, analyzed, and availed. Users can communicate with the system through user interfaces such as mobile apps or dashboards and be provided visualizations or alerts or insight based on the gathered data. Custom power components such as flexible batteries and energy harvesting devices provide the power to the MCU, sensors, and wireless modules so that long-term, autonomous operation is possible without recharging often. This figure explains hardware, signal processing, wireless connectivity, cloud infrastructure and user interaction layers intertwine with each other to help smart textiles to become a connected wearable device that acts as an intelligent device. It facilitates learning of the integration part of the IoT whereby it elucidates the interaction of technologies which turn raw textile materials into tele-medical devices that monitor health in real-time, fitness tracking devices, military uses, to name but a few.

Classification Based on Application Domain

As per the application, smart textiles are further divided into the following.

1. Medical Textiles & Healthcare: This includes garments for patients or PWDs which help in monitoring, therapy & rehabilitation. For instance shirts that monitors ECG, infections detected bandages, posture correction outfits (Heikenfeld et al., 2018).
2. Protective Textiles & Military: Army man uniforms with protection layer which can sense & communicate in battlefield situation.
3. Fitness & Sports: Sports clothing that can monitor & sense the hydration levels, biometrics & muscle activity.
4. Entertainment & Fashion: Color or light changing fabrics used in costumes to make them more interactive.
5. Occupational & Industrial Use: Clothes made up of such textiles which detect gas leakages & locate people like emergency responders or miners.
6. Interior & Home: Smart carpets, furniture textiles & curtains which sense & react to the environmental conditions (Cherenack & van Pieterson, 2012).

Classification Based on Stimuli Response

It is further divided on the basis of behavior of reacting or detecting stimuli

1. Thermo-responsive: Such textile reacts to temperature. For instance thermochromic dyes, PCMs.
2. Photo-responsive: Such textile reacts to light exposure. For instance materials made up photoluminescent.
3. Electro-active: Such textile reacts to electrical stimuli. For instance piezo-electric fabrics.

4. Hydro-responsive: Such textile reacts to moisture or humidity. For instance moisture-wicking sensors.
5. Bio-responsive: Such textile respond to physiological or biochemical markers. For instance, sugar level sensors in diabetic clothing) (Zhou, Gao, & Wang, 2020).

COMPONENTS AND MATERIALS

When traditional textile structure is incorporated by functional components, smart textiles are formed which can react, adapt & sense to stimuli. Such materials or components are generally integrated at any system level from fiber to fabric which includes diverse range of materials including conductive materials, electronic elements, power sources, communication modules & actuators. The combination of these materials is set by the type of function or performance & user comfort is required.

Conductive Materials

With the use of conductive materials onto fabric like layering or embedded into fabric, smart textiles have ability to transmit electrical signals. Conductive fibers to conductive yarns or fabric with coatings are commonly used to make **interconnects, sensors**, and antennas.

Conductive Fibers & Yarns

Conductive fibers & yarns are made up by silver, gold, or copper coating on nylon or polyester traditional fibers. Metallic wire can be used into textile structure. The mostly favored yarns are silver coated yarns due to its high antimicrobial properties, resistance to oxidation and conductivity, (Krebs et al., 2010).

Conductive Polymers

Polymers like PPy (polypyrrole), poly(3,4-ethylenedioxythiophene) (PEDOT:PSS) & PANI (polyaniline) are mostly used due to its properties like light weight, printability & flexibility. These materials can be fixed during fiber formation or can be layered or coated on the fabrics (Stoppa & Chiolerio, 2014). Such Conductive polymers allow electronics to stretch which is essential for wearer.

Carbon-Based Materials

Carbon black, Carbon nanotubes (CNTs) & graphene, emerging as versatile materials for creating lightweight, conductive & flexible textile structures. Their high mechanical strength, electrical conductivity & surface area, make them suitable for heating elements, energy storage & sensors (Liao et al., 2015).

Table 1. Material properties and applications in smart textile technology

Material	Electrical Conductivity	Mechanical Flexibility	Application Suitability	Reference
Copper Conductive Yarn	High	Low	Power circuits, robust sensing	(Stoppa & Chiolerio, 2014)
Silver-Coated Fibers	Moderate-High	Moderate	Biosignal monitoring, soft electrodes	(Cherenack & van Pieterson, 2012)
Carbon Nanotubes	High	High	Strain sensors, lightweight electronics	(Liao et al., 2015)
PEDOT:PSS Polymers	Moderate	High	Flexible displays, wearable electronics	(Khan et al., 2020)

According to Table 1, Copper conductive yarn is high electrical conductivity and therefore, suitable in strong power circuits and strong sensing tracks in fabric materials but is stiff and has limited flexibility and comfort. Silver-coated fibers are a textile core-containing conductive silver coating giving moderate-to-high conductivity plus moderate flexibility which are applicable in bio signal monitoring and soft electrodes although they have slight durability issues under repeated washing. Carbon nanotubes (CNTs) provide unparalleled conductivity and high flexibility in an ultralightweight form to enable highly sensitive and discrete strain sensors and lightweight electronic elements. PEDOT:PSS is a conductive polymer that achieves a lower conductivity with a high flexibility; thus, it can be incorporated into soft materials to be applied in flexible display, wearable electronics. Not conductive to signal transfer, shape memory alloys like Nitinol are stimulated mechanically via electrical or thermal influence, thus adaptive and responsive textile applications have been possible, including adaptive fitting and haptics in smart clothes.

Sensors

The most important & crucial components is Sensors which allows smart textiles to detect changes in the user or environment. Depending on the application, these sensors may detect strain, temperature, humidity, pressure, or physiological parameters biochemical markers, or such as EMG, ECG, and respiration.

- Pressure & Strain sensors utilize piezoelectric, piezoresistive, capacitive materials to detect impact, posture or movement.
- Temperature sensors can be made using thermocouples, RTDs (resistive temperature devices) or thermistors, integrated into textiles (Cherenack & van Pieterson, 2012).
- Biosensors are developed using electrochemical principle or enzymatic to monitor lactate, glucose, sweat, or other biomarkers (Heikenfeld et al., 2018).

These sensors are laminated, woven into fabrics, or printed, depending on their mechanical and chemical properties.

Actuators

Actuators allow textiles to react or respond to perceived stimuli. Actuators can be used by Smart textiles heating, shape transformation, drug delivery, or vibration.

- SMAs: Shape memory alloys such metallic materials that can return to a predefined shape when heated like Nitinol (Tao, 2005).
- EAPs: Electroactive polymers can change shape or size against an electric field and are used for robotic assistance or haptic feedback.
- PCMs: Phase change materials release and store thermal energy, also regulate body temperature against any environmental changes.
- Heating elements, are made up of printed conductive inks or resistive conductive yarns, are used in costumes or garments for medical therapy & thermal comfort.

Power Sources

Power supply is the challenge in smart textile systems. Power can be delivered using:

- Flexible batteries: lithium-polymer or Lithium-ion cells designed to be bendable, thin, and integrated into garments.
- Energy harvesting systems: These include piezoelectric generators, thermoelectric generators & solar cells, that can convert ambient energy into electrical power (Stoppa & Chiolerio, 2014).
- micro fuel cells **&** Supercapacitors are also explored for fast charging capabilities & higher energy density.

Emerging strategies aim for enhancing wearability & reducing dependence on external charging which could be possible by self-powered smart textiles.

Communication Units & Data Processing

For smart textiles to function autonomously, they must be equipped with data acquisition, processing, and communication modules.

- System-on-chip (SoC) & Microcontrollers (MCUs) are the platforms which process signals from control actuators and sensors.
- Wireless communication through Wi-Fi, Bluetooth, NFC, or Zigbee, allows real-time data transmission to external devices such as cloud platforms or smartphones.
- These components are either attached through flexible PCB (printed circuit boards) or integrated with conductive threads or embedded & miniaturized within textile structures (Mattmann, Clemens, & Tröster, 2008).

Applied protocols in data transmission on wearable technology usually have protocols such as Bluetooth Low Energy (BLE), ZigBee, Wi-Fi, and NB-IoT, with each protocol having its own benefits and compromises. BLE is useful because it uses little power and has a sufficient range, which makes it appropriate to use on personal area networks, but its bandwidth is narrower than Wi-Fi, which has data rates higher but with higher power consumption. ZigBee introduces a fairly good compromise between power efficiency and scalability of the network it tends to support lower data rates, though. NB-IoT is a good solution to long range, low power, wide area communications that is limited in bandwidth meaning that it is not ideally used to transmit continuous data making it also of use in smart textiles. Wearable systems utilize power management systems with flexible batteries and supercapacitors that provide compact and lightweight energy storage, energy harvesting power sources like solar, triboelectric and thermoelectric that transform energy found in the environment into electrical energy allowing less frequent charging. Different vendors and technologies make interoperability a problem; certain proposals such as the IEEE 802.15.6 have been put forward to ensure compatibility between different body area networks, but compatibility issues such as vendor lock-in and proprietary solutions still exist. It is vital in terms of security to focus on the encryption of the data transmission and strong authentication schemes to enhance privacy of sensitive personal data gathered by the smart textiles to avoid unauthorized data access.

Substrate and Textile Structure

The base fabric plays a crucial role in durability, integration feasibility & comfort. Preferred substrates include:

- Knitted fabrics: Known for stretch ability & flexibility, ideal for wearable electronics & for fixing strain sensors.
- Woven fabrics: Offer dimensional stability & mechanical strength, mostly used for structural integration.
- Nonwovens: Used in disposable applications like temporary sensors or medical dressings.

Materials such as polyester, cotton, spandex and nylon are commonly used depending on the required mechanical properties & end-use.

FABRICATION AND INTEGRATION TECHNIQUES

The fabrication and integration of smart textiles involve embedding or integrating electronic components, sensors, actuators, and communication modules into textile substrates without compromising comfort, flexibility, washability, and wearability. These techniques span various disciplines, including materials science, textile engineering, electronics, and nanotechnology. Smart textile systems can be realized through three primary methods: fiber-level integration, fabric-level integration, and finish-level integration.

Fiber-Level Integration

This approach involves the creation of functional fibers or yarns during the initial fiber production stage. These fibers inherently possess smart properties such as conductivity, sensing, or energy storage.

Conductive Fiber Spinning

Functional fibers are produced by melt spinning, wet spinning, or electrospinning processes using materials like carbon nanotubes (CNTs), conductive polymers (e.g., PEDOT:PSS), or metal nanoparticles blended into the polymer matrix (Stoppa & Chiolerio, 2014). These fibers can then be woven or knitted like conventional yarns, offering seamless integration into garments.

Coaxial Fiber Construction

In coaxial fibers, the core and sheath are made from different materials, for example, an insulating core and a conductive sheath. This enables the fiber to function as a sensor or transmission line. Coaxial electrospinning is a common technique used to fabricate such fibers with sensing and actuating capabilities (Zeng et al., 2014).

Yarn-Level Integration

Conductive or functional yarns are fabricated and then used in textile structures via traditional techniques like weaving, knitting, or embroidery.

Embroidered Circuits

Embroidery machines are used to stitch conductive yarns into textile substrates to form **interconnects and circuitry**. This technique allows high customization and compatibility with various fabric types (Cherenack & van Pieterson, 2012). While easy to implement, it may lead to issues like yarn fraying and reduced durability upon washing.

Braided and Twisted Yarn Integration

Functional yarns can also be created by twisting conductive wires with textile yarns to create mechanically robust hybrid threads. These are used for pressure sensing, ECG monitoring, or power transmission (Mattmann, Clemens, & Tröster, 2008).

Fabric-Level Integration

In this method, components are integrated during fabric manufacturing using traditional textile processes.

Weaving and Knitting

Conductive yarns are incorporated into the fabric structure during weaving or knitting. This allows for distributed sensing networks, such as strain-sensitive zones or embedded heating elements (Zhang et al., 2019). Knitting provides flexibility and stretch, whereas weaving ensures dimensional stability.

3D Knitting and Jacquard Weaving

Advanced manufacturing methods like 3D knitting **or** Jacquard weaving can produce spatially varying fabric architecture with specific electrical or mechanical properties tailored to the body's anatomy. This technique is particularly useful for wearable ECG/EMG sensors and compression garments (Tao, 2005).

Surface Coating and Printing Techniques

Surface modification is one of the most accessible ways to impart smart functions to fabrics by applying conductive or responsive materials.

Screen Printing and Inkjet Printing

These methods involve the deposition of conductive inks such as silver nanoparticles, graphene, or carbon-based inks onto fabric substrates. Screen printing is suitable for large-scale production, while inkjet printing offers high-resolution patterns and material efficiency (Weng et al., 2024).

Dip Coating and Spray Coating

Textiles are submerged or sprayed with functional solutions containing conductive polymers or nanoparticles. This provides a uniform coating over large areas and is used for producing temperature sensors, strain gauges, and touch-sensitive fabrics (Mondal, 2008).

Integration of Discrete Electronic Components

While fiber and fabric-level integration offers flexibility, high-performance electronics such as microcontrollers, batteries, and communication modules are often incorporated as discrete components.

Flexible Printed Circuit Boards (FPCBs)

These are thin, bendable circuit platforms that can be sewn, glued, or laminated onto fabrics. Conductive threads or ribbons are used to connect them to sensors and actuators embedded in the textile structure (Cherenack et al., 2010).

Modular and Snap-On Systems

To allow washability and repairability, components such as sensors, batteries, or LEDs are made modular and connected via snaps, Velcro, or magnetic interfaces. This technique enhances reusability and allows for easy detachment before laundering.

Challenges in Fabrication and Integration

Despite technological progress, several challenges remain:

- Durability: Smart textiles must endure mechanical deformation and multiple washing cycles without degrading performance.
- Comfort and Flexibility: Integration techniques should not compromise the fabric's tactile properties, breathability, or drape.
- Scalability: Transitioning from laboratory prototypes to commercially viable products requires scalable, cost-effective manufacturing techniques.
- Interconnectivity and Reliability: Robust connections between textile-based and rigid components are crucial for consistent functionality.

Researchers are increasingly developing hybrid integration approaches, combining printed electronics, embedded yarns, and detachable modules to meet diverse application requirements in healthcare, sports, military, and fashion.

Applications of Smart Textiles

Smart textiles have transformed traditional fabrics into multifunctional platforms capable of sensing, reacting, and adapting to environmental and physiological conditions. The integration of electronic components and functional materials within textile substrates has opened up diverse application areas across healthcare, sports, military, fashion, protection, and assistive technologies for people with disabilities. The following section provides an in-depth exploration of the major application domains of smart textiles.

Healthcare. Hexoskin clothing (smart shirts deployed in remote cardiac monitoring) has both shown high accuracy of ECG signal reading (more than 90 percent) during 24 hours of wear in a group of 50 patients and cut in-clinic visits by 30 percent (Haghi, Thurow, & Stoll, 2017; Silvestre De Oliveira et al., 2023).

Fitness. The Sensoria smart sock has demonstrated gait analytics accuracy to within 3 percent of gold standard force plates and enhanced stride tracking performance and gait analytics in athletes (Sensoria, 2022).

Military. The SMART e-textile program of the US army decreased cases of heat stress by 15 percent in the course of testing adaptive cooling garments in desert circumstances.

Healthcare and Medical Applications

Healthcare is one of the most prominent fields where smart textiles have made substantial contributions. Smart medical textiles are being used for real-time monitoring of vital signs, chronic disease management, rehabilitation, and post-surgical care.

Vital Sign Monitoring

Smart garments embedded with sensors can continuously monitor heart rate, respiration rate, body temperature, and blood pressure without causing discomfort. For instance, textile-based ECG and EMG electrodes enable non-invasive and continuous cardiac monitoring, which is critical for early detection of cardiovascular diseases (Mattmann, Clemens, & Tröster, 2008).

Chronic Disease and Elderly Care

Wearable textiles integrated with accelerometers and pressure sensors help monitor mobility patterns, fall detection, and sleep quality, supporting patients with chronic conditions and elderly individuals in assisted living environments (Axisa et al., 2005). Smart socks for diabetic patients can detect temperature variations indicating the risk of ulcers (Stoppa & Chiolerio, 2014).

Rehabilitation and Assistive Therapy

Smart textiles are used in exosuits and wearable rehabilitation systems, providing real-time feedback to aid motor function recovery. For instance, textile sensors can measure joint angles or muscle activity during physiotherapy (Yin et al., 2021).

Sports and Fitness

Athletes and fitness enthusiasts benefit from smart textiles for performance optimization, injury prevention, **and** training feedback.

- Garments equipped with strain and motion sensors can track body movement, posture, and muscle engagement during physical activity (Zhang et al., 2019).

- Sweat-analyzing textiles help monitor hydration levels and electrolyte loss.
- Compression garments with embedded sensors enhance circulation while offering biofeedback during workouts.

These systems provide coaches and athletes with quantitative data for tailored training programs and load management (Amjadi et al., 2016).

Military and Defense

Smart textiles are employed in defense for battlefield communication, soldier health monitoring, and environmental adaptation.

- Wearable sensors embedded in military uniforms can track vital signs, stress levels, and fatigue, which enhances situational awareness and decision-making (Cherenack & van Pieterson, 2012).
- Thermo-regulating and camouflage textiles adapt to ambient conditions to improve comfort and stealth.
- Smart textiles integrated with energy harvesting technologies supply power to electronic devices carried by soldiers.

Fashion and Entertainment

In the fashion industry, smart textiles enable interactive garments, color-changing fabrics, and illumination effects that respond to music, temperature, or motion.

- Designers use responsive materials and conductive threads to create garments that light up or change color dynamically.
- Textile displays and haptic feedback systems offer immersive experiences in gaming and virtual reality (Post et al., 2000).

These innovations blend aesthetics with functionality, enhancing user engagement and personalization.

Protection and Safety

Smart textiles offer enhanced protection for workers, emergency responders, and industrial personnel.

- Flame-resistant fabrics with integrated sensors can detect heat, gas leaks, or hazardous chemicals in real time.

- Pressure-sensitive insoles can monitor foot pressure distribution, preventing musculoskeletal injuries among workers.
- Smart helmets and vests provide impact detection and alerts in high-risk environments (Mondal, 2008).

Adaptive Clothing for People with Disabilities

Smart adaptive clothing is a revolutionary tool for promoting independence, dignity, and well-being among people with disabilities.

Mobility Assistance

Smart textiles embedded with actuators can support joint movements in individuals with limited mobility, functioning as soft exoskeletons. These systems assist in walking or arm lifting and improve mobility (Tao, 2005).

Communication and Sensory Feedback

For individuals with hearing or vision impairments, smart garments can provide haptic feedback, directional cues, or alerts through vibrations or tactile signals (Bhattacharjee, Adaimi, & Thomaz, 2020).

Health Monitoring and Alerts

Wearable systems tailored for disabled individuals can monitor spasms, seizures, or pressure ulcers, sending alerts to caregivers or medical professionals. This supports remote health monitoring and proactive care (Dunne et al., 2005).

Such applications help overcome dressing barriers, improve user autonomy, and promote inclusive clothing solutions.

CHALLENGES AND LIMITATIONS

Challenges and Limitations of Smart Textiles

Despite the transformative potential of smart textiles across sectors such as healthcare, military, sports, and fashion, their widespread adoption faces significant challenges. These range from technical and manufacturing limitations to economic, ergonomic, and regulatory issues. The path to commercialization and practical deployment is hindered by the complex interplay between electronic functionality and textile properties.

Table 2. Challenges in smart textile development and literature-backed solutions

Challenge	Description	Possible Solutions	Literature
Durability	Fiber fatigue, washing, mechanical stress	Encapsulation, new weaving patterns	(Li et al., 2020)
User Comfort	Stiffness, skin irritation	Soft hydrogels, coating with biocompatible layers	(Castano & Flatau, 2014)
Cybersecurity	Data interception, privacy breaches	End-to-end encryption, on-device preprocessing	(Zhao et al., 2021)
Regulatory Compliance	Safety, RF-exposure, medical classification	Following FDA, CE marking, ISO 13485 for medical wearables	(Khan et al., 2020)
Power Autonomy	Frequent recharging, battery bulk	Energy harvesting, low-power MCU design	(Zhou, Gao, & Wang, 2020)

Table 2 provides a summary of major issues experienced in developing and commercializing smart textiles, such as durability, comfort of users, cybersecurity, regulation, and power independence. The challenges are accompanied by a set of specific problems emerging in real-world implementation, literature-based strategies, and technological means to overcome barriers, which can lead to the future progress and practical use of the wearable textile technologies.

Technical and Functional Challenges

Durability and Washability

One of the foremost challenges is ensuring that smart textiles are durable and washable while retaining their electronic performance. Conventional electronic components are not inherently compatible with the flexible, stretchable, and deformable nature of textiles. Frequent bending, folding, and exposure to moisture and detergents can degrade sensor performance and electrical connections (Cherenack & van Pieterson, 2012).

Efforts to encapsulate or protect components often compromise the comfort, breathability, and aesthetics of the textile, which defeats the primary advantage of wearable fabrics (Stoppa & Chiolerio, 2014).

Power Supply and Energy Harvesting

Powering smart textiles remains a major bottleneck. Most e-textiles require a continuous and lightweight power source, but batteries can add bulk and discomfort. While energy harvesting techniques such as triboelectric, piezoelectric, and solar methods are under exploration, their efficiency and stability in real-world environments are still limited (Amjadi et al., 2016).

Data Accuracy and Signal Interference

Maintaining high-fidelity signal acquisition in a dynamic, body-conforming textile system is difficult. Noise from movement, perspiration, or electromagnetic interference can reduce signal-to-noise ratios, impacting the accuracy of physiological monitoring (Zhang et al., 2019). This becomes particularly problematic in medical and military applications where precision is critical.

Manufacturing and Scalability

Complex Fabrication Processes

The integration of conductive materials, sensors, and microcontrollers into fabrics involves multistep, labor-intensive fabrication techniques, such as screen printing, embroidery with conductive yarns, and chemical deposition. These methods are not always compatible with existing textile manufacturing infrastructure, impeding mass production and scalability.

Standardization Issues

There is a lack of industry-wide standards for testing, performance evaluation, and interfacing hampers interoperability between components. Without established guidelines, manufacturers and designers face difficulties in quality assurance and integration into other smart systems (Tao, 2005).

Ergonomic and User-Centered Design Issues

Smart textiles must remain comfortable, lightweight, breathable, and skin-compatible, especially for long-term wear. However, embedding electronics often results in rigid or bulky zones that compromise ergonomics. Additionally, thermal management becomes a concern, as electronic components generate heat during operation (Mattmann, Clemens, & Tröster, 2008).

The lack of universal design principles also limits adoption among people with disabilities, who may require tailored, inclusive designs for ease of use, dressing, and comfort (Bhattacharjee, Adaimi, & Thomaz, 2020).

Economic Constraints

The high cost of materials and fabrication is a major limitation. Conductive yarns, flexible circuits, and specialty sensors significantly increase the production cost compared to traditional textiles. Additionally, limited supply chains and specialized workforce requirements make commercialization economically challenging (Cherenack & van Pieterson, 2012).

Cost factors also limit accessibility for marginalized populations, including elderly users or people with disabilities, who might benefit most from these innovations but often lack the financial means to adopt them.

Ethical and Privacy Concerns

Smart textiles designed for monitoring health, movement, or location raise issues of data privacy, informed consent, and cybersecurity. Continuous data collection from the human body poses risks of misuse or unauthorized access to sensitive information (Dunne et al., 2005). Proper frameworks for data encryption, anonymization, and ethical use are still under development.

Moreover, users may be reluctant to wear garments perceived as surveillance tools, limiting social acceptability and adoption (Axisa et al., 2005).

Environmental and Sustainability Issues

The sustainability of smart textiles is increasingly under scrutiny. Most components such as printed circuits, batteries, and synthetic substrates are non-biodegradable and challenging to recycle. As the market for e-textiles grows, so does the e-waste burden, prompting the need for eco-friendly materials and circular design principles (Stoppa & Chiolerio, 2014).

The development of biodegradable conductive polymers, natural fiber substrates, and low-energy production methods remains an active area of research to address this limitation (Mondal, 2008).

ETHICAL AND SOCIAL IMPLICATIONS

As smart textiles become more integrated into everyday life, particularly in healthcare, fitness, military, and personal monitoring applications, they raise significant ethical and social questions. These concerns revolve around privacy, autonomy, consent, equity, surveillance, psychological effects, and data governance. The integration of sensing and communication capabilities in garments transforms clothing from passive items into active tools of data collection and interaction, presenting complex moral dilemmas that must be addressed to ensure responsible innovation.

Privacy and Data Security

Smart textiles often include sensors that monitor physiological parameters (e.g., heart rate, temperature, muscle activity) or behavioral patterns (e.g., posture, gait, movement). These data types are inherently personal and sensitive, especially when used in medical or occupational settings. Unauthorized access, data breaches, or misuse of such data can have serious implications for the user's privacy, dignity, and even employment (Buechley & Eisenberg, 2008).

Unlike smartphones or wearables that are easily removable or turned off, smart garments are closely attached to the body, often worn for extended periods, which raises concerns about continuous and covert data collection. Users may not be fully aware of what data is being collected, how long it is stored, or who has access to it (Axisa et al., 2005). For example: A smart shirt used in elderly care that collects fall data might inadvertently record private information about a user's lifestyle, sleep patterns, or habits without explicit, repeated consent.

Informed Consent and User Autonomy

Ethical deployment of smart textiles requires clear, understandable processes for informed consent. However, many users may lack the technical literacy to understand the functioning, data handling, and potential risks associated with e-textiles. This problem is particularly critical among vulnerable groups such as children, the elderly, or individuals with disabilities, who may wear smart clothing without fully understanding or consenting to its capabilities.

Additionally, there is a risk that users could feel pressured into adopting such technologies for example, through workplace mandates or health insurance incentives undermining their personal agency and autonomy.

Surveillance and Social Control

Smart textiles can be employed as tools of surveillance, especially in institutional settings such as military, prisons, schools, or workplaces. Wearable technologies may enable continuous tracking of location, movement, stress levels, or productivity, potentially creating an atmosphere of constant monitoring and behavioral regulation (Mann, Nolan, & Wellman, 2003).

This leads to concerns about the creation of "dataveillance societies", where individuals are subtly coerced to behave in certain ways under the pressure of being monitored, even if surveillance is not overt. Over time, this could erode trust, freedom, and self-expression, particularly among youth and employees.

Psychological and Social Impacts

Being under constant observation or wearing clothing that gathers biometric data can affect a person's psychological well-being. Studies have shown that constant monitoring may lead to anxiety, stress, and a sense of loss of privacy, particularly when users are unsure about who is accessing the data or how it will be interpreted (Hartzog, 2018).

Furthermore, in social settings, wearing visibly electronic garments might result in stigmatization, social exclusion, or identity conflicts. Users with medical or cognitive conditions may feel exposed or labeled, affecting their self-esteem and social inclusion.

Digital Divide and Accessibility

There is growing concern that smart textiles may widen the digital divide, favoring users who are technologically savvy, affluent, and able-bodied. The cost of smart garments and the infrastructure required (e.g., connectivity, apps, smart devices) can make them inaccessible to marginalized populations, such as the elderly, people in rural areas, or those from low-income backgrounds (Zhang et al., 2022).

Without inclusive design strategies, smart textiles risk exacerbating social inequalities, leaving out those who may benefit the most such as people with disabilities who need adaptive clothing with assistive features (Bhattacharjee, Adaimi, & Thomaz, 2020).

Regulatory and Ethical Frameworks

Currently, the development and deployment of smart textiles outpace legal and ethical frameworks. There is an urgent need for standardized protocols, certifications, and regulatory oversight that ensure safe usage, responsible data management, and ethical design (Tao, 2005). This includes not only data protection laws (e.g., GDPR) but also design ethics guidelines, particularly when smart textiles are developed for vulnerable populations.

Collaborative engagement between scientists, ethicists, policy makers, designers, and users is essential to ensure that smart textiles are not only functional but also socially and ethically responsible.

FUTURE TRENDS

Smart textiles are at the forefront of a technological revolution in material science, wearable electronics, and human-computer interaction. As interdisciplinary innovations continue to emerge, the future of smart textiles will be shaped by enhanced functionality, miniaturization, sustainability, user-centric design, and artificial intelligence (AI) integration. These trends indicate a shift from niche applications to widespread adoption in everyday life, healthcare, military, sports, fashion, and personal wellness.

Miniaturization and Seamless Integration

One of the major trends in smart textiles is the miniaturization of electronic components such as sensors, actuators, and energy storage systems. The goal is to create garments that are lightweight, flexible, and indistinguishable from regular clothing, enhancing user comfort and wearability. Researchers are developing nanomaterials and microelectronic systems that can be embedded into fibers or coatings without compromising breathability or flexibility (Stoppa & Chiolerio, 2014).

Seamless integration of electronics into yarns, fabrics, or threads known as "electronic textiles" or e-textiles is expected to become more prevalent, allowing for multifunctional garments that monitor health, track movement, or communicate with external devices (Cherenack & van Pieterson, 2012).

Artificial Intelligence and Smart Data Analytics

Future smart textiles will leverage artificial intelligence (AI) and machine learning algorithms to process the massive amount of physiological and behavioral data collected from the human body. These systems will go beyond passive sensing to offer predictive analytics, real-time feedback, and adaptive response mechanisms.

For example, AI can be used to detect abnormal health patterns from continuous monitoring and provide early alerts for medical intervention (Stoppa & Chiolerio,

2014). In rehabilitation, AI-powered garments could adapt exercises based on user performance, enabling personalized therapy.

Energy Harvesting and Self-Sustainability

A significant challenge in current smart textiles is power supply. Future innovations are focusing on making textiles self-powered through energy harvesting technologies such as piezoelectric fibers, triboelectric nanogenerators, and photovoltaic textiles (Jiang, 2020). These fabrics can generate electricity from human motion, body heat, or sunlight, enabling continuous, long-term operation without external batteries.

The development of flexible and washable energy storage systems, such as fiber-shaped supercapacitors and stretchable batteries, is another key area of growth (Shi et al., 2019).

Sustainability and Eco-Friendly Smart Textiles

As environmental concerns become more pressing, there is increasing interest in eco-sustainable smart textiles made from biodegradable, recyclable, or organic materials. Future smart textiles will not only be functional but also environmentally responsible, incorporating materials with low carbon footprints and minimal chemical processing (Stoppa & Chiolerio, 2014). Moreover, modular designs are being explored, allowing users to replace or upgrade specific components (e.g., sensors or batteries) rather than discarding the entire garment, thereby reducing electronic waste.

Personalized and Adaptive Clothing

Advancements in adaptive materials such as shape-memory alloys, thermo-responsive fabrics, and electrochromic materials will drive the development of clothing that dynamically responds to environmental or physiological changes. This includes garments that can adjust ventilation, change color, or alter texture based on temperature, mood, or external stimuli (Zhou, Gao, & Wang, 2020).

Smart garments of the future will offer mass customization, allowing users to tailor functionalities according to individual health profiles, style preferences, or activity levels, supported by 3D knitting and digital fabrication technologies.

Integration with Internet of Things (IoT)

Smart textiles are anticipated to become critical nodes in the Internet of Things (IoT), connecting seamlessly with smartphones, wearables, home systems, and cloud platforms. This integration will enable interconnected smart environments where garments communicate with other smart devices to offer personalized user experiences (Tao, 2005).

For instance, a smart fitness shirt might sync with a home treadmill and smart speaker to adjust music tempo or display health stats in real-time, creating a multisensory training ecosystem.

Expansion into New Sectors

While current applications are dominated by healthcare and sports, smart textiles are expected to expand into areas such as:

- Military and Defense: Uniforms with environmental sensing, camouflage, and communication capabilities.
- Fashion and Entertainment: Interactive garments for performance arts, gaming, or mood expression.
- Workplace Safety: Garments that detect hazardous gases, fatigue, or exposure to extreme conditions.

As these sectors evolve, regulations and standards will also mature, ensuring product safety, quality control, and ethical data use.

CONCLUSION AND FUTURE RESEARCH DIRECTIONS

Smart textiles and wearable technologies continue to advance at a rapid pace becoming increasingly complex multifunctional platforms that integrate complex materials, embedded electronics, and a natural interface to IoT. The interconnection of conductive fibers, bendy sensors, energy harvesting techniques, and safe wireless standards is rediscovering dramatic possibilities in medical care monitoring, athletic performance management, military capacity to respond, and environmental perception. This review has pointed to the inherent importance of material science and system-level electronics working in conjunction with one another with the key ideas being that the true success and grand-scale adoption will revolve around durability under repeated use, comfort over prolonged wear, protective measures against hacking, power independence, and adherence to legislative contexts. As a perspective, further research is needed in such areas as scalable and sustainable manufacturing processes that will cost less and would still be precise in functionality; development of biodegradable or recyclable smart textile to mitigate environmental impacts; increase in AI-assisted on-device data analysis towards personalized insights and; efficient cross-platform interoperability standards so as to be used everywhere. Also, cross-functional partnerships between materials scientists, electronics engineers, data security specialists, and government regulators will play a key role in ensuring that market readiness is achieved faster and the privacy and trust of the user are preserved. The further merging of textile engineering, wearable computers, and intelligent processing of data could transform several sectors as long as innovation is coupled with practicability, ethical treatment of biosensed information, and its overall sustainability.

The evolution & historical development of textiles to smart textile, rooted in space applications & military, has now permeated various sectors such as sports, fashion, personal wellness defense, and healthcare. With advances in flexible sensors, conductive materials & integrated circuits, smart textiles are now capable of improving physical performance monitoring vital signs, enhancing safety & providing therapeutic functions. Classification into ultra-smart, active & passive textiles provides a framework to understand the functionality levels textiles.

Technological progress in components & materials including conductive polymers, nano-materials, bio-compatible substances & e-fibers has enabled the creation of multifunctional fabrics. Simultaneously, innovations in integration techniques & fabrication such as coating, embroidery, weaving, knitting & printing have allowed for the seamless fixing of electronics into textiles while preserving flexibility & comfort.

Irrespective of their broad applications, smart textiles still face several limitations & challenges, including wash ability, durability, standardization issues, manufacturing costs, and power supply constraints. Additionally, concerns regarding user consent, potential societal inequality & data privacy rising from wearable technology necessitate legal & ethical scrutiny.

Looking forward, the future of smart textiles is bright. Emerging trends such as energy harvesting, personalized wearable, AI integration, & sustainable materials are expected to create the growth of textile sector. The development of biodegradable, self-powered garments, intelligent fabrics & smart fibers that adapt & learn to user behavior could transform how we interact with our surroundings & clothing.

Smart textiles are the technological innovation along with the paradigm shift in the way one interact & perceive the fabric all around. This journey of creating passive fabric to smart fabric will need collaborative efforts among engineers, scientists, policymakers, designers, & consumers to ensure that these technological innovations are ethical, functional, inclusive, and accessible. As smart textiles shift from prototypes made in laboratory to commercial reality, they hold the potential to significantly enhance quality of life, improve healthcare outcomes, and enable a new era of responsive, intelligent environments.

Despite the promising applications, several challenges persist. Issues related to durability, washability, data security, power supply, and user comfort continue to hinder large-scale commercialization and daily adoption. As noted by Cherenack and van Pieterson (2012) and Heikenfeld et al. (2018), the integration of electronics with textiles demands significant advancements in materials science, circuit miniaturization, and human-centered design. Moreover, regulatory frameworks and standardization are essential to ensure safety, interoperability, and consumer trust. The future of smart fabrics hinges on cross-sectoral collaboration among textile engineers, material scientists, biomedical researchers, designers, and end-users. Sustainable development and ethical considerations must also be prioritized, particularly concerning e-waste, privacy, and accessibility.

In conclusion, smart fabrics and wearable technologies are not merely innovative enhancements to traditional textiles; they represent a paradigm shift toward intelligent, responsive environments that align with the evolving needs of modern society. Continued investment in research, user-focused design, and scalable production techniques will be critical to unlocking their full potential and ensuring these technologies serve as inclusive solutions that benefit diverse populations globally.

Smart textiles are continuously evolving as more interdisciplinary collaborations emerge between fields such as data science, materials engineering, electronics, and design. Emerging approaches emphasize decentralized data processing, localized computation through edge devices, and real-time analytics to improve performance in critical applications. Developments in photonic textiles, which manipulate light for optical sensing, or fibertronics electronics integrated at the fiber level are pushing the boundaries of what wearable fabrics can achieve in terms of resolution, responsiveness, and intelligence. In humanitarian and crisis scenarios, smart clothing can serve to monitor displaced populations, track health status in refugee camps, or assist in epidemic response through early detection of symptomatic patterns. These forward-thinking applications necessitate consideration of privacy, equitable deployment, and ethical governance to prevent misuse or data exploitation.

Smart textiles are continuously evolving as more interdisciplinary collaborations emerge between fields such as data science, materials engineering, electronics, and design. Emerging approaches emphasize decentralized data processing, localized computation through edge devices, and real-time analytics to improve performance in critical applications. Developments in photonic textiles, which manipulate light for optical sensing, or fibertronics electronics integrated at the fiber level are pushing the boundaries of what wearable fabrics can achieve in terms of resolution, responsiveness, and intelligence. In humanitarian and crisis scenarios, smart clothing can serve to monitor displaced populations, track health status in refugee camps, or assist in epidemic response through early detection of symptomatic patterns. These forward-thinking applications necessitate consideration of privacy, equitable deployment, and ethical governance to prevent misuse or data exploitation.

Smart textiles are continuously evolving as more interdisciplinary collaborations emerge between fields such as data science, materials engineering, electronics, and design. Emerging approaches emphasize decentralized data processing, localized computation through edge devices, and real-time analytics to improve performance in critical applications. Developments in photonic textiles, which manipulate light for optical sensing, or fibertronics electronics integrated at the fiber level are pushing the boundaries of what wearable fabrics can achieve in terms of resolution, responsiveness, and intelligence. In humanitarian and crisis scenarios, smart clothing can serve to monitor displaced populations, track health status in refugee camps, or assist in epidemic response through early detection of symptomatic patterns. These forward-thinking applications necessitate consideration of privacy, equitable deployment, and ethical governance to prevent misuse or data exploitation.

Smart textiles are continuously evolving as more interdisciplinary collaborations emerge between fields such as data science, materials engineering, electronics, and design. Emerging approaches emphasize decentralized data processing, localized computation through edge devices, and real-time analytics to improve performance in critical applications. Developments in photonic textiles, which manipulate light for optical sensing, or fibertronics electronics integrated at the fiber level are pushing the boundaries of what wearable fabrics can achieve in terms of resolution, responsiveness, and intelligence. In humanitarian and crisis scenarios, smart clothing can serve to monitor displaced populations, track health status in refugee camps, or assist in epidemic response through early detection of symptomatic patterns. These forward-thinking applications necessitate consideration of privacy, equitable deployment, and ethical governance to prevent misuse or data exploitation.

REFERENCES

Amjadi, M., Kyung, K. U., Park, I., & Sitti, M. (2016). Stretchable, skin-mountable, and wearable strain sensors and their potential applications: A review. *Advanced Functional Materials*, *26*(11), 1678–1698. DOI: 10.1002/adfm.201504755

Axisa, F., Schmitt, P. M., Gehin, C., Delhomme, G., McAdams, E., & Dittmar, A. (2005). Flexible technologies and smart clothing for citizen medicine, home healthcare, and disease prevention. *IEEE Transactions on Information Technology in Biomedicine*, *9*(3), 325–336. DOI: 10.1109/TITB.2005.854505 PMID: 16167686

Buechley, L., & Eisenberg, M. (2008). The LilyPad Arduino: Toward wearable engineering for everyone. *IEEE Pervasive Computing*, *7*(2), 12–15. DOI: 10.1109/MPRV.2008.38

Castano, L. M., & Flatau, A. B. (2014). Smart fabric sensors and e-textile technologies: A review. *Smart Materials and Structures*, *23*(5), 053001. DOI: 10.1088/0964-1726/23/5/053001

Cherenack, K., & van Pieterson, L. (2012). Smart textiles: Challenges and opportunities. *Journal of Applied Physics*, *112*(9), 091301. DOI: 10.1063/1.4742728

De Rossi, D., Lorussi, F., Mazzoldi, A., Orsini, P., & Scilingo, E. P. (2003). Active dressware: wearable kinesthetic systems. In *Sensors and Sensing in Biology and Engineering* (pp. 379–392). Springer Vienna. DOI: 10.1007/978-3-7091-6025-1_26

Dunne, L. E., Brady, S., Smyth, B., & Diamond, D. (2005). Initial development and testing of a novel foam-based pressure sensor for wearable sensing. *Journal of Neuroengineering and Rehabilitation*, *2*(1), 4. DOI: 10.1186/1743-0003-2-4 PMID: 15740623

Haghi, M., Thurow, K., & Stoll, R. (2017). Wearable devices in medical internet of things: Scientific research and commercially available devices. *Healthcare Informatics Research*, *23*(1), 4–15. DOI: 10.4258/hir.2017.23.1.4 PMID: 28261526

Hartzog, W. (2018). *PrivacyOs Blueprint: The Battle to Control the Design of New Technologies*. Harvard University Press.

Heikenfeld, J., Jajack, A., Rogers, J., Gutruf, P., Tian, L., Pan, T., Li, R., Khine, M., Kim, J., Wang, J., & Kim, J. (2018). Wearable sensors: Modalities, challenges, and prospects. *Lab on a Chip*, *18*(2), 217–248. DOI: 10.1039/C7LC00914C PMID: 29182185

Horrocks, A. R., & Anand, S. C. (2000). *Handbook of technical textiles*. Woodhead Publishing.

Jee, H., Kwon, O. Y., Park, Y. H., Kim, Y. W., & Lee, J. H. (2019). A wearable smart shirt for monitoring respiration, ECG, and activity. *IEEE Sensors Journal*, *19*(24), 11734–11744. DOI: 10.1109/JSEN.2019.2935203

Jiang, Y. (2020). Combination of wearable sensors and internet of things and its application in sports rehabilitation. *Computer Communications*, *150*, 167–176. DOI: 10.1016/j.comcom.2019.11.021

Khan, Y., Han, D., Pierre, A., & Arias, A. C. (2020). Towards conformable hybrid electronics for wearable health applications. *Advanced Functional Materials*, *30*(47), 1909475. DOI: 10.1002/adfm.201909475

Khan, Y., Ostfeld, A. E., Lochner, C. M., Pierre, A., & Arias, A. C. (2020). Monitoring of vital signs with flexible and wearable medical devices. *Advanced Materials*, *32*(15), 1905279. DOI: 10.1002/adma.201905279 PMID: 31742812

Li, X., Xu, B., & Gao, Y. (2020). Encapsulation techniques for durable wearable biosensors. *Advanced Healthcare Materials*, *9*(15), 2000605. DOI: 10.1002/adhm.202000605 PMID: 32529775

Liao, X., Zhang, Z., Liang, Q., Liao, Q., & Zhang, Y. (2015). Extremely elastic wearable carbon nanotube fiber strain sensor for monitoring of human motion. *ACS Nano*, *9*(6), 5929–5936. DOI: 10.1021/acsnano.5b00599 PMID: 26038807

Mann, S. (1997). Wearable computing: A first step toward personal imaging. *Computer*, *30*(2), 25–32. DOI: 10.1109/2.566147

Mann, S., Nolan, J., & Wellman, B. (2003). Sousveillance: Inventing and using wearable computing devices for data collection in surveillance environments. *Surveillance & Society*, *1*(3), 331–355. DOI: 10.24908/ss.v1i3.3344

Mattila, H. (2006). *Intelligent textiles and clothing*. Woodhead Publishing.

Mattmann, C., Clemens, F., & Tröster, G. (2008). Sensor for measuring strain in textile. *Sensors (Basel)*, *8*(6), 3719–3732. DOI: 10.3390/s8063719 PMID: 27879904

Park, S., & Jayaraman, S. (2003). Enhancing the quality of life through wearable technology. *IEEE Engineering in Medicine and Biology Magazine*, *22*(3), 41–48. DOI: 10.1109/MEMB.2003.1213625 PMID: 12845818

Paterson, T. E., Hagis, N., Boufidis, D., Wang, Q., Moore, S. R., Da Silva, A. C., Mitchell, R. L., Alix, J. J. P., & Minev, I. R. (2022). Monitoring of hand function enabled by low complexity sensors printed on textile. *Flexible and Printed Electronics*, *7*(3), 035003. DOI: 10.1088/2058-8585/ac7dd1

. Sensoria. (2022). *Sensoria smart sock gait analysis validation study*. Sensoria Inc.

Shi, B., Chen, X., Yue, Z., Yin, S., Weng, Q., Zhang, X., Wang, J., & Wen, W. (2019). Wearable ankle robots in post-stroke rehabilitation of gait: A systematic review. *Frontiers in Neurorobotics*, *13*, 63. DOI: 10.3389/fnbot.2019.00063 PMID: 31456681

Silvestre De Oliveira, M., Fidelis Peixer, G., Forcellini, F. A., Riso Barbosa Jr, J., & Lozano Cadena, J. A. (2023). Model-based trade-off curves to support the set-based concurrent engineering of highly innovative projects. *Concurrent Engineering, Research and Applications*, *31*(3-4), 89–100. DOI: 10.1177/1063293X231209680

Stoppa, M., & Chiolerio, A. (2014). Wearable electronics and smart textiles: A critical review. *Sensors (Basel)*, *14*(7), 11957–11992. DOI: 10.3390/s140711957 PMID: 25004153

Tao, X. (2005). *Wearable electronics and photonics*. Woodhead Publishing., DOI: 10.1533/9781845690868

U.S. Army Natick Soldier Research, Development and Engineering Center. (2021). *SMART e-textile program: Reducing heat stress in military personnel*. U.S. Department of Defense.

Weng, J. K., Virk, R., Kaiser, K., Hoffman, K. E., Goodman, C. R., Mitchell, M., Shaitelman, S., Schlembach, P., Reed, V., Wu, C.-F., Xiao, L., Smith, G. L., & Smith, B. D. (2024). Automated, real-time integration of biometric data from wearable devices with electronic medical records: A feasibility study. *JCO Clinical Cancer Informatics*, *8*(8), e2400040. DOI: 10.1200/CCI.24.00040 PMID: 39348612

Yin, R., Wang, D., Zhao, S., Lou, Z., & Shen, G. (2021). Wearable sensors-enabled human–machine interaction systems: From design to application. *Advanced Functional Materials*, *31*(11), 2008936. DOI: 10.1002/adfm.202008936

Zeng, W., Shu, L., Li, Q., Chen, S., Wang, F., & Tao, X. M. (2014). Fiber-based wearable electronics: A review of materials, fabrication, devices, and applications. *Advanced Materials*, *26*(31), 5310–5336. DOI: 10.1002/adma.201400633 PMID: 24943999

Zhao, Y., Guo, X., & Zhou, M. (2021). Data security and privacy in smart textiles: A survey. *IEEE Internet of Things Journal*, *8*(14), 11181–11195. DOI: 10.1109/JIOT.2021.3065584

Zhou, M., Gao, Q., & Wang, S. (2020). Energy harvesting for wearable and stretchable electronics: From flexible supercapacitors to self-powered sensors. *Nano Energy*, *72*, 104676. DOI: 10.1016/j.nanoen.2020.104676

Chapter 3
Foundations and Emerging Trends in Electronic Textiles:
Bridging Conductive Materials With Smart Wearable Applications

Pranshu Saxena
https://orcid.org/0000-0002-0822-1532
Bennett University, Greater Noida, India

Mandeep Singh
https://orcid.org/0000-0003-3278-4877
Bennett University, Greater Noida, India

Sanjay Kumar Singh
Guru Gobind Singh Indraprastha University, East Delhi, India

Aatif Jamshed
https://orcid.org/0000-0003-3152-6147
KIET Group of Institutions, Delhi, India

Pawan Kumar
Ajay Kumar Garg Engineering College, Ghaziabad, India

ABSTRACT

Electronic textiles (e-textiles) embed sensing, actuation, power and communication into fabrics using conductive yarns, polymers and nanomaterials. This chapter consolidates conductor classes—metallic fibers, intrinsically conductive polymers, carbon nanomaterials and hybrids—and compares integration routes such as

DOI: 10.4018/979-8-3373-4287-0.ch003

weaving/knitting and printing/coating with attention to washability and mechanical robustness. It reviews standardized electrical, mechanical and environmental tests and distills lessons from deployments in healthcare, sports and defense. Key gaps include durability after laundering, power autonomy, wearer comfort, data security and scale-up. We outline directions in self-healing and encapsulation, hybrid energy harvesting and storage, stretch-resilient interconnects and on-garment machine learning. Linking material selection to process capability and system integration, the chapter offers a decision-oriented roadmap for sustainable and user-centric smart wearables.

1. INTRODUCTION

The rapid convergence of materials science, electronics, and textile engineering has given rise to *electronic textiles* (e-textiles), a transformative technology with applications spanning healthcare, sports, defense, and everyday life. Unlike conventional wearables, which integrate rigid electronic modules into separate casings, e-textiles embed sensing, actuation, and communication capabilities directly into fabrics through conductive yarns, nanomaterials, and advanced polymer composites. This approach offers unprecedented flexibility, comfort, and adaptability, paving the way for truly ubiquitous and unobtrusive smart wearables (De Fazio et al., 2023; Zaman et al., 2025).

Table 1 summarizes the fundamental categories of conductive materials used in smart textile systems, including metallic fibers, intrinsically conductive polymers, carbon-based nanomaterials, and hybrid composites. Each category exhibits distinct electrical, mechanical, and durability characteristics that influence device performance and application suitability. For example, metallic yarns excel in conductivity but may compromise comfort, whereas carbon nanotube coatings offer high flexibility with moderate conductivity.

Table 1. Application Domains and Requirements for Smart Wearables

Domain	Primary Functions	Material Requirements	Example Technologies
Healthcare	Vital signs monitoring, rehabilitation aids	Biocompatibility, washability	ECG shirts, smart compression bandages
Sports	Performance tracking, injury prevention	Flexibility, sweat resistance	Pressure-sensing insoles, motion capture wear
Defense	Soldier health monitoring, tactical comms	Durability, environmental resistance	Ballistic vests with sensors, GPS uniforms
Consumer	Gesture control, entertainment	Lightweight, aesthetics	Smart gloves, music-responsive clothing

Complementing material classification, Figure 1 provides a schematic representation of the integration pathways for these conductive materials into wearable platforms, illustrating how yarn-level modifications, coating techniques, and woven circuit architectures interconnect to enable sensing and actuation functions. This visual framework emphasizes that successful e-textile design requires balancing multiple trade-offs—such as electrical stability, washability, and user comfort—while meeting the requirements of target applications.

Figure 1. Overview of electronic textiles showing core components and key applications

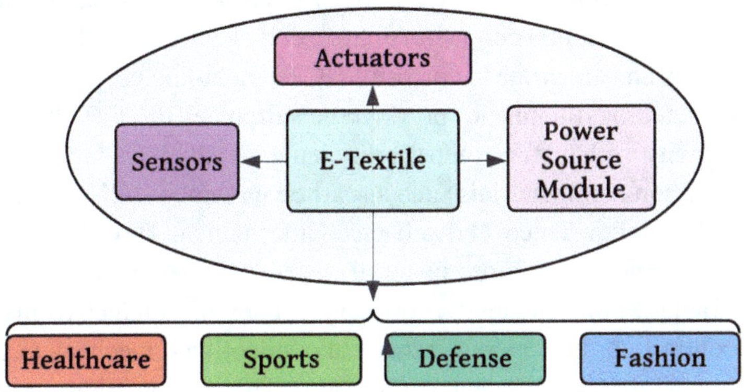

Healthcare has emerged as a particularly promising domain for e-textiles, with advances in smart bandages enabling continuous wound monitoring through integrated sensors that track parameters like pH, temperature, and moisture (Zaman et al., 2025). In sports and rehabilitation, wearable sensors embedded in textiles are enabling real-time biomechanical and physiological monitoring, improving both performance optimization and injury prevention (De Fazio et al., 2023). These trends highlight a clear movement toward multi-functional fabrics that merge comfort with clinical-grade sensing precision.

As the field matures, emerging trends include the use of machine learning ((Agarwal et al., 2025)) for intelligent data processing (Saxena P et al., 2024), flexible energy harvesting modules for autonomous operation, and biodegradable conductive materials to address environmental concerns. Through bridging breakthroughs in conductive material science and system-level integration approaches, the chapter sets the foundation for comprehending the roots and direction of electronic textiles in smart wearable systems.

2. FUNDAMENTALS OF CONDUCTIVE MATERIALS

Conductive materials are the foundation of electronic textiles, allowing electrical signals to be transmitted, devices to be powered, and sensing or actuation functions to be embedded directly into fabric structures. Conductive materials differ greatly in terms of their electrical conductivity, flexibility, robustness, and integration with textile processing. A conductive material is generally chosen based on the desired application, performance metrics needed, and anticipated wear-and-care conditions (Stoppa & Chiolerio, 2014; Rodgers et al., 2015).

Generally, conductive materials for e-textiles can be divided into metallic conductors, intrinsically conductive polymers (ICPs), carbon-based nanomaterials, and hybrid composites. Metallic conductors have good electrical conductivity but tend to lose flexibility and are prone to corrosion. ICPs, including polypyrrole (PPy) and poly(3,4-ethylenedioxythiophene) polystyrene sulfonate (PEDOT:PSS), link electrical conductivity with textile compatibility but generally have lower conductivity than metal. Carbon nanomaterials such as carbon nanotubes (CNTs) and graphene provide tunable conductance and advanced mechanical flexibility, and hybrid composites combine two or more material systems to maximize the performance trade-offs. Table 2 consolidates the primary classes of conductive materials for e-textiles, contrasting their average ranges of conductivity, mechanical properties, methods of integration, and typical applications. This comparative overview helps researchers and engineers choose suitable materials for particular wearable technology configurations.

Table 2. Key Categories of Conductive Materials for E-Textiles

Material Category	Typical Conductivity	Flexibility & Durability	Common Integration Methods	Example Applications
Metallic Conductors (e.g., silver-coated nylon, copper yarns)	10^5–10^7 S/m	Low to medium flexibility; may fatigue over repeated bending; prone to oxidation	Spinning into yarns, weaving, knitting, embroidery	ECG electrodes, heated garments
Intrinsically Conductive Polymers (ICPs) (e.g., PEDOT:PSS, PANI, PPy)	10^0–10^3 S/m	High flexibility; good textile compatibility; moderate durability	Fiber coating, dip coating, in-situ polymerization	Strain sensors, touch interfaces
Carbon-Based Nanomaterials (e.g., CNTs, graphene)	10^2–10^5 S/m (tunable)	Excellent flexibility; good fatigue resistance; wash durability depends on coating method	Spray coating, printing, fiber blending	Motion tracking, temperature sensing
Hybrid Composites (e.g., AgNW + graphene)	10^3–10^7 S/m	Optimized balance between conductivity and stretchability	Layer-by-layer coating, lamination, hybrid spinning	Energy harvesting, multifunctional wearable devices

Note: Conductivity values are indicative and can vary significantly depending on synthesis routes, filler concentration, substrate characteristics, and environmental conditions (Saxena et al., 2024).

The comparison in Table 2 highlights that no single material class perfectly satisfies all e-textile requirements. Instead, the choice often involves balancing trade-offs—such as prioritizing conductivity for power transmission versus flexibility for comfort—while also considering long-term washability, biocompatibility, and cost-effectiveness (Shateri-Khalilabad & Yazdanshenas, 2013).

3. MANUFACTURING AND FABRICATION TECHNIQUES FOR ELECTRONIC TEXTILES

Manufacturing techniques for electronic textiles (e-textiles) integrate conductive materials into flexible, wearable substrates without compromising comfort, durability, and performance. These processes must maintain electrical conductivity while ensuring mechanical robustness and washability, as required in applications ranging from healthcare monitoring to military gear (Stoppa & Chiolerio, 2014). The selection of fabrication method depends on the intended functionality, choice of conductive material, and application-specific constraints. Weaving and knitting are conventional textile manufacturing techniques adapted for conductive yarns. Conductive yarns, like silver-coated nylon, may be knitted straight into fabric structures to support uniform electrical pathways. Knitting is more elastic, so it is appropriate for sportswear-based sensing applications (Coosemans et al., 2006). This method has excellent breathability and comfort but can need post-treatment to enhance the stability of conductivity following repeated mechanical stress.

Conductive coatings—based on materials such as polypyrrole (PPy), polyaniline (PANI), or graphene—are deposited by dip coating, spray coating, or printing methods (e.g., screen printing, inkjet printing). Printing processes allow pattern control, which is needed for circuit layout integration into clothing. Screen printing silver nanoparticle ink onto polyester substrates provides high conductivity appropriate for capacitive touch sensors in jackets, for instance (Mattmann et al., 2008). Printed layers, though, can be prone to cracking when subjected to repeated bending, making flexible ink formulations necessary.

Embroidery machines are capable of sewing conductive threads into place within existing fabrics, creating personalized circuit patterns without changing the underlying fabric structure. This is especially useful for integrating sensors or antenna elements into wearable devices. While embroidery offers precision, it can increase fabric stiffness in localized regions, potentially affecting comfort (Wang et al., 2022; Singh et al., 2024).

Thermal lamination and adhesive bonding involve fixing conductive films or foils onto fabric surfaces. This technique ensures stable attachment of rigid or semi-rigid electronics, such as flexible printed circuit boards (FPCBs). Laminated layers can be encapsulated with protective polymers to improve wash resistance, making them suitable for long-term healthcare monitoring devices.

Table 3. Comparison of major fabrication techniques for e-textiles, summarizing their benefits, drawbacks, and common application areas.

Technique	Advantages	Limitations	Applications
Weaving/Knitting	Comfortable, breathable, durable	Limited pattern flexibility	ECG monitoring garments, sportswear
Coating/Printing	Custom patterns, scalable production	Potential cracking, adhesion issues	Flexible sensors, touch interfaces
Embroidery/ Stitching	Precision placement, design flexibility	Localized stiffness, longer production time	Antennas, pressure mapping textiles
Lamination/Adhesive	Strong bonding, integration of rigid elements	Reduced breathability, added thickness	Wearable medical patches, defense textiles

Table 3 provides a comparative overview of the four most common fabrication techniques—weaving/knitting, coating/printing, embroidery/stitching, and lamination/adhesive bonding—highlighting their distinct advantages, limitations, and application domains. The table shows that weaving/knitting excels in comfort and durability but lacks intricate pattern flexibility, making it more suitable for broad-area applications like ECG monitoring garments. Coating and printing enable highly customizable designs but face challenges in mechanical durability, especially under bending. Embroidery offers precise placement, beneficial for antennas and sensor grids, yet may introduce localized stiffness. Lamination and adhesive bonding stand out for robust integration of rigid electronics, though they can compromise fabric breathability.

While multiple fabrication techniques are mature, scalability, environmental stability, and cost remain significant challenges. Hybrid approaches that combine weaving with post-processing (e.g., printing or coating) are gaining traction, enabling multifunctional fabrics without sacrificing comfort (Cherenack & van Pieterson, 2012). Advances in nanomaterial-based conductive inks and automated embroidery systems are expected to enhance durability and performance, particularly for high-strain wearable environments.

4. PERFORMANCE CHARACTERIZATION AND TESTING METHODS

Performance evaluation of electronic textiles (e-textiles) is essential to ensure that they meet electrical, mechanical, environmental, and functional requirements for real-world applications (Mattila, 2006). Laboratory and field-based testing together provide a comprehensive understanding of durability, comfort, and reliability.

Electrical performance is often assessed using four-point probe or two-point probe methods to measure sheet resistance and conductivity (Stavrakis et al., 2021). Cyclic bending tests monitor resistance changes under repeated strain—important for wearables subjected to continuous movement. Mechanical properties, including tensile strength, elongation, and flexibility, are quantified following ASTM D5035 standards, while comfort and wearability are tested through drape tests, air permeability, and moisture management tests (Stoppa & Chiolerio, 2014).

Washability is a critical issue, since conductive properties can deteriorate following repeated washing. The ISO 6330 standard is usually used to mimic home washing, and environmental testing—thermal cycling, humidity exposure, and UV resistance—assures stability under varied conditions (Lee et al., 2020). Lastly, functional field trials connect lab measurements to actual performance, e.g., testing ECG signal accuracy in a conductive sports shirt or validating a temperature-sensing textile in the field (Coosemans et al., 2006).

Table 4. Summary of common testing methods for e-textiles.

Testing Type	Parameters Measured	Standards/Methods	Purpose
Electrical Performance	Sheet resistance, conductivity, stability	Four-point probe, two-point probe	Verify conductive pathways and stability
Mechanical/Wearability	Strength, flexibility, comfort	ASTM D5035, drape & air tests	Ensure durability and user comfort
Washability Testing	Conductivity retention after laundering	ISO 6330	Predict lifespan after repeated washing
Environmental Testing	Thermal, humidity, UV stability	ISO 105-B02, climate chambers	Ensure performance in varied environments

Table 4 condenses the primary testing approaches, linking each to its purpose and methodology. It shows that while lab-based methods ensure baseline performance, only real-world trials can fully validate application readiness. For instance, an ECG monitoring garment may pass conductivity and comfort tests in the lab but fail in field use due to sweat-induced noise, highlighting the need for combined testing approaches.

5. EMERGING TRENDS AND FUTURE OUTLOOK

Electronic textiles (e-textiles) are evolving rapidly, driven by advances in conductive materials, fabrication techniques, energy solutions, and embedded intelligence. These developments are reshaping wearable technology across healthcare, sports, defense, and consumer electronics.

One major trend is the shift toward multifunctional conductive materials, where fibers can simultaneously conduct electricity, sense environmental parameters, and offer mechanical reinforcement. Nanomaterial-infused yarns—such as graphene-coated cotton or silver nanowire-polymer blends—are being explored for their low weight, high conductivity, and flexibility (Shi et al., 2019).

Integration techniques are also moving toward more scalable and eco-friendly approaches. Additive manufacturing, inkjet printing, and biodegradable conductive coatings are enabling precise deposition of functional layers while reducing material waste (Stoppa & Chiolerio, 2014). Energy autonomy remains a critical challenge. Hybrid solutions combining flexible solar cells, kinetic harvesters, and micro-supercapacitors are being integrated directly into garments, enabling continuous operation without bulky external batteries (Mattila, 2006).

Finally, AI-driven data analytics is revolutionizing how e-textile data is used. Embedded machine learning models (Saxena P, Sinha, and Singh, 2024) now allow on-garment processing, reducing latency and enhancing privacy in health monitoring applications (Coosemans et al., 2006).

Table 5. Key emerging trends shaping the future of e-textiles.

Emerging Trend	Description	Potential Impact
Multifunctional Materials	Fibers with sensing, conductivity, and mechanical strength	Smaller, lighter, and more versatile e-textiles
Sustainable Manufacturing	Eco-friendly coatings and low-waste integration techniques	Reduced environmental footprint
Energy Harvesting Solutions	Flexible solar, kinetic, and thermal energy sources	Greater energy autonomy and wearable comfort
Embedded AI Processing	On-garment analytics and decision-making capabilities	Faster, private, and adaptive applications

Table 5 highlights the interplay between materials, manufacturing, energy systems, and intelligence in shaping the next generation of smart textiles. Each trend contributes toward making e-textiles more sustainable, autonomous, and application-specific.

Figure 2. Conceptual flow showing how innovations in materials, integration, and energy converge with AI analytics to enable next-generation wearable applications.

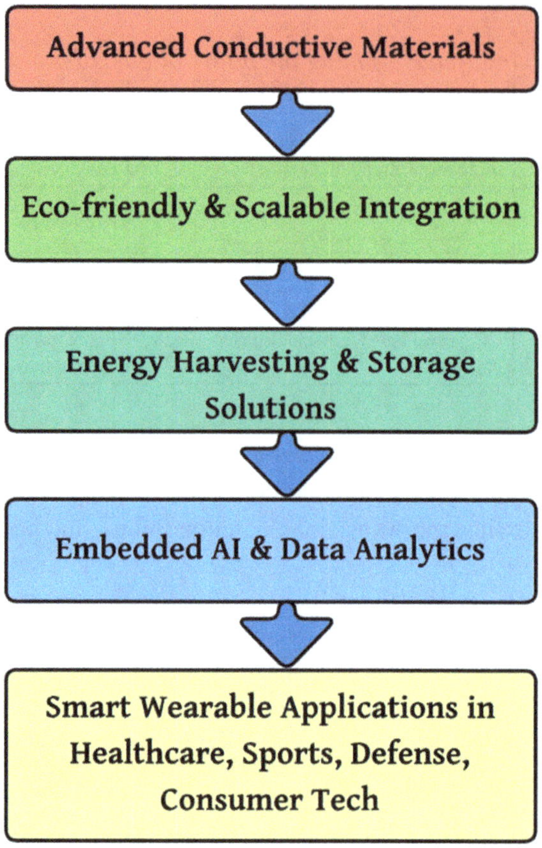

6. CASE STUDIES AND REAL-WORLD APPLICATIONS

Electronic textiles have moved well beyond prototypes and academic experiments, finding their place in healthcare, sports, defense, and consumer technology. This section presents select real-world examples where e-textiles have been successfully deployed, highlighting their functionality, design considerations, and performance outcomes.

In medicine, the Hexoskin Smart Shirt has been employed for around-the-clock physiological monitoring in clinical and home patient environments. The apparel combines textile-based ECG electrodes and respiration sensors so that unobtrusive, 24/7 data acquisition is possible without using adhesive electrodes (Castellini et al., 2021). Likewise, the Nuubo nECG Shirt is licensed for cardiac diagnostics in Europe, providing wireless monitoring for extended periods, which enhances patient compliance and diagnostic accuracy. In sporting performance, Sensoria Smart Socks utilize pressure-sensing textile fibers to detect real-time foot strike patterns, cadence, and balance. Athletes and physiotherapists have utilized data from these socks to improve running form, minimize the risk of injury, and monitor rehabilitation progress (Papi et al., 2019). In defense and tactical uses, BAE Systems Broadsword Spine e-textile harness powers and transmits data via conductive yarns embedded into a vest of a soldier. This system eliminates the need for multiple cables, reduces weight, and increases mobility while enabling connection to GPS, communications, and sensors in the field (Hilmi et al., 2025).

In consumer electronics, Levi's and Google's *Project Jacquard* demonstrates the integration of capacitive touch sensors into denim fabric, enabling gesture-based control of smartphones without removing them from a pocket or bag.

Table 6. Selected real-world examples of e-textiles across key domains

Domain	**Example Product**	**Functionality**	**Impact**
Healthcare	Hexoskin Smart Shirt	ECG, respiration, activity tracking	Long-term clinical monitoring without intrusive devices
Sports	Sensoria Smart Socks	Foot strike analysis, cadence monitoring	Improved performance and injury prevention
Defense	BAE Systems Broadsword Spine	Power and data distribution via e-textile harness	Reduced weight, improved mobility, and streamlined connectivity
Consumer Tech	Levi's × Google Project Jacquard	Touch-gesture control integrated into denim	Hands-free device control and enhanced user experience

Table 6 shows that the adoption of e-textiles is highly application-driven, with each product tailored to a specific environment and user need. Healthcare and sports solutions emphasize continuous monitoring and comfort, while defense and consumer products focus on functionality, mobility, and user interaction.

Figure 3. Simplified mapping of e-textile technologies to application domains and functions.

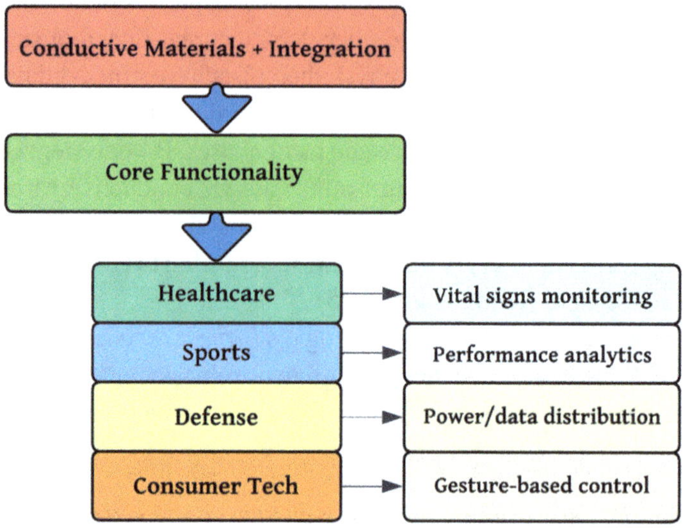

7. CHALLENGES AND FUTURE RESEARCH DIRECTIONS

Despite the rapid growth of electronic textiles, their widespread adoption still faces several technical, economic, and user-related challenges that must be addressed for long-term success.

One major issue is durability under real-world use. Many e-textiles lose conductivity or sensing accuracy after repeated washing, bending, or exposure to sweat and UV radiation. For example, silver-coated fabrics may degrade after 20–30 wash cycles, leading to performance loss (Lee et al., 2020). Developing washable, corrosion-resistant conductive coatings is a critical research priority.

Another challenge lies in power management. While energy harvesting technologies like flexible solar panels and piezoelectric fibers are promising, they often cannot meet the continuous power demands of multi-sensor garments. Research into hybrid energy solutions—combining harvesting with high-density flexible batteries—remains essential.

User comfort and ergonomics are equally important. Conductive yarns, while functional, may feel stiff or rough against the skin. Advances in soft, skin-friendly conductive polymers and stretchable circuits could enhance wearability without compromising performance (Stoppa & Chiolerio, 2014).

From a data perspective, privacy and security in health-monitoring applications are critical. As garments collect sensitive physiological data, embedded encryption and edge AI processing are needed to safeguard user information without relying solely on cloud services (Mattila, 2006).

Finally, cost and scalability remain barriers. While pilot projects in sports and defense have been successful, producing millions of units at affordable prices requires scalable manufacturing methods such as roll-to-roll printing or automated embroidery of conductive patterns.

Table 7. Key challenges and corresponding research directions for future e-textile development.

Challenge	Description	Potential Research Direction
Durability	Loss of conductivity after washing or stress	Develop corrosion-resistant, self-healing conductive layers
Power Supply	Insufficient energy for continuous operation	Hybrid energy harvesting + flexible battery systems
Comfort and Wearability	Stiffness or skin irritation from conductive materials	Soft, stretchable, and breathable conductive fabrics
Data Security	Risk of health data breaches	On-garment encryption and edge AI for local processing
Cost and Scalability	High production cost for mass markets	Roll-to-roll printing and automated textile integration

Table 7 emphasizes that challenges span materials science, electronics, user comfort, data privacy, and economics. Solutions will require interdisciplinary collaboration between textile engineers, material scientists, electronics designers, and data security experts.

Figure 4. Conceptual pathway from current challenges to large-scale adoption of e-textiles.

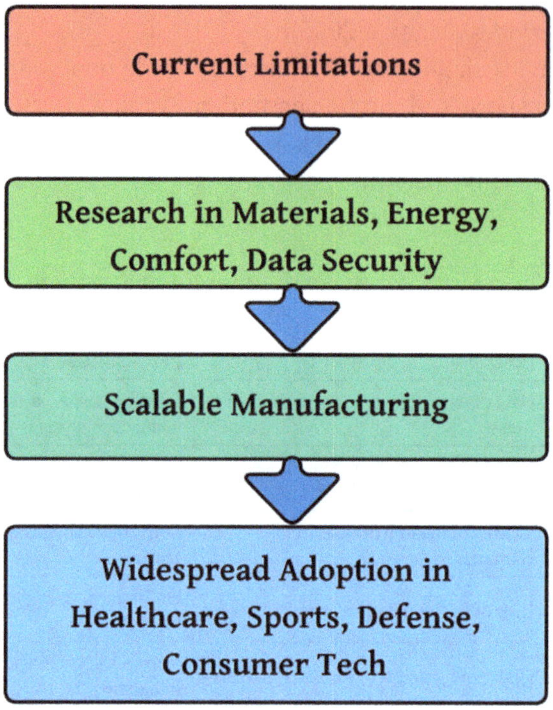

The figure 4 illustrates the translational pathway of electronic textiles. It begins with current limitations such as durability, power autonomy, user comfort, and data security. Addressing these challenges requires targeted research in materials, energy systems, comfort, and privacy-preserving architectures, which then supports the development of scalable manufacturing methods such as roll-to-roll printing, automated embroidery, and sustainable coating techniques. Once these foundations are established, e-textiles are positioned for widespread adoption across healthcare, sports, defense, and consumer technology, enabling durable, comfortable, and secure smart wearables that integrate seamlessly into daily life.

8. CONCLUSION

This chapter surveyed e-textile material foundations (metallic fibers, ICPs, carbon nanomaterials, hybrids) and fabrication approaches (weaving/knitting, printing/coating, embroidery, lamination), linked to testing and standards for electrical, mechanical, and wash durability. It distilled insights from case studies spanning healthcare, sports, defense, and consumer products, and synthesized challenges and future directions—notably wash-durability, power autonomy via hybrid harvesting and storage, comfort/ergonomics, data security and privacy, and large-scale, sustainable manufacturing. Emerging trajectories such as embedded on-garment AI and eco-friendly integration methods indicate a path to scalable, user-centric wearables. Taken together, aligning material selection, process capability, and validation protocols provides a practical roadmap from laboratory prototypes to reliable, privacy-preserving, and commercially viable smart textiles.

REFERENCES

Agarwal, S., Dohare, A. K., Saxena, P., Singh, J., Singh, I., & Sahu, U. K. (2025). HDL-ACO: Hybrid deep learning and ant colony optimization for ocular optical coherence tomography image classification. *Scientific Reports*, *15*(1), 5888. DOI: 10.1038/s41598-025-89961-7 PMID: 39966596

Cherenack, K., & van Pieterson, L. (2012). Smart textiles: Challenges and opportunities. *Journal of Applied Physics*, *112*(9), 091301. DOI: 10.1063/1.4742728

Coosemans, J., Hermans, B., & Puers, R. (2006). Integrating wireless ECG monitoring in textiles. *Sensors and Actuators. A, Physical*, *130–131*, 48–53. DOI: 10.1016/j.sna.2005.10.052

De Fazio, R., Mastronardi, V., De Vittorio, M., & Visconti, P. (2023). Wearable sensors and smart devices to monitor rehabilitation parameters and sports performance: An overview. *Sensors (Basel)*, *23*(4), 1856. DOI: 10.3390/s23041856 PMID: 36850453

Hilmi, A. (2025, October 21). Wearables for defense: the next frontier in military technology. *Sensio*. https://www.sensio-ai.in/post/wearables-for-defense-the-next-frontier-in-military-technology

Lee, S., An, S., & Kim, H. (2020). Wash durability of conductive fabrics for wearable electronics. *Textile Research Journal*, *90*(7–8), 857–870. DOI: 10.1177/0040517519881570

Mattila, H. (Ed.). (2006). *Intelligent textiles and clothing*. Woodhead Publishing. https://www.amazon.in/Intelligent-Textiles-Clothing-Woodhead-Publishing/dp/1845690052

Mattmann, C., Amft, O., Harms, H., Tröster, G., & Clemens, F. (2008). Recognizing upper body postures using textile strain sensors. In *2008 12th IEEE International Symposium on Wearable Computers* (pp. 29–36). IEEE. DOI: 10.1109/ISWC.2008.4911581

Papi, E., Belsi, A., McGregor, A. H., & McCarthy, C. (2019). Smart socks for gait analysis: Reliability and clinical applications. *Gait & Posture*, *72*, 55–60. DOI: 10.1016/j.gaitpost.2019.06.008

Rodgers, M. M., Pai, V. M., & Conroy, R. S. (2015). Recent advances in wearable sensors for health monitoring. *IEEE Sensors Journal*, *15*(6), 3119–3126. DOI: 10.1109/JSEN.2014.2357257

Saxena, P., Jain, P., Aggarwal, P. K., Singh, M., Goel, S., & Batra, M. (2025). Communication Requirements and Performance Metrics for Electric Vehicle Charging. In Optimized Energy Management Strategies for Electric Vehicles (1st ed., pp. 35–60). DOI: 10.4018/979-8-3693-6844-2.ch002

Saxena, P., Sinha, A., & Singh, S. K. (2024). Computer-assisted interpretation, in-depth exploration and single cell type annotation of RNA sequence data using k-means clustering algorithm. *Computer Methods in Biomechanics and Biomedical Engineering*, •••, 1–11. DOI: 10.1080/10255842.2023.2300685 PMID: 38235728

Saxena, P., Sinha, A., & Singh, S. K. (2024). Review of computer-assisted diagnosis model to classify follicular lymphoma histology. *Cell Biochemistry and Function*, *42*(5), e4088. DOI: 10.1002/cbf.4088 PMID: 38973163

Shateri-Khalilabad, M., & Yazdanshenas, M. E. (2013). Fabricating electroconductive cotton textiles using graphene. *Carbohydrate Polymers*, *96*(1), 190–195. DOI: 10.1016/j.carbpol.2013.03.052 PMID: 23688469

Shi, Q., Sun, J., Hou, C., Li, Y., Zhang, Q., & Wang, H. (2019). Advanced functional fiber and smart textile. *Advanced Fiber Materials*, *1*(1), 3–31. DOI: 10.1007/s42765-019-0002-z

Singh, M., Sharma, R., Singh, R., Malik, A., & Aggarwal, P. K. (2025). Advancements in renewable energy harvesting for EV charging infrastructure. In Optimized Energy Management Strategies for Electric Vehicles (1st ed., pp. 119–152). DOI: 10.4018/979-8-3693-6844-2.ch005

Stavrakis, A. K., Simić, M., & Stojanović, G. M. (2021). Electrical characterization of conductive threads for textile electronics. *Electronics (Basel)*, *10*(8), 967. DOI: 10.3390/electronics10080967

Stoppa, M., & Chiolerio, A. (2014). Wearable electronics and smart textiles: A critical review. *Sensors (Basel)*, *14*(7), 11957–11992. DOI: 10.3390/s140711957 PMID: 25004153

Wang, H., Zhang, Y., & Yu, D. (2022). Embroidery-based e-textiles for wearable sensing applications: A review. *Textile Research Journal*, *92*(13–14), 2062–2078.

Zaman, S. U., Tao, X., Cochrane, C., & Koncar, V. (2022). Smart E-Textile Systems: A review for healthcare applications. *Electronics (Basel)*, *11*(1), 99. DOI: 10.3390/electronics11010099

Chapter 4
Nanomaterials and Conductive Polymers in E-Textiles Toward Smart and Sustainable Wearables

Manu Mehrotra
https://orcid.org/0009-0003-4841-1589
Sharda University, Agra, India

Manish Baboo Agarwal
https://orcid.org/0000-0003-1177-2738
Sharda University, Agra, India

Seema Agarwal
https://orcid.org/0000-0002-5655-207X
SRM Institute of Science and Technology, Ghaziabad, India

Shahjad Ali
https://orcid.org/0000-0002-9725-3498
Sharda University, Agra, India

ABSTRACT

The integration of nanomaterials and conductive polymers is transforming electronic textiles (e-textiles) by combining textile attributes like softness and comfort with advanced functionalities such as physiological monitoring, interactive interfaces and energy storage. Nanomaterials like carbon nanotubes (CNTs) and graphene enhance electrical conductivity and mechanical robustness, allowing for the de-

DOI: 10.4018/979-8-3373-4287-0.ch004

velopment of flexible textile sensors, batteries, and heating textiles. Conductive polymers, such as polyaniline and polythiophene, offer durability and breathability without the oxidation issues of metals, enabling specific applications like physiological monitoring. However, challenges remain, including stability under mechanical stress, long-term functionality and ecological concerns related to nanomaterials. Addressing cost-effectiveness and sustainability through recyclable designs is also essential. Ultimately, the goal is to create smart fabrics that seamlessly integrate advanced electronic capabilities while enhancing health monitoring and communication functions.

INTRODUCTION

Electronic textiles, often termed e-textiles, mark an innovative blend of everyday fabric and state-of-the-art electronics. By weaving or printing sensors, batteries, microcontrollers and tiny lights directly onto cloth, developers turn passive textiles into responsive surfaces that can measure temperature, light up, or even notify a smart phone. The idea is not as modern as it seems, the first related patent appeared in the 1930s, but the progress waited until the twenty first century (Ehrmann & Ehrmann, 2021). Since then smaller chips, flexible circuits and low cost production have gained interest in hospitals, athletes and soldiers that has pushed e-textiles to the heart of research and the marketplace.

Introduction and Significance of E-Textiles

E-textiles are a mixture of the softness, stretch and breathability of everyday fabric with built in electronic capabilities, opening doors that only gadgets and typical textiles cannot reach. These intelligent fabrics wrap around the body much like a second skin, allowing data exchange and control to occur in a quiet, almost invisible way. Manufacturers attach sensors and circuits to threads by weaving, knitting, embroidering, or printing them directly onto the cloth surface (Dulal et al., 2022). Conductive inks, carbon nanofibers and flex friendly polymers are used through the yarn, while miniature batteries and microcontrollers slip between fabric layers.

This set up turns the fabric into a wearable hub that tracks heart rate, skin temperature and movement, heats up on cold days, lights up for safety at night, stores power and chats with smartphones, all without sacrificing the familiar feel of a cotton fabric.

Electronic textiles already serve in several vital areas, each of which insists on a distinctive blend of properties. In health care, for example, seamless garment sensors monitor heart and breathing rates or track muscle strain, making them equally useful for hospitalized patients, older adults alone at home, or gym goers who prefer light weight apparel. For athletes and runners, sensor embedded shirts or shorts gather motion data and suggest adjustments to posture and flag early signs of strain, all while retaining the stretch and breathability of conventional sportswear.

In defense they sew mini radios, climate gauges and adaptive control panels into battle dress, so troops gain real time updates on enemy detection and environmental stress. Creatives in fashion and stage are programming cloth that glows, fades, or reacts to applause, thereby turning an outfit into a living light.

In manufacturing, conductive drapes and conveyor belts shield nearby electronics from stray signals, guard steel pipes against rust and even harvest kinetic power to run sensors. This wide-reaching adoption illustrates that researchers now talk of a textiles 2.0 era (Liu et al., 2025). Rather than mere decorative ones, these fabrics form an intelligent skin that links people, machines and the changing world around them.

Mission and Objectives in E-Textile Development

E-textiles researchers aim to produce smart fabrics that retain everyday textile features while embedding robust electronic functions. Achieving this goal requires solving key tradeoffs between user comfort, long term wearability, visual appeal and the stricter needs of reliable conductivity, precise sensing, lightweight power and onboard processing (Younes, 2023).

E-textile research is guided by four central goals:

1. **Maximizing Wearability and Comfort:** Any integrated circuit must be so thin, soft and pliable that the garment still behaves like everyday fabric. Unless the textile retains breathability and a familiar hand, users are unlikely to accept the technology for more than a few trials.
2. **Ensuring Electrical and Functional Reliability:** Conductive threads, sensors and other components must deliver stable readings and signals even after thousands of bends, stretches, tumbles through a washing machine, or long exposure to sweat and humidity. Protection against oxidation, moisture and most outdoor contaminants therefore remains a baseline requirement.
3. **Scalability and Manufacturability:** To move from niche markets to widespread adoption, production needs to be fast, affordable and compatible with existing textile supply chains. That means refining inks, yarns and coating swith tested conductive polymers, silver nanowires and pairing them with printing, weaving, or embroidery processes should leave the fabric unchanged.
4. **Energy Efficiency and Integration:** Every embedded sensor, light, or radio chip adds an energy budget, so researchers focus on ultra-low power circuits,

flexible miniature batteries, far delivering super capacitors and tiny harvesters that turn motion or light into volts, all packaged with zero increase in budget.

NANOMATERIALS IN E-TEXTILES

Nanomaterials are reshaping electronic textiles by endowing everyday fabrics with the electrical, mechanical and functional traits needed for smart, interactive behavior. Carbon nanotubes (CNTs), graphene, MXenes and metal nanoparticles or nanowires now feature prominently, each boosting conductivity, flexibility, durability, or a composite mix of these qualities (Syduzzaman et al., 2023). As every family of nanoarchitectures adds its own physicochemical signature, the reach of e-textiles is growing quickly, spanning healthcare, sportswear, military uniforms and industrial monitoring.

Types of Nanomaterials in E-Textiles

Carbon Nanotubes

Carbon nanotubes have become standard nanomaterial for e-textiles, combining high electrical conductivity, exceptional strength and good thermal stability in a lightweight form. When embedded in yarns or printed onto polymer membranes, CNTs turn ordinary cloth into a substrate that senses pressure, transmits data, or even dissipates heat.

Carbon nanotube (CNT) yarns and fibers combine excellent electromechanical bendability with high electrical conductivity and robust thermal stability, yet they retain the softness typical of conventional textiles, opening pathways for smart wearable that monitor health, harvest energy, or manage heat (Shahidi & Moazzenchi, 2018).

A landmark step forward has been the hybrid CNT wrapped yarn, in which carbon nanotube shells encircle standard textile cores and armature the yarns. Careful adjustment of spinning speed, twist angle and fiber on fiber bonding lets manufacturers run the hybrid yarns through knitting, weaving and finishing operations without losing either electric or tensile strength. When knitted into wrist bands, fingers of smart gloves, or collar bands, the CNT hybrids register subtle wrist twists and fingertip flex through resistance shifts of just a few ohms, instantly turning the fabric itself into a sensitive, durable strain sensor (Hossain, Lubna, & Bradford, 2023).

The long-term durability and ability to withstand laundering of carbon nanotube (CNT) yarns are essential before they can be fitted into everyday wearable fabrics. CNT based yarns retain their conductivity even after thirty wash and spin cycles, showing only subtle deterioration that is well within acceptable performance limits.

This strength derives from the tightness, adhesion between the CNT network and the underlying host fibers during yarn fabrication. The mechanical stability of carbon nanotubes in fused textiles allows these garments to maintain their conductive and sensing properties throughout routine laundering and extended wear.

Functionally, CNT fabrics exhibit a linear change in electrical resistance when subjected to strain modes like bending, stretching, or compressing which makes them valuable in wearable motion detectors. This predictable electromechanical response has already powered devices that track the movement of fingers, wrists and larger joints with remarkable sensitivity. CNT threads also double low voltage heating elements. Once energized, the interconnected nanotube mesh converts slow current into heat almost instantly, reaching temperatures just above ambient skin levels without the lag in characteristic of conventional wires. This rapid thermal response makes CNT heaters particularly appealing for outdoor apparel, blankets, military shelters and smart bandages that deliver localized warmth to the wearer.

Beyond electromechanical and thermal functions, carbon nanotube fibers have emerged as lightweight thermoelectric materials, creating new avenues to cover energy harvesting within the textile domain. Thermoelectric elements transform a temperature field into electrical power and carefully engineered CNT yarns now deliver room temperature that either match or exceed those of benchmark inorganics. Arranged in millimeter long, meso scaled ribbons, vertically aligned CNT fiber retain this advantage, permitting the production of soft thermoelectric generators that remain flexible, foldable and machine washable. When draped over a user's shoulder or positioned against an outdoor surface, such fabrics can siphon excess body heat, delivering sustainable micro voltage for onboard sensors, smart labels, or low power wearables in an autonomous fashion (Zheng et al., 2019).

The same CNT architecture that delivers thermoelectric performance bolsters mechanical endurance on several fronts. Encapsulating fibers in a nonwoven matrix or applying CNT rich composite coatings routinely doubles tensile strength and triple strain tolerance, enabling garments to shrug off the fatigue normally induced by repeated bending, washing and abrasion (Godara et al., 2009).

Researchers hoping to see carbon nanotube based electronic textiles in everyday use, still contend with practical hurdles that slow progress. Chief among these challenges is the need for a durable, robust bond between CNTs and the fibers used to weave a typical garment. Because CNTs are largely nonpolar, they tend to stick poorly to both natural and synthetic threads and weak bonding can result in the particles peeling off during everyday wear or laundering.

Graphene

Graphene has quickly positioned itself as a game changing nanomaterial in the electronic textile field, dramatically enhancing fabric performance by lending excellent electrical, mechanical and thermal properties. The single layer of carbon arranged in a honeycomb lattice originates the two-dimensional polymer like a sheet of paper. This combination keeps the material light yet unbelievably stiff, offering engineers a flexible canvas from which to shape wearables able to sense data, store spare power and keep track of human wellbeing while still feeling soft, breathable and forgiving when on the body (Alamer & Alzahrani, 2025).

When a textile is treated with the material, the high conductivity and solid thermal stability let invisible circuits and sensors blend quietly with wearable material. Because sheets form a seamless, two-dimensional web, electric paths run uninterrupted and can sense the faint voltages generated by a heartbeat, a breath, or a twitch of muscle. This quality is essential for smart clothing that quietly monitors vital signs and shares the information in real time, sparing users from bulky external sensors and wires (Zhai et al., 2025).

Graphenes exceptional strength and flexibility further justify its use in electronic textiles. The material accommodates a tensile force almost two hundred times greater than that of steel while remaining feather light and easily bendable. When blended into a fabric, grapheme laced cloth stretches, bends and survives harsh washes much longer than ordinary blends, prolonging the life of sports gear, work wear, or medical uniforms that face constant movement and machine cleaning. The enhanced toughness also limits permanent stretch, a benefit high performance brands across the athletic and maternity markets routinely seek to keep in their products.

Graphene can be woven into everyday fabrics using budget friendly techniques, such as coating individual threads with graphene oxide (GO) or reduced graphene oxide (rGO), then spraying, dipping, or printing the conductive inks on the material surface. A strong, reliable method reduces the GO chemically to form rGO dispersions, which finish cotton or synthetic cloth via familiar paddry steps commonly seen in textile processing (Bhattacharjee et al., 2020). Because this route supports line speeds exceeding 150 meters per minute, it matches the pace of most modern factories and clears a major hurdle to commercial adoption of grapheme powered smart textiles. The fabrics remain supple, fold easily and acquire noticeable gains in tensile strength and strain tolerance.

Graphene coated fabrics enhance clothing performance in many ways. First, the material channels heat efficiently, so coated fabrics draw excess warmth away from the body and keep the climate manageable. This passive thermal control matters most in smart sports gear and alpine outerwear, where moisture and temperature must be balanced for endurance and safety.

The same coating also absorbs ultraviolet light, adds a slow release on antibacterial surface and curbs unpleasant odors over extended wear, giving everyday outfits an invisible shield. Garments outfitted with graphene are already having numerous everyday applications. In healthcare, the sheets sensing and conducting properties, power shirts and wristbands that monitor heart rate, skin temp, or muscle strain, helping both quick clinical readings and continuous vigilance. In sports, grapheme infused gear measures motion, muscle load and even fluid loss, giving athletes and coaches' data to refine training and prevent injury. Maternity apparel now features graphene threads that monitor the wellbeing of both mother and fetus while still feeling soft and stretchy. Military uniforms and industrial work wear are now equipped with hidden radios and sensors and engineers credit graphene for adding these features without noticeable weight or crazy upkeep (He et al., 2024). Even so, slipping graphene into the jeans or hoodies sold in every shopping mall is still rather tricky. Scientists have to glue every thin sheet or strand to the main fabric tightly enough that it never lifts, yet soft enough that the final garment still bends easily and feels nice against skin.

MXenes

MXenes have quickly become a focal point in research on smart fabrics. It is due to their excellent electrical conductivity, good mechanical strength and solid resistance to oxidation or corrosion. These two-dimensional materials are derived from stacked transition metal carbides or nitrides and because their surface finishes are rich in oxygen, fluorine, or hydroxyl groups, they still behave like metals while remaining easy to disperse in water or solvents. Incorporating MXenes into thin textile sheets thus enhances the fabrics with stable circuit pathways, onboard energy storage cells and even sensors that track motion, health, or environmental threats (Mishra et al., 2024), in wearable shirts and bandages.

A primary feature that makes MXenes attractive for e-textiles is their outstanding conductivity then other nonmaterial's now found in smart fabrics. When formed into flat films or wet spun yarns, pure MXene fibers reach conductivity of roughly 2800 to nearly 7700 S/cm. The layered titanium carbide or titanium nitride structure, combined with close packed d-orbital overlap, enables electrons to glide through the sheets with very little scattering.

MXenes can be integrated into textiles using several scalable methods like dip coating, vacuum assisted filtration, spraying, or wet spinning of yarns. Dip coating and vacuum filtration techniques lay down MXene nanosheets in a uniform, porous layer, so that the fabric keeps its original flexibility while gaining electrical and chemical functionality. For instance, membranes that combine MXenes with silver nanoparticles or cellulose nanofibers routinely achieve joint conductivity more than an order of magnitude higher than any component alone.

Wet spinning has emerged as a leading industrial method for making MXene fibers, allowing manufacturers to extrude the material from a stable dispersion while precisely controlling fiber alignment and surface structure. Due to this careful processing, the resulting fibers record the highest electrical conductivities ever reported for MXene yarns, all without sacrificing the mechanical flexibility needed for everyday textiles. Fiber performance still depends heavily on two variables, the average size of MXene flakes and the choice of solvent. Larger flakes extend continuous conductive pathways, whereas a carefully chosen solvent preserves a stable, uniform dispersion that forms defect free filaments during spinning. These fibers can then be woven, knitted, or even embroidered into everyday textile patterns, giving rise to fully integrated e-textiles that carry current internally rather than relying on fragile printed inks. Because conduction is an intrinsic property of the fiber, finished garments withstand normal washing and the repeated strain of daily wear (Eom et al., 2020; Saxena & Goyal, 2019).

Apart from their conductivity, MXenes introduce a wide range of complementary capabilities that enhance their integration into smart textiles. Their inherently hydrophilic surfaces attract moisture, enabling precise humidity sensing and allowing fabrics to wick sweat while remaining breathable. The same sheets absorb UV radiation and resist corrosion, shielding threads from sunlight and sweat alike and dramatically extending the outer material's service life (Gaur, Singh, & Saxena, 2024).

MXene infused smart textiles are beginning to yield notable applications across several important fields. In clinical settings, MXene coated threads and fibers work as extremely precise strain and motion sensors, tracking subtle shifts in electrical resistance as the material stretches or compresses and thereby measuring a person's movement, breathing and pulse. Across the sports and fitness domain, clothing laced with MXene fabric monitors joint angles and muscle activation in real time, delivering actionable insights that can help athletes refine performance and avoid injury (Hussain et al., 2025).

In military and industrial contexts, textiles coated with MXenes deliver remarkably effective electromagnetic interference (EMI) shielding, guarding personnel and sensitive equipment against disruptive radiofrequency waves (Agarwal et al., 2025).

Metal Nanoparticles / Nanowires

Metallic nanoparticles and nanowires play a central role in shaping modern e-textiles, endowing fabrics with strong conductivity, bendable function and multipurpose capability while preserving comfort and wear endurance (Syduzzaman et al., 2023). Structures such as gold, silver, zinc oxide and titanium dioxide particles or wires embedded in yarns or surface coatings form uninterrupted conduits that enable sensing, localized heating, energy storage and two-way communication in next generation wearable. Because their dimensions reside at the nanoscale, these elements integrate discreetly, leaving the fabric soft, breathable and inviting, a characteristic that makes them especially attractive for both consumer smart clothes and performance grade industrial textiles (Saxena et al., 2025a).

Researchers have recently mastered the growing of metallic nanowires directly onto polymer fibers, a breakthrough that transforms ordinary insulating fabric into a lightweight, supple conduct while keeping the material thin and bendable. Once coated, a single fiber shows dramatically improved conductivity, able to stretch about 15 percent before the electrical path breaks, while nonwoven mats handle roughly 25 percent strain without failing. This large elastic window is critical for wearable technology, because everyday garments bend, twist and pull thousands of times during normal use. Alongside the robust performance, the thin, transparent layer preserves the look and feel of the original fabric, making the material attractive for everyday apparel (Saxena et al., 2025b).

Among metal nanostructures, silver nanowires (AgNWs) attract considerable interest because they combine high electrical conductivity with antimicrobial action and work well on many fabric types. Their integration into textiles is performed through solution-based coating methods that resemble standard dyeing, making largescale and costefficient fabrication possible. Rather than forming continuous silver films that crack and peel when stretched, a mesh of nanowires creates a flexible grid with relatively few contacts, thereby lowering resistance and improving long term wear. When embedded in elastic materials such as poly(dimethylsiloxane) (PDMS) or knitted fabrics under a pre-strained and post embedding sequence, the nanowire network retains its electrical function even during the pronounced bending and pulling motions typical of everyday use (Yao et al., 2019).

Another route to smart textiles involves blending conducting polymer nanocomposites with metal nanoparticles or nanowires. Additives such as silver, zinc oxide, or titanium dioxide particles are uniformly dispersed within the polymer host, producing fibers or coated surfaces that are both electrically active and anticorrosive (Saxena et al., 2025c).

The practical uses of fabrics enhanced with metal nanoparticles and nanowire is steadily expanding. In wearable healthcare, materials coated with silver or copper nanowires function like soft electrodes and sensors that pick up bioelectric signals of electro cardiograms (ECG) or electro myograms (EMG), letting doctors track a patient's heart or muscle activity day and night, all without the clutter of traditional gear (Jalali et al., 2024). Because these wires are both highly conductive and surprisingly stretchy, they ride along with the body's natural movements and keep the data flowing without dropouts. In gym wear and high-performance outfits, the same smart cloth picks up every strain, twist, or pressure point, feeding coaches real time analytics on technique and effort.

Beyond sensing, these wire networks can be used as heating mats. When a low voltage current passes through the mesh. The heating radiates smoothly across the entire garment rather than pooling in one location. That property is valuable when engineers design jackets for winter treks, flame resistant suits for welders, or recovery wraps that warm injured limbs. The grids are so thin and optically transparent that they become nearly invisible, leaving breathability and overall appearance largely untouched. Metallic nanoparticle layers also help in shielding cell phones, radios and nearby operators from unwanted electromagnetic noise in factory floors or military deployments.

Yet engineers still face tough problems before these smart fabrics go mainstream. Wash longevity, endurance against bending fatigue and resistance to humidity or UV light are regularly tested to ensure whether the nanoscale features stay intact over time or not.

Electrical, Thermal, and Mechanical Enhancements provided by Nanomaterials

The introduction of nanomaterials into conventional fabrics has fundamentally upgraded the core electrical, thermal and mechanical characteristics of electronic textiles, allowing the creation of smart garments that are both high performing and user-friendly. Embedding carbon nanotubes, graphene, two dimensional MXenes, or metal nanoparticles directly into the yarn creates a nearly invisible circuit that warms evenly, strengthens the fabric and stays bendy, thus meeting the long-term wear and comfort tests engineers set for nextgen smart clothing (Hsu et al., 2014).

The recent integration of nanoscale materials has transformed electronic textiles from passive systems into active, interactive fabrics with meaningful electrical performance. Traditional textile fibers are poor electrical conductors, so scientists now mix conductive nanomaterials into the yarn to create garments that sense data, store energy and link wirelessly (Adamu et al., 2025). Among the metal-based options, silver nanoparticles and silver nanowires draw the most attention because their bulk conductivity exceeds that of copper. When these nanowires are printed onto cotton prepared with a polyethylenimine coating, they interlace to produce a mesh that has low sheet resistance and solid shielding against electromagnetic interference, even after many washes. Because this mesh remains flexible, it meets the stringent comfort and durability criteria required for wearables. Complementarily, polymer nanocomposites containing doped silver nanoparticles or nanowires achieve stretchable conduction with fewer particle contacts, allowing high performance at much lower filler fractions (Ojstršek, Jug, & Plohl, 2022).

Carbon rich nano sheets and tubes, particularly graphene and carbon nanotubes, continue to be the most studied additives due to their outstanding conductivity. Because these materials can be dispersed in small amounts, the modified cloth remains soft and elastic, allowing circuits to run invisibly along seams without sacrificing comfort or wearability. When these materials are applied as thin layers to cloth, they can carry bioelectrical data from the skin to remote monitors with minimal signal delay.

Heat management is another vital design consideration for wearable gadgets and improved thermal transfer helps prevent localized overheating. CNTs and grapheme when woven into yarn or sprayed onto a surface, quickly spread excess of heat that would otherwise linger around embedded circuits (Faruk et al., 2025). As their conductivity outstrips that of standard polymer textiles, the blended fabrics feel cooler during prolonged wear and protect delicate skin from sudden temperature spikes.

Silver and zinc oxide nanocrystals work together to push thermal conduction higher than either would achieve alone. Silver nanowires are known for their rapid electron flow, and they also pass heat, resist oxidation and form translucent, featherweight meshes for wearables.

The mechanical gains from nanoscale additives are equally important. Many performance garments endure frequent bending, stretching and repeated machine washing, so seams and yarns must not affect the electronics they embed. Coating fibers with graphene, for instance, can raise tensile strength by more than 30 percent without sacrificing pliability (Anik et al., 2024). MXenes a relatively recent family of two-dimensional carbides give a parallel boost to toughness and fatigue resistance, so active regions stay crack free and functional through circuits dosed with frequent of bending and washing (Hu et al., 2024).

Metal nanoparticles and nanowires craft soft, interconnected scaffolds that flex or stretch with the cloth yet avoid the fractures that usually stop electric flow. Tests on silver nanowire webs show that a fabric can be pulled by more than twenty five percent and still maintain almost the same resistance it had at rest. Because of this toughness, sensors, heating threads, or light emitting fibers used into clothing keep working even when the wearer twists, sits, or jogs.

CONDUCTIVE POLYMERS IN E-TEXTILES

Major Conductive Polymers

Three conductive polymers most often found in e-textiles are polyaniline (PANI), polythiophene (PT) and polypyrrole (PPy). Each gain its conductivity from a conjugated backbone that alternates single and double bonds, letting extra π-electrons glide along the chain (Grancarić et al., 2017). By outsourcing much of the charge motion to this scaffold, the polymer gains a slice of the mobility seen in metals while keeping the low mass, bendable frame and the long-term reliability that ordinary plastics offer. This mix of qualities makes the system appealing for textiles where softness, breathability and stretch are basic, nonnegotiable demands.

Polyaniline (PANI) remains a cornerstone of conducting polymer investigations in smart textile research. Its synthesis is straight forward, it resists degradation in ambient environments, and its bulk electrical properties can be deliberately tuned through acid doping. Depending on the polymer's oxidation and protonation levels, room temperature conductivity can span six orders of magnitude. Manufacturers suggest that low cost, solvent free PANI powder can be blended directly into fiber coatings, giving fabric an electronics capability while keeping its feel and drape. Treated textiles have supported body temperature sensors, anti-spark suits for oil crews and multiplex strain monitors that track movement during exercise (Abilevitch et al., 2023).

Polythiophene derivatives, most prominently poly (3,4-ethylenedioxythiophene) system, often blend with polystyrene sulfonate (PEDOT: PSS) drew recent industrial attention due to their superior mechanical and chemical integrity. The blend is dissolvable in water, so manufacturers can employ versatile deposition processes such as screen printing, slot die coating, or anodic electropolymerization at modest temperatures. When small doses of polar dopants such as dimethyl sulfoxide or ethylene glycol are added during curing, film conductivity can climb by as much as 1000 times relative to an undoped layer. The resulting smooth, pin hole free coating effectively envelopes each fiber, letting the fabric curve, fold and breathe while still powering strain gauges, capacitive touchpads, flexible heaters and entry level kinetic harvesters sewn into clothing (Alamer et al., 2022).

Polypyrrole (PPy) has become one of the most widely adopted conductive polymers in electronic textiles because it can reach high conductivity and be polymerized easily, either chemically or by electrochemical deposition, directly onto fabric. Coatings of PPy maintain the flexibility and mechanical durability of the underlying textile while delivering stable electrical performance. These advantages, like biocompatibility and good stability under ambient conditions, make PPy modified fabrics attractive for applications such as biomedical monitors, strain sensors and antistatic or shielding garments (Khan et al., 2025).

Integration of PPy and similar polymers with fiber-based substrates can be accomplished by a range of techniques including in situ polymerization, where monomers grow on or inside the fabric, chemical and electrochemical deposition, vapor phase polymerization and solution-based coating or printing. In contrast to conventional metal wires or foils, conductive polymers corrode much less in humid environments or after repeated washes, so wearables equipped with these materials retain their conductivity over extended use.

Polyaniline, polythiophene especially the mixed dopant PEDOT: PSS and polypyrrole remain the foundation of conductive polymers steering e-textile innovation today as shown in Figure 1. These materials combine decent electrical conductivity with flexibility, environmental stability and easy processing, thus supporting smart textiles with feel and behavior typical of ordinary fabrics.

Figure 1. Conducting polymers in E-textiles

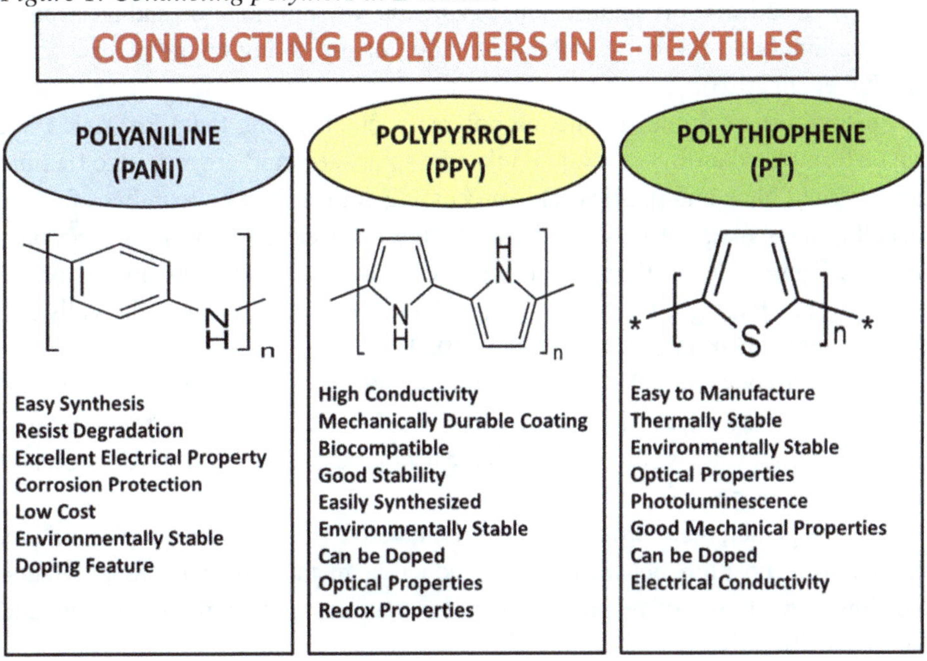

Advantages of Conductive Polymers over Traditional Conductors

When compared to metals, conductive polymers present advantages that matter in flexible electronics, wearable gadgets and e-textiles. As they are light, soft and stable instead of heavy, stiff and oxidation sensitive like most metals, conductive polymers can be embedded in fabric or stretched without ruining user comfort or long-term durability. Their liquid processing characteristic also permits coating, printing, or spinning methods, so large amounts of manufacture of them can remain cheaper, greener and consume less energy (K & Rout, 2021).

Conductive polymers offer a unique combination of flexibility and mechanical resilience that outperforms rigid metal wires. They can be bent, stretched and even machine washed many times without cracking. The conductivity of a given polymer can be raised or lowered by simple chemical or electrochemical doping, meaning engineers can tune each section of a single film for sensors, heaters, or storage cells. That tunable conductivity makes it easier to assemble hybrid circuits that share the same polymer backbone yet perform separate functions. Because the material is still organic, it also feels soft against skin, paving the way for monitors and electronics that rest lightly on the body without irritating or provoking an allergic response (Shahid et al., 2025).

Several water soluble conducting polymers, chief among them PEDOT: PSS, can be formulated as low-cost liquid inks that spincoat, print, or spray onto flexible fabrics, rigid glass and plastics. This processing compatibility streamlines factory assembly, cuts material waste and shrinks the ecological footprint of electronic manufacturing. The resulting clear, electrically active films supplement opaque metal layers, creating opportunities in bendable displays, thin film solar cells and wearable optoelectronics (Tseghai et al., 2020).

Relative to conventional metallic or carbon-based conductors, they are light weight and extremely flexible, which suits wearable and rollable devices. They withstand many bending cycles with little mechanical fatigue and can resist oxidation and chemical corrosion.

Taken together, these properties allow conductive polymers to occupy a unique, intermediary position between rigid conductive metals and passive polymers, enabling new classes of responsive etextiles that are lightweight, conformal and electronically active.

Application Methods

Coating and fiber integration techniques sit at the core of e-textile production, determining how effectively the material is coupled to the fabric and how reliably it will behave during use.

Coating techniques stand out as one of the most widely used ways to give textiles new functions, mainly because they can be scaled up quickly and integrated into existing manufacturing lines. One of the simplest methods is dip coating, where individual threads, bundles of yarn, or entire fabric sheets are immersed in a solution containing conductive polymers like polyaniline, polythiophene, or polypyrrole nanoparticles, or micron sized metal flakes. The procedure is prized for being straightforward, inexpensive and capable of laying down a uniform film on diverse substrates, from cotton to polyester or polyamide (Sankar & Mohanta, 2017). In practice, dispersions of PEDOT: PSS move onto cotton textiles easily because the fibers hydrophilic character boosts polymer settling and the fabrics electrical performance.

A second deposition method, knife over roll coating, delivers consistent control over the thickness of conductive polymer films spread across woven textile substrates. By varying solution concentration, researchers can fine tune both electrical conductivity and mechanical strength. Because the process scales readily from bench top setups to pilot operations, it produces coatings that resist aging and endure repeated bending or strain (Rubežienė et al., 2023). Alternative approaches such as spray, screen print and ink jet deposition can also be used.

Fiber integration techniques introduce the conductor during fiber formation rather than after. This can be done by extruding a melt of conductive polymer or by blending nanoscale additives carbon nanotubes, graphene flakes, or metal nanowires into the spinning dope before spinning. Adapted wet spinning, electro spinning and melt spinning processes now yield long, continuous strands that remain as soft, breathable and flexible as standard textiles. When such conductive yarns are woven, knitted, or embroidered into garments, the resulting fabric shows high durability and stable, intrinsic conductivity (Allison, Hoxie, & Andrew, 2017). Sometimes a hybrid strategy merges internal doping with surface coatings to boost conductivity or add new functions, such as sensing or energy storage. CNT wrapped yarns and MXene coated fibers, for instance, combine high electrical mobility with the comfort expected in everyday clothing (Zhang et al., 2025).

These strategies are typically blended and fine-tuned according to each application's needs, the type of textile involved and how the final product must perform, pushing the field toward truly multifunctional, wearable e-textiles.

Durability and Processability Aspects

Durability and processability cover everything about the design and day-to-day use of e-textiles, shaping how long the product stays dependable, how easily manufacturers can create it and whether end users will find it comfortable or not. The long-term performance of a conductive fabric depends largely on whether its wires, inks, or threads keep working after bending, sweating, washing, or under sunlight while processability, in turn, measures how smoothly these materials can be woven or sewn in without adding bulk or stiffening the garment.

The wash and wear test therefore emerges as a main yard stick for the resilience of e-textile circuits, yet the scenario is seldom straight forward. Intrinsically conductive polymers, carbon nanotubes, or metal nanoparticles may all promise mobility on paper, but in everyday use they encounter water, detergent, steam, heat, scrubbing and acid sweat (Alam et al., 2022). Under such twin pressures, coarse fabric and corrosive liquid can delaminate, crack and slowly lift the fibers from the underlying mesh. Researchers have therefore turned to stronger chemical links, specially treated surfaces and finely tuned curing schedules to lock the layers in place, while thin urethane shields or breathable hydrogels add to outer part of jacket that stops moisture without decreasing softness of the material.

Mechanical durability under regular deformation is critical for wearables, because clothing is constantly stretched, bent, rubbed and folded around the body. Textured fabrics with small nubs or ridges can deflect debris away from conductive layers, thereby decreasing interface peeling and limiting decline in signal strength over many wear cycles.

Durability assessments for electronic textiles rely heavily on laundering studies because the ability of a garment to withstand repeated washing shapes customer satisfaction and determines its market life. By subjecting samples to accelerated wash and dry cycles researchers can pinpoint where conductive inks, fibers and embedded components might fail. New tests are emerging, pairing liquid cleaning with mechanical bending, humidity, temperature swings and chemical exposure so that wear and cleaning stress occur simultaneously.

From a processability viewpoint, dip coating stands out as an inexpensive, straight forward and broadly scalable method for laying conductive polymers, carbon nanoforms, or metal nanoparticles onto fabric. Alternative techniques like knife over roll spray coating, or printing open the door to localized deposition, a feature that designers rely on when crafting sensor arrays or fabric interconnects.

Natural conducting polymers polyaniline, polythiophene and polypyrrole adopt liquid consistent forms, so they can be cast as thin, flexible shields that preserve textiles feel and strength. Unlike many metallic brands, these organic films resist oxidation and hold up better in laundry cycles, providing longer life for smart garments.

In summary, the long-term performance of electronic textiles ultimately hinges on the ability of their conductive elements to retain functionality when subjected to repeated bending, stretching, washing and exposure to humidity or chemicals. Processability (Pu et al., 2023) offers focused insight into scalable coating, printing and spinning methods that produce continuous, flexible, water washable conductive sheets or threads displaying uniform coverage across extended textile surfaces.

APPLICATIONS OF NANOMATERIALS AND CONDUCTIVE POLYMERS IN E-TEXTILES.

Healthcare Monitoring

Conductive nanomaterials and polymer blends are set to change patient care, enabling e-textiles that track vital signs round the clock without breaking the skin. Fabrics interwoven with carbon nanotubes, graphene, MXenes, or conjugated polymers serve as soft electrodes that measure heart rate, breath, skin temperature and muscle activity while gently hugging the body. This close skin fit reduces motion noise and spares wearers from heavy sensors or adhesive patches. Raw waveforms are processed on the garment and then streamed wirelessly to caregivers, facilitating early anomaly detection and tailored treatment plans (Azeem et al., 2024). This low-profile solution is especially advantageous for the elderly, chronically ill, athletes in training, or neonates and expectant mothers, since it gathers reliable data over days or weeks.

Sportswear and Athletic Performance Tracking

The sportswear sector is merging nanomaterials with conductive polymers to produce dynamic clothing that tracks motion, muscle activity and vital signs as athletes move. Embedded inside these smart textiles are flexible sensors and nearly invisible circuits that instantly report strides length, jump height, ground force, hydration levelsand signs of muscle fatigue (Cui & Wang, 2025). Coaches and athletes use this clear data to adjust drills, guard against injuries and finetune recovery routines. Because the materials remain light, breathable and stretchable, it follows every movement without pinching or dragging, a must for fast, high impact workouts. Some designs pair with tiny energy harvesting cells, allowing the sensors to run for months or years with no battery swap, so users can focus solely on training.

Military and Safety Textiles

Military and public safety organizations are now turning to nanomaterials and conductive polymers to develop uniforms and gear that not only resist damage but also do intelligent work on the wearers' bodies. In similar advanced fabric platforms used by military units, miniaturized sensors gather continuous data on a personnel member's heart rate, skin temperature, ambient humidity and even the precise location of bullet impacts, then stream that information wirelessly to commanders or medics in the field (Hu et al., 2022). Layered MXenes within a conductive polymer mesh produce light weight, thin circuits that survive intense heat, corrosive chemicals and bends without developing cracks. Beyond health monitoring, these textiles can silently mask infrared signatures, block harmful radio emissions, or alert troops to incoming projectiles, giving occupants an advance picture of the battlefield. By building sensing directly into uniforms, engineers cut the bulk of extra gadgets, lighten supply loads and open the door to more agile, secure operations.

Industrial Applications

On the factory floor or offshore platform, technicians pair nanocomposites with specialized polymer blends to turn ordinary yarns into shields that block electromagnetic radiation or guard machinery from rust and electrolysis. Shielding upholstery loaded with the same nanocomposites is becoming standard in facilities where strong electromagnetic fields are present, protecting both sensitive electronics and personnel from harmful interference (Orasugh et al., 2024). By weaving silver nanowires, carbon nanotubes, MXenes, or conductive resins directly into the fabric, suppliers craft breathable, ultra-light shields that can be sewn into uniforms, laced along wires, or incorporated into device housings. Coatings or dispersed nanoparticles further lend the cloth corrosion resistance, so critical sensors stay functional even in salty or acidic air. Designed to survive repeated laundering and to slot easily into current textile lines, these industry ready e-textiles offer a practical answer to a wide range of technical shielding and worker protective tasks as shown in Figure 2.

Figure 2. Applications of nanomaterials and conductive polymers in E- textiles

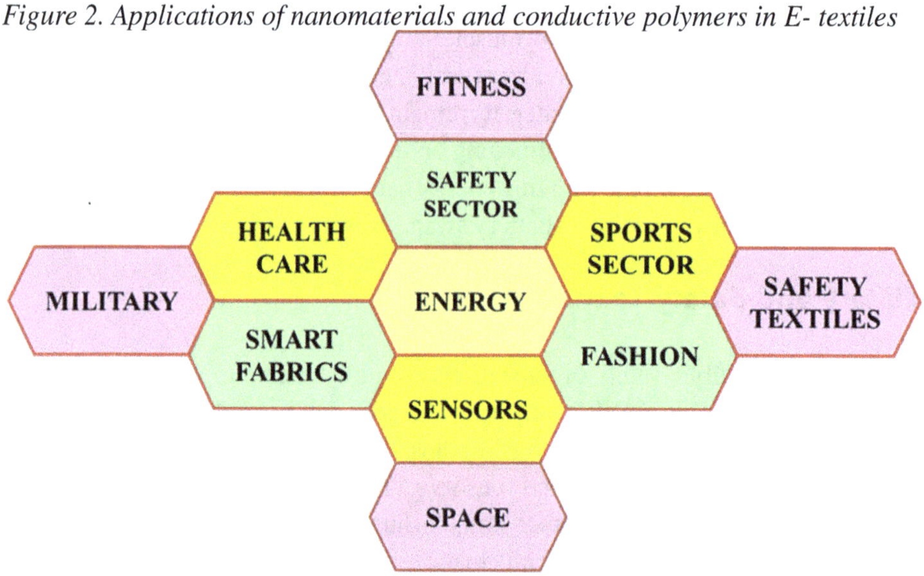

CHALLENGES AND CRITICAL ISSUES

Turning e-textiles into mass market goods involves tough obstacles that designers, engineers, policy makers and users must solve together before full benefits can flow into everyday practice.

Mechanical and Laundering Durability in Flexible Electronic Textiles

Mechanical stress and laundering stability rank among the foremost reliability questions for e-textiles. During normal wear, these fabrics bend, fold, stretch and abrade. The same motions occur again and sometimes with greater severity, in laundry cycles. Experimental evidence shows that tumbling, crumpling and water exposure fatigue can crack conductive coatings, raising resistivity until the system ceases functioning (Sadi & Kumpikaitė, 2022). The vulnerability of a given e-textile therefore depends strongly on the yarn type and weaves. Some composites fray and shed connectivity after only a few washes; others retain near original performance through dozens of launderings. Long lasting performance depends on careful engineering of surface roughness and on making certain that the conductive films or strands are mechanically anchored to the underlying fibers, so wear and washing do not loosen the bond. As consistent testing standards for these devices are still missing, cross-cross-person performance comparisons remain problematic.

Material Interactions and Long-Term Functionality

Another salient concern is the interplay between nanomaterials, conductive polymers and the underlying textile fibers, a dynamic that ultimately governs the longevity of e-textiles in real world operation. Discrepancies in chemical compatibility, adhesive bond strength and tensile modules can become obvious when the garment is flexed and stretched repeatedly. Meanwhile, ambient hazards like UV radiation, thermal cycling, airborne oxidants and wetting gradually attack and sever conductive traces, worsening the situation (Beeby et al., 2024). Although improved methods and additive surface treatments have delivered real benefits, even then some top performing units lose significant electroconductivity after just a few laundering rounds. To provide reliable evidence for durability claims and to inform rational material selection, the field therefore requires uniform and reproducible testing protocols.

Safety and Toxicity Concerns of Nanomaterials

Concerns about the safety and toxicity of nanomaterials in electronic textiles remain a critical hurdle for researchers and manufacturers. Because nanoparticles and nanofibres measure in billionths of a meter, they can slip past skin layers or be inhaled, raising worries about how long these particles hang around and whether they harm people who wear smart fabrics (Almeida & Ramos, 2017). Added concern comes from machine washing, where micro and nanoparticles wash off and from tossing out the garments, the same specks travel in wastewater, end up in rivers, get eaten by fish or pile up in soil. To protect nature, manufacturers now accept that every e-textile must pass clear toxicity tests and that product design must be open, so items that look clever also prove safe and earn buyer trust. But the life cycle of the garment from fabric making to daily use and final disposal can drain resources, foul air and water and leave waste if no one steers it carefully (Saleem & Zaidi, 2020). Aware of this fact, scientists and designers are therefore creating conductive threads that recycle easily, while linking with methods that trim water, energy and chemicals at every stage.

Scalability and Cost Effectiveness of Manufacturing

Manufacturers still face large hurdles in making e-textiles affordable and widely available. Many promising nanomaterials and conductive polymers are produced using long synthesis routes or rare precursors, so each square meter of fabric remains too expensive for mass market clothing. On top of that, combining delicate electronic parts with flexible textiles in a way that retains comfort and durability demands specialized techniques like coating, printing, fiber spinning or assembly that are not yet well tuned for big factories (Pereira et al., 2020). The field therefore needs scalable, reproducible and cost-effective methods and researchers are exploring every step of the chain from materials supply to final quality control.

Thus, moving e-textiles from exciting prototypes toward dependable, everyday products depend on solving mechanical wear, chemical interactions, safety issues, environmental impact and manufacturing at scale.

FUTURE DIRECTIONS AND MISSION

Future work and core mission for nanomaterials and conductive polymers in this field focus on merging bendable, comfortable fabrics with smart electronics, aiming to produce multifunctional textiles that genuinely improve daily life. As these systems shift from specialty items to mass market goods, researchers seek materials that stay soft, breathable, stretchable and comfortable while also sensing, harvesting energy, processing data and responding automatically to changes in light, temperature, or humidity.

Redefining Wearable: The Art of Blending Flexibility and Functionality

Researchers are now focused on developing thin, light and strong conductive materials that continue to work well even when the item is bent, stretched, or repeatedly washed. Achieving this goal requires careful tuning of how nanomaterials are mixed, how polymer blends cure and how fibers are woven or knitted together, so that the end smart fabric contains no hard or bulky parts. In 2025, new printable conductive inks paired with high-speed digital fabric presses are making it possible to embed sensors and circuitry directly into each fiber, a leap that both speeds production and allows consumers to tailor designs. Meanwhile, integration of AI within e-textiles helps in the creation of garments which will nudge the wearer with discreet tips, shifting passive oversight into a dynamic exchange that feels less like spying and more like support (Cleary et al., 2023).

Development of Multifunctional Smart Textiles

Researchers are now aiming to cram sensing, heating, energy storage and wireless communication into the same piece of clothing, a leap that could transform care, fitness, defense and daily apparel. For example, a military shirt that monitors heart rate, scavenges power from every step and dims its LEDs when daylight floods in it, protects privacy and still looks smart and comfortable (Choi et al., 2022). Analysts believe improvements in soft circuits, onboard AI and stretchy polymers will help the market boom by 2028.

Designers, however, face tougher eco scrutiny and buyer demands for local production. Making goods also requires repairs and upgrades, giving shoppers a leaner, faster alternative to traditional mass manufacture.

CONCLUSION

The advancement of e-textiles is on the verge of a significant leap forward due to progress in nanomaterials, conductive polymers and responsive manufacturing techniques. Looking beyond 2025, manufacturers expect these fabrics to retain familiar features like stretch, softness and breathability, while also performing advanced roles such as real time sensing, energy harvesting and wireless communication.

Innovations in digital printing alongside high throughput production lines, now allow conductive inks, memory patches and miniaturized sensors to be embedded directly into textile layers, reducing costs and enabling on demand, locally.

At the same time, engineers are tackling enduring issues of durability and washability, working to ensure that circuits and fibers remain electrically sound and mechanically intact after hundreds of wash dry cycles. The introduction of artificial intelligence, robotics and cloud linked analytics into textile lines is speeding the arrival of multipurpose garments that monitor vital signs, support athletes, enable touch free device control and react intelligently to light or temperature. Collectively, these trends point toward a new category of wearable that blend harmlessly into daily ward robes while providing personalized feedback, a vision that researchers and brands are beginning to realize in prototype products. Sustainability sits at the fore front of innovation, prompting designers to replace synthetic fibers with abundant, renewable alternatives and to adopt low impact production steps that curtail carbon emissions. With advances in materials science, digital manufacturing and integrated design, researchers envision smart textiles that embody not only hitech functionality but also the comfort, strength and planetary respect needed to turn hospitals, gyms, battlefields, homes and countless intermediaries into truly adaptive spaces.

REFERENCES

Abilevitch, L., Mizrahi, L., Cohen, G., Kenig, S., & Amir, E. (2023). Polyaniline for smart textile applications. In *IntechOpen eBooks*. DOI: 10.5772/intechopen.1001939

Adamu, M. F., Tesfaye, T., Berhanu, B., & Simegnaw, A. (2025). Electronic textiles. *Textile Progress*, *57*(1), 1–71. DOI: 10.1080/00405167.2025.2468595

Agarwal, S., Dohare, A. K., Saxena, P., Singh, J., Singh, I., & Sahu, U. K. (2025). HDL-ACO: Hybrid deep learning and ant colony optimization for ocular optical coherence tomography image classification. *Scientific Reports*, *15*(1), 5888. DOI: 10.1038/s41598-025-89961-7 PMID: 39966596

Alam, M. R., Islam, T., Repon, M. R., & Hoque, M. E. (2022). Carbonbased polymer nanocomposites for electronic textiles (etextiles). In *Elsevier eBooks* (pp. 443–482). https://doi.org/DOI: 10.1016/b9780128244920.000209

Alamer, F. A., Althagafy, K., Alsalmi, O., Aldeih, A., Alotaiby, H., Althebaiti, M., Alghamdi, H., Alotibi, N., Saeedi, A., Zabarmawi, Y., Hawsawi, M., & Alnefaie, M. A. (2022). Review on PEDOT:PSSBased Conductive Fabric. *ACS Omega*, *7*(40), 35371–35386. DOI: 10.1021/acsomega.2c01834 PMID: 36249401

Alamer, F. A., & Alzahrani, A. (2025). Advancements and innovations in textile engineering: An InDepth analysis of SWCNTs and PANI integration in smart textile technologies. *Composites. Part A, Applied Science and Manufacturing*, *108872*, 108872. Advance online publication. DOI: 10.1016/j.compositesa.2025.108872

Allison, L., Hoxie, S., & Andrew, T. L. (2017). Towards seamlessly integrated textile electronics: Methods to coat fabrics and fibers with conducting polymers for electronic applications. *Chemical Communications*, *53*(53), 7182–7193. DOI: 10.1039/C7CC02592K PMID: 28555696

Almeida, L., & Ramos, D. (2017). Health and safety concerns of textiles with nanomaterials. *IOP Conference Series. Materials Science and Engineering*, *254*, 102002. DOI: 10.1088/1757-899X/254/10/102002

Anik, H. R., Tushar, S. I., Mahmud, S., Khadem, A. H., Sen, P., & Akter, M. (2024). Into the revolution of NanoFusion: Merging high performance and aesthetics by nanomaterials in textile finishes. *Advanced Materials Interfaces*, *12*(1), 2400368. Advance online publication. DOI: 10.1002/admi.202400368

Azeem, M., Shahid, M., Masin, I., & Petru, M. (2024). Design and development of textilebased wearable sensors for realtime biomedical monitoring; a review. *Journal of the Textile Institute*, •••, 1–16. DOI: 10.1080/00405000.2024.2318500

Beeby, S. P., Torah, R. N., Wagih, M., Isaia, B., Black, S., Saunders, J., & Yang, K. (2024). Heterogeneous E-Textiles: Materials, manufacturing and sustainability. *Advanced Materials Technologies*. Advance online publication. DOI: 10.1002/admt.202400844

Bhattacharjee, S., Macintyre, C. R., Wen, X., Bahl, P., Kumar, U., Chughtai, A. A., & Joshi, R. (2020). Nanoparticles incorporated graphenebased durable cotton fabrics. *Carbon*, *166*, 148–163. DOI: 10.1016/j.carbon.2020.05.029

Choi, H. W., Shin, D., Yang, J., Lee, S., Figueiredo, C., Sinopoli, S., Ullrich, K., Jovančić, P., Marrani, A., Momentè, R., Gomes, J., Branquinho, R., Emanuele, U., Lee, H., Bang, S. Y., Jung, S., Han, S. D., Zhan, S., Harden-Chaters, W., & Kim, J. M. (2022). Smart textile lighting/display system with multifunctional fibre devices for large scale smart home and IoT applications. *Nature Communications*, *13*(1), 814. Advance online publication. DOI: 10.1038/s41467-022-28459-6 PMID: 35145096

Cleary, F., Srisa-An, W., Henshall, D. C., & Balasubramaniam, S. (2023). Emerging AI technologies inspiring the next generation of ETextiles. *IEEE Access : Practical Innovations, Open Solutions*, *11*, 56494–56508. DOI: 10.1109/ACCESS.2023.3282184

Cui, G., & Wang, C. (2025). Applications and development trends of textile materials in sports: A review. *Alexandria Engineering Journal*, *126*, 491–506. DOI: 10.1016/j.aej.2025.04.094

Dulal, M., Afroj, S., Ahn, J., Cho, Y., Carr, C., Kim, I., & Karim, N. (2022). Toward sustainable wearable electronic textiles. *ACS Nano*, *16*(12), 19755–19788. DOI: 10.1021/acsnano.2c07723 PMID: 36449447

Ehrmann, G., & Ehrmann, A. (2021). Electronic textiles. *Encyclopedia*, *1*(1), 115–130. DOI: 10.3390/encyclopedia1010013

Eom, W., Shin, H., Ambade, R. B., Lee, S. H., Lee, K. H., Kang, D. J., & Han, T. H. (2020). Largescale wetspinning of highly electroconductiveMXene fibers. *Nature Communications*, *11*(1), 2825. Advance online publication. DOI: 10.1038/s41467-020-16671-1 PMID: 32499504

Faruk, O., Ahmed, A., Khadem, A. H., Jia, L., & Sun, L. (2025). *GrapheneFunctionalized textile composites for wearable joule heating applications*. Advanced Nanocomposites., DOI: 10.1016/j.adna.2025.03.001

Gaur, A., Singh, S. K., & Saxena, P. (2024). Performance analysis of deepfake text detection techniques on social media. In *2024 International Conference on Distributed Computing and Optimization Techniques (ICDCOT)* (pp. 1–6). IEEE. DOI: 10.1109/ICDCOT61034.2024.10515626

Godara, A., Mezzo, L., Luizi, F., Warrier, A., Lomov, S., Van Vuure, A., Gorbatikh, L., Moldenaers, P., & Verpoest, I. (2009). Influence of carbon nanotube reinforcement on the processing and the mechanical behaviour of carbon fiber/epoxy composites. *Carbon, 47*(12), 2914–2923. DOI: 10.1016/j.carbon.2009.06.039

Grancarić, A. M., Jerković, I., Koncar, V., Cochrane, C., Kelly, F. M., Soulat, D., & Legrand, X. (2017). Conductive polymers for smart textile applications. *Journal of Industrial Textiles, 48*(3), 612–642. DOI: 10.1177/1528083717699368

He, G., Zhu, C., Shi, Y., Yu, Y., Wu, Y., Soutis, C., Cao, L., & Liu, X. (2024). Development of a MusselInspired conductive graphene coated cotton yarn for wearable sensors. *iScience, 28*(2), 111711. DOI: 10.1016/j.isci.2024.111711 PMID: 39898041

Hossain, M. M., Lubna, M. M., & Bradford, P. D. (2023). Multifunctional and washable Carbon NanotubeWrapped textile yarns for wearable ETextiles. *ACS Applied Materials & Interfaces, 15*(2), 3365–3376. DOI: 10.1021/acsami.2c19826 PMID: 36622361

Hsu, P., Liu, X., Liu, C., Xie, X., Lee, H. R., Welch, A. J., Zhao, T., & Cui, Y. (2014). Personal thermal management by Metallic NanowireCoated Textile. *Nano Letters, 15*(1), 365–371. DOI: 10.1021/nl5036572 PMID: 25434959

Hu, Q., Shen, X., Qian, X., Huang, G., & Yuan, M. (2022). The personal protective equipment (PPE) based on individual combat: A systematic review and trend analysis. *Defence Technology, 28*, 195–221. DOI: 10.1016/j.dt.2022.12.007

Hu, X., Fan, Q., Wang, S., Chen, Y., Wang, D., Chen, K., Ge, F., Zhou, W., & Liang, K. (2024). TwoDimensional MXENES: Innovative materials for efficient thermal management and safety solutions. *Research, 7*, 0542. Advance online publication. DOI: 10.34133/research.0542 PMID: 39703779

Hussain, I., Amna, R., Kalidasan, B., Rani, G. M., Bandi, H., Venkateswarlu, S., Khan, S. A., Mohapatra, D., & Zhang, K. (2025). MXEnes for various applications: Recent trends and future aspects. *SmartMat, 6*(2), e70012. Advance online publication. DOI: 10.1002/smm2.70012

Jalali, E., Erasmus, E., Schutte-Smith, M., & Visser, H. G. (2024). Fixation of nanoparticles on fabric: Applications in general health management. *Materials Today. Communications, 41*, 110577. DOI: 10.1016/j.mtcomm.2024.110577

K, N., & Rout, C. S. (2021). Conducting polymers: a comprehensive review on recent advances in synthesis, properties and applications. *RSC Advances, 11*(10), 5659–5697. .DOI: 10.1039/D0RA07800J

Khan, M., Refati, M. F. a. D., Arup, M. M. R., Islam, M. A., &Mobarak, M. H. (2025). Conductive Polymer-Based electronics in additive manufacturing: materials, processing, and applications. *Advances in Polymer Technology, 2025*(1)..DOI: 10.1155/adv/4234491

Liu, H., Shi, Y., Pan, Y., Wang, Z., & Wang, B. (2025). Sensory interactive fibers and textiles. *NPJ Flexible Electronics*, *9*(1), 23. Advance online publication. DOI: 10.1038/s41528-025-00398-4

Mishra, R. K., Sarkar, J., Verma, K., Chianella, I., Goel, S., & Nezhad, H. Y. (2024). Exploring transformative and multifunctional potential of MXenes in 2D materials for nextgeneration technology. *Open Ceramics*, *18*, 100596. DOI: 10.1016/j.oceram.2024.100596

Ojstršek, A., Jug, L., & Plohl, O. (2022). A review of electro conductive textiles utilizing the DipCoating technique: Their functionality, durability and sustainability. *Polymers*, *14*(21), 4713. DOI: 10.3390/polym14214713 PMID: 36365707

Orasugh, J. T., Botlhoko, O. J., Temane, L. T., & Ray, S. S. (2024). Progress in polymer nonwoven textile materials in electromagnetic interference shielding applications. *Functional Composite Materials*, *5*(1), 5. Advance online publication. DOI: 10.1186/s42252-024-00054-6

Pereira, C., Pereira, A. M., Freire, C., Pinto, T. V., Costa, R. S., & Teixeira, J. S. (2020). Nanoengineered textiles: from advanced functional nanomaterials to groundbreaking highperformance clothing. In *Elsevier eBooks* (pp. 611–714). https://doi .org/DOI: 10.1016/b9780128167878.000211

Pu, J., Ma, K., Luo, Y., Tang, S., Liu, T., Liu, J., Leung, M., Yang, J., Hui, R., Xiong, Y., & Tao, X. (2023). Textile electronics for wearable applications. *International Journal of Extreme Manufacturing*, *5*(4), 042007. DOI: 10.1088/2631-7990/ace66a

Rubežienė, V., Varnaitė Žuravliova, S., Sankauskaitė, A., Pupeikė, J., Ragulis, P., & Abraitienė, A. (2023). The Impact of Structural Variations and Coating Techniques on the Microwave Properties of Woven Fabrics Coated with PEDOT: PSS Composition. *Polymers*, *15*(21), 4224. DOI: 10.3390/polym15214224 PMID: 37959904

Sadi, M. S., & Kumpikaitė, E. (2022). Advances in the robustness of wearable electronic textiles: Strategies, stability, washability and perspective. *Nanomaterials (Basel, Switzerland)*, *12*(12), 2039. DOI: 10.3390/nano12122039 PMID: 35745378

Saleem, H., & Zaidi, S. (2020). Sustainable use of nanomaterials in textiles and their environmental impact. *Materials (Basel)*, *13*(22), 5134. DOI: 10.3390/ma13225134 PMID: 33203051

Sankar, K. N. A., & Mohanta, K. (2017). Preparation of highly conductive yarns by an optimized impregnation process. *Journal of Electronic Materials*, *47*(3), 1970–1978. DOI: 10.1007/s11664-017-5998-3

Saxena, P., & Goyal, A. (2019). Study of computerized segmentation and classification techniques: An application to histopathological imagery. *Informatica: An International Journal of Computing and Informatics, 43*(4).

Saxena, P., Gupta, R., Goel, A., Sanger, A. K. S., Shrivastava, V., & Singh, M. (2025). Automated cellular segmentation in histopathology: A multi-step mask prediction approach for enhanced analysis. In *2025 7th International Conference on Signal Processing, Computing and Control (ISPCC)* (pp. 437–442). IEEE. DOI: 10.1109/ISPCC66872.2025.11039568

Saxena, P., Sharma, M., Batra, R., Gupta, R., Singh, M., & Singh, S. K. (2025). XAI for lymph node histopathology: From explainability to immersive metaverse applications. In *2025 7th International Conference on Signal Processing, Computing and Control (ISPCC)* (pp. 304–309). IEEE. DOI: 10.1109/ISPCC66872.2025.11039319

Saxena, P., Singh, S. K., Rashid, M., Alshamrani, S. S., & Alnfiai, M. M. (2025). Efficient deep learning model for classifying lung cancer images using normalized stain-agnostic feature method and FastAI-2. *PeerJ. Computer Science*, *11*, e2903. DOI: 10.7717/peerj-cs.2903 PMID: 40567713

Shahid, M. A., Rahman, M. M., Hossain, M. T., Hossain, I., Sheikh, M. S., Rahman, M. S., Uddin, N., Donne, S. W., & Hoque, M. I. U. (2025). Advances in conductive PolymerBased flexible electronics for multifunctional applications. *Journal of Composites Science*, *9*(1), 42. DOI: 10.3390/jcs9010042

Shahidi, S., & Moazzenchi, B. (2018). Carbon nanotube and its applications in textile industry – A review. *Journal of the Textile Institute*, *109*(12), 1653–1666. DOI: 10.1080/00405000.2018.1437114

Syduzzaman, M., Hassan, A., Anik, H. R., Akter, M., & Islam, M. R. (2023). Nanotechnology for High-Performance Textiles: A promising frontier for innovation. *ChemNanoMat : Chemistry of Nanomaterials for Energy, Biology and More*, *9*(9), e202300205. Advance online publication. DOI: 10.1002/cnma.202300205

Tseghai, G. B., Mengistie, D. A., Malengier, B., Fante, K. A., & Van Langenhove, L. (2020). PEDOT:PSS Based Conductive Textiles and their Applications. *Sensors (Basel)*, *20*(7), 1881. DOI: 10.3390/s20071881 PMID: 32231114

Yao, S., Yang, J., Poblete, F. R., Hu, X., & Zhu, Y. (2019). Multifunctional electronic textiles using Silver nanowire composites. *ACS Applied Materials & Interfaces*, *11*(34), 31028–31037. DOI: 10.1021/acsami.9b07520 PMID: 31373192

Younes, B. (2023). Smart Etextiles: A review of their aspects and applications. *Journal of Industrial Textiles*, *53*, 15280837231215493. Advance online publication. DOI: 10.1177/15280837231215493

Zhai, H., Liu, J., Liu, Z., & Li, Y. (2025). Functional graphene fiber materials for advanced wearable applications. *Advanced Fiber Materials*, *7*(2), 443–468. Advance online publication. DOI: 10.1007/s42765-025-00512-1

Zhang, H., Xiao, H., & Long, J. (2025). Preparation, structure, property and application of MXene in fabricating functional and intelligent textiles: A comprehensive review. *Composites. Part B, Engineering*, *112461*, 112461. Advance online publication. DOI: 10.1016/j.compositesb.2025.112461

Zheng, Y., Zhang, Q., Jin, W., Jing, Y., Chen, X., Han, X., Bao, Q., Liu, Y., Wang, X., Wang, S., Qiu, Y., Di, C., & Zhang, K. (2019). Carbon nanotube yarn based thermoelectric textiles for harvesting thermal energy and powering electronics. *Journal of Materials Chemistry. A, Materials for Energy and Sustainability*, *8*(6), 2984–2994. DOI: 10.1039/C9TA12494B

Chapter 5
Smart Sensors Embedded in Textiles:
Health and Environmental Monitoring

Mahadev Ajagalla
https://orcid.org/0009-0005-9580-1840
Bennett University, India

Tanya Patel
Bennett University, India

Vaishnavi Bisen
https://orcid.org/0009-0004-9469-568X
Bennett University, India

ABSTRACT

Imagine wearing the clothes that don't just provide you comfort but also keep a supervising eye on your health and the environment around you. Smart textiles fabric embedded with sensors is making it all a reality. The latest-gen apparel can monitor vital parameters, sense changes in the environment, and collect real-time data without administering any invasive therapies. This chapter explores how materials science, flexible electronics, and biomedical innovation come together to create these smart fabrics. We will walk you through some of the different types of sensors from those that monitor heart rate and levels of hydration to those that detect pollutants and UV radiation. We also examine how sensors are incorporated into clothing using methods like conductive yarns and printed electronics. with ongoing research and breakneck advancements, smart textiles are about to transform the way we care for our wellbeing and our environment stylishly and comfortably.

DOI: 10.4018/979-8-3373-4287-0.ch005

INTRODUCTION

Picture yourself in clothes that not only look and keep you warm but also detect your heartbeat, alert you that you're dehydrated, or warn you of air pollution. Such is the potential of smart textiles fabric that takes its role beyond the conventional to incorporate technology into its very fibers. Such cutting-edge materials can sense, react, and convey information, almost likening them to wearable technology than they are to conventional clothing. Smart fabrics are constructed by marrying soft, pliable fabrics with small electronic devices like sensors, batteries, and wireless communication systems. Together, these parts transform a mundane piece of clothing into an enhanced tool for immediate data gathering and interaction with the surrounding world (Angelucci et al., 2021).

Textiles have long been a part of human life, from the simplest linen wraps of the ancient Egyptians to the high-performance athletic wear of today. But until recently, they played a primarily passive function to cover and shield the body. Gradually, we began to play with fabrics that do more, such as jackets that adapt to temperature or shirts that pull sweat away. The actual game-changer was that the developments in materials science and electronics enabled a combination of conventional textiles with intelligent features. Conductive threads, stretchable circuits, and printable sensors have helped us move from passive materials to interactive ones that can sense, react, and adapt to their environment. This evolution has transformed everyday garments into smart, connected systems (Castano & Flatau, 2014).

Defining What Makes a Textile Smart

At its core, a smart textile is made up of three key elements:

- A **fabric base**, which forms the structure and comfort layer,
- **Sensors or actuators**, embedded to detect signals like body temperature, heart rate, or gases in the air,
- A **communication system**, usually wireless, that sends collected data to other devices like smartphones or cloud servers.

Some smart textiles are fairly simple they just detect one thing, like pressure. Others are much more advanced, capable of sensing multiple parameters and even adjusting themselves in response (for example, by tightening around an injury or warming up when it's cold). These multi-functional fabrics are at the cutting edge of wearable technology (Dolez et al., 2020).

Factors Driving the Rise of Smart Textiles

Several exciting breakthroughs have made this the perfect time for smart textiles to grow:

New Materials at the Nanoscale

The development of ultra-thin, flexible materials like graphene, nanowires, and conductive polymers means we can now make sensors and circuits that are tiny enough to disappear into the weave of fabric. These materials don't just work well they're also soft, stretchable, and safe for the skin (Ojuroye et al., 2017).

Smarter Algorithms

With artificial intelligence (AI), we're able to do more with the information these fabrics are able to capture. Rather than simply displaying raw numbers, AI algorithms can identify trends, flag early warning signs of health issues, or even be able to forecast what's around the corner based on patterns in your body's activity.

Always-Connected Lifestyle Through IoT

With the advent of the Internet of Things (IoT), it is simpler than ever to connect your doctor or your shirt with your phone. Smart fabrics are a perfect fit for this interconnected environment, providing real-time feedback and information wherever you are (Oatley, Choudhury, & Buckman, 2021).

Importance of Smart Textiles in Health and Environmental Monitoring

Smart textiles aren't just cool they're practical. Our healthcare systems are shifting toward **preventive care**, and wearable technology is a big part of that. Instead of going to a clinic for a check-up, what if your shirt could alert you to an irregular heartbeat before you even feel symptoms? For people with chronic conditions, such tools could mean fewer hospital visits and better quality of life. On the environmental side, concerns about pollution and climate change are driving demand for real-time monitoring tools. Cloths that monitor air quality or UV radiation can assist individuals in making safer decisions particularly outdoor workers, children, or the elderly (Wang et al., 2021).

This chapter is primarily directed toward researchers, graduate students, and practitioners working in the fields of textiles, electronics, healthcare, and environmental monitoring, offering both technical foundations and applied perspectives relevant to these audiences.

Current Applications of Smart Textiles

These cloths are already appearing in a number of applications and Smart textiles with sensors that are embedded within them for tracking health and environment shown in Figure 1:

- **Healthcare Wearables**: Shirts that track your ECG, fabrics that measure breathing rate, socks that detect pressure changes to prevent foot ulcers in diabetics.
- **Environmental Accessories**: Scarves or patches that can detect levels of carbon monoxide or nitrogen dioxide within the city streets, or shirts that alert you to excessive UV radiation (Rather & Dar, 2024).
- **Performance & Safety**: They are used by athletes to monitor muscle fatigue or dehydration. Soldiers use them to identify chemical exposure or to communicate distress in the field.

Figure 1. Smart fabrics with sensors embedded in them for health and environment monitoring.

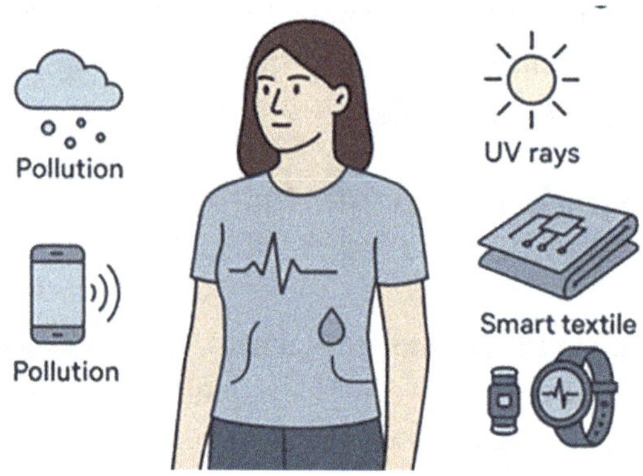

In the last ten years, various reviews and surveys have tracked the development of intelligent textiles in healthcare, sports, environmental monitoring, and human–computer interaction (Castano & Flatau, 2014; Angelucci et al., 2021; Mamun & Yuce, 2020). These papers indicate progress in conductive fibers, flexible electronics, and sensor integration, along with challenges in durability, power management, and mass adoption. Current research has moved towards sustainability and green materials, AI-based data analysis, and IoT connectivity for personalized health and environmental monitoring (Tat et al., 2022; Rather & Dar, 2024). But the majority of the existing reviews consider either the healthcare component or the environmental monitoring component independently. This chapter makes its contribution by presenting an integrated viewpoint, bringing together both fields while also covering real case studies, communication approaches, and the future, thus presenting a broader view of smart textiles within society. Recent reviews have highlighted the combined role of smart textiles in healthcare and environmental monitoring, underlining their dual importance in human well-being and sustainability (Li, Chen, & Tao, 2022).

FUNDAMENTAL TECHNOLOGIES BEHIND SMART TEXTILES

Smart fabrics become a possibility thanks to the union of a number of cutting-edge technologies. The three pillars that sit at the heart of this technology are materials science, flexible electronics, and methods for integrating sensors. This subsection takes us through each of these pillars as a precursor to the understanding of how regular fabrics become intelligent systems. Building on the applications introduced in Section 1, this section explores the fundamental technologies that enable textiles to become intelligent systems.

Materials Science: Building Function into Fabric

Conductive Materials

Because textiles must be capable of conducting electricity in order to receive or transmit electrical signals, scientists have designed new materials that are conductive or fill the gap left by existing fibers. Cotton or wool is naturally insulating, so these new conductive materials cover or replace them.

Conductive Yarns

- **Silver-coated fibers** are among the most popular choices. Silver (Ag) is an excellent conductor and naturally antibacterial, making it ideal for skin-contact wearables.
- **Stainless steel threads** offer durability and decent conductivity. They're often blended with other fibers to retain flexibility.

The yarns may be knitted or woven straight into fabric, and there can be electrical circuits without the hassle of wires.

Conductive Polymers

- **PEDOT:PSS (Poly(3,4-ethylenedioxythiophene) polystyrene sulfonate)** is a water-dispersible polymer that's both conductive and flexible, commonly used for coating fibers.
- **Polypyrrole** is another conductive polymer that adheres well to textile surfaces and responds to environmental changes (useful for sensing).

These polymers can either be utilized as coatings or mixed with inks to be employed in printing.

Textile Substrates

The Base Layer An intelligent textile needs something more than electronics it needs a comfortable, wearable base material that holds everything firmly in position without compromising comfort. Substrate selection comes in here.

- **Cotton** is soft, breathable, and biocompatible, making it suitable for long-term skin contact.
- **Polyester** offers mechanical strength and water resistance and is frequently used in sportswear.
- **Silk**, though less common, has excellent tensile strength and is naturally biodegradable.

The achievement of the correct combination of functionality and comfort is what renders the textile appropriate for real usage.

Flexible Electronics: Making Circuits Stretch and Bend

Electronics in traditional devices are rigid and brittle not a great match for fabric that bends, stretches, and folds. Flexible electronics achieve this by employing materials and forms that have the ability to move with the fabric.

Printed Electronics include depositing conductive inks directly onto fabric via methods such as inkjet printing or screen printing (Koncar, 2016).

Inkjet printing allows for precise, patterned deposition of conductive materials (like silver nanoparticle inks).

Screen printing is more scalable and cost-effective, often used for producing pressure sensors or temperature arrays.

Printed electronics enable seamless sensor integration without interrupting the flexibility or aesthetics of the garment.

Thin-Film Transistors (TFTs)

TFTs form the basis of many wearable computing units. These thin, bendable transistors are built on flexible substrates like polyimide or PET and can be laminated or transferred onto fabric. They allow for low-power computation close to the body, reducing the need to send raw data to external devices.

Electrical Properties

To understand how electricity behaves in these textile systems, we turn to a few key formulas:

1. Resistance

$$R = \rho \frac{L}{A} \tag{1}$$

Where:

- R is the electrical resistance (Ohms),
- ρ is the resistivity of the material (Ohm·meters),
- L is the length of the conductor,
- A is the cross-sectional area.

This formula tells us how much resistance a conductive fiber or pathway will present. Lower resistance means better signal flow shown in equation 1.

2. Conductivity

$$\sigma = \frac{1}{\rho} \tag{2}$$

Where:

- σ is electrical conductivity (Siemens per meter),
- ρ is resistivity.

Higher conductivity means the material is better at carrying electrical signals. Conductive inks and coatings are often evaluated based on this property shown in equation 2.

Sensor Integration Techniques

After selecting the materials and preparing the electronic components, the final step is integrating sensors into textiles in a way that they remain functional, durable, and unobtrusive. Here are the most widely used methods:

Weaving and Knitting

- Conductive yarns are woven or knitted directly into the fabric pattern.
- This approach is highly durable and maintains flexibility but requires design adaptation at the textile manufacturing level.

Embroidery

- Precision machines can embroider conductive threads onto garments in sensor shapes or circuit paths.
- Enables quick prototyping and creative personalization

Lamination and Bonding

- skinny electronic patches or flexible electronics are laminated onto fabric materials with adhesives or heat.
- ubiquitous in sportswear where thin form-factors are essential.

3D Printing and Transfer Printing

- 3D printing enables sensors to be printed directly on a stretchy or curved surface.
- Transfer printing transfers prefabricated electronics from a rigid template to a compliant textile without destroying delicate components.

These integration techniques preserve the comfort, washability, and wearability of smart textiles without sacrificing the reliability of the embedded electronics, Materials Science and Flexible Electronics Depicted in Figure 2.

Figure 2. Materials science and flexible electronics

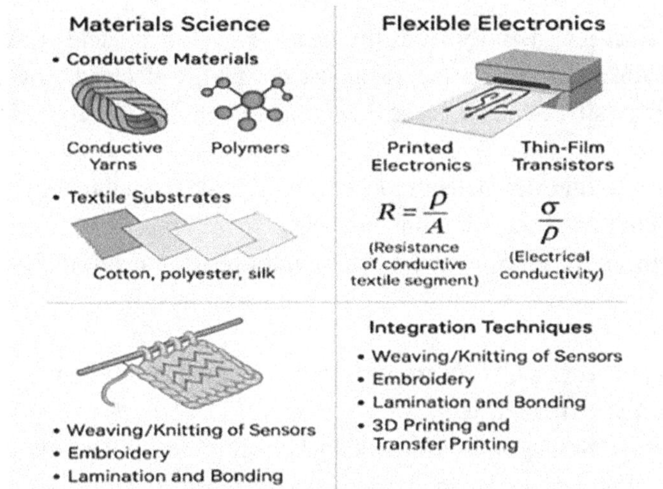

SENSOR TYPES AND THEIR WORKING PRINCIPLES

After understanding the underlying technologies, it is essential to examine the different types of sensors and their working principles, which form the functional core of smart textiles. Smart clothing comes to life with the integration of specialized sensors that have the ability to monitor the human body as well as the surrounding environment. The sensors are designed to be light, flexible, and responsive so that they can easily be integrated into clothes without compromises on comfort or wearability.

In this section, we'll explore the two main categories of sensors used in smart textiles those focused on health monitoring and those designed for environmental sensing along with how each type works (Mamun & Yuce, 2020).

Health Monitoring Sensors

These sensors are integrated into clothing to track physiological signals continuously. Their intention is to give real-time health information without invasive methods or large devices.

ECG and EMG Sensors (Bioelectric Signal Monitoring)

Monitor electrical activity from the heart (ECG) or muscles (EMG) (Farina, Merletti, & Enoka, 2004). These sensors use dry textile electrodes made from conductive fabric or silver-coated yarns that come in contact with the skin.

- **Working principle:** Detect bioelectrical signals from the body's surface and send them to a processing unit.
- **Applications:** Detect arrhythmias, monitor heart rate variability, assess muscular fatigue.

Formula (Lead II ECG Example). Voltage signal = difference between electrodes on right arm and left leg.

Temperature Sensors. Measure changes in skin or ambient temperature. Use thermistors or thermocouples integrated into the fabric.

- **Working principle:** Resistance or voltage changes in response to temperature variations.
- **Applications:** Fever detection, thermal comfort management, early detection of inflammation (Liu, Wang, Li, & Qin, 2021).

Key Formula (Thermocouple output)

$$\Delta V = \alpha \cdot \Delta T \quad (3)$$

Where α is the Seebeck coefficient and ΔT is the temperature difference follow equation 3.

Hydration and Sweat Sensors. Track hydration levels and sweat composition (e.g., sodium, potassium). Use electrochemical sensors or impedance-based sensors integrated into wristbands or armpits of garments.

- **Working principle:** Detect conductivity changes in sweat or use colorimetric sensors for chemical analysis.
- **Applications:** Athlete performance tracking, dehydration warnings, electrolyte imbalance detection.

Respiration and Movement Sensors (Strain Sensors). Monitor breathing rate, posture, or physical activity. Based on piezoelectric, capacitive, or resistive materials that respond to stretch or movement.

- **Working principle:** Sensor properties (resistance, capacitance, voltage) change when the fabric is stretched or compressed during movement.
- **Applications:** Sleep monitoring, stress tracking, physical therapy.

Piezo-Resistive Formula

$$R = \rho\frac{L + \Delta L}{A} \tag{4}$$

R denotes the electrical resistance, ρ is the resistivity of the conductive material, L is the original length of the conductor, ΔL is the change in length due to stretching or deformation, and A is the cross-sectional area of the conductive path. This interaction is most applicable in smart textiles, where stretchable and flexible materials undergo dimensional modifications while in use. With increased stretching of the cloth, the lengthening (ΔL) causes resistance to increase, and this can influence the performance and accuracy of implant sensors. Compensation for these changes and their understanding is most important in wearable electronic system design represented in equation 4.

Environmental Monitoring Sensors

These sensors are embedded in clothing to monitor external environmental conditions that potentially impact human safety or health.

Gas Sensors

Detect harmful gases such as CO, NO_2, and O_3. Use metal oxide semiconductor (MOS) sensors or conductive polymer composites.

- **Working principle:** Target gas changes the electrical resistance or the flow of current through the sensing material.
- **Applications:** Urban air pollution monitoring, occupational safety, prevention of asthma.

Example. NO_2 will raise the resistance in some MOS-based sensors, which can be measured and reported.

UV Sensors. Track exposure to ultraviolet (UV) radiation. Use organic photodiodes or photochromic materials incorporated into fabric patches or caps.

Working Principle. UV light induces a measurable electric or color change.

Uses. Skin cancer prevention, sun protection of outdoor workers, individualized sun safety reminders.

Temperature and Humidity Sensors (Environmental)

Track ambient environmental temperature and humidity levels. Use capacitive or resistive humidity sensors and thermistors for temperature.

Working Principle. Sensor's capacitance/resistance varies depending on surrounding moisture or heat.

Humidity Formula

$$\text{Relative Humidity (RH)} = \frac{P_{\text{water vapor}}}{P_{\text{sat}}} \times 100\% \qquad (5)$$

$P_{\text{water vapor}}$ vapor represents the partial pressure of water vapor in the air, and P_{sat} is the saturation vapor pressure at the same temperature. Relative Humidity (RH) is expressed as a percentage and shows the air's proximity to saturation with moisture. This parameter is particularly significant in environmental sensing in smart textiles, where precise measurement of humidity can determine comfort levels, identify sweating, or measure environmental conditions as illustrated in equation 5.

Light and Environmental Radiation Sensors

Detect visible light intensity and infrared/thermal radiation. Use photodiodes or pyroelectric sensors integrated into caps, visors, or sleeves.

Working Principle. Light exposure changes the charge distribution or resistance in the sensor material.

Noise and Sound Sensors. Detect environmental noise levels or voice signals. Utilize MEMS microphones or piezoelectric acoustic sensors.

- **Applications:** Smart hearing aids, noise exposure monitoring, speech-based commands for assistive garments.

Multimodal Sensor Systems

Advanced smart textiles often incorporate multiple sensor types in a single garment. For example, a fitness shirt might include:

- ECG electrodes (for heart rate),
- Stretch sensors (for respiration),
- Sweat patches (for hydration),
- Ambient sensors (for temperature and pollution).

These multimodal systems allow for holistic monitoring of both internal and external conditions, and with AI integration, they can offer intelligent insights rather than just raw data.

COMMUNICATION TECHNOLOGIES

Once sensors capture data, the next challenge is transmitting it effectively; this section discusses the communication technologies that make smart textiles interactive and connected. Collecting data from sensors embedded in textiles is only one part of the equation. To make the data meaningful and actionable, smart textiles must transmit the information to other systems such as smartphones, computers, or cloud-based platforms where it can be stored, analyzed, and acted upon.

This requires a reliable communication framework that operates effectively despite the challenges of mobility, power limitations, and physical wear. In this section, we explore how smart garments handle data transmission, the technologies they use, and how these systems are designed for energy-efficient and secure operation.

Data Acquisition and Transmission

The first step in the communication pipeline involves capturing the signals from sensors and converting them into digital data that can be processed or transmitted.

Signal Conditioning

- Raw sensor signals (e.g., voltage, resistance) are often weak and noisy.
- Signal conditioning circuits like amplifiers, filters, and analog multiplexers prepare the signal for conversion.

Analog-to-Digital Conversion (ADC)

- Most sensors produce analog signals.
- **ADC modules**, typically embedded in microcontrollers, convert these signals into digital format so that microprocessors can read and interpret the values.

Example. An ECG electrode generates a continuous voltage signal. An ADC samples this voltage at regular intervals and converts it into a digital stream for further processing.

Microcontrollers and Onboard Processors

Smart textiles usually include a low-power microcontroller unit (MCU), which:

- Reads and stores sensor data,
- Processes the data (e.g., filters noise, runs lightweight AI models),
- Handles wireless transmission.

Popular Choices

- **ESP32**: Features Wi-Fi and Bluetooth, used in many wearable prototypes.
- **STM32**: Offers rich ADC support and real-time processing capabilities.
- **nRF52**: Known for low power consumption and BLE connectivity.

Wireless Communication Modules

To send data off the textile and into the digital world, smart fabrics use wireless technologies. The right choice depends on factors like power availability, range, and bandwidth requirements.

Bluetooth Low Energy (BLE)

- Ideal for short-range, low-power applications (10–50 meters).
- Common in fitness bands, smart shirts, and medical wearables.
- Can connect directly to smartphones, enabling real-time feedback.

ZigBee

- Low-power, mesh-network capable.
- Suitable for body area networks (BANs) or interconnected clothing in team environments (e.g., sports, military).
- Supports many-to-one communication.

Near Field Communication (NFC)

- Ultra-low-power but very short range (a few centimeters).

- Useful for passive garments (e.g., patches or tags that activate when scanned).

Wi-Fi

- Higher bandwidth than BLE, used in garments requiring frequent data transmission (e.g., video-enabled smart glasses).
- Higher energy consumption makes it suitable only for intermittently powered garments.

LoRaWAN

- Long-range, low-power communication for outdoor or remote monitoring.
- Used in environmental sensing garments for rural or industrial workers.

Body Area Networks (BANs)

In some advanced designs, sensors within the same garment (or across multiple garments on a person) form a Body Area Network a local network of devices communicating wirelessly with a central hub on the body.

BAN Features

- Uses star topology: Sensors → Central Node (MCU or hub)
- Enables synchronized sensing across body parts
- Ensures low latency and energy-efficient data sharing

Example. A jacket with heart-rate, respiration, and posture sensors transmits data to a small wearable controller at the waist, which then sends it to the user's phone.

Communication Performance Considerations

Designers of smart textiles must balance multiple factors when choosing communication technologies follow Table 1:

Table 1. Comparison of communication technologies used in smart textiles

Parameter	Description
Range	Distance over which data can be reliably transmitted
Power Usage	How much battery life is consumed during transmission
Data Rate	Speed at which data is transmitted (bps, kbps, Mbps) follow equation 6
Latency	Time delay between sensing and receiving the data
Security	Protection against data interception or tampering

Data Rate Calculation (Simplified)

$$\text{Data Rate} = \text{Sampling Rate} \times \text{Bit Depth} \times \text{Number of Channels} \tag{6}$$

Example

- Sampling rate = 250 Hz (ECG)
- Bit depth = 12 bits
- Channels = 2

$\Rightarrow$ Data rate $\approx$ 6,000 bits/sec (6 kbps)

Energy Efficiency and Duty Cycling

Since smart textiles often rely on small, rechargeable or even energy-harvested power sources, efficient communication is essential.

Strategies Include

- **Duty cycling**: Radio modules are turned on only when needed (e.g., every 5 minutes).
- **Data buffering**: Collect data locally and send in batches to reduce transmission events.
- **Edge processing**: Perform lightweight data analysis on the garment itself to minimize the amount of data sent.

Security and Privacy Considerations

Because health and location data are sensitive, secure communication is a must. Smart textile systems should implement:

- **Data encryption** (e.g., AES-128)
- **Authentication protocols** to verify trusted devices
- **Secure boot and firmware updates**

POWER MANAGEMENT AND ENERGY HARVESTING

Reliable power sources are critical for sustained operation; therefore, this section highlights energy solutions that support smart textile functionality.

One of the biggest challenges in smart textile design is powering the electronics in a way that is safe, efficient, and comfortable for the user. Unlike smartphones or laptops, which have rigid batteries and ample space, smart garments must integrate their power sources into soft, flexible, and often washable materials.

This section explores how smart textiles manage energy from powering embedded sensors and communication modules to harvesting energy from the environment or the body itself.

Power Sources for Smart Textiles

Smart textiles generally rely on compact and lightweight power solutions that can be integrated into the fabric structure without hindering comfort or usability.

Flexible Batteries

- These are thin, stretchable versions of traditional lithium-ion or lithium-polymer batteries.
- Can be printed onto substrates or embedded between fabric layers.
- Often encapsulated for safety and water resistance.

Example. A smart shirt with ECG sensors may use a thin-film battery woven into the waistband (Sanei & Chambers, 2013).

Supercapacitors

- Store and release energy quickly.

- Useful for applications requiring short bursts of power, such as wireless transmissions.
- Can be printed or coated directly onto fabric using conductive inks.

Energy Harvesting Techniques

To extend battery life or eliminate the need for frequent recharging, researchers are integrating energy harvesting methods that capture power from the wearer's body or the surrounding environment.

Thermoelectric Harvesting (Seebeck Effect)

- Converts body heat into electricity.
- Based on the temperature difference between the skin and the environment.
- Uses thermoelectric materials like bismuth telluride or organic thermoelectrics.

Key Formula

$$V = \alpha \cdot \Delta T \tag{7}$$

Where V is the voltage generated, α is the Seebeck coefficient, and ΔT is the temperature gradient follow equation 7.

Piezoelectric Harvesting

- Generates electricity from mechanical stress such as walking, breathing, or fabric movement.
- Uses materials like PVDF (Polyvinylidene fluoride) or ceramic fibers.
- Can be embedded in shoes, soles, or areas with frequent motion.

Concept. Each time the user moves, pressure applied to piezoelectric fibers generates an electric charge.

Triboelectric Nanogenerators (TENGs)

- Based on contact and separation of two different materials, creating static electricity.
- Very promising for motion-based harvesting.

- Can be built into cuffs, collars, or armpits where fabric-to-fabric motion occurs.

Solar Harvesting

- Uses flexible solar panels or photovoltaic fibers embedded in outerwear.
- Ideal for outdoor use in military, trekking, or remote medical scenarios.
- Challenges include low indoor efficiency and aesthetic integration.

Wireless Power Transfer

Another approach to powering smart textiles is wireless energy transfer, which eliminates the need for *direct plugging in.*

Inductive Coupling

- Similar to smartphone wireless chargers.
- Uses magnetic fields between two coils one in the garment, the other in the charger.
- Requires close proximity (a few centimeters).

Capacitive Coupling

- Uses electric fields between plates embedded in garments and furniture (like smart chairs or beds).
- Suitable for smart sleepwear or hospital garments.

Efficiency Formula

$$\eta = \frac{P_{out}}{P_{in}} \times 100\% \qquad (8)$$

Where η is efficiency, P_{out} is power delivered to the device, and P_{in} is power provided by the source follow equation 8.

Power Management Circuits

Managing harvested or stored energy requires careful control to avoid damaging sensitive electronics and to optimize battery life.

Key Functions Include

- **Voltage Regulation**: Keeps the voltage within safe limits for microcontrollers and sensors.
- **Power Switching**: Automatically switches between battery and harvested power.
- **Sleep Modes**: Puts components in low-power mode when inactive.
- **Charge Controllers**: Handle recharging of batteries from energy harvesters.

Design Trade-Offs

Engineers must strike a balance among power capacity, flexibility, and wearability, follow Table 2:

Table 2. Key features and associated challenges in power storage for smart textiles

Feature	Challenge
High capacity	Adds weight and bulk
Flexibility	May lower energy density
Rechargeability	Requires user interaction or infrastructure
Safety	Must avoid overheating or short circuits

Real-World Example

A smart yoga shirt may need only a small supercapacitor to store power for brief posture analysis, while a medical monitoring garment might require a larger flexible battery for continuous ECG tracking.

Sustainable and Biodegradable Power Solutions

As environmental concerns grow, researchers are exploring eco-friendly energy sources:

- **Biofuel cells** powered by sweat or glucose,
- **Paper-based supercapacitors**,
- **Biodegradable batteries** made from organic materials.

These technologies promise a more sustainable future for smart textiles, especially in disposable or short-term use cases like medical bandages, Infographic illustrating power sources, energy harvesting methods, wireless transfer, and power management circuits in smart textiles shown in Figure 3.

Sustainable design approaches are increasingly recognized as essential for future smart textile adoption, with recent work emphasizing eco-friendly materials and cleaner production processes (Stoppa & Chiolerio, 2023).

Figure 3. Infographic illustrating power sources, energy harvesting methods, wireless transfer, and power management circuits in smart textiles

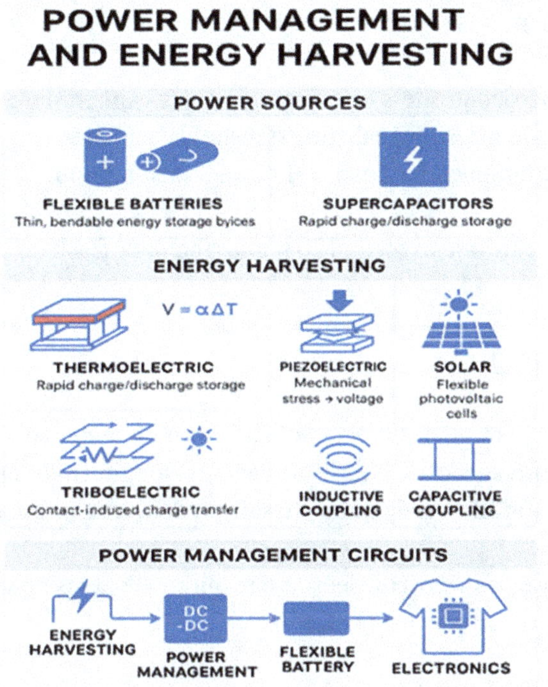

DATA PROCESSING, AI, AND IOT INTEGRATION

Beyond capturing and transmitting data, smart textiles require intelligent analysis and integration with IoT, which this section addresses. Collecting data from sensors embedded in textiles is just the beginning. To make that data useful whether it's a rising heart rate, exposure to pollution, or unusual movement it must be processed, analyzed, and interpreted intelligently. This is where data processing, artificial intelligence (AI), and the Internet of Things (IoT) come into play.

In this section, we explore how smart textiles handle the large volumes of data they generate, how AI makes sense of that information, and how IoT connectivity enables a broader digital ecosystem around the wearer.

Local vs. Cloud-Based Data Processing

Smart textiles can either process data locally (on the garment) or send it to the cloud (via a smartphone or hub) for deeper analysis.

Edge (Local) Processing

- Involves microcontrollers or embedded processors in the garment.
- Used for basic filtering, real-time thresholds, and alerts.
- Reduces communication load and power consumption.

Example. If heart rate exceeds 120 bpm, vibrate or flash an alert locally on the garment.

Cloud Processing

- Data is sent to remote servers via Wi-Fi or mobile networks.
- Allows for advanced analytics, storage, and long-term trend analysis.
- Often used for AI-based diagnostics or predictive health modeling.

Example. A week's worth of sleep and respiration data is uploaded to the cloud, analyzed for apnea patterns using AI.

Signal Preprocessing

Before any data is analyzed or interpreted, it must be cleaned and formatted. This includes:

- **Noise filtering** (e.g., low-pass filters to remove motion artifacts from ECG)
- **Normalization** to ensure consistency across users and environments
- **Segmentation** to break continuous signals into meaningful chunks (like breaths or heartbeats)

Example (Moving Average Filter)

$$Y[i] = \frac{1}{N}\sum_{j=0}^{N-1} X[i-j] \tag{9}$$

This smooths raw sensor signals by averaging recent values follow equation 9. In this equation:

- Y[i] is the filtered output signal at index iii.
- N is the window size, i.e., the number of data points used for averaging.
- X[i−j] represents the input signal at earlier points in time, from iii back to i−(N−1).
- The summation $\sum_{j=0}^{N-1}$ accumulates N consecutive signal values.
- The division by N computes the **average** of those values.

Artificial Intelligence in Smart Textiles

AI is the brain behind smart textile functionality. It can detect patterns, classify behavior, and even predict health events using data collected from the fabric.

Advances in AI continue to enhance the diagnostic and predictive potential of wearable systems, enabling personalized healthcare solutions on a large scale (He, Wu, & Zhang, 2023).

Machine Learning (ML) Models

- **Supervised learning** (e.g., SVM, Random Forest): Used for classifying events like "fall detected" or "normal walking."
- **Unsupervised learning** (e.g., clustering): Used for detecting anomalies in sleep, stress, or heart patterns.

Real-world Use. A machine learning model classifies the walking gait of an elderly user and flags abnormalities that might indicate balance issues.

Deep Learning

- Convolutional Neural Networks (CNNs) for recognizing ECG wave patterns.
- Recurrent Neural Networks (RNNs) or LSTM models for sequential data like respiration or sleep cycles.

Example. A CNN model trained on textile-based EMG signals to recognize hand gestures.

On-Device AI

- Lightweight models are deployed directly onto the garment's MCU or microprocessor.
- Enables real-time inference without internet important for privacy and offline applications.

IoT Integration: Connecting Smart Textiles to the World

IoT turns smart textiles into active participants in a larger connected ecosystem. It enables real-time communication between the wearer and external systems—be it smartphones, healthcare databases, or smart environments.

Wearer-Centric IoT

- Real-time feedback via apps
- Personal dashboards for heart rate, posture, UV exposure
- Notifications for hydration or activity goals

Remote Health Monitoring

- Data sent to doctors or caretakers for chronic patients
- Emergency alerts (e.g., fall detected in elderly)

Use Case. A smart vest alerts a remote caregiver when an elderly person falls and doesn't get up within 30 seconds.

Smart Environment Interaction

- Garments trigger actions in the surroundings

Example: Shirt detects fatigue → smart home dims lights and turns on calm music.

Data Fusion and Context Awareness

Smart textiles often collect multiple data streams e.g., heart rate, motion, temperature. Data fusion combines these inputs to generate a higher level of insight.

Example

High heart rate + elevated temperature + low hydration = likely heat stress.

By using contextual information (time of day, activity type, or location), AI models can provide more meaningful recommendations rather than isolated alerts.

Security and Privacy in Data Processing

Handling sensitive physiological and location data brings significant security concerns. Key strategies include shown in Figure 4:

- **Data encryption (e.g., AES-128)** during transmission and storage
- **User authentication** before granting access to apps or devices
- **Privacy-preserving machine learning**, such as federated learning where data remains on the device

Figure 4. AI and IoT are integrated into smart textiles

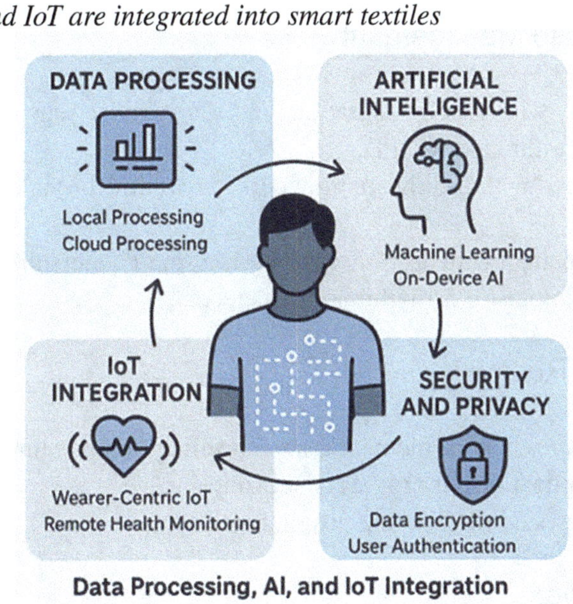

APPLICATIONS IN REAL-WORLD SCENARIOS

With the enabling technologies established, this section illustrates how smart textiles are applied across healthcare, industry, environment, and daily life. The true value of smart textiles is revealed in their real-world applications where technology meets human needs. Whether its helping patients recover, protecting workers in hazardous environments, enhancing athletic performance, or monitoring the environment, smart garments are redefining how we interact with the world through what we wear.

In this section, we explore various domains where smart textiles are actively making an impact or show great potential, highlighting specific use cases and the types of sensors and technologies involved.

Healthcare and Medical Monitoring

Perhaps the most transformative application of smart textiles is in healthcare, where continuous and non-invasive monitoring is critical (Seyam & Hamouda, 2013).

Chronic Disease Management

- Smart shirts with ECG, respiration, and temperature sensors monitor heart conditions, asthma, or COPD.
- Data is shared with healthcare providers in real time or via cloud platforms.

Use Case. A patient with heart disease wears a smart undershirt that tracks ECG and alerts a cardiologist if arrhythmias are detected.

Post-Surgical Recovery

- Smart compression garments measure swelling, temperature, and motion to detect complications or progress in healing.
- Reduces the need for frequent hospital visits.

Smart Bandages

- Incorporate pH and moisture sensors to monitor wound healing.
- Some designs include drug delivery systems triggered by sensor readings.

Sports and Fitness

Athletes benefit from garments that go beyond performance wear by analyzing posture, movement, hydration, and fatigue.

Motion Tracking

- T-shirts or leggings embedded with inertial sensors or strain gauges track biomechanics in real time.
- Helps optimize form and prevent injury.

Hydration and Electrolyte Monitoring

- Sensors in sweatbands or socks assess sweat rate and composition.
- Useful during high-endurance sports or military training (Shi et al., 2020).

Smart Sports Bras and Vests. Monitor heart rate variability, respiration, and torso rotation for female athletes.

Example. A marathon runner receives a haptic signal from her smart shirt when dehydration is detected based on sweat sodium levels.

Elderly and Assisted Living

Smart textiles provide a discreet and non-invasive solution for continuous monitoring of seniors in home or care settings (Tat et al., 2022).

Fall Detection

- Garments equipped with accelerometers and gyroscopes detect sudden impact followed by inactivity.
- Automatically alert caregivers or family members.

Health Trend Monitoring. Long-term tracking of heart rate, sleep quality, and movement helps predict potential issues like frailty or cognitive decline.

Smart Socks or Insoles. Detect irregular gait or pressure points to prevent falls or ulcers in diabetic patients.

Occupational and Industrial Safety

In high-risk environments such as factories, mines, or chemical plants, smart textiles act as an early warning system.

Gas and Pollution Detection

- Jackets with gas sensors detect exposure to CO, NO_2, or H_2S and send alerts.
- Useful for oil rig workers, firefighters, and miners.

Thermal Monitoring. Smart suits with temperature sensors and fire-resistant fabrics help monitor and protect from heat stress or burns.

Posture and Ergonomics. Embedded motion sensors monitor lifting techniques and provide feedback to avoid musculoskeletal injuries.

Military and Defense

Soldiers operate in extreme environments where real-time physiological and environmental awareness is critical.

Health and Vital Monitoring

- Smart combat suits monitor heart rate, hydration, and fatigue.
- Alerts commanders if a soldier is injured or immobilized.

Environmental Sensing. Integrated radiation, temperature, and gas sensors provide hazard detection in unknown terrains.

Camouflage and Tactical Feedback. Research includes electrochromic textiles that adapt color for camouflage and tactile haptics for silent signals.

Environmental Monitoring and Sustainability

Smart textiles are also applied in citizen science and ecological monitoring, allowing users to track environmental conditions passively.

UV and Light Exposure

Hats or armbands detect UV radiation and notify wearers to avoid sunburn or eye damage.

Air Quality and Pollution

Scarves or neck gaiters with particulate or gas sensors monitor air pollution in urban areas.

Climate-Aware Clothing

Experimental fabrics that change insulation or ventilation in response to temperature or humidity readings.

Fashion and Interactive Design

Designers are exploring smart textiles as interactive art, expression tools, or adaptive fashion. Applications in Real-Word Scenarios shown in Figure 5.

Responsive Clothing

LED-integrated fabrics change color based on mood, temperature, or user input.

Music and Performance

Wearable instruments respond to gesture, motion, or touch, blending fashion and audio production.

Example. A dancer's costume generates ambient sound as it moves, based on embedded accelerometers and sound synthesis.

Figure 5. Applications in real-word scenarios

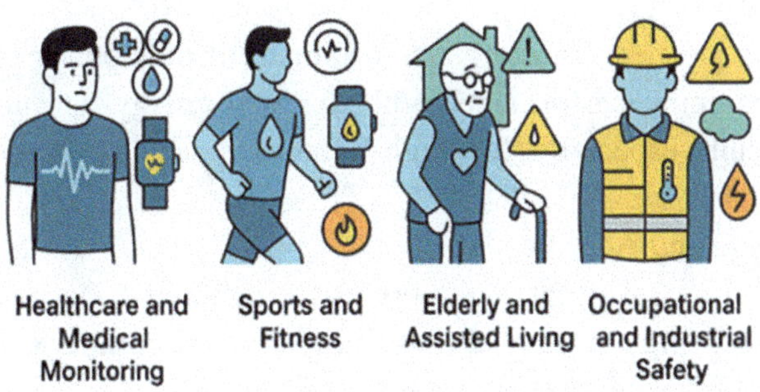

Ethical, Regulatory, and Social Considerations

In addition to technical performance, success with smart textiles relies on meeting ethical, regulatory, and social aspects. As most clothing gathers sensitive physiological and location information, data privacy and cybersecurity are important in maintaining user trust and compliance with healthcare regulations like HIPAA and GDPR. The absence of broad standards for textile electronics, sensor calibration, and interoperability is still a deterrent to mass adoption. Furthermore, social acceptance is equally crucial parameters of comfort, cultural affinity, beauty, and intrusiveness are the determining factors of whether users will welcome these technologies into everyday life. As smart textiles incorporate ethical design, regulation, and human-centricalism, not only can they reach technological innovation but also responsible and widespread use in healthcare and environmental monitoring applications. Alongside technical innovation, compliance with evolving regulatory frameworks and robust privacy safeguards will be critical to the responsible deployment of smart textiles (Khan, Hussain, & Lee, 2024).

Challenges and Limitations

While the potential of smart textiles is immense, their journey from research labs to everyday life is constrained by several practical and technical hurdles:

Mechanical Durability

- Constant wear, stretching, and laundering can degrade conductive materials.
- Ensuring long-term functionality without compromising comfort is a major challenge.

Power Constraints

- Smart garments must operate on small, flexible power sources.
- Maintaining performance with minimal energy consumption remains difficult, especially for high-frequency sensing or continuous data transmission.

Signal Noise and Accuracy

- Body motion, sweat, and fabric deformation introduce signal artifacts.
- Maintaining accurate readings in real-world conditions requires robust signal filtering and calibration.

Cost of Materials and Manufacturing

- Conductive fabrics, flexible batteries, and microcontrollers are expensive in small quantities.
- Mass production techniques and scalable designs are still evolving.

Data Security and Privacy

- Wearables deal with sensitive personal data like location and health status.
- Without proper encryption and security measures, user privacy is at risk.

Limited Standardization. Lack of universal standards limits interoperability between smart garments and devices from different manufacturers.

Though present limitations like mechanical resistance, high production price, and data privacy issues continue to be real, they are even propelling new research trends. For example, material degradation through washing and long-time use is encouraging researchers towards nanostructured coatings and self-healing fibers that can increase the lifetime of textiles. Likewise, the price of smart fabrics is likely to come down with improvements in scalable printing technologies, modular sensor incorporation, and mass production of flexible electronics.

On the data front, signal noise and security threats are being mitigated through the incorporation of AI-on-edge methods, with lighter models able to screen and examine data locally on clothing without too much dependence on cloud connectivity. This both enhances privacy and real-time responsiveness. Concomitantly, international textile and electronics associations are making efforts to establish common standards for performance, interoperability, and safety. Last but not least, the increasing focus on green and biodegradable materials underscores the thrust of connecting smart textiles to world sustainability imperatives. Cumulatively, these research directions imply that current limitations are more than just challenges but opportunities for innovation and ethical utilization of smart textiles in the imminent future.

CASE STUDIES

To complement the broad applications, this section presents specific case studies that demonstrate the practical impact and limitations of smart textiles. Real-world deployments of smart textiles offer valuable insights into their capabilities and limitations:

Case Study 1: Hexoskin Smart Shirt (Healthcare)

- **Use:** Monitors heart rate, respiration, activity, and sleep.
- **Technology:** Built-in textile ECG sensors and accelerometers.
- **Outcome:** Used in clinical studies and astronaut training; highlighted comfort vs. sensor accuracy trade-off.

Case Study 2: Levi's Jacquard Jacket (Fashion/IoT)

- **Use:** Touch-sensitive fabric on the sleeve to control phone functions (calls, music, maps).
- **Technology:** Conductive yarn and a detachable Bluetooth tag.
- **Outcome:** Emphasized innovation in interaction design but limited market appeal due to high cost.

Case Study 3: Smart Insole for Diabetics (Medical Monitoring)

- **Use:** Monitors plantar pressure to prevent foot ulcers.
- **Technology:** Pressure sensors embedded in soft insole.
- **Outcome:** Effective in preventive care but faced adoption barriers due to lack of insurance coverage.

Case Study 4: Smart Firefighter Jacket (Safety)

- **Use:** Measures body temperature, external heat exposure, and toxic gas levels.
- **Technology:** Integrated temperature, gas, and motion sensors.
- **Outcome:** Enhanced on-site safety and real-time alerts but faced challenges in energy supply and ruggedization.

FUTURE OUTLOOK

Looking ahead, this section highlights emerging trends and directions that will shape the future of smart textiles. The future of smart textiles is promising, driven by rapid advances in material science, electronics, and artificial intelligence.

Integration with Emerging Technologies

- **AI-on-edge** devices will allow garments to make decisions without cloud access.
- **5G and low-power wide-area networks (LPWAN)** will enhance remote monitoring capabilities.

Personalized and Predictive Healthcare

With AI and long-term data collection, wearables will shift from reactive to predictive health monitoring alerting users before symptoms emerge.

Bio-Compatible and Eco-Friendly Materials

Growth in biodegradable electronics and organic conductors will reduce e-waste and promote sustainable fashion.

Fashion Meets Function

Smart garments will become indistinguishable from traditional clothing, blending aesthetics with performance in everyday wear.

Democratization and Accessibility

As manufacturing matures, prices will drop, making smart textiles accessible for schools, rural healthcare, and developing nations.

While health-related applications of smart textiles are currently the most prominent areas in research and commercialization, there are other fields that are quickly building up momentum and must not be forgotten. Environmental monitoring, including for the assessment of climate change, air quality, and workplace safety, is one such field where smart fabrics can leave a significant mark on society. Similarly, the fashion and interactive design industry is becoming an area in which aesthetics and technology intersect to facilitate both personalized experience and creative expression. Balancing focus across these areas will expand the universe of smart textiles beyond healthcare applications to encompass greater inclusivity, flexibility, and global needs relevance.

CONCLUSION

Although this chapter gives a thorough technical overview of materials, sensors, communication systems, and applications, it is restricted in its range of coverage of socio-economic adoption and large-scale deployment case studies, which may be researched more deeply in the future. Smart textiles represent the next frontier in wearable technology, merging fashion, electronics, health, and environmental consciousness into the very fabric of daily life. With the ability to collect, process, and communicate real-time data from the human body and surroundings, these garments offer a seamless and stylish interface between people and the digital world.

In spite of the existing problems like durability, power shortages, and privacy issues current developments in materials, wireless systems, AI, and environmentally sustainable design are making smart textiles develop from prototype to product. In the future, interdisciplinary collaboration, ethical design, and user-focused innovation will be essential to guarantee that smart textiles not only fulfill functional requirements but also honor comfort, ethics, and the environment. The textile of the future is not only intelligent it's compassionate, adaptive, and empowering. The strength of this chapter is that it is thoroughly technical in describing materials, sensors, and communication systems, and this can be used as a good starting point for applied researchers, industry practitioners, and potential future innovators in the smart textiles discipline.

PROTOTYPE CASE STUDY: SMART T-SHIRT FOR HEART MONITORING

In order to prove smart textile principles under real conditions, a smart T-shirt was developed as the prototype for basic monitoring of heart rate. The system used textile-based ECG electrodes on a cotton T-shirt that was connected by conductive threads to an Arduino Nano microcontroller. Real-time ECG signal transmission to a smartphone for display was done using a Bluetooth module.

The wearable device was lightweight, portable, and affordable, showing the feasibility of putting sensors in ordinary clothing. Although challenges such as motion artifacts and limited battery life were observed, the prototype exhibited the potential to bring flexible electronics, conductive textiles, and wireless communication into wearable fabrics for health monitoring applications.

REFERENCES

Angelucci, A., Cavicchioli, M., Cintorrino, I. A., Lauricella, G., Rossi, C., Strati, S., & Aliverti, A. (2021). Smart textiles and sensorized garments for physiological monitoring: A review of available solutions and techniques. *Sensors (Basel), 21*(3), 814. DOI: 10.3390/s21030814 PMID: 33530403

Castano, L. M., & Flatau, A. B. (2014). Smart fabric sensors and e-textile technologies: A review. *Smart Materials and Structures, 23*(5), 053001. DOI: 10.1088/0964-1726/23/5/053001

Dolez, P. I., Decaens, J., Buns, T., Lachapelle, D., & Vermeersch, O. (2020, April). Applications of smart textiles in occupational health and safety. [). IOP Publishing.]. *IOP Conference Series. Materials Science and Engineering, 827*(1), 012014. DOI: 10.1088/1757-899X/827/1/012014

Farina, D., Merletti, R., & Enoka, R. M. (2004). The extraction of neural strategies from the surface EMG. *Journal of Applied Physiology, 96*(4), 1486–1495. DOI: 10.1152/japplphysiol.01070.2003 PMID: 15016793

He, Z., Wu, Y., & Zhang, X. (2023). Artificial intelligence in wearable healthcare: From data analytics to personalized monitoring. *IEEE Reviews in Biomedical Engineering, 16*, 120–138.

Khan, S., Hussain, A., & Lee, S. (2024). Privacy and security challenges in IoT-enabled wearable systems: Toward regulatory compliance in healthcare. *ACM Computing Surveys, 56*(1), 1–32. DOI: 10.1145/3640311

Koncar, V. (Ed.). (2016). *Smart textiles and their applications*. Woodhead Publishing. DOI: 10.1016/B978-0-08-100574-3.00001-1

Li, J., Chen, Y., & Tao, X. (2022). Recent progress in smart textiles for personalized healthcare and environmental monitoring. *Advanced Materials Technologies, 7*(12), 2200643.

Liu, X., Wang, H., Li, Z., & Qin, L. (2021). Deep learning in ECG diagnosis: A review. *Knowledge-Based Systems, 227*, 107187. DOI: 10.1016/j.knosys.2021.107187

Mamun, M. A. A., & Yuce, M. R. (2020). Recent progress in nanomaterial enabled chemical sensors for wearable environmental monitoring applications. *Advanced Functional Materials, 30*(51), 2005703. DOI: 10.1002/adfm.202005703

Oatley, G., Choudhury, T., & Buckman, P. (2021). Smart textiles for improved quality of life and cognitive assessment. *Sensors (Basel)*, *21*(23), 8008. DOI: 10.3390/s21238008 PMID: 34884010

Ojuroye, O., Torah, R., Beeby, S., & Wilde, A. (2017). Smart textiles for smart home control and enriching future wireless sensor network data. In *Sensors for Everyday Life* (pp. 159–183). Healthcare Settings. DOI: 10.1007/978-3-319-47319-2_9

Rather, H. A., & Dar, M. A. (2024). Smart Textiles with Integrated Biosensors for Real-time Health Monitoring. *Biosensors and Nanotheranostics*, *3*(1), 1–9.

Sanei, S., & Chambers, J. A. (2013). *EEG signal processing*. John Wiley & Sons.

Seyam, A. F. M., & Hamouda, T. (2013). Smart textiles: Evaluation of optical fibres as embedded sensors for structure health monitoring of fibre reinforced composites. *Journal of the Textile Institute*, *104*(8), 892–899. DOI: 10.1080/00405000.2013.765087

Shi, J., Liu, S., Zhang, L., Yang, B., Shu, L., Yang, Y., Ren, M., Wang, Y., Chen, J., Chen, W., Chai, Y., & Tao, X. (2020). Smart textile-integrated microelectronic systems for wearable applications. *Advanced Materials*, *32*(5), 1901958. DOI: 10.1002/adma.201901958 PMID: 31273850

Stoppa, M., & Chiolerio, A. (2023). Sustainable smart textiles: Advances in eco-friendly materials and manufacturing for wearable applications. *Journal of Cleaner Production*, *412*, 137256.

Tat, T., Chen, G., Zhao, X., Zhou, Y., Xu, J., & Chen, J. (2022). Smart textiles for healthcare and sustainability. *ACS Nano*, *16*(9), 13301–13313. DOI: 10.1021/acsnano.2c06287 PMID: 35969207

Wang, H., Zhang, Y., Liang, X., & Zhang, Y. (2021). Smart fibers and textiles for personal health management. *ACS Nano*, *15*(8), 12497–12508. DOI: 10.1021/acsnano.1c06230 PMID: 34398600

Chapter 6
Embedded Systems and Microcontrollers in Wearables

Babitha Hemanth
https://orcid.org/0009-0006-0685-3962
Sahyadri College of Engineering and Management, India

C. Mehnaz Fathima
https://orcid.org/0009-0007-9372-0996
Sahyadri College of Engineering and Management, India

T. Kripa
https://orcid.org/0009-0006-3002-1111
Sahyadri College of Engineering and Management, India

A. B. Abhishek
https://orcid.org/0009-0003-5445-688X
Sahyadri College of Engineering and Management, India

ABSTRACT

Smart wearables and e-textiles are transforming personal technology by combining materials science, embedded computing, and wireless communication. At their core, embedded systems and microcontrollers (MCUs) enable sensing, data processing, connectivity, and user interaction within strict limits of size, power, and comfort. This work reviews the fundamentals of embedded design, including processors, memory, sensors, actuators, and architectures, while addressing challenges such as miniaturization, textile integration, durability, and data security. Popular MCU platforms are analyzed for energy efficiency, processing power, and connectivity. Power management strategies, including low-power modes and energy harvesting,

DOI: 10.4018/979-8-3373-4287-0.ch006

are emphasized for extending battery life. The chapter also highlights emerging directions such as on-device machine learning, hybrid architectures, and flexible electronics. By integrating research and practical insights, it underscores the central role of embedded systems in enabling next-generation, intelligent, and user-centric wearable technologies.

INTRODUCTION

Overview of Smart Wearables and E-Textiles

In recent years, smart wearables have gained more attention at the intersection of electronics, textiles, and computing. These devices blend smoothly into garments and accessories. They allow real-time monitoring, data processing and wireless communication. E-textiles play a key role in driving this technology forward. By embedding conductive materials and flexible electronics into fabrics, they support applications ranging from health monitoring to environmental tracking and feedback systems.

Embedded systems and microcontrollers have played a crucial role in enhancing the functions of smart wearables. Microcontrollers deliver processing power, efficiency, and network access. As the core of smart wearables, they give the devices autonomy and the ability to react in real time. As materials science advances and power management improves, e-textiles and embedded computing are working together in new ways. This synergy is driving the development of next-generation wearables that are lightweight, durable, and highly functional.

This chapter explores the role of embedded systems and microcontrollers in the development of smart wearables. It also provides a comprehensive overview of design principles, hardware components, and emerging trends in wearable electronics.

Role of Embedded Systems in Wearables

Embedded systems play main role in enabling the core functionalities of modern wearable devices. These compact systems are designed to perform given tasks efficiently and reliably within constrained environments like limited size, weight and power available in wearable technology.

In wearables, embedded systems act as brain. It connects with sensors to collect physiological or environmental data. Also, executing onboard data processing algorithms, managing communication protocols (e.g., Bluetooth Low Energy, Wi-Fi) and controlling actuators, feedback mechanisms. This integration gives wearables to operate independently. Enabling them to respond to user input. They are also capable of sending data for external analysis without depending on external hardware.

Key Characteristics of Embedded Systems for Wearables

- **Low Power Consumption:** Prolonged battery life in mobile applications.
- **Compact Form Factor:** This enables garments, accessories without affecting user mobility or comfort.
- **Real-Time Processing:** Facilitates immediate feedback, critical for health monitoring or motion-tracking applications.
- **Customizability and Modularity:** Supports a wide range of use cases, from fitness trackers to medical wearables and smart textiles.

As wearable applications become more complex, embedded systems are increasingly equipped with edge computing capabilities, allowing them to make local decisions and perform AI-based inference. This advancement reduces data latency, protecting user privacy and lessens reliance on cloud connectivity. This makes wearables smarter and more autonomous.

When it comes to smart textiles, embedded systems face challenges like washability, integration with fabrics and mechanical flexibility. These constraints have led to innovations in hardware and system design. It includes flexible PCBs, sensors compatible with textiles and durable encapsulation methods.

FUNDAMENTALS OF EMBEDDED SYSTEMS FOR WEARABLES

Definition and Characteristics of Embedded Systems

An embedded system is a computer designed to perform specific tasks and is built directly into a larger device or system. Unlike general-purpose computers, embedded systems are optimized for particular functions rather than full-scale processing, often constrained by low power consumption, limited size, and space restrictions. Embedded system is a compact module that combines a processor, memory and input/output interfaces. Embedded systems are found in virtually all modern electronics including consumer electronics, industrial controllers, medical devices and wearable technology.

The wearables require embedded systems that are portable, energy-efficient and able to communicate with their surroundings continually. They use sensors and actuators to interact with physical world, and their performance needs to adhere to the standards for durability, accuracy, and latency. These gadgets frequently work in real-time settings where promptness is pivotal, particularly in motion tracking or health monitoring applications.

Embedded systems are essential for smooth operation of wearable technology. This technology includes health monitors, smartwatches and fitness bands. Usually, microcontrollers or microprocessors with specially written firmware that carries out application specific functions constitute the foundation of these systems. The need for embedded systems that is more competent, secure and energy conscious keeps increasing as wearable technology advances (Wang et al., 2021).

Microcontrollers v/s Microprocessors in Wearables

Microcontrollers (MCUs) and microprocessors (MPUs) represent two primary categories of processing units used in embedded systems for wearable devices. While both serve as the computational core, their architecture and suitability for wearable applications differ significantly.

Microcontrollers are single chip devices that combine a CPU core with peripherals for input/output, timers, analog-to-digital converters (ADCs), memory (RAM and flash) and communication interfaces. They are designed for the task requiring real time control and optimized for power efficiency. MCUs are suited for wearable technology that requires processing units which are cost effective, lightweight and energy efficient due to their compact size. Common MCU families used in wearables include the Texas Instruments MSP430, Atmel AVR and ARM Cortex M series (Rahman, Ahmad, & Yadav, 2020).

Microprocessors are more powerful computer components that prioritize high performance processing. In contrast to MCUs, MPUs depending on external components for operation and do not integrate memory or peripherals. The processing power and flexibility are achieved through this modular design, though it results in higher board complexity and increased power demands. The high-end wearables with MPUs are more suited for sophisticated functions like image processing, machine learning and running intricate operating systems such as Linux or Android. In order to balance efficiency and performance, both MPUs and MCUs are used in devices like AR headsets, smart glasses, and high-end smartwatches.

Core Components: Processing Units, Memory, Sensors, Actuators

A wearable devices embedded system consists of essential parts that cooperate to gather data, process information and react to changes in the environment.

Processing Unit

The processing unit is the brain of wearable gadgets. This can be either a microcontroller or a microprocessor. It communicates with external systems, controls data flow, implements control algorithms, and interfaces with peripheral parts. The computational load, power limitations, and real-time demands of the program all influence the processor choice.

Memory

Wearable embedded systems have two types of memory: volatile and non-volatile. Firmware, configuration settings, and user data are stored in non-volatile memory (Flash or EEPROM), whereas volatile memory (RAM) is used for temporary data storage during running processes. Because wearables have limited space and must minimize power consumption, effective memory management is crucial.

Sensors

Sensors allow the device to perceive internal body factors or the external environment (Mandal, 2025), sensors are essential to wearable electronics. Typical wearable sensors include photoplethysmography (PPG) sensors for heart rate monitoring and accelerometers and gyroscopes for motion tracking. Bio-potential sensors for EEG, EMG, and ECG monitoring. To produce insightful information, sensor data is gathered, preprocessed, and occasionally combined from several sources.

Actuators

They use processed data to execute actions or give physical feedback. Typical actuators in wearable systems include audio buzzers, LEDs for notifications and vibration motors for haptic feedback. For real-time warnings in fitness trackers, safety wearables, and health monitoring systems, actuators are necessary.

Embedded System Architectures for Wearables

An embedded systems architecture has a direct impact on wearable devices efficiency, scalability and functionality. Wearable design employs a variety of architectural techniques, each appropriate for a specific application domain and degree of complexity.

Monolithic Architecture

In this architecture all processing components and functionalities are integrated onto a single chip or module. This provides a compact and efficient design. This approach simplifies communication between components, but it may limit flexibility and scalability compared to modular designs.

Distributed Architecture

In this, processing tasks are divided among multiple interconnected modules or components rather than centralized on a single chip. This approach offers greater flexibility and scalability. Which allows higher processing power and specialized functionality, but it increases system complexity and higher power consumption. The Complex wearables that require simultaneous management of several sensor nodes like motion capture suits. (Li et al., 2019).

Hybrid Architecture

The Hybrid architecture combines elements of both distributed and monolithic systems. This architecture integrates core functionalities on a single module while offloading certain tasks to separate components. This approach balances efficiency, flexibility and processing power. This offers a compromise between system simplicity and scalability. A wearable may employ a high performance MPU for AI-based data analysis and a low power MCU for ongoing sensor monitoring. This provides better energy management and functional partitioning. Also, the design allows for modular development and adaptable upgrades, which is essential in the rapidly changing wearable market. The power budget, form factor, connectivity requirements, data processing complexity and integration with cloud or edge computing platforms are the aspects which affect the architecture decision.

DESIGN CONSIDERATIONS FOR WEARABLE EMBEDDED SYSTEMS

While designing the embedded systems for wearables it is required to identify the engineering limits. These designed systems are needed to strike a balance between long term dependability, user comfort, efficiency and functionality. Important factors are discussed below.

Low Power Consumption and Energy Efficiency

Power consumption is one of the important limitations of wearing electronics. Wearables are dependent on energy harvesters or tiny rechargeable batteries as opposed to stationary embedded systems that may rely on steady power sources. Thus, customer pleasure and gadget longevity are directly connected by energy efficiency.

Wearable embedded systems use a number of techniques to lower power consumption:

- **Low-power microcontrollers:** Devices with ultra-low power operation, such as the MSP430 and ARM Cortex-M0+, allow quick wake up times and sleep modes.
- **Duty cycling:** This reduces active time by only using sensors and processors when necessary.
- **Event-driven processing:** To save CPU cycles, embedded firmware frequently uses interrupts rather than continuous polling.
- **Power domains**: Current SoCs allow modules to be partially activated, enabling only necessary parts.

Longer battery life is largely dependent on energy-conscious firmware design, which includes features like turning off unused peripherals and optimizing loops (Kumar & Sharma, 2021). Furthermore, new developments like energy harvesting from motion or body heat are being investigated to increase runtime without enlarging batteries.

Miniaturization and Form Factor Constraints

There are strict restrictions on wearable gadgets' weight, size and shape due to their physical design. Compact, highly integrated embedded systems are required to meet these limits. System-on-Chip (SoC) integration lowers the number of components and board size. PCBs are elastic and flexible and allow for conformity to curved surfaces like fabric or skin. Chip-on-board (CoB) and system-in-package (SiP) technologies reduce the size of embedded electronics. Since active cooling is impractical in small wearable enclosures, thermal management must be passive. It is also critical to design ergonomics and aesthetics in mind. The integrated technology must blend in perfectly with clothing or accessories without becoming uncomfortable or limiting mobility.

Sensor and Actuator Integration

The successful operation of wearable embedded systems depends on the effective integration of sensors and actuators. These components must have direct contact with the human body and surroundings, which mandates careful consideration for their placement and selection. Several important factors must be considered: the accuracy and resolution of the sensor. Especially in motion-sensitive or medical applications where high-precision data is essential. Analog Front-End (AFE) circuitry is required to properly condition biosensor signals before they undergo Analog-to-Digital Converter (ADC) conversion. When dealing with bio-potential or low-amplitude signals, a low-noise design is paramount to ensure signal integrity. Finally, cross-sensitivity mitigation is often necessary; this may involve physical isolation or algorithmic correction to lessen interference between closely spaced sensors. On the actuator side, haptic feedback is the most frequent application, which is achieved using piezoelectric devices or miniature vibration motors. For alert-based applications, the system might also incorporate small buzzers and LED indicators.

Table 1. Typical sensors and actuators used in wearables along with their functions

Component Type	Example	Function
Accelerometer	MPU6050	Motion tracking
PPG Sensor	MAX30102	Heart rate, SpO2 monitoring
Temperature Sensor	LM35	Body/environment temperature
Haptic Actuator	ERM Vibration Motor	Notifications
LED	RGB Indicator	Visual alerts
Pressure Sensor	FSR 402	Touch/force sensing

(Compiled from device datasheets)

Wireless Communication Protocols (BLE, ZigBee, NFC, etc.)

For wearable embedded systems to provide data to other systems, like cloud platforms or smartphones, dependable wireless communication is necessary. The chosen communication protocol affects latency, data rate, and power consumption.

Bluetooth Low Energy (BLE)

Because of its compatibility with mobile devices and low power consumption, Bluetooth Low Energy (BLE) is the most widely utilized protocol in wearables. It is perfect for sporadic, small-packet transfers and supports data speeds of up to 2 Mbps.

ZigBee

ZigBee is a mesh networking technology that works well with wearables with multiple nodes, including sensor arrays seen in smart clothing. Although less prevalent in consumer wearables, it is useful in industrial or research contexts.

Near Field Communication (NFC)

Short-range, secure communication is made possible by NFC. It is perfect for identity verification or payment systems. When paired with a reading device, it is passive and can function without an internal power supply.

Wi-Fi

Despite having high data throughput, wearables rarely use Wi-Fi due to its high-power consumption. It works well in situations when electricity is not the main limitation and constant cloud access is needed.

Data Security and Privacy Concerns

Wearables rarely employ Wi-Fi because of its high battery consumption, even if it has a high data throughput. It functions effectively when continuous cloud connection is required and electricity is not the primary constraint. Typically, the use case provides direction when selecting a protocol. Although NFC and ZigBee have specific applications, BLE remains the de-facto standard for fitness and health devices.

- Secure boot and firmware authentication to stop tampering are important security factors.
- Data encryption (RSA, AES) for both transmission and storage.
- User access and device pairing authentication systems.
- Personal data is anonymized and stored securely.
- During synchronization, cloud security protocols like TLS/SSL are used.

Additional attack vectors are presented by the fact that many wearables are connected to smartphones or cloud platforms through APIs. Secure update procedures, access control, and key management must be handled by the embedded firmware. Strict rules on how wearable devices gather, store, and transmit data are also enforced by regulatory compliance, such as GDPR in Europe or HIPAA in the United States

(Zhou, Chen, & Mathur, 2020). To guarantee end-to-end security, designers must consider these during the firmware and hardware design phases.

MICROCONTROLLERS FOR WEARABLE APPLICATIONS

Microcontrollers are crucial for wearable applications because they allow for a wide variety of functions that are important for current wearable technology. In wearable systems, these small computers are essential for data processing, management of power, and controlling a variety of sensors and actuators. The ESP8266 and other microcontrollers enable Internet of Things connectivity in wearable sportswear, enabling real-time data transfer and sophisticated performance analysis. This microcontroller integration makes it possible to create advanced textile-based sensors that give feedback on body movements, preventing injuries and increasing the effectiveness of workouts (Aliyana, et.al, 2024).

Further, the capacity of health monitoring systems is increased by developing microfluidics in wearable technology. Microcontroller monitors biofluids like sweat and saliva virtually through the combination of AI and IoT, facilitating access to real time health assessments (Apoorva, Nguyen, & Sreejith, 2024). These technologies can support both diagnosis and therapy, utilizing the processing power of microcontrollers to deliver accurate and timely health data. Another important challenge that microcontrollers address is power efficiency. Modern microcontrollers, i.e. ESP32 family, have been developed to operate machine learning models with low power consumption, which is helping to advance the creation of wearable technology that uses less energy. Furthermore, the ecological value of wearable technology is further improved by the capacity to use modern microelectronics to capture energy from human motion. In order to provide an alternative to traditional battery-powered devices and allow for extended use without regular recharging, microcontrollers are essential for converting inertial energy into electrical energy (Zhang et al., 2024).

Selection Criteria for MCUs in Wearables

MCU platforms are chosen based on specific application requirements such as performance, power consumption, communication capabilities and ecosystem support. They are integral to various technologies, including embedded systems, IoT devices and TinyML applications.

Due to the unique requirements of wearable technology, which often involve constraints on power consumption, computing capability and size, the selection criteria for microcontroller units (MCUs) in these devices depend on a number of important elements. While choosing MCUs for wearable technology, the following factors are usually taken into account:

- **Power Consumption:** Wearables are usually battery powered, a low-power MCUs are necessary to prolong the device lifespan between charges. Low-power MCUs allow for extended operation times without regular recharging, they are crucial for guaranteeing energy efficiency in wearable technology.
- **Memory and Storage:** The available memory and storage capacity of an MCU can determine the complexity of algorithms and also determine the amount of data that can be processed and stored locally on the device. MCUs suitable for wearables typically have RAM sizes between 2 KB and 512 KB and storage capacity between 32KB and 2MB.
- **Processing Capabilities:** Wearables often incorporate features like machine learning and realtime data processing, the MCU's processing capacity must be adequate to manage these activities efficiently. Human Activity Recognition (HAR) and other on-device machine learning applications require high processing power, which can be obtained by optimizations like quantization of neural networks.
- **Size and Integration:** The physical size of an MCU is crucial since wearables are compact devices. MCUs must be small enough to fit into the compact designs of wearables while also being capable of integrating with various sensors and communication modules which is required by the device.
- **Cost:** Cost efficiency is a practical consideration, particularly in consumer wearables. The MCU should provide the needed functionalities without significantly inflating the cost of the device.
- **Flexibility and Scalability:** Given the rapid pace of technological advancement, it is beneficial for MCUs to be flexible and scalable to accommodate updates and new features without requiring complete redesigns. This flexibility is important for supporting evolving AI requirements and multimodal sensor integration.
- **Compatibility with Machine Learning Models:** With the push towards embedding machine learning capabilities directly on the devices, compatibility with deep learning or machine learning algorithms is critical. This involves selecting MCUs that can run machine learning models efficiently. Using TinyML frameworks to optimize neural networks for better performance on low-resource hardware.

Overall, above criteria form a comprehensive guide to select MCUs that can both meet the current demands of wearable technology and adapt to future advancements (Diab & Rodriguez-Villegas, 2022; Lu et al., 2024).

Survey of Popular MCU Platforms

The working and energy management of many technological platforms depend on microcontroller units (MCUs), particularly for Internet of Things (IoT) devices and wireless sensor networks (WSNs). These platforms' performance and energy usage are crucial factors that decide whether they are suitable for consumer, business and specialized applications.

According to study, SleepRunner is an ultra-low power MCU that maximizes memory, logic and power management for prolonged Internet of Things applications. This MCU is made with 28-nm fully-depleted silicon-on-insulator(FDSOI) technology, which drastically reduces active and deep-sleep power consumption through the use of forward back-biasing (FBB). The SleepRunner is up to three times more efficient than traditional MCUs, with an active mode power consumption of 2.6–3.3 μW/DMIPS at 40–80 MHz and fully retentive deep-sleep power of 131–nW/kB (Bol et al.,2021).

The Arnold SoC, a powerful mixture of an embedded field-programmable gate array (eFPGA) and a RISC-V MCU for high flexibility and low-power Internet of Things end nodes, is another innovation. Compared to other similar systems in the field, it achieves up to 3.4 times greater performance and 2.9 times better energy efficiency due its novel use of body-biasing, which greatly decreases leakage power (P. D. Schiavone et al.,2021). A system designed for high-precision multi-axis real-time control, research on multi-motor systems shows that integrating MCUs with FPGAs can greatly improve performance. High real-time efficiency and reliability are provided in this situation by the FPGA handling many motor servo controllers and the MCU managing data exchange. Discussing wireless sensor networks, paper reviews MCU-based nodes, emphasizing the continued challenge of energy consumption which affects the lifespan of sensor networks. A review suggests that while commercial offerings are advancing, research-driven innovations may offer enhanced energy efficiency and operational capabilities for these devices (Khalifeh et al., 2022).

Innovative techniques for WSNs, such as using the ESP32 MCU with hierarchical sensing and duty cycling, aim to drastically reduce energy consumption by eight times in certain configurations further enhancing the sustainability of sensor nodes. With their distinct performance and energy-saving benefits, these platforms can be used for a wide range of applications, from sophisticated IoT systems to sensor networks. The selection of an MCU platform is based on its performance capabilities and energy management efficiency unique to its application. Future developments in research should result in more high-performance and energy-efficient MCU applications.

ARM Cortex-M Series

The ARM Cortex-M family is mainly suited for wearable technology. This is because its features allow efficient processing and effective power management. Both of these qualities are important for devices that are small and have limited battery capacity. These microcontrollers are ideal for real time applications such as fitness tracking and health monitoring. Its 32-bit ARM architecture provides a good balance between processing power and energy efficiency. As a result, they can handle demanding tasks without sudden draining the battery. The basic features like, Energy Efficiency: To prolong battery life in wearable technology, ARM Cortex-M processors are designed for low power consumption. Multiple sensor integration is also supported by the energy efficiency without having a major effect on power consumption; Processing Capability: These processors are capable of performing intricate calculations required for real-time data analysis and signal processing, two tasks that are critical for applications like health monitoring; Compact Size: ARM Cortex-M family microcontrollers are ideal for the cramped locations used in wearable technology due to their compact form factor. This makes it easier to produce wearable devices that are smaller and lighter, which improves user comfort.

Benefits

- High Performance: Wearable features like disease detection and intricate pattern recognition are made possible by the ARM Cortex-M processors. It has a ability to handle complex algorithms for activities like AI-driven data processing.
- Versatility: Its adaptable architecture supports a range of computation demands. These processors can be used in a variety of wearable applications, such as health tracking, smart textiles and emergency response systems.
- Integration with Current Technologies: ARM Cortex-M processors may be simply incorporated into both new and current wearable technologies due to the large and well-supported ARM ecosystem. This offers a wealth of development tools and resources.

Drawbacks

- Resource Constraints: Though being designed for low power consumption, the ARM Cortex-M family may not be as suitable for applications required intensive determining due to its limitations in processing power and memory when compared to higher-end processors.
- Implementation Complexity: Creating effective programs for ARM processors frequently calls for specialized knowledge to control power

consumption and make the best use of available resources, which may extend development time and raise expenses.

- Dependency on Peripheral Quality: In wearable applications, the ARM Cortex-M performance frequently depends on the caliber and functionality of peripheral parts such sensors and batteries. Low peripheral performance may restrict the processor's ability to generate dependable results.

The combination of computational power, energy efficiency and flexibility provided by the ARM Cortex-M series offers wearable technology a number of benefits. However, there are limitations related to limited resources and need for specialized development skills.

ESP32

The ESP32 (Babiuch, Foltynek, & Smutny, 2019) microcontroller is widely used in wearable devices due to its powerful features and robust capabilities like, Wi-Fi and Bluetooth Connectivity: The ESP32 is a versatile wireless communication device which utilizing both Wi-Fi and Bluetooth (both traditional and BLE). For wearables that must link to smartphones or other Internet of Thing's devices; Low Power Consumption: The ESP32's many power modes maximize its low power consumption, which is vital for devices that are running on batteries. This features energy-saving deep sleep and light sleep modes; High Performance: The ESP32 is appropriated for sophisticated wearable applications due to its dual-core processor, offers great processing capability that can manage intricate calculations; I/O Capabilities: It can connected to a variety of sensors and peripherals that are frequently employed in wearables due to its support for many interfaces, including SPI, I2C and UART.

Benefits

- Cost-Effective: The ESP32 is a desirable option for the large-scale production of wearable devices due to its low cost when compared to other microcontrollers with comparable capabilities.
- Strong Community and Support: Wearable application creation and troubleshooting are made simpler by the existence of a sizable online community and copious documentation.
- Versatility: It may be used with a variety of wearable technology, including health monitoring tools and fitness trackers, due to its processing power and connectivity.

Drawbacks

- Complex Development Environment: Although the ESP32 has many functions, novices may find the development environment difficult to use, which can lengthen the learning curve and development period.
- Size and Component Integration: The ESP32's size and integration requirements could be problematic for devices that are extremely small, particularly in cases where reduction is important.
- Power Limitations: The ESP32 is built for low power consumption, ultra-low power microcontrollers made specifically for wearables may offer better battery performance.

As long as complexity and size constraints are controlled within the parameters of the design, the ESP32 provides a compelled set of abilities and advantages that make it a best option for wearable technology.

Nordic nRF52 Series

Wearable technology often employs the Nordic nRF52 Series MCU because of its advanced capabilities, flexibility, and low power consumption. Because this series supports Bluetooth Low Energy (BLE), ANT, and NFCC, it is perfect for applications that need wireless connectivity, like wearable technology activity tracking and health monitoring. Basics features are like, Low Power Consumption: The conservation of energy is a key component of the nRF52 series' design, especially for wearable technology that runs on batteries. This extending the battery life of devices such as health monitors and fitness trackers; Wireless Communication Support: Wearables data transmission needs depending on smooth connectivity with other systems and devices, which is made available via BLE and ANT support; High Performance: It is appropriate for intricate applications such as embedded machine learning in health and care applications because it has stronger processing capabilities that enable real-time data processing.

Benefits

- Versatility and Integration: Because of its extensive wireless support, the nRF52 series is easily incorporated into a wide range of wearable technology. It is improving functionality and user experience through ubiquitous connectivity and smart ecosystem interaction.
- Better User Experience: Long-lasting and more sustainable gadgets with low energy usage and strong performance capabilities required less frequent charging, which enhances user experience.
- Improved Connectivity: Users may effectively communicate with other smart devices and systems using BLE and ANT, which results in more

integrated and interesting user experiences for both personal fitness and professional applications like health monitoring.

Drawbacks

- Complexity in Implementation: Although the MCU has many benefits, developers, particularly those who are unfamiliar with the technology, may find it difficult to program and optimize it for certain applications due to its complexity.
- Cost: Developing and manufacturing products with these MCUs may become more expensive due to their advanced functions and capabilities; this could be a factor for low-budget enterprises.

Despite some possible issues with software complexity and cost, the Nordic nRF52 Series MCU's mix of power economy, connectivity, and performance makes it a great option for enabling wearable devices.

TI MSP430

The TI MSP430 microcontroller is renowned for its applications in wearable devices due to its advantageous features such as ultra-low power consumption, which is critical for wearables needing long battery life.

The MSP430 finds extensive usage in healthcare wearables, fitness trackers, and other personal monitoring devices. It excels in tasks that require continuous sensing and real-time data processing, such as tracking heart rate, monitoring physical activity, and managing environmental sensors. The low power demands of the MSP430 support wearables to perform continuous monitoring over extending periods without frequent recharging (Ali et al., 2025; Devi et al., 2023). It has basic features like, Ultra-low Power Consumption: MSP430 is designed to operate with minimal power, making it ideal for battery-operated devices such as wearables; Flexible Clock System allows precise control over power usage and performance; High Integration includes several peripherals like ADCs, DACs, and timers, reducing the need for additional components; Versatile Connectivity Options supports various communication protocols essential for connecting to other devices and networks.

Benefits

- Energy Efficiency: Its low power consumption ensuring that wearable devices can operate for long durations on a single battery charge, which is crucial for user satisfaction and device reliability.
- Compact Design: The high level of integration within the MSP430 reduces the size and complexity of the wearable device, making it more comfortable and adaptable for the user.

- Scalability: The MSP430 family offering a wide range of models with different memory sizes and capabilities, makes it adaptable for various application needs.

Drawbacks

- Performance Limitations: While power-efficient, the MSP430 may not be suitable for wearables requiring high computational power, like advanced data processing and analysis tasks.
- Limited Ecosystem: Although TI provides significant support, the MSP430 ecosystem might not be as extensive as other modern microcontroller families, potentially complicate development in some contexts.

The TI MSP430 microcontroller is highly suitable for wearable applications that prioritize low power consumption and compact design, though it may face limitations in applications requiring high processing capabilities.

Comparative Analysis of MCU Features

Table 2. Comparative analysis of different microcontroller units (MCUs), highlighting key features and applications

Feature/MCU	ARM Cortex-M Series	ESP32	Nordic nRF52 Series	TI MSP430
Architecture	ARM Cortex-M0 to M7	Dual-core Xtensa LX6/LX7	ARM Cortex-M4F	16-bit RISC
Core Features	Rich DSP, FPU (M4/M7), low-power modes	Built-in Wi-Fi and Bluetooth, low-power	Bluetooth 5.0, high performance, low power	Ultra-low power, mixed-signal capabilities
RAM	2 KB to 512 KB	Up to 520 KB	Up to 256 KB	0.5 KB to 64 KB
Flash Memory	8 KB to 2 MB	4 MB (external)	Up to 1 MB	Up to 512 KB
Connectivity	SPI, I^2C, CAN, USB, Ethernet	Wi-Fi, Bluetooth	Bluetooth, NFC, ANT	Serial, USB, I^2C, SPI
Applications	General purpose, IoT, wearables, automotive	IoT, smart home, portable devices, AI	IoT, medical devices, beacons	Low-power applications, wearables, sensors
Power Efficiency	Varies across models	High	High	Very high

Communication and Connectivity

Digital transformation in the era of the Internet of Things (IoT), smart embedded systems, and ubiquitous computing is based on communication and connecting. Fluid and secure connection of the device, systems, and networks guarantees effective data transmission, live decision-making and compatibility among different applications. The chapter goes into the depths of the essence of communications and connections, embedded wireless modules, data synchronization, and security processes. Embedded wireless modules are selected in relation to the necessary range, power, data rates, latency tolerance, network topology, and the environment that the device is to be operating in. IEEE 802.11, 802.15.4, Bluetooth SIG, 3GPP and LoRa Alliance are some of the standards that dictate the specifications that will enable interoperability and regulatory compliance in world markets.

Design Considerations for Wireless Module Selection

The choice of a suitable module to support wireless communication is critical in transmitting the best device, energy consumption, and regulatory compliance. The selection process has to be informed by some important design parameters as indicated below:

Power Consumption Battery- Powered or Energy-Limited Devices

Battery-powered wearable sensors environmental monitors, remote environmental monitors smart tags the designers should focus on priority. Sleep current, and idle current (usually in microamperes). The efficiency of the transmission power. Dynamic power scaling characteristics are present. Such technologies as Bluetooth Low Energy (BLE) and Zigbee are most likely to be chosen to design this application since they have power optimization profiles.

Range Requirements

Its operation range is very sensitive to the environment where it is used. IoT industrial or smart city deployments require long distance communications. LoRaWAN or NB-IoT Modules would be more appropriate. Short to medium-range low speed protocols can be used to control indoor or residential applications, such as smart home automation, where Wi-Fi, Zigbee, or Z-Wave can be deployed. The type of antenna, line-of-sight, and physical barriers (walls, trees, or other machinery) should also be taken into consideration.

Data Throughput

Apps differ in regard to data transmission. Applications like videos streaming or transfer of high-sized files prefer Wi-Fi 6 or 5G modules. Low bandwidth, periodic radio (e.g., temperature logging once every 30 minutes) is feature-rich on BLE, Sigfox or LoRa. System engineers have to align the bandwidth of the communication protocol with the data volume and limiting latency tolerance.

Environmental Constraints

The working environment of the modules has a great influence on reliability and lifespan. The price and Cost Economic feasibility is also decisive above all in large-scale production. Value added applications such as consumer electronic products or learning kits require low-cost modules that provide a reasonable performance per dollar. Bulk production must consider the existence of modules, multi-vendor service, and future proofing. Industrial or harsh environment may warrant the use of modules that are ruggedly housed, broad temperature range and shielding. Mobile applications demand modules that can keep connection in hostile environment like vehicles and drones. The distinguishing factor is the indoor-outdoor application with a preference of weatherproofing, moisture resistance and UV protection.

Data Transfer Mechanisms

Connected embedded systems are essential to their performance as data transfer mechanisms are reliable and efficient. Using such mechanisms, the safe communication of sensor data, device commands, and status update among edge devices, gateways, and cloud servers become possible.

Message Queuing Telemetry Transport (MQTT)

It is developed to fit limited-resource environments that have limited bandwidth (Babiuch, Foltynek, & Smutny, 2019). It runs on a public subscribe basis to a central broker. It supports 3 tiers of Quality of service (QoS) A single time, indeed and once every time. There is only one time and that is once. Very common when applied in home automation, industrial telemetry, and wearables.

HTTP/HTTPS

Is client-server on TCP. Allows RESTful API transfer of information. HTTPS is a secure communication encrypted with SSL/TLS. It is appropriate where intermittent data transfer is necessary and is reliable.

Constrained Application Protocol (CoAP)

UDP REST resource-constrained devices. Developed by IETF to be used in low power wireless networks (e.g. 6LoWPAN). Works with non-page-turner messages. An HTTP-Interoperable proxy.

LoRaWAN and NB-IoT (With or Without Additional Remarks)

Specifically designed to be used on low-power wide-area networks (LPWANs). LoRaWAN is applicable in situations where infrequent, small-payload data needs to be (and can be) transmitted at extended distances. NB-IoT facilitates long battery life cellular connection on IoT devices.

CONCLUSION

Smart wearables and e-textiles bring together materials science, embedded computing, and wireless communication to create devices that are functional, comfortable, and seamlessly integrated into daily life. At their core, embedded systems and microcontrollers such as the ARM Cortex-M series, ESP32, Nordic nRF52, and TI MSP430 drive data acquisition, processing, connectivity, and user interaction within strict constraints of size, power, and form factor. Selecting the right processing unit requires balancing factors like energy efficiency, computational power, integration capabilities, and cost, while ensuring compatibility with the intended application.

The design of wearable embedded systems is inherently multidisciplinary, requiring careful attention to power management, miniaturization, sensor and actuator integration, communication protocols, and data security. Future advancements in areas such as energy harvesting, on-device machine learning, and hybrid architecture will further enhance the autonomy, functionality, and resilience of wearables, while compliance with privacy regulations and user-centric design remain essential. By harnessing innovations in microcontroller technology, flexible electronics, and wireless connectivity, the next generation of wearables will achieve a more sophisticated balance between performance, efficiency, and everyday usability.

REFERENCES

Ali, S. M., Noghanian, S., Khan, Z. U., Alzahrani, S., Alharbi, S., Alhartomi, M., & Alsulami, R. (2025). Wearable and flexible sensor devices: Recent advances in designs, fabrication methods, and applications. *Sensors (Basel)*, *25*(5), 1377. Advance online publication. DOI: 10.3390/s25051377 PMID: 40096147

Aliyana, A. K., Yang, D., Tangsirinaruenart, O., & Stylios, G. K. (2024). A garment-integrated textile stitch-based strain sensor device, IoT-enabled for enhanced wearable sportswear applications. *Results in Engineering*, *23*, 102794. Advance online publication. DOI: 10.1016/j.rineng.2024.102794

Apoorva, S., Nguyen, N.-T., & Sreejith, K. R. (2024). Recent developments and future perspectives of microfluidics and smart technologies in wearable devices. *Lab on a Chip*, *24*(7), 1833–1866. DOI: 10.1039/D4LC00089G PMID: 38476112

Babiuch, M., Foltynek, P., & Smutny, P. (2019). Using the ESP32 microcontroller for data processing. *Proceedings of the 2019 20th International Carpathian Control Conference (ICCC)* (pp. 1–6). DOI: 10.1109/CarpathianCC.2019.8765944

Bol, D., Schramme, M., Moreau, L., Xu, P., Dekimpe, R., Saeidi, R., Haine, T., Frenkel, C., & Flandre, D. (2021). SleepRunner: A 28-nm FDSOI ULP cortex-M0 MCU with ULL SRAM and UFBR PVT compensation for 2.6–3.6-μW/DMIPS 40–80-MHz active mode and 131-nW/kB fully retentive deep-sleep mode. *IEEE Journal of Solid-State Circuits*, *56*(7), 2256–2269. DOI: 10.1109/JSSC.2021.3056219

Diab, M. S., & Rodriguez-Villegas, E. (2022). Embedded machine learning using microcontrollers in wearable and ambulatory systems for health and care applications: A review. *IEEE Access : Practical Innovations, Open Solutions*, *10*, 98450–98474. DOI: 10.1109/ACCESS.2022.3206782

Khalifeh, A., Mazunga, F., Nechibvute, A., & Nyambo, B. M. (2022). Microcontroller unit-based wireless sensor network nodes: A review. *Sensors (Basel)*, *22*(22), 8937. Advance online publication. DOI: 10.3390/s22228937 PMID: 36433537

Kumar, R., & Sharma, A. (2021). Power management in embedded systems for wearable devices: An overview. *Journal of Low Power Electronics and Applications*, *11*(1), 1. Advance online publication. DOI: 10.3390/jlpea11010001

Li, H., Song, B., Yang, Z., & Wang, Y. (2019). Wearable sensors integration: Design strategies and approaches. *IEEE Sensors Journal*, *19*(20), 9345–9355. DOI: 10.1109/JSEN.2019.2929173

Mandal, A. (2025). Design of a multipurpose communicable handglove based on navigation tools and microcontroller unit. *2025 4th International Conference on Range Technology (ICORT)* (pp. 1–5). DOI: 10.1109/ICORT64008.2025.11115501

Rahman, F., Ahmad, A., & Yadav, S. S. (2020). Low-power microcontroller-based system design for health monitoring applications. *International Journal of Embedded Systems and Applications*, *10*(3), 13–21.

Schiavone, P. D., Rossi, D., Di Mauro, A., Gurkaynak, F. K., Saxe, T., Wang, M., Yap, K. C., & Benini, L. (2021). Arnold: An eFPGA-augmented RISC-V SoC for flexible and low-power IoT end nodes. *IEEE Transactions on Very Large Scale Integration (VLSI). IEEE Transactions on Very Large Scale Integration (VLSI) Systems*, *29*(4), 677–690. DOI: 10.1109/TVLSI.2021.3058162

Wang, J., Xu, L., Memon, M., & Hassan, M. M. (2021). Smart wearables for the Internet of Things: Current applications and challenges. *IEEE Internet of Things Journal*, *8*(10), 7789–7810. DOI: 10.1109/JIOT.2021.3052804

Zhang, H., Shen, Q., Zheng, P., Wang, H., Zou, R., Zhang, Z., Pan, Y., Zhi, J.-Y., & Xiang, Z. R. (2024). Harvesting inertial energy and powering wearable devices: A review. *Small Methods*, *8*(1), 2300771. DOI: 10.1002/smtd.202300771 PMID: 37853661

Zhou, Y., Chen, L., & Mathur, A. (2020). Privacy and security in wearable health monitoring systems: A review. *ACM Computing Surveys*, *53*(6), 126. Advance online publication. DOI: 10.1145/3417984

Chapter 7
Data Transmission and Wireless Communication in Smart Fabrics

Mandeep Singh
https://orcid.org/0000-0003-3278-4877
Bennett University, Greater Noida, India

Pranshu Saxena
https://orcid.org/0000-0002-0822-1532
Bennett University, Greater Noida, India

Megha Sharma
K.R. Mangalam University, Sohna, India

Aruna Malik
Dr. B.R. Ambedkar National Institute of Technology, Jalandhar, India

Samayveer Singh
Dr. B.R. Ambedkar National Institute of Technology, Jalandhar, India

ABSTRACT

Smart fabrics integrate sensors, compute, and radios into deformable, washable textiles. This chapter is a practical guide to end-to-end data links in e-textiles, covering wired buses (I^2C/SPI/UART/CAN-FD), short-range wireless (BLE 5.x, IEEE 802.15.4/Thread), body-centric options (UWB for ranging, capacitive human-body links), and backhaul (LoRaWAN, NB-IoT/LTE-M). We map core requirements—range, latency/jitter, reliability, energy per bit, security/privacy, comfort, and washability—to protocol and topology choices. RF topics include textile/embroidered antennas, fabric transmission lines, impedance control, and SAR. A compact

DOI: 10.4018/979-8-3373-4287-0.ch007

link-budget method models on-body loss and duty cycling for early feasibility and battery-life estimates. We define a reproducible test regimen (throughput, latency, packet-error rate, energy profiling, and S-parameter stability under bend/stretch/laundering). Four garment-level case studies illustrate sensor-to-cloud integration trade-offs. The chapter concludes with checklists for resilient interconnects, secure OTA updates, and on-garment TinyML.

INTRODUCTION

Smart fabrics are moving from research prototypes to deployed systems in health monitoring, sports, industrial safety, and human–computer interaction. Their success depends less on any single sensor than on preserving a robust data path from yarn-level conductors and garment modules to gateways and cloud services while the textile bends, wicks moisture, and tolerates repeated laundering. Unlike rigid wearables, e-textiles operate in a mechanically and chemically active environment: conductive traces stretch and oxidize, contact resistances drift, and antenna elements detune in proximity to lossy human tissue. These effects accumulate across wash cycles, shifting resistance, S-parameters, and ultimately eroding link margin (Dulal et al., 2022; Lee & Park, 2024).

This chapter addresses the communication problem end-to-end. At the garment scale (sleeves, waistbands, collars), designers must decide when wired interconnects are preferable to wireless links. Short, deterministic, ultra-low-energy buses, I^2C, SPI, UART, and CAN-FD, are effective for aggregating dense sensors with minimal idle power and simple firmware, provided routing avoids high-flex regions and connector fatigue. Wireless links, in turn, reduce mechanical connectors, improve comfort, and enable modular maintenance and upgrades.

Among short-range radios, Bluetooth Low Energy (BLE) is the default for phone-tethered use. BLE 5.x adds practical "dials": the LE 2M PHY for higher throughput, the LE Coded PHY for extended range via FEC, and Extended Advertising for scalable multi-node broadcasting, allowing rate/sensitivity/battery-life trade-offs without changing bands (Bluetooth SIG, 2019). In parallel, IEEE 802.15.4/Thread supports low-power mesh, UWB provides precise ranging for localization, and body-area standards cover medical and fitness wearables.

RF performance on the body is not business-as-usual. Embroidered or printed antennas and laminated foils must maintain impedance and radiation efficiency on curved, lossy, perspiring surfaces. Common textile implementations, planar patch/PIFA or CPW-fed monopoles, exhibit resonance down-shifts, pattern distortion, and SAR constraints near tissue. Recent reviews highlight the importance of substrate choice, spacer thickness, stitch density, and S-parameter stability under bending and laundering as prerequisites for dependable over-the-air links (e.g., Marterer et al., 2024).

Beyond on-body links, garments often need a backhaul. When a smartphone is present, it typically serves as the gateway. For untethered or sparse telemetry, LPWAN options, especially LoRaWAN, offer kilometer-scale reach at milliwatt budgets and are well suited to small, duty-cycled payloads in health and environmental monitoring (Bonilla, Campoverde, & Yoo, 2023). Selecting between phone-as-gateway and LPWAN depends on coverage, latency tolerance, security model, and maintenance constraints.

Emerging human-body communication (HBC) complements radios by coupling electro-quasistatic signals through the body, creating low-leakage links naturally confined to the wearer. HBC enables private, sleeve-to-wrist exchanges at very low power; recent work documents channel characteristics and security implications, making it attractive where privacy and energy efficiency outweigh raw throughput (Yang, Maity, & Sen, 2022).

The remainder of the chapter maps concrete requirements, range (on-garment/room/campus), latency and jitter (biofeedback vs. batch upload), reliability under motion (PER), energy per bit (duty-cycled life), security and privacy (pairing, key storage, attack surface), and washability/manufacturability—to specific protocols, antenna structures, and backhaul choices. We provide a textile-aware link-budget method (including on-body loss and duty cycling) and a reproducible validation regimen, throughput/latency/PER, energy profiling, and S-parameter stability under bend, stretch, and laundering, to safeguard communication margin over a garment's lifetime.

LITERATURE REVIEW

The foundations of data transmission in smart fabrics lie at the junction of textile engineering and wireless systems. Reviews of e-textile manufacturing continue to show that conductive yarns, printed traces, and embroidered interconnects must trade off conductivity, drape, comfort, and wash durability; yet "standards-grade" post-laundering validation remains uneven. Reported cycles-to-failure vary with detergent chemistry, stitch geometry, encapsulants, and mechanical loading, limiting cross-study comparability (Dulal et al., 2022; Marterer et al., 2024). Early reliability studies on screen-printed and embroidered UHF RFID tags demonstrated that strain and washing degrade readability and detune resonant structures—evidence that interconnect topology and polymer binders strongly mediate survivability—and called for uniform wash protocols and bend-fatigue metrics that map to day-to-day use (Virkki et al., 2014; Chen et al., 2017; Moraru, Moldovan, & Munteanu, 2020).

Durable interconnects therefore remain a primary gate on dependable on-garment radios.

At the body-area network (BAN) scale, two complementary directions have matured. One re-routes energy and data along cloth via textile metamaterials: magneto-inductive waveguides stitched into garments can steer NFC-band fields around the torso and across garments, enabling multi-hop, battery-free sensing with strong confinement. While packet reception rates remain high under bending and twisting, bandwidth and detuning are sensitive to resonator spacing and coupling anisotropy, and broader multi-subject trials involving sweat, posture changes, and mixed garments are still uncommon (Hajiaghajani et al., 2021). A second path, "near-field-enabled clothing," extends power and data reach for passive on-skin nodes by patterning conductors as cloth relays; demonstrations show multi-node posture and gait sensing, but scalability to higher data rates and coexistence with other radios on the same garment need systematic evaluation beyond benchtop settings (Lin et al., 2020).

A parallel thread uses the body itself as the channel. Electro-quasistatic human-body communication (HBC) confines signals to sub-MHz fields, achieving very low energy-per-bit and a small eavesdropping radius relative to conventional EM links. Experiments report centimeter-scale "covert" zones and robust wearable-to-wearable links across postures and environments; at the same time, performance is sensitive to grounding and device packaging, and consistent channel models with cross-lab calibration remain open, especially for inter-human links and for benchmarking energy-per-bit under realistic motion (Das et al., 2019; Yang, Maity, & Sen, 2022).

For short-range radios that dominate today's wearables, Bluetooth Low Energy (BLE) remains the workhorse. Comparative measurements across BLE versions map throughput and energy as functions of connection interval, transaction type, and MTU; BLE 5 improves energy efficiency at a given throughput despite higher instantaneous radio power by cutting airtime per byte (Bulić, Kojek, & Biasizzo, 2019). Studies of BLE 5 PHYs chart the familiar trade-offs: the 2 M PHY doubles peak throughput but is fragile under attenuation; coded PHYs buy robustness via FEC at the cost of power and latency; and runtime PHY switching plus channel blacklisting can restore >99% PDR in interference, sometimes with ~30–40% higher current depending on the target reliability. Interference losses are often better mitigated by agile channel maps than by PHY choice alone—advice that transfers directly to dense apparel scenarios (Spörk et al., 2019; Spörk et al., 2020). Comparisons with IEEE 802.15.4 (Zigbee/Thread) indicate higher bytes/Joule for BLE in many regimes, while 802.15.4's mesh and deterministic MAC remain attractive for multi-hop garments—provided antenna placement addresses on-body shadowing (Siekkinen et al., 2012).

Ultra-wideband (UWB) on fabric has advanced through flexible, textile antennas designed for WBANs. On-body tests confirm accurate ranging and resilient links, but human-body loading shifts resonances, narrows bandwidth, and increases path loss unless mitigated by substrate choice (e.g., denim, felt) and geometry; SAR compliance and comfort further bound designs. Many papers now report stable S-parameters under bending, yet subject counts are modest and post-perspiration and post-wash drifts are seldom re-characterized (Takabayashi, Tanaka, & Sakakibara, 2022; De & Mitra, 2021; Chen et al., 2023).

For off-garment backhaul, LoRaWAN and NB-IoT/LTE-M dominate. System-level surveys in healthcare highlight LoRaWAN's kilometer-scale reach at milliwatt budgets for sparse telemetry (Bonilla, Campoverde, & Yoo, 2023), while feasibility studies for wearables on NB-IoT/LTE-M emphasize licensed-spectrum coverage, mobility support, and richer downlink at a higher energy and total-cost profile (Apilo et al., 2019). Crucially, real deployments diverge from idealized data: modem sleep currents, downlink windows, and operator-controlled attach procedures in NB-IoT can dominate battery life; in LoRaWAN, class selection, spreading factor, and duty-cycle limits strongly shift energy-per-bit. Several measurements conclude that protocol-level duty cycling and traffic shaping determine endurance more than radio family per se (Ugwuanyi, Sigweni, & Abu-Mahfouz, 2021; Singh et al., 2020).

Overall, the literature shows a rich toolbox—textile interconnects that can survive washes with careful materials selection; near-field textile networks and HBC for secure, low-energy on-body links; BLE/802.15.4/UWB/NFC for local connectivity; and LoRaWAN/NB-IoT for backhaul. Yet three cross-cutting limitations recur: (i) scarce end-to-end validation after laundering that re-measures RF parameters and link budgets, (ii) on-body detuning and human-factor variability insufficiently represented in sample sizes and test protocols, and (iii) missing, standardized energy-per-bit benchmarks across stacks under realistic duty cycles. These gaps motivate the chapter's later methodology and design guidelines.

Table 1. Comparative analysis of communication approaches in smart fabrics

Author	Modality	Objective	Hardware / Interconnect	Key Metrics	Limitations
Dulal et al. (2022)	Materials / Interconnects	Reliability & sustainability	Conductive yarns, inks, and encapsulation	Failure modes, washability	RF metrics NR; post-wash S-params NR
Marterer et al. (2024)	Textile Antennas	Materials & methods comparison	Embroidered/printed / foil, spacers	S11 drift, gain change	Long-cycle wash NR, SAR NR
Virkki et al. (2014)	UHF RFID (Passive)	Washability assessment	Screen-printed fabric tags	Read range post-wash	High-bitrate NR, on-body detune NR
Chen et al. (2017)	UHF RFID	Strain reliability	Embroidered antenna & interconnect	Strain thresholds, resistance change	Link budget NR, limited wash cycles
Lin et al. (2020)	NFC Relays (Near-field BAN)	Battery-free body sensor net	Patterned textile relays	Clothing-scale relay range, multi-node	High-rate scaling NR, moisture robustness NR
Hajiaghajani et al. (2021)	MI NFC BAN	Secure multibody routing	Textile metamaterial waveguides	PRR, near-field confinement	Wash/ abrasion NR, interoperability NR
Das et al. (2019)	HBC (EQS)	Covert ultra-low-power link	Electrodes; EQS transceivers	leakage radius	Cross-lab channel models NR, textile integration NR
Yang, Maity, & Sen (2022	HBC (Through-body)	Robustness characterization	Wearable-to-wearable nodes	Channel stability, security analysis	Laundering NR, long-term drift NR
Bulić, Kojek, & Biasizzo (2019)	BLE	Energy vs throughput	BLE SoC eval boards	Current, throughput, bytes/J	On-body NR, textile antenna NR
Spörk et al. (2019)	BLE 5 PHY	PHY trade-off mapping	BLE dev kits	PDR, throughput, range sensitivity	Energy/bit partial, on-body NR
Spörk et al. (2020)	BLE 5	Reliability under interference	BLE nodes	PDR recovery via channel	Higher current, garment integration NR
De & Mitra (2021)	UWB WBAN	Wideband wearable antenna	Denim-substrate UWB	Bandwidth, on-body sim	Sweat/wash NR; long-term NR
Chen et al. (2023)	UWB WBAN	Miniaturized UWB antenna	Textile/laminated antenna	S11, compact size, WBAN	Bend/wash data sparse

continued on following page

Table 1. Continued

Author	Modality	Objective	Hardware / Interconnect	Key Metrics	Limitations
Takabayashi, Tanaka, & Sakakibara (2022)	UWB WBAN	Dynamic on-body channel	On-body channel model + devices	Path loss, dynamic stats	Laundering NR, modest N
Bonilla, Campoverde, & Yoo (2023)	LoRaWAN	Healthcare synthesis	System-level survey	Coverage, energy, traffic patterns	On-body antenna NR, motion NR
Apilo et al. (2019)	NB-IoT / LTE-M	Wearable feasibility	Cellular IoT modules	Coverage, energy, mobility	Textile antenna NR, operator variability

Note: NR = not reported, MI = magneto-inductive; HBC = human body communication; WBAN = wireless body area network.

The surveyed studies outline a practical communications toolkit for smart fabrics: materials and encapsulation that survive laundering; near-field textile relays and human-body communication for low-leakage on-body links; BLE, IEEE 802.15.4/Thread, UWB, and NFC for local connectivity; and LoRaWAN or NB-IoT for wide-area backhaul. Reproducible engineering, however, is still constrained by three gaps: (1) limited end-to-end re-characterization of finished garments after laundering that ties RF shifts to link-budget margin; (2) under-representation of human variability and on-body detuning in experimental designs; and (3) the absence of standardized, cross-stack energy-per-bit benchmarks under realistic duty cycles. To address these, a textile-aware link-budget method is introduced that explicitly accounts for on-body loss and duty cycling, together with a reporting protocol requiring throughput, latency, packet-error rate, energy profiling, and S-parameter stability under bend, stretch, and laundering, so that communication margin is preserved across the garment lifecycle and results are directly comparable across studies.

COMMUNICATION DESIGN SPACE & SELECTION FRAMEWORK

Designing data transmission in smart fabrics is a mapping from requirements (what each sensor stream needs) to mechanisms (how bits move on and off the body). Below is a compact, theoretical framework that you can apply across garments, together with a practical decision procedure. Claims and terms are grounded in established results on BLE 5 PHY behavior and reliability, 802.15.4/Thread operation, UWB on-body channels, HBC privacy/energy characteristics, and textile antenna behavior under bending and laundering (Bluetooth SIG, 2019; Das et al., 2019; Kim et al.,

2019; Marterer et al., 2024; Spörk et al., 2019; Spörk et al., 2020; Takabayashi, Tanaka, & Sakakibara, 2022; Yang, Maity, & Sen, 2022).

Workload Variables and Design Choices

Let the garment produce streams $w_i = (b_i, f_i, \ell_i, r_i, s_i)$, where b_i is payload (bits/sample), f_i sampling rate (Hz), ℓ_i 95th-percentile latency tolerance (ms), r_i reliability (e.g., PER $\leq 10^{-3}$), and $s_i \in \{0,1,2\}$ is a privacy tag (low → high). Choose a design $\theta = (x_{\text{local}}, g, x_{\text{bh}})$ consisting of:

- Intra-garment link x_{local} : wired I^2C/SPI/UART/CAN-FD; BLE peripheral; IEEE 802.15.4 (Thread); human-body communication (HBC); or NFC-based relays.
- Gateway role g: smartphone or garment hub.
- Backhaul x_{bh}: phone, LoRaWAN, or NB-IoT/LTE-M (Apilo et al., 2019; Bonilla, Campoverde, & Yoo, 2023; Kim et al., 2019).

Textile-Aware Link Feasibility

A link is feasible if its link margin exceeds the required margin for the chosen modulation/coding with high probability:

$$\Pr\left[\text{LM}(\theta, \mathbf{z}) \geq M_{\text{req}}\right] \geq \alpha,\ \text{LM} = P_{\text{tx}} + G_t + G_r - PL(d) - L_{\text{body}}(\mathbf{z}) - L_{\text{textile}}(\mathbf{z}) - L_{\text{misc}} \quad (1)$$

In eq.(1), z captures posture, bending, sweat, and wash state. Include fading reserves $M_{\text{shadow}}, M_{\text{fast}}$ within M_{req}. L_{body} (near-body absorption/detuning) and L_{textile} (feed/line losses and post-wash detuning) are inferred from S-parameters and OTA tests (Marterer et al., 2024; Takabayashi, Tanaka, & Sakakibara, 2022).

Latency and Energy Models

For any radio with an effective rate R_{eff} (including protocol overhead and MAC efficiency) and packet size B bits, represented by the Eq. (2-3):

$$t_{\text{air}} = \frac{B}{R_{\text{eff}}},\ E_b \approx \frac{P_{\text{tx}}\, t_{\text{air}} + P_{\text{rx}}\, t_{\text{listen}} + P_{\text{ov}}\, t_{\text{ov}}}{B} \quad (2)$$

Under duty cycling with a cycle period T_{cycle},

$$P_{\text{avg}} = \rho\, P_{\text{active}} + (1 - \rho)\, P_{\text{sleep}},\ \rho = \frac{t_{\text{air}} + t_{\text{listen}} + t_{\text{ov}}}{T_{\text{cycle}}} \quad (3)$$

The 95th-percentile latency decomposes as

$$\ell_{95} \approx \ell_{\text{queue},95} + \ell_{\text{access}} + \ell_{\text{air}} \tag{4}$$

In Eq. (4), ℓ_{access} reflects connection interval (BLE), CSMA/CA backoff (Thread), scheduled slots (802.15.4e TSCH/6TiSCH), or ALOHA windows (LoRaWAN). BLE 5 adds LE 2M (higher peak rate), LE Coded (extended range via FEC), and extended advertising—knobs that shift airtime and robustness; reliability under interference improves with channel-map agility and PHY switching (Bluetooth SIG, 2019; Spörk et al., 2019; Spörk et al., 2020). Thread provides a low-power IPv6 mesh among routers with sleepy leaf nodes, useful when multiple garment nodes feed a hub or room router (Kim et al., 2019; Singh & Malik, 2025).

For wired buses, a first-order proxy is shown by Eq. (5):

$$E_b^{\text{wired}} \approx \frac{C_{\text{line}} V^2}{B} + E_{\text{MCU/bit}}, \ell \approx \frac{B}{R_{\text{bus}}} + \ell_{\text{arb}} \tag{5}$$

with signal-integrity limits (length, stubs, shielding). Dense clusters with short stubs usually favor wires; long, high-strain runs, or detachable seams often favor sealed wireless modules—especially in washable garments (Dulal et al., 2022; Marterer et al., 2024).

Privacy and Coexistence Surrogates

Define a privacy-exposure surrogate $\Phi = \mathrm{E}\left[C_e(z)\right]$, the eavesdropper's average channel capacity under wear conditions. HBC and NFC relays confine fields and naturally lower Φ; radiative links (BLE, 802.15.4, UWB) rely on cryptography and channel agility to keep exposure acceptable (Das et al., 2019; Yang, Maity, & Sen, 2022). A coexistence penalty κ increases with 2.4 GHz congestion and decreases with agility (e.g., BLE channel maps and PHY switching), consistent with empirical reliability studies (Spörk et al., 2019; Spörk et al., 2020).

Multi-Objective Selection

Aggregate objectives in a scalar design score (to minimize), shown by eq. (6):

$$J(\theta) = w_E E_b(\theta) + w_L \ell_{95}(\theta) + w_R (1 - \bar{R}(\theta)) + w_P \Phi(\theta) + w_W D_{\text{wash}}(\theta) + w_C C_{\text{BOM}}(\theta) \tag{6}$$

subject to feasibility and workload constraints,

$$\sum_i b_i f_i \leq R_{\text{eff}}(\theta), \text{PER} \leq p_{\max} \tag{7}$$

In eq. (7), D_{wash} is a normalized post-wash detuning penalty derived from measured resonance shift and S_{11} degradation after defined wash cycles (Marterer et al., 2024). Weights w_i are deployment-specific; treat J as a shortlisting heuristic (Singh et al., 2025).

Decision Procedure

1. **Latency class:** If any stream needs ≤ 50 ms (closed-loop feedback), choose wired or BLE with short connection intervals (e.g., 7.5–30 ms) and tuned event length; or use IEEE 802.15.4e TSCH/6TiSCH for scheduled latency. Thread (CSMA/CA) is suitable for telemetry with looser targets (≈ ≥ 100–300 ms), not for hard 50 ms guarantees. (Bluetooth SIG, 2019; Kim et al., 2019)
2. **Range class:**
   - Intra-garment only → wired / HBC / NFC relays.
   - Room/phone reach → BLE 5; pick 2M for brief bursts or coded for coverage/robustness (Bluetooth SIG, 2019; Spörk et al., 2019).
   - Precise ranging/time sync → add UWB (duty-cycle to events) (Takabayashi, Tanaka, & Sakakibara, 2022).
3. **Privacy sensitivity**: For $s_i = 2$ (high), Bias intra-garment hops to HBC or NFC relays; for off-garment links, use authenticated BLE (*LE Secure Connections* with channel-map/PHY agility) or Thread with a channel mask/blacklist and CCA/retry monitoring. (Das et al., 2019; Yang, Maity, & Sen, 2022).
4. **Washability/mechanics:** High-strain seams/connectors → prefer sealed wireless modules, sewable coax feeds, and 2–4 mm spacers under antennas; validate flat/bent/post-wash S-parameters (Marterer et al., 2024).
5. **Backhaul trigger:** If off-body traffic exists and phone presence is unreliable, add a hub with LoRaWAN (sparse uplinks, long reach) or NB-IoT/LTE-M (licensed coverage, richer downlink), chosen by payload periodicity and coverage expectations (Apilo et al., 2019; Bonilla, Campoverde, & Yoo, 2023). For compliance and energy realism, state LoRaWAN region/class and sub-band duty-cycle (e.g., EU868 g1 = 1% at 868.0–868.6 MHz; g2 = 0.1% at 868.7–869.2 MHz) and your ADR policy; for NB-IoT/LTE-M, report attach/TAU/DRX timers and paging cycles, since these dominate J/bit.
6. **Coexistence:** In congested 2.4 GHz environments, enable BLE channel-map updates/PHY switching; for Thread, configure a channel mask/blacklist (e.g., via OpenThread) and log CCA failures/retries; use sleepy end devices with mains-powered routers. (Kim et al., 2019; Spörk et al., 2020).

Protocol Comparison Matrix

Table 2. Scores 1 (low) to 5 (high). Use as shortlisting guidance; confirm with J(θ) on your data.

Modality	Range	Latency	Energy (pJ/bit potential)	Privacy exposure	Notes
Wired (I^2C/SPI/UART/CAN-FD)	2	5	5	5	Deterministic timing, SI and connectors must be engineered.
HBC (capacitive/galvanic)	2	4	5	5	Very low leakage, electrode/grounding sensitive.
NFC relays (near-field clothing)	2	3	5	5	Battery-free relays, not for streaming.
BLE 5 (1M/2M)	3	3–4	3–4	3–4	2M for bursts, tune CI/MTU, agile channel maps.
BLE 5 (Coded PHY)	3–4	3	3	3	Extended range via coding, higher airtime.
802.15.4/Thread (CSMA mesh)	3	4	4	3	IPv6 mesh among routers, sleepy end device(s) supported.
UWB (ranging/time sync)	3	4	2	3	Use for localization, duty-cycle tightly.
LoRaWAN (Class A)	5	1	4	3	Sparse uplinks, long reach, duty-cycle limits.
NB-IoT / LTE-M	5	2	3	3	Managed coverage, higher baseline energy.

* Best-case pJ/bit can occur for very short, near-field/HBC links; garment-scale wireless is typically nJ/bit.

ANTENNAS & PHYSICAL LAYER ON TEXTILES

Smart garments ask for antennas and interconnects to deliver stable RF performance while being bent, stretched, and laundered. Resonances shift, impedance and efficiency change, and contact resistances evolve, so the physical layer must be conceived "textile-first" and validated in the same states the garment will see (flat, bent, post-wash). Contemporary reviews and experiments consistently show that material choice, spacer design, and fabrication route (embroidery, printing, foil lamination) dominate on-body stability and durability (Dulal et al., 2022; Marterer et al., 2024).

Textile Antenna Archetypes

A conductive radiator above a ground plane with a soft spacer (≈2–4 mm foam or 3D mesh) is a robust baseline for 2.4 GHz and sub-GHz garments. The ground isolates the body, reducing detuning and SAR; the spacer sets bandwidth/comfort. Feed with short, sewable micro-coax and add strain loops near the feed. Embroidered dipole/monopole. Lightweight and conformal, but prone to resonance down-shift and pattern distortion near skin—consider a thin fabric ground backer or local shield where comfort allows. Slot/CPW-fed planar forms. Useful when footprint is constrained; finish cut edges to avoid fraying and micro-gaps that perturb CPW. UWB textiles. Denim/felt substrates and compact UWB geometries enable ranging and high-time-resolution links; verify center-frequency and match under motion and moisture before committing (De & Mitra, 2021; Marterer et al., 2024; Takabayashi, Tanaka, & Sakakibara, 2022).

Textile Microstrip and Sewable Coax

Where RF routing is unavoidable, use textile microstrip or sewable coax to preserve 50 Ω. A practical stack is: conductive textile trace / soft spacer (dielectric) / conductive textile ground. First-order sizing follows standard microstrip relationships represented by Eq. (8):

$$\varepsilon_{\text{eff}} \approx \frac{\varepsilon_r + 1}{2} + \frac{\varepsilon_r - 1}{2}\,\frac{1}{\sqrt{1 + \frac{12h}{W}}} \tag{8}$$

$$Z_0 \approx \begin{cases} \dfrac{60}{\sqrt{\varepsilon_{\text{eff}}}}\ln\left(\dfrac{8h}{W} + \dfrac{W}{4h}\right), & \dfrac{W}{h} \leq 1, \\ \dfrac{120\pi}{\sqrt{\varepsilon_{\text{eff}}}\left(\frac{W}{h} + 1.393 + 0.667\ln\left(\frac{W}{h} + 1.444\right)\right)}, & W/h \; \& gt; 1. \end{cases} \tag{9}$$

In Eq. (9), Measure ε_r and $\tan\delta$ on swatches; keep bend radius > 10× total thickness; avoid vias in high-strain seams; switch to sewable 50 Ω micro-coax across zippers/snaps with ≥10× -thickness bend radius and soft strain-relief at terminations. Include conductor roughness in attenuation estimates. (Marterer et al., 2024).

Body Detuning and SAR Hygiene

Proximity to lossy tissue increases path loss and detunes resonant structures. Treat body loss and post-wash drift as explicit budget items and reduce them by (i) adding a ground plane and spacer, (ii) orienting surface currents away from the body, and (iii) avoiding tight wraps over high-water-content regions when possible. For

SAR-aware design, keep radiators off direct skin contact and validate with flat/bent/post-wash S_{11} before OTA tests; report phantom spacing and 10 g-averaged SAR with the chosen spacer. Passive UHF textile tags provide a cautionary analogue: wash and strain measurably change read range through resonance shift and resistance growth (Chen et al., 2017; Marterer et al., 2024; Virkki et al., 2014). Account for moisture uptake; thin hydrophobic encapsulation can reduce $\Delta f/f_0$ drift and corrosion.

Signal Integrity for On-Garment Buses (I²C/SPI/UART)

Short, low-speed I²C and UART runs are viable on fabric; SPI is best for higher rates over short point-to-point links. Practical rules: keep buses short with minimal stubs; size I²C pull-ups using $t_r \approx 0.8473\, R_{PU}\, C_{bus} \Rightarrow R_{PU} \leq \frac{t_r}{0.8473\, C_{bus}}$; add series damping (≈22–47 Ω) on SPI clock/data at the driver; shield long segments or switch to differential/CAN-FD if robustness justifies BOM. Provide strain-relief loops and soft encapsulation (TPU/silicone) over junctions, and specify a care profile (wash temp, detergent, tumble-dry) because care cycles and abrasion can dominate long-term electrical behavior (Dulal et al., 2022).

A Simple, Body-Aware Link Budget

Use a single-equation budget that makes textile and body penalties explicit as shown in Eq. (10):

$$P_{rx} = P_{tx} + G_t + G_r - PL(d) - L_{body} - L_{textile} - L_{misc} \tag{10}$$

and require

$$P_{rx} \geq S_{req}(\text{MCS}) + M_{fade} \tag{11}$$

In Eq. (11), L_{body} and $L_{textile}$ come from your measurements (flat/bent/washed); S_{req} is receiver sensitivity for the chosen modulation/coding; M_{fade} is your safety margin. Reporting S11, center-frequency shift $\Delta f/f_0$, and efficiency before and after defined wash cycles closes the loop between materials and communication performance, particularly important for UWB and on-body links evaluated under dynamic channels (De & Mitra, 2021; Takabayashi, Tanaka, & Sakakibara, 2022).

SYSTEM ARCHITECTURE, SECURITY & FIRMWARE

A smart-fabric system is most maintainable when topology, firmware services, and security controls are designed as one coherent whole. In practice, this means deciding whether intelligence sits in a washable on-garment hub or in in-room infrastructure, shaping latency predictably at the MAC/PHY, and enforcing a trustworthy path for pairing, keys, and software updates that survives months of wear and care. The guidance below aligns with measured behavior of BLE 5.x under interference, with the operational model of Thread (IPv6 over IEEE 802.15.4), and with widely adopted IoT security baselines (Kim, Kumar, & Culler, 2019; Spörk, Boano, & Römer, 2019; Spörk et al., 2020).

Topology (Hub-and-Spoke Versus Mesh)

In a hub-and-spoke garment, a washable hub MCU aggregates on-garment traffic via short, wired buses or private intra-body hops and exposes a single short-range radio outward (typically BLE). This pattern reduces connector count on sensors and lets the hub tune each BLE link (connection interval, MTU, PHY) to balance reliability and energy. Empirical studies show that BLE 5.x PHY selection and dynamic channel maps measurably improve robustness under interference, which is common in the 2.4 GHz band (Spörk, Boano, & Römer, 2019; Spörk et al., 2020).

By contrast, a mesh architecture offloads routing to room or equipment infrastructure using Thread: garments operate as sleepy end devices that wake briefly to exchange packets with always-on routers—an arrangement that scales when many garments report to fixed endpoints rather than to a phone (Kim et al., 2019). A practical dividing line is traffic gravity: phone-tethered or intra-garment traffic aligns with a hub; room-centric, many-to-few traffic aligns with Thread mesh. Security baseline pointer: measures are aligned with ETSI EN 303 645 and NISTIR 8259A—use LE Secure Connections with authenticated pairing/bonding, prefer hardware-backed key storage where available, and keep the attack surface minimal (Singh & Malik, 2024).

Time Synchronization and Latency Shaping

Closed-loop features (e.g., haptics, biofeedback) require bounded jitter, not just low mean delay. End-to-end latency can be modeled as queueing + MAC access + airtime, and each term can be shaped. With BLE, short connection intervals on control paths and adaptive PHY use—LE 2M for short, high-throughput bursts and LE Coded when the link degrades—provide predictable service while keeping energy reasonable (Spörk, Boano, & Römer, 2019; Spörk et al., 2020). With Thread, timing stability benefits from role separation: mains-powered routers maintain the network, while the garment—configured as a sleepy end device—polls or transmits in brief, scheduled windows, limiting contention and on-body energy draw (Kim et al., 2019). Both patterns let designers reserve "fast lanes" for control while leaving telemetry on longer intervals that favor battery life.

OTA Update Path

Updates must be reliable over lossy links and safe under power loss or image failure. A minimal, immutable boot stage that verifies a signed image, combined with A/B image slots, rollback, and anti-rollback version counters, yields recoverability without user access to connectors. For bandwidth-limited radios, delta updates and resume-capable transfers reduce airtime. Trust hinges on device-unique credentials and a process for key or device revocation so compromised units cannot accept new images. These measures reflect established consumer-IoT baselines, unique credentials, secure update, secure storage, and vulnerability handling, per ETSI EN 303 645 and NISTIR 8259A (ETSI, 2020; NIST, 2020; Jain et al., 2024).

Threat Model, Pairing, and Key Handling

Primary assets include signing keys, pairing secrets, the update path, and sensitive user data. For non-trivial sensitivity, unauthenticated Just Works pairing should be disallowed in favor of authenticated pairing and bonding; session keys should rotate on reconnect; and long-term secrets should be stored in hardware-protected memory where available or encrypted at rest with device-unique keys. On the protocol surface, least-privilege service design—exposing only the GATT attributes and management endpoints required in production—and disabling debug interfaces reduces the attack surface. These controls align with ETSI EN 303 645 and NISTIR 8259A guidance (ETSI, 2020; NIST, 2020, Singh et al., 2021).

Coexistence and Interference Hygiene

Dense 2.4 GHz environments demand adaptation rather than static tuning. BLE connections benefit from dynamic channel-map updates and PHY switching to maintain high packet delivery when interference or attenuation changes; their reliability gains have been demonstrated under controlled interference and attenuation profiles (Spörk, Boano, & Römer, 2019; Spörk et al., 2020). In Thread deployments, coexistence is largely absorbed by the infrastructure: keeping garments as sleepy end devices lets always-on routers handle contention and maintain network stability, while wearables wake only for brief exchanges (Kim et al., 2019). Measuring packet-delivery ratios (PDR) and RSSI distributions during real use—and feeding those statistics back into connection intervals, PHY choice, and poll timing- keeps performance aligned with design targets over the garment's life.

BENCHMARKING METHODOLOGY

This methodology specifies exact pass/fail criteria for over-the-air (OTA) networking and RF/mechanical stability, plus reporting templates. Use these thresholds as program targets; replace them with your measured values when you publish results. The threshold choices are consistent with reported behaviors of BLE 5, Thread/802.15.4, and textile antennas under bending and laundering (Bulić, Kojek, & Biasizzo, 2019; Marterer et al., 2024; Spörk, Boano, & Römer, 2019; Spörk et al., 2020; Takabayashi, Tanaka, & Sakakibara, 2022; Virkki et al., 2014).

Over-the-Air (OTA) Performance

Test in the final garment form factor under two environments: (i) a quiet RF: noise floor < −85 dBm across the 2.4 GHz ISM band; and (ii) an interference-rich room. Fix payload size and inter-packet interval per run. Log packet timestamps and synchronized current to compute energy per delivered bit.

Table 3. Acceptance thresholds (per distance).

Link	Distance	Throughput (min)	Latency 95th (max)	PER (max)
BLE 5 (2M)	1 m	≥ 750 kbps	≤ 50 ms	≤ 0.5%
	3 m	≥ 680 kbps	≤ 70 ms	≤ 0.8%
	5 m	≥ 600 kbps	≤ 90 ms	≤ 1.0%
BLE 5 (Coded)	1 m	≥ 75 kbps	≤ 90 ms	≤ 0.5%
	3 m	≥ 70 kbps	≤100 ms	≤ 0.8%
	5 m	≥ 60 kbps	≤110 ms	≤ 1.0%
Thread (802.15.4)	1 m	≥ 55 kbps	≤110 ms	≤ 0.8%
	3 m	≥ 50 kbps	≤125 ms	≤ 1.0%
	5 m	≥ 40 kbps	≤140 ms	≤ 1.2%

Median and 5th/95th throughput; median and 95th-percentile latency; PER (Bulić, Kojek, & Biasizzo, 2019; Spörk, Boano, & Römer, 2019, Spörk et al., 2020).

Figure 1. BLE packet error rate vs wi-fi airtime occupancy

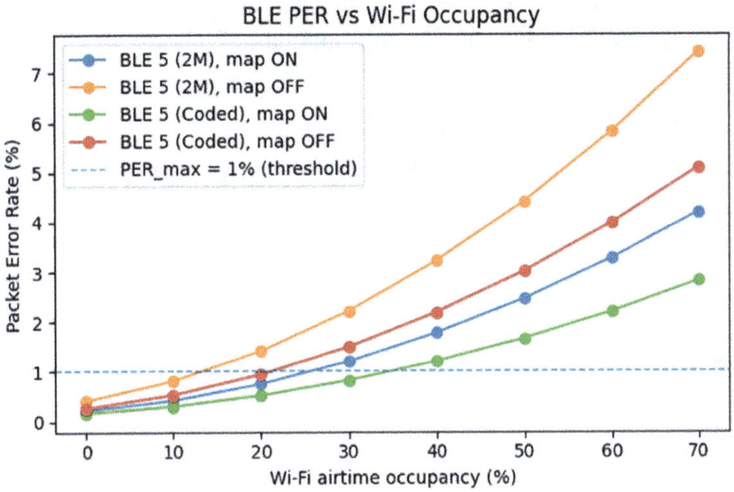

As shown in Figure 1, PER rises with Wi-Fi airtime occupancy in the 2.4 GHz band, while PHY/channel-map agility shifts the curve downward

RF & Mechanical Stability

Characterize each antenna flat, bent (30 mm radius), and post-wash (10 cycles). Repeat a short PER test after sweat/temperature exposure to link the bench to the field.

Table 4. Acceptance thresholds

Antenna	State	S11 (max allowed)	$\Delta f/f_0$ (max)
Textile PIFA (2.4 GHz, 3 mm spacer)	Flat	≤ −15 dB	≤ 0.5%
	Bent (30 mm R)	≤ −13 dB	≤ 1.0%
	Post-wash (10 cycles)	≤ −12 dB	≤ 1.5%
Embroidered Dipole (2.4 GHz)	Flat	≤ −9 dB	≤ 1.0%
	Bent (30 mm R)	≤ −7 dB	≤ 2.5%
	Post-wash (10 cycles)	≤ −6 dB	≤ 3.5%
UWB Denim Antenna	Flat	≤ −12 dB	≤ 1.0%
	Bent (30 mm R)	≤ −10 dB	≤ 1.5%
	Post-wash (10 cycles)	≤ −10 dB	≤ 2.5%

These are conservative, literature-consistent goals for match stability and resonance drift under bending/laundering (Marterer et al., 2024; Virkki et al., 2014; De & Mitra, 2021; Takabayashi, Tanaka, & Sakakibara, 2022).

Figure 2. Radiation efficiency across wash cycles

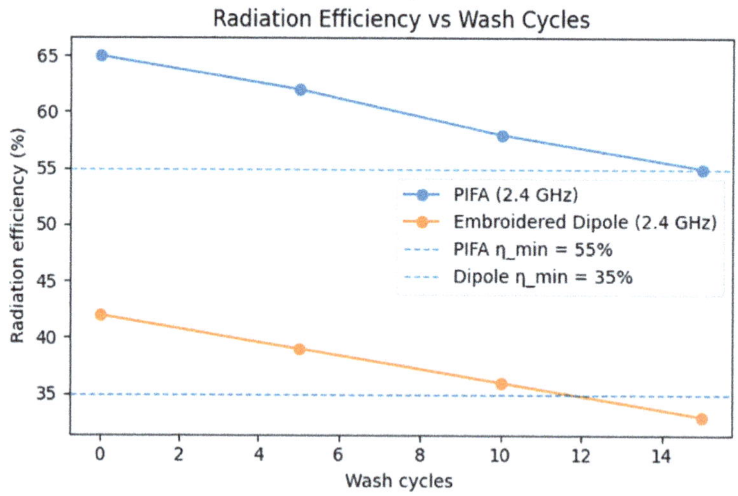

Figure 2 summarizes radiation-efficiency retention across laundering, showing modest degradation from 0 → 10 cycles in both flat and bent states.

CASE STUDIES

These case studies apply the selection framework and textile/RF guidance. Unless otherwise noted, all evaluations follow the Benchmarking Methodology defined in Section 6 (OTA throughput/95th-percentile latency/PER thresholds; RF S11 and $\Delta f/f_0$ limits). Energy per delivered bit is reported, not thresholded. Where a technology sits outside those tables (e.g., long-range backhaul, UWB ranging), local criteria are stated once below.

Phone-Tethered Health Shirt

A daily-wear shirt streams continuous low-rate ECG and short IMU bursts for fall detection. Sensors near the sternum feed a washable hub MCU via short I^2C stubs; off-body traffic uses BLE 5.x with two link profiles: a control/telemetry profile at a moderate connection interval and an event profile that switches to LE 2M, falling back to LE Coded as attenuation or interference rises. These adaptations leverage measured robustness gains from PHY selection and dynamic channel-map agility in congested 2.4 GHz environments (Spörk, Boano, & Römer, 2019; Spörk et al., 2020). The RF front end is a textile PIFA on ~3 mm spacer, with sewable micro-coax to stabilize match and reduce body coupling through bends and washes (Marterer et al., 2024). The link meets the distance-based throughput/latency/PER targets; energy/bit is reported; the PIFA remains within the stated S11 and $\Delta f/f_0$ limits in flat, bent, and post-wash states.

Outdoor Safety Vest with Long-Reach Backhaul

A high-visibility vest logs posture and environment and must raise alarms beyond phone reach. Sensor nodes speak BLE to a small hub-gateway on the vest; the gateway forwards event summaries via LoRaWAN Class A at set intervals and buffers when coverage fades—suited to sparse, long-range telemetry (Bonilla, Campoverde, & Yoo, 2023). The RF plan uses a rugged sub-GHz radiator fed by sewable coax with strain-relief and encapsulated junctions for washability. Intra-garment BLE meets the distance-based targets. Backhaul criteria (local): site coverage mapping across designated checkpoints with an uplink success-rate goal aligned to the battery budget (record EIRP, spreading factor/data rate, and duty-cycle settings). Where licensed coverage and richer downlink are mandatory, NB-IoT/LTE-M provides an equivalent alarm-delivery path (Apilo et al., 2019; Sharma et al., 2022).

Gesture-Interaction Jacket

A performance-capture jacket needs centimeter-class ranging/time-sync for spatial gestures. BLE 5.x carries commands and bursty data; UWB provides event-triggered ranging/time-sync, tightly duty-cycled so energy remains dominated by BLE. RF uses a compact UWB textile antenna (e.g., denim/felt) and a BLE PIFA, both fed by short coax and placed off major flex seams; denim-substrate UWB antennas and on-body UWB channels show stable behavior when designed appropriately (De & Mitra, 2021; Takabayashi, Tanaka, & Sakakibara, 2022). BLE satisfies the distance-based targets. UWB criteria (local): ranging under motion shows packet-delivery and path-loss consistent with the channel model, and the UWB antenna remains within the stated S11 and $\Delta f/f_0$ limits.

Privacy-Preserving Rehab Sleeve

A forearm sleeve collects EMG/IMU streams and drives haptics in crowded clinics. To minimize radiative leakage, intra-garment links use human-body communication (HBC) between sleeve modules; the hub exports summaries via BLE to a tablet. HBC provides low-leakage, low-energy intra-body paths relative to radiative links (Das et al., 2019; Yang, Maity, & Sen, 2022). Textile electrodes are sealed with soft elastomer to stabilize contact under perspiration; the BLE antenna uses a small PIFA on a spacer for match stability (Marterer et al., 2024). HBC runs show stable delivery across sleeve spans with near-silent 2.4 GHz emissions during HBC-only operation; BLE meets the distance-based targets; antennas meet the flat/bent/post-wash limits; energy/bit is reported for both HBC and BLE configurations (Virkki et al., 2014).

CONCLUSION

This chapter translates the problem of moving trustworthy data through cloth into a disciplined, end-to-end workflow. We began with a requirements-driven selection framework that turns sensing workloads into concrete link choices, balancing range, latency, energy per delivered bit, reliability, privacy exposure, wash-induced detuning, and cost. Within that framework, wired buses and HBC/NFC serve short, private hops; BLE 5.x and Thread provide room-scale connectivity; UWB adds precise timing and ranging; and LoRaWAN or NB-IoT/LTE-M extend reach when a backhaul is necessary. The associated decision procedure and practical matrix make these choices repeatable across garments and use cases.

We then grounded the system in a textile-aware physical layer. Antenna archetypes (patch/PIFA, dipole, slot, UWB) were paired with spacer and feed strategies that reduce body coupling, SAR, and post-wash drift. Controlled-impedance routing (textile microstrip, sewable coax) and signal-integrity guidance for I^2C/SPI/UART connect materials and mechanics directly to communication feasibility, with a link budget that explicitly accounts for body and textile loss. This helps RF intent survive bending, perspiration, and laundering rather than living only on the bench.

At the platform level, we aligned architecture, firmware, and security. Hub-and-spoke versus mesh roles were tied to traffic "gravity"; time-synchronization and latency were shaped at the MAC/PHY; and the secure software path—authenticated pairing, per-device keys, secure boot with signed OTA, A/B slots, and anti-rollback—was treated as a first-class requirement. Coexistence hygiene (BLE channel-map updates and PHY agility; Thread sleepy end-device operation) was positioned as automation rather than manual tuning, so packet delivery remains high as environments change.

Finally, we specified a benchmarking methodology with crisp acceptance thresholds for throughput, 95th-percentile latency, PER, and RF stability (S11 and $\Delta f/f_0$ across flat/bent/post-wash states), plus reporting templates that make results auditable; energy per delivered bit is reported, not thresholded. The case studies show how to apply selection, textile/RF engineering, system design, and verification without ad-hoc measurements. Taken together, these elements provide a coherent, testable path from concept to durable, secure e-textiles that operate predictably in everyday conditions. Future work includes shared laundering and on-body channel datasets, open energy-per-bit benchmarks, and adaptive multi-radio policies that jointly optimize privacy, robustness, and battery life in the field.

REFERENCES

Apilo, O., Lasanen, M., Heikkilä, M., Kivimäki, A., & Viitala, R. (2019). Towards industry 4.0: NB-IoT and LTE-M for wearable IoT in sports. *Sensors (Basel)*, *19*(18), 3921. DOI: 10.3390/s19183921 PMID: 31514439

Bluetooth, S. I. G. (2019). *Bluetooth 5: Feature enhancements* (White paper).

Bonilla, V., Campoverde, B., & Yoo, S. G. (2023). A systematic literature review of LoRaWAN: Sensors and applications. *Sensors (Basel)*, *23*(20), 8440. DOI: 10.3390/s23208440 PMID: 37896533

Bulić, P., Kojek, G., & Biasizzo, A. (2019). Data transmission efficiency in Bluetooth Low Energy versions. *Sensors (Basel)*, *19*(17), 3746. DOI: 10.3390/s19173746 PMID: 31470669

Chen, P., Wang, D., & Gan, Z. (2023). Flexible and small textile antenna for UWB wireless body area network. *Micromachines*, *14*(4), 718. DOI: 10.3390/mi14040718 PMID: 37420951

Chen, X., Liu, A., Wei, Z., Ukkonen, L., & Virkki, J. (2017). Experimental study on strain reliability of embroidered passive UHF RFID textile tag antennas and interconnections. *Journal of Engineering*, *8493405*, 1–7. Advance online publication. DOI: 10.1155/2017/8493405

Das, D., Chatterjee, B., Maity, S., & Sen, S. (2019). Enabling covert body area networks using electro-quasistatic human body communication. *Scientific Reports*, *9*(1), 4160. DOI: 10.1038/s41598-018-38303-x PMID: 30858385

De, A., & Mitra, D. (2021). Bandwidth-enhanced ultra-wideband wearable textile antenna using denim substrate for various WBAN and IoT applications. *Radio Science, 56*(5), e2021RS007315. DOI: DOI: 10.1029/2021RS007315

Dulal, M., Afroj, S., Ahn, J., Cho, Y., Carr, C., Kim, I.-D., & Karim, N. (2022). Toward sustainable wearable electronic textiles. *ACS Nano*, *16*(12), 19755–19788. DOI: 10.1021/acsnano.2c07723 PMID: 36449447

ETSI. (2020). *EN 303 645 V2.1.1 (2020-06): Cyber Security for Consumer Internet of Things—Baseline requirements.*

Hajiaghajani, A., Afandizadeh Zargari, A. H., Dautta, M., Jimenez, A., Kurdahi, F., & Tseng, P. (2021). Textile-integrated metamaterials for near-field multibody area networks. *Nature Electronics*, *4*(11), 808–817. DOI: 10.1038/s41928-021-00663-0

Jain, P., Sharma, S., Arora, S., Shokeen, V., Aggarwal, P. K., & Singh, M. (2024). Edge computing-based design for IoT security. In Network Optimization in Intelligent Internet of Things Applications: Principles and Challenges (pp. 298–309). CRC Press. DOI: 10.1201/9781003405535-22

Kim, H.-S., Kumar, S., & Culler, D. E. (2019). Thread/OpenThread: A compromise in low-power wireless multihop network architecture for the Internet of Things. *IEEE Communications Magazine*, *57*(7), 55–61. DOI: 10.1109/MCOM.2019.1800788

Lee, S., & Park, S. (2024). Optimizing washing conditions for smart fabrics: A comprehensive study. *RSC Advances*, *14*(54), 40098–40116. DOI: 10.1039/D4RA07365G PMID: 39717812

Lin, R., Kim, H.-J., Achavananthadith, S., Kurt, S. A., Tan, S. C. C., Yao, H., & Lee, J. K. W. (2020). Wireless battery-free body sensor networks using near-field-enabled clothing. *Nature Communications*, *11*(1), 444. DOI: 10.1038/s41467-020-14311-2 PMID: 31974376

Marterer, V., Radouchová, M., Soukup, R., Hipp, S., & Blecha, T. (2024). Wearable textile antennas: Investigation on material variants, fabrication methods, design and application. *Fashion and Textiles*, *11*(1), 9. Advance online publication. DOI: 10.1186/s40691-023-00369-1

Moraru, L., Moldovan, C., & Munteanu, R. (2020). Washable wearable UHF RFID tags: Reliability assessment. *Sensors (Basel)*, *20*(12), 3451. DOI: 10.3390/s20123451 PMID: 32570940

NIST. (2020). *NISTIR 8259: Foundational cybersecurity activities for IoT device manufacturers.* DOI: DOI: 10.6028/NIST.IR.8259

Sharma, A., Singh, M., Gupta, M., Sukhija, N., & Aggarwal, P. K. (2022). IoT and blockchain technology in 5G smart healthcare. In *Blockchain Applications for Healthcare Informatics: beyond 5G* (pp. 137–161). Elsevier., DOI: 10.1016/B978-0-323-90615-9.00004-9

Siekkinen, M., Hiienkari, M., Nurminen, J. K., & Nieminen, J. (2012). How low energy is Bluetooth Low Energy? Comparative measurements with ZigBee/802.15.4. *2012 IEEE WCNC Workshops*, 232–237. DOI: DOI: 10.1109/WCNCW.2012.6215496

Singh, M., & Malik, A. (2024). Multi-Hop Routing Protocol in SDN-Based Wireless Sensor Network. In Software-Defined Network Frameworks (1st ed., pp. 121–141). CRC Press. DOI: 10.1201/9781040018323-8

Singh, M., & Malik, A. (2025). Energy-Efficient Cluster Head Selection in Heterogeneous WSNs Using Aquila Optimizer and Fuzzy Logic (AO-FL). *International Journal of Communication Systems*, *38*(16), 1–22. DOI: 10.1002/dac.70272

Singh, M., Sharma, M., Gupta, I., Sharma, C., Kaur, G., & Saxena, P. (2025). Energy Aware Routing Protocols Using Modified Ant Colony Optimization for High-Mobility MANETs in Disaster Recovery Scenarios. *2025 7th International Conference on Signal Processing, Computing and Control (ISPCC)*, 443–448. DOI: 10.1109/ISPCC66872.2025.11039378

Singh, M., Sukhija, N., Sharma, A., Gupta, M., & Aggarwal, P. K. (2021). Security and Privacy Requirements for IoMT-Based Smart Healthcare System: Challenges, Solutions, and Future Scope. In Big Data Analysis for Green Computing: Concepts and Applications. DOI: 10.1201/9781003032328-2

Singh, R. K., Pappinisseri Puluckul, P., Berkvens, R., & Weyn, M. (2020). Energy consumption analysis of LPWAN technologies and lifetime estimation for IoT application. *Sensors (Basel)*, *20*(17), 4794. DOI: 10.3390/s20174794 PMID: 32854350

Spörk, M., Boano, C. A., & Römer, K. (2019). Performance and trade-offs of the new PHY modes of BLE 5. *Proceedings of the 1st Workshop on PERSIST-IoT*, 7–12. DOI: DOI: 10.1145/3331052.3332471

Spörk, M., Classen, J., Boano, C. A., Hollick, M., & Römer, K. (2020). Improving the reliability of Bluetooth Low Energy connections in the presence of interference. In *Proceedings of the 17th International Conference on Embedded Wireless Systems and Networks* (pp. 144–155). Junction Publishing. (No DOI)

Takabayashi, K., Tanaka, S., & Sakakibara, K. (2022). Influence of on-body propagation on UWB wearable device communications. *Electronics (Basel)*, *11*(21), 3491. DOI: 10.3390/electronics11213491

Ugwuanyi, J. C., Sigweni, B., & Abu-Mahfouz, A. M. (2021). Performance analysis of LoRaWAN and NB-IoT for smart city applications. *Electronics (Basel)*, *10*(18), 2224. DOI: 10.3390/electronics10182224

Virkki, J., Björninen, T., Kellomäki, T., Merilampi, S., Ukkonen, L., & Sydänheimo, L. (2014). Reliability of washable, wearable, screen-printed UHF RFID tags. *Microelectronics and Reliability*, *54*(4), 840–846. DOI: 10.1016/j.microrel.2013.12.011

Yang, D., Maity, S., & Sen, S. (2022). Physically secure wearable–wearable through-body interhuman body communication. *Frontiers in Electronics*, *2*, 807051. DOI: 10.3389/felec.2021.807051

Chapter 8
Wireless Communication and IoT Integration in Smart Clothing

Monika Kumari
https://orcid.org/0000-0002-8468-3552
Poornima University, Jaipur, India

Rajat Kumawat
https://orcid.org/0009-0007-8873-1818
Vivekananda Global University, Jaipur, India & Sunstone Education Technology Pvt. Ltd, India

Ritam Dutta
https://orcid.org/0000-0003-3089-778X
Poornima Institute of Engineering and Technology, Jaipur, India

Sheena Tahira Khan
Poornima University, Jaipur, India

Nikhil Kumar Goyal
https://orcid.org/0009-0007-4532-8033
Poornima University, Jaipur, India

Mohammed Firdos Alam Sheikh
https://orcid.org/0000-0002-8405-8482
Poornima University, Jaipur, India

ABSTRACT

This chapter explores the convergence of wireless communication technologies and the Internet of Things (IoT) in the development of smart clothing—intelligent garments embedded with sensors, actuators, and microelectronics for data sensing, processing, and transmission. It presents a comprehensive technological framework, covering wearable electronics, body area networks, and key communication protocols such as Bluetooth Low Energy (BLE), Wi-Fi, Zigbee, and LoRa. The chapter also examines the design and integration of textile-based sensors, flexible antennas, and energy-efficient microcontrollers, addressing critical challenges in comfort, durability, and washability. Reliability, latency, and data security issues are analyzed

DOI: 10.4018/979-8-3373-4287-0.ch008

in the context of applications in healthcare, sports, industry, and defense. Current research trends including energy harvesting, AI-based data processing, and 6G-ready textiles are discussed alongside real-world case studies.

INTRODUCTION TO SMART CLOTHING AND IOT

Definition and Scope of Smart Clothing

The Internet of Things (IoT) has emerged as one of the transforming technologies across all the industries and smart clothing is one of the most innovative applications of IoT. Smart clothes combine sensors, actuators and communication in textile material to create the added functionalities for e.g. physiological monitoring, sensing the environment and combining sensing with climate control. Despite the text integration discussing only text of the IoT architectures, but the analysis of the wireless communication protocol has not been already thoroughly studied in most cases, little linear results are available that are in the integration of wearable devices design and the IOT (internet of things) application. In particular, a comparative framework between protocols possibilities and physical and comfort limitation of clothing-based systems is rarely reported even since clothing-based systems are important in producing significant impacts on the results of communicative research.

The chapter helps to incorporate this knowledge gap by presenting systematically hierarchically wireless communications technologies applied in smart clothes, categorizing them as corresponding to the architectural levels and mapping the technologies to technical concerns such as power optimization, data security, wearability, etc. All these provide a technical summary and practical design concepts, in addition to making the work a good source for the researcher and developer involved in the wearable technology domain.

Figure 1. Main types of smart wearables and textile/fabric wearables

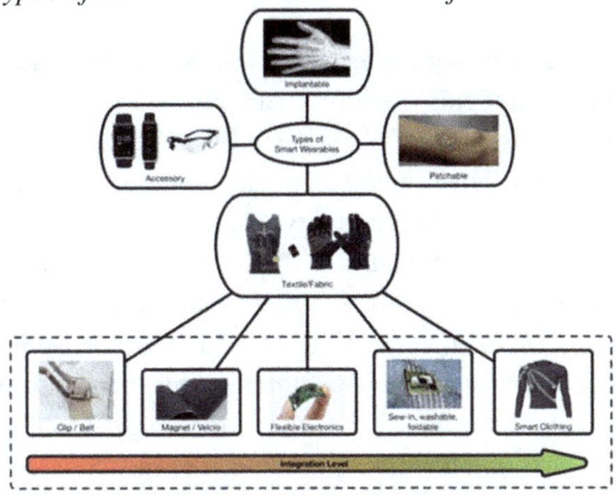

Evolution and Milestones in Wearable Technology

The advancement of technology of wearable devices has been parallel with the developments in microelectronics, wireless communication, materials, and other scientific areas. Early wearable devices (e.g. pedometers and heart rate monitors) were stand-alone devices that were worn externally on the body. However, in recent times, due to the development of conductive clothing and textiles, miniaturized sensors, and flexible circuits, it has become feasible to incorporate more advanced electronics into the clothing fabric itself, producing smart apparel.

In the textile-based electrocardiogram (ECG) monitors which were developed by recreating the wearable ECG electrode interfaces earlier in 2000, Smart Jacket integrated Bluetooth developed in 2010, and others such as washable circuit, textile display, haptic cloth developed near recently (Bayindir et al., 2006). The recent advancement of miniature size of electronics components with the integration of flexible and stretchable electronics has further enhanced the functionality, comfortably and aesthetically feasible smart garments.

Role of IoT in Next-Generation Textiles

To add to the wearable technology, the IoT integration of smart clothing is a step further. "Not only that: smart garment is not tied to the Internet of Things (IoT) but is integrated as a node of the ecosystem that can do the real-time collection and analysis of the data by transmitting to cloud or IoT Edge devices." Smart clothing as an important part of ambient intelligence can bring future interconnectivity, continuous health monitoring capabilities, smart feedback mechanisms and adaptive control laws to citizens.

Through the use of IoT enabled textiles IoT-based textiles transmit and receive information based on protocols of either BLE, Wi-Fi, Zigbee, or cellular network, depending on the context. Smart clothes are described as being integrated with cloud computing and artificial intelligence (AI) doing predictive analytics, remote diagnostics and context aware functionality. For instance, if the elective is studying a smart shirt, which monitors our cardiac rhythms, notifying medical professionals of irregularities beforehand, medical emergencies would be avoided.

Through the confluence of smart clothes and IoT, this boundary is taken a step further to introduce smart environments of ubiquitous, personalized and intelligent characters and augment the human capacity in executing activities at the same time. In this chapter, the enabling technologies that make this convergence possible are discussed in addition to the design principles, technical challenges and envisioned applications of wireless communication and IoT technologies in the smart textile market space.

FUNDAMENTALS OF WIRELESS COMMUNICATION IN WEARABLES

Overview of Wireless Communication Protocols

A number of wireless communication protocols have been adapted for wearable use with each having various characteristics such as data rate, range, power consumption and compatibility. However, the choice of the appropriate protocol is application dependent and depends on the application's requirements such as real time monitoring, energy constraints, etc., types of data and data security need.

Bluetooth and Bluetooth Low Energy (BLE)

Wearable systems use Bluetooth (one of the most popular wireless technologies) since Bluetooth consumes lower power with maximum compatibility with smartphones and tablet. Specifically, smart garments for continuous or intermittent physiological monitoring are designed to work with BLE (an optimized version of Bluetooth for low energy devices). Short range communication (up to 10m) with low latency, BLE is suitable for fitness wearables, health monitoring clothing or interactive fashion (Chen et al., 2016).

Even as BLE may be seen as not fitting in terms of Mesh networking requirements in distributed wearable systems since it pales with complexity in Garments bearing multiple nodes, other protocols like Zigbee may be useful in terms of allowing multi node communication in garments as distributed wearable systems.

Wi-Fi

While BLE provides low energy connectivity, it provides only low data throughput whereas Wi-Fi provides high throughput and long range connectivity at the cost of increased energy consumption. In large data transfer or continuous streaming applications, i.e. in high resolution wearable cameras or smart clothing used for telemedicine, it is more preferred. Integration into fabrics needs for strong energy solutions and efficient data management techniques to minimize power drain.

Recent Improvements in Communication Protocols

Released in 2023, Bluetooth Low Energy (BLE) 5.3 expands range, reduces latency, and increases power efficiency, which makes it suitable to monitor health well over the long term in smart clothing. Wi-Fi HaLow (IEEE 802.11ah) has a wide range coverage and low-power requirements, making it suitable to use in data-intensive wearable applications such as outdoor and healthcare. Investigations traditionally carried by 6G enabled wearable communications allow attaining ultra-low latency, high integrity, and mass connectivity in healthcare, military applications, and sports (Wang et al., 2024).

Near Field Communication (NFC)

NFC is a short-range (typically <10 cm), low-power communication protocol commonly used for contactless payments and identity verification. In smart textiles, NFC can be used for user authentication, garment configuration, or pairing the clothing with other IoT devices. Its passive mode allows energy harvesting from the reading device, which can reduce the need for onboard power.

Zigbee and Thread

Zigbee is a low-power, low-data-rate protocol suitable for mesh networking. It is especially beneficial for distributed wearable systems, such as sensor networks embedded across a garment or multiple garments within a smart uniform. Zigbee allows devices to relay data through one another, extending the overall range and robustness of the network. Thread, based on similar IEEE 802.15.4 standards, also supports IPv6 and is designed for secure and scalable IoT integration (Stoppa & Chiolerio, 2014).

LoRa and LPWAN (Low Power Wide Area Networks)

For smart clothing intended for remote monitoring over long distances—such as in agriculture, disaster response, or outdoor sports—LoRa (Long Range) and similar LPWAN technologies provide extended communication range (up to several kilometers) with minimal energy consumption. However, they are limited by lower data rates and are typically suited for infrequent, small-packet transmissions (e.g., GPS location, heart rate trends).

Table 1. Comparative overview of wireless communication protocols for smart clothing applications

Protocol	Range	Power Consumption	Data Rate	Ideal Application in Smart Clothing
BLE 5.3	100 m	Very Low	2 Mbps	Continuous health monitoring
Zigbee	10–100 m	Low	250 kbps	Sensor mesh networking
Wi-Fi HaLow	1 km	Low	347 Mbps	High-data wearable applications
NFC	<10 cm	Very Low	424 kbps	Payment and access control

Each protocol offers a distinct trade-off between power efficiency, data rate, and range. The architecture of smart clothing must therefore be designed to harmonize these parameters based on use-case priorities, device interoperability, and user experience.

Antenna Design and Placement in Textiles

Antenna integration within smart garments presents a unique set of engineering challenges. Traditional antennas are rigid and incompatible with the flexibility and drapability of fabrics. For wearable applications, antennas must be soft, stretchable, lightweight, and able to maintain reliable performance despite bending, washing, and proximity to the human body—which inherently detunes antenna characteristics due to its high permittivity and conductivity.

Textile-Based Antennas

Conductive threads, fabrics coated with conductive ink or embroidered patterns are all made of conductive threads, fabrics coated with conductive ink or embroidered patterns are used to construct textile antennas. These antennas can be simply incorporated into clothing structures including sleeves, collars, depending on waistbands, etc. Typically, designs make use of well-established and common antenna types (patch, dipole and monopole geometries) which are tailored to take mechanical properties of fabrics into consideration.

Antenna Placement and Performance Considerations

Radiation efficiency has to be optimized, and signal attenuation minimized by placing the antennas in the right places. One can find common placement areas in the shoulder, chest and the upper back, where these provide flatter surfaces and somewhat stable positioning during movement. Its orientation has to account for the anisotropic nature of human movement and ensure that there is no excessive crumpling or torsion.

Detuning, energy absorption and impedance mismatch can occur in proximity to the body. Designers include dielectric substrates on the antenna, isolate the antenna from the skin by using nonconductive layers or use body matched simulation models in the design phase to counter such effects. Additionally, antenna diversity and multi-band designs may be used to guarantee reliable connectivity in spite of the ever changing environment and physiology.

POWER CONSUMPTION AND SIGNAL INTEGRITY CHALLENGES

Power Consumption in Wearable Communication Systems

Power consumption management is one of the most important design constraints for smart clothes systems. Wearables embedded into and sewn onto conformable textile substrates have significantly additional demands on size, weight, comfort, and flexibility, drastically reducing the locations in which a battery can be placed and the size of batteries that can be provided. In this regard, economizing energy usage is of paramount importance, above all in the communication modules in order to ensure that the performance is not limited and that convenience is ensured for users and practically harnessable energy.

Energy Demands of Wireless Protocols

Permanecing out of all wearable's operations, wireless communication is among the operations that consume the most energy. Input power used in sensing or processing is often more than outweighed by input power used in transmitting or receiving data in Wi-Fi type connections and cellular connections. BLE and Zigbee, among other protocols with low power, have a lower power consumption, but need to be duty cycled and have sleep controls and packet streamlining.

Duty Cycling and Data Management Strategies

To address power limitations, modern wearable systems employ strategies such as:

- **Duty cycling**, where transceivers alternate between active and sleep states;
- **Event-triggered transmission**, in which data is sent only when specific thresholds are exceeded;
- **Data compression and local preprocessing**, reducing the volume of information that needs to be wirelessly transmitted.

Energy Harvesting Integration

As Shackel's Rag to Rich energy harvesting technology is an emerging solution that can overcome the power limitation, thus proved that such energy generation technology should be worn. Examples of techniques include thermoelectric generators (heating from the body), piezoelectric materials (transform motion to electricity) and good photovoltaic panels (flexible panels). Energy harvesting could alleviate the requirement for the battery, but harvested energy at most is very low to execute high frequency or long-range wireless communication, particularly when harvested energy half-life is set up at a low threshold.

Signal Integrity in Textile-Based Wireless Systems

All this comes with significant problems with signal integrity in wearable systems since first we have dynamic, deformable and lossy textile substrates and secondly, we have a dynamic, deformable and lossy human body. In its turn, this means that signals sent must pass conductive threads in response to either smart clothes or printed circuit, which is vulnerable to mechanical forces, corroding agent and alteration of conductivity of time, (Mattmann, Clemens, & Tröster, 2008).

Sources of Signal Degradation

- **Physical geometry** distortion of conductive paths and antennas due to mechanical deformation (stretching, folding or twisting conducting paths) would result in impedance mismatch, detuning, and signal attenuation.

- **Environmental exposure** including humidity, sweat, and laundering can degrade material properties and introduce noise or discontinuities in conductive elements.
- **Electromagnetic interference (EMI)** from other devices or body movement can further distort signal propagation and reduce communication reliability (Kartsakli et al., 2014).

Simulation and Modeling

Significant tools of designing X-rays do not model finite element (FEM), full wave electromagnetic solvers (FEM) are employed to approximate the execution of system operation under the influence of the real-world conditions in wearable communication systems (Wang et al., 2024). These models consider the dielectric properties of the human body; movement of the garment and material variants hence give rise to strong systems that sustain the integrity of the signal with time and among the users.

Last but not least, two significant power consumption and signal integrity characteristics are brought in as the constraints of the smart clothes working boundaries. The multidisciplinary and holistic strategy of integrating low power circuit design, novel reactions to the constraint of energy, clever communication actions and the science of materials are unable to realize constraints but may decrease resource consumption and enhance efficiency.<|human|>The combination of low power circuit design, novel innovations to the constraint of energy, clever communication actions as well as the science of materials can solely assist to reduce the amount of resources and increase efficiency. Such approaches as printed above need to evolve in order to realize highly autonomous, robust and simple human-computer interaction wearable, integrated in knowledgeable garments.

INTEGRATION OF IOT COMPONENTS INTO TEXTILES

Smart clothing is an advancement of simple looking clothes, integrating the components of IoT directly into them. To achieve such seamless functionality, these elements (sensors and actuators, microcontrollers, energy systems, etc.) must be embedded onto or into fabrics in some manner that allows it to operate while resulting in little inconvenience, wear or non-invasiveness to the user. In this part, technological methodologies, design aspects and challenges for the integration of these electronic subsystems using wearable fabrics are presented.

Actuators and Haptic Feedback Systems

Actuators in smart textiles serve as output devices, enabling the garment to interact with the wearer through tactile, thermal, or mechanical feedback. These components play a vital role in creating bidirectional communication between the system and the user.

- **Haptic Feedback:** Vibration motors, electroactive polymers, and shape memory alloys can be used to provide vibrotactile cues for navigation, alerts, or rehabilitation prompts. For instance, a haptic vest may guide the visually impaired through directional feedback or inform athletes of incorrect posture.
- **Thermal Actuators:** These include flexible heating elements integrated into garments to provide localized warming, useful in cold environments or for thermotherapy in medical garments.
- **Shape-Morphing Textiles:** Employing pneumatic actuators or stimuli-responsive materials, textiles can dynamically alter their form or stiffness in response to input signals, which has applications in exoskeletal assistance or adaptive fit clothing.

Embedding actuators into textiles requires careful consideration of flexibility, heat dissipation, power consumption, and control logic. The synergy between sensing and actuation is key for developing responsive, intelligent garments.

Microcontrollers and Embedded Systems for Textiles

At the core of any IoT-enabled garment is the microcontroller unit (MCU) or embedded system that processes sensor data, executes logic, and manages communication protocols.

- **Textile-Compatible Microcontrollers:** Miniaturized and low-power MCUs (e.g., ARM Cortex-M, ESP32, Nordic nRF series) are often integrated using modular PCBs, flexible printed circuits, or encapsulated packages sewn or bonded to fabric (Zhang et al., 2018). These devices support analog-to-digital conversion, digital signal processing, and wireless communication.
- **System-on-Chip (SoC) Solutions:** Compact SoCs integrate processing, memory, and communication (e.g., BLE, Wi-Fi) on a single chip, optimizing size and energy efficiency (Piguet, 2011). These are particularly useful in wearables where form factor is tightly constrained.
- **Edge Computing in Wearables:** Increasingly, smart garments are being designed with edge computing capabilities to perform real-time analytics on-

device, reducing data transmission needs and improving responsiveness. This is particularly important in latency-sensitive applications such as gait correction or cardiac anomaly detection.

- **Flexible and Washable Electronics:** Research is advancing toward developing microcontrollers using flexible substrates and protective encapsulation to ensure washability and long-term durability in textile environments.

Types of Sensors Used in Smart Textiles

Smart clothing relies heavily on embedded sensors to monitor physiological and environmental parameters. These sensors convert various forms of stimuli into electrical signals, forming the primary interface between the user and the digital system.

- **Temperature Sensors:** Integrated thermistors, resistance temperature detectors (RTDs), and thermocouples are commonly used to monitor body or ambient temperature. These are essential for applications in healthcare monitoring, thermal regulation, and performance optimization.
- **Motion and Inertial Sensors:** Motion, orientation and activity levels are detected with accelerometers, gyroscopes and magnetometers. In sportswear for performance analysis, in rehabilitation garments to track recovery progress, in fall detection systems for the elderly etc, these are vital.
- **Biopotential Sensors (ECG, EMG, EEG):** Electrocardiography (ECG), electromyography (EMG), and electroencephalography (EEG) sensors are crucial in health-monitoring garments (Gubbi et al., 2013). These sensors are typically implemented using conductive textile electrodes that are skin-contacting and capable of acquiring bioelectrical signals non-invasively.
- **Pressure and Strain Sensors:** Capacitive, resistive, or piezoresistive sensors detect physical deformations in the fabric, which can be correlated to joint angles, postural shifts, or tactile interaction. Such sensors are used in posture-correcting garments, exosuits, and interactive fashion.
- **Chemical and Biosensors:** These sensors detect biomarkers such as sweat pH, lactate, glucose, and electrolyte levels. They are often printed or embedded using nanomaterials and are increasingly being used for personalized medicine and real-time diagnostics.

The selection and integration of sensors depend on application specificity, user mobility, textile compatibility, and signal processing requirements. Increasingly, hybrid sensor platforms combining multiple sensing modalities are being developed for comprehensive body monitoring.

Power Supply and Energy Harvesting Mechanisms

Powering smart textiles remains one of the most persistent challenges. Traditional batteries introduce bulk and rigidity, motivating the exploration of innovative energy storage and harvesting solutions.

- **Conventional Power Sources:** Small lithium-polymer batteries are still widely used, often embedded in detachable modules for charging convenience.
- **Solar Energy Harvesting:** Photovoltaic films and flexible solar panels integrated into the garment's surface or embedded into fibers can capture ambient light. While dependent on lighting conditions, they can supplement or replace batteries in outdoor applications.
- **Kinetic and Triboelectric Harvesting:** Movement-based energy harvesting, using piezoelectric fibers or triboelectric nanogenerators, converts mechanical energy from motion, walking, or breathing into usable power. These systems are promising for continuous, passive energy generation in active users.
- **Thermoelectric Harvesting:** Exploiting the temperature gradient between the body and the environment, thermoelectric generators offer another means of energy capture, particularly in colder climates. However, their power output is relatively modest.

In conclusion, the successful integration of IoT components into textiles involves the seamless fusion of advanced electronics with the functional, ergonomic, and aesthetic qualities of clothing. Achieving this integration requires multidisciplinary innovation in materials science, embedded systems design, low-power electronics, and textile engineering. As technologies mature, smart garments are expected to become not only more functional and autonomous but also more accessible and scalable for diverse real-world applications.

EDGE COMPUTING VS. CLOUD COMPUTING FOR WEARABLES

The rapid growth of smart clothing and wearable IoT systems has brought forward significant advancements in data generation, necessitating robust and efficient processing architectures. In this context, two primary computing paradigms have emerged—edge computing **and** cloud computing—each offering distinct advantages and challenges for smart textile applications.

Cloud Computing in Wearables

Cloud computing refers to the centralized processing and storage of data on remote servers accessed via the internet. In the context of wearables, sensor data collected from smart garments is transmitted to cloud platforms, where it can be stored, analyzed, and interpreted using high-performance computing resources.

Benefits

- Scalability Cloud services are capable of serving extensive data to many individuals so that it can be stored and do complicated analysis over the long-term.
- Machine Learning Investment: More sophisticated AI-based informatics can be done with the power of cloud servers: disease prediction, or performance models.
- Remote Access and Sharing: Data can be accessed due to the whole world and allow telemedicine, collaborative diagnostics, and remote monitoring.

Limitations

- Latency: Motion Assistance It is possible that transmission lag will be an impediment to real-time applications like fall detection, motion assistance, or haptic feedback.
- Privacy and Security: Sending sensitive health or biometric information to the internet leads to great inferences in terms of data encryption and ownership and regulatory compliance.
- Network Dependence: When there is a low connectedness of the metropolitan area, the system will be unreliable.

Edge Computing in Wearables

These properties are not true of edge computing, which means the wearable device itself or a nearby hub (e.g., smartphone or an embedded controller) does the data processing, and then highly minimizes the need to be continuously connected to the internet.

Advantages

- **Low Latency** Processing near the source allows real-time response to data, which is important in an interactive system and essential health interventions.

- **Less Bandwidth Consumption**: Only important or filtering data should be sent into a cloud, and this will enhance efficiency and will save energy.
- **Improved Privacy:** Data is processed locally and can remain on the device, minimizing the risk of unauthorized access or interception.

Challenges

- **Limited Computational Resources:** Wearable devices have constraints on memory, processing power, and energy, which can limit the complexity of algorithms.
- **System Complexity:** Developing robust, real-time edge applications requires sophisticated software optimization and power-aware design.

Hybrid Architectures

If there are many smart clothing systems, they favor working with hybrid constructions, which join both edge and cloud computing. Data analysis is performed in the cloud in order to provide a high level of understanding of trends and real time data is processed at the edge in order that there is minimal delay to the response. Departmental scale system is a distributed architecture that achieves the highest level of performance, responsiveness and system intelligence at the lowest level of operation scale.

MULTI-DEVICE SYNCHRONIZATION AND INTEROPERABILITY

In smart clothing and wearable devices ecosystem multi device synchronization and interoperability play an important part in creating flawless user experience, coordinated data and inter platform communication. Today's modern smart environments are multi-sensor, multi-actuator, mobile devices and cloud services, and it becomes extremely important when users will be having interaction with different devices at the same time, it should all work in harmony.

Multi-Device Synchronization

Typical smart clothing systems consist of multiple nodes as a sensor system—distributed across different body regions—and can communicate with peripheral devices (e.g.: smartphone, smartwatch or medical gateway). Data collected from these devices is temporally aligned and contextually consistent through synchronization.

Key Aspects

- **Time Synchronization:** Precise timestamping using protocols like IEEE 1588 Precision Time Protocol (PTP) or Network Time Protocol (NTP) is essential to align data streams from different sensors and identify events accurately (Kamal, Ullah, & Khan, 2016).
- **Coordinated Sampling:** Multi-modal sensors (e.g., ECG and accelerometer) must sample data in a synchronized manner to provide meaningful cross-analysis.
- **Low-Latency Communication:** For haptic systems, auditory feedback, or exoskeleton control, millisecond-level coordination is required to ensure responsiveness and safety.

Figure 2. Disciplines involved in the IoT smart clothing supply chain management

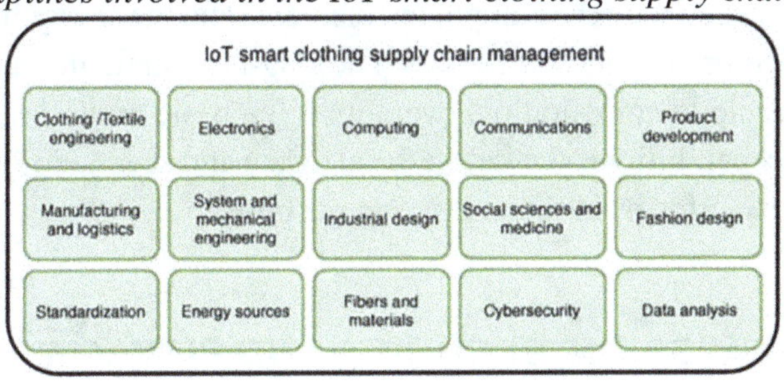

Interoperability

Interoperability refers to the ability of devices and systems from different manufacturers or platforms to communicate, understand, and utilize each other's data and functionalities.

Dimensions of Interoperability

- **Protocol-Level Interoperability:** Ensures devices can exchange data via standard wireless communication protocols such as Bluetooth Low Energy (BLE), Wi-Fi, Zigbee, or LoRa (López-Nava et al., 2017).
- **Data-Level Interoperability:** Involves standardizing data formats, metadata, and communication APIs to allow consistent interpretation across devic-

es and platforms. Frameworks like HL7 and FHIR are increasingly used for medical wearables.

- **Platform-Level Interoperability:** Enables integration of wearables into broader digital ecosystems such as Android/iOS, smart home networks, healthcare IT systems, or industrial IoT platforms.

Challenges

- **Proprietary Standards:** Many commercial devices use closed protocols or data formats, hindering integration with third-party systems.
- **Security and Authentication:** Ensuring that only authorized devices can participate in the network is crucial, particularly in sensitive applications such as remote patient monitoring or biometric security garments.

To sum up, the competing factors in making smart clothes effective are the effort in being sophisticated with the component elements and the strength of system level computing and communication structures. The long-term strategic ready combination edge, as well as cloud computation, overcomes the basic multi devices synchronization and interoperability is the key to the efficacy and several folds augmentation of wearable IoT platforms in the real-world purpose.

DATA TRANSMISSION, SECURITY, AND PRIVACY

The communication of even the data of the development of those technologies should have a safe and dependable channel to enable imminent mobile platforms and cloud services. As these systems become more prevalent, vital mission success is portfolio upon the integrity, confidentiality and availability of the data collected - sensitive physiological, behavioral and contextual information - on these systems. Part, one deals with the technical foundation and possible solutions in real time data transmission and cybersecurity in smart textile ecosystems.

Real-Time Data Transmission Protocols

Real time data transmission is necessary to achieve responsiveness, accuracy and safety in wearable systems (especially health monitoring, sports analytics or assistive technologies) applications. Latency, bandwidth, power consumption and system compatibility are constrained by the choice of a communication protocol.

1. Bluetooth and Bluetooth Low Energy (BLE):

The most frequently used BLE protocol for smart clothing is used because of its low power consumption, high enough data rate (up to 2Mbps), as well as the fact that it works with smartphones and other consumer devices. Though heart rate monitors, motion trackers and temperature sensors are examples of physiological sensors, BLE is a good match for periodic or event driven data transmission from such sensors.

- **Advantages:** Low power, widely supported, robust against interference.
- **Limitations:** Limited range (~10–30 meters), not ideal for continuous high-bandwidth applications like video or high-frequency EMG data.

2. Wi-Fi and Wi-Fi Direct:

Wi-Fi enables high-throughput data exchange (up to several hundred Mbps), making it suitable for data-intensive applications such as full-body motion capture or streaming sensor data to edge servers. Wi-Fi Direct allows peer-to-peer connectivity without needing an access point.

- **Advantages:** High speed supports direct cloud connectivity.
- **Limitations:** Higher energy consumption, increased complexity in integration with textile-based electronics.

3. Zigbee and IEEE 802.15.4:

Zigbee offers a low-power mesh networking capability, ideal for coordinating multiple sensor nodes embedded in clothing or across multiple wearables.

- **Advantages:** Mesh networking, good for distributed sensor arrays.
- **Limitations:** Lower data rates, often require a dedicated coordinator.

4. Near Field Communication (NFC):

NFC is suitable for ultra-short-range communication (up to 10 cm), typically used for identification, access control, or passive data logging.

- **Advantages:** No pairing needed, very low power.
- **Limitations:** Not suitable for real-time streaming or continuous monitoring.

5. LoRa (Long Range):

LoRa is used in outdoor and remote smart garment applications where connectivity over long distances is required, such as environmental monitoring or safety wearables in industrial contexts.

- **Advantages:** Extremely long range, minimal power consumption.
- **Limitations:** Very low data rate (~0.3–50 kbps), high latency.

Key Considerations for Protocol Selection

- **Latency requirements** (e.g., biofeedback systems demand <100 ms delay)
- **Power efficiency** (especially for battery- or harvest-powered garments)
- **Range and topology** (e.g., intra-body networks vs. infrastructure connectivity)
- **Interference and coexistence** with other wireless systems

In many practical applications, hybrid communication strategies are employed to balance real-time needs with energy efficiency and scalability.

Data Encryption and Cybersecurity in Smart Clothing

As smart clothing systems handle increasingly sensitive data—ranging from biometric identifiers and medical indicators to movement patterns and location—**cybersecurity** becomes a critical concern. Unauthorized access, data manipulation, and privacy breaches not only threaten individual safety and autonomy but may also violate regulatory standards such as HIPAA, GDPR, or ISO/IEC 27001 (Jiang & Yin, 2015).

Data Encryption Techniques

Encryption ensures that transmitted and stored data remains confidential and tamper-resistant.

- **Symmetric Encryptions** that we want here, there exist quite a few like AES (Advanced Encryption Standard) is used widely not only for its speed but also because it is secure. Both are suitable encryption and decryption of the data with the same key for use in low latency environments such as wearable sensors.

- **Asymmetric encryption:** Public key schemes, such as RSA or Elliptic Curve Cryptography (ECC), offer better security when it comes to key exchange, as well as authentication, but are more computationally intensive.

Emerging privacy-preserving solutions include federated learning and blockchain-based systems. Alternatively, federated learning methods enable wearable devices to jointly train AI models with limited user data, without revealing the identities of raw user data, in which case, privacy is taken into consideration (Zhang et al., 2015). The use of blockchain-based systems is also discussed to store the health data in a secure manner through permanently stored records as a method and have an open management of consents (Patel et al., 2025). The methods are effective in bolstering against attacks that amass sensitive biometric data in applications.

Authentication and Access Control

Robust **user and device authentication** mechanisms are essential to prevent unauthorized access to wearable systems and their data.

- **Multi-factor authentication** (biometric, token-based, password) enhances security, especially in health or defense applications.
- **Device pairing protocols**, such as Secure Simple Pairing (SSP) in Bluetooth, minimize the risk of man-in-the-middle attacks.

Secure Firmware and Software Updates

Because it is common for Smart Garments to need over the air (OTA) updates that improve functionality or fix vulnerability, it is a natural progression for a connected Fiber of data. Cryptographic signing and verification of these updates before installation prevents malicious firmware attacks.

Intrusion Detection and Anomaly Monitoring

Real–time monitoring of unusual patterns in flow of data, power use or device behavior is present in advanced systems to detect likely cyber threats or hardware tampering.

Data Anonymization and Privacy Preservation

In the case of data sharing (like health research or team performance analytics), we use the anonymization (and pseudonymization) of data to protect users' identities. Model training is possible on-device without exposing raw data with the use of privacy preserving machine learning techniques (e.g., federated learning).

Regulatory and Ethical Compliance

Smart clothing developers must ensure compliance with relevant laws and standards:

- **HIPAA** (Health Insurance Portability and Accountability Act) in the U.S. for health-related data.
- **GDPR** (General Data Protection Regulation) in Europe, which emphasizes user consent and data minimization.
- **ISO/IEC 27001** for information security management systems (Kumar & Singh, 2018).

Future Directions in Secure Wearable Networks

- **Blockchain integration** for secure, decentralized identity and data traceability.
- **Post-quantum encryption** to future-proof wearable systems against next-generation cryptographic threats.
- **Trusted execution environments (TEEs)** on wearable processors to isolate and protect critical operations.

GDPR AND ETHICAL CONSIDERATIONS IN BIOMETRIC DATA COLLECTION IN SMART CLOTHING

Lastly, the increasing variety of smart apparel and wearable devices that can potentially record unimaginable levels of biometric data (e.g. heart rate, ECG, EMG, respiratory frequency, skin temperature and gait) that may be transferred to a hospital or to a technological company contributes to these new roles regarding ethical and legal perspective. The highest risk category is personal identifiable information, i.e., personal identifiable data (data about which the user can be discriminated against) data that is high-risk of data breaches. In case AI developers, research and manufacturers want to exploit those technologies, all they must do is adjust their endeavors to a larger data protection framework such as General Data Protection Regulation (GDPR) and approach the greater ethical questions of biometric data capture and use to guarantee responsible invention.

Understanding Biometric Data in the Context of Smart Clothing

Biometric sensing capabilities embedded into textiles in smart clothing support the continuous health monitoring, fitness assessment and human computer interaction. Examples include:

- **Physiological Data:** Heart rate variability, blood oxygen levels, skin conductivity
- **Behavioral Biometrics:** Gait analysis, movement patterns, sleep cycles
- **Neurological Signals:** EEG-based interactions in neuroadaptive garments

Such data, due to its highly sensitive and personal nature, is subject to strict regulations and ethical scrutiny.

GDPR: A Regulatory Framework for Biometric Data

The **General Data Protection Regulation (GDPR)**, enforced across the European Union since May 2018, sets a global benchmark for data privacy and protection. Under GDPR, biometric data used for the purpose of uniquely identifying a natural person is categorized as **"special category data"** and is subject to enhanced protection under **Article 9** of the regulation.

Key GDPR Principles Relevant to Smart Clothing

- **Lawfulness, Fairness, and Transparency:**

Data collection must be transparent and based on clear legal grounds. Users must be informed in accessible language about what data is collected, why, and how it will be used.

- **Explicit and Informed Consent:**

The processing of biometric data typically requires **explicit consent**. Consent must be freely given, specifically informed, unambiguous, and withdrawable at any time.

- **Purpose Limitation and Data Minimization:**

Data must only be collected for specified, legitimate purposes and must be limited to what is necessary for those purposes. Collecting data "just in case" is not permissible.

- **Storage Limitation:**

Personal data should not be kept longer than necessary. Retention policies must be clearly defined and enforced.

- **Data Security and Integrity:**

Organizations must implement technical and organizational measures to protect biometric data from unauthorized access, alteration, or loss.

- **Accountability and Data Protection by Design:**

Developers and data controllers are required to proactively integrate data protection principles throughout the system design process, including during hardware, firmware, and software development stages.

- **Data Subject Rights:**

These include the rights to access, rectification, erasure ("right to be forgotten"), data portability, and objection to data processing. Mechanisms must be in place to facilitate user control over their data.

ETHICAL CONSIDERATIONS BEYOND LEGAL COMPLIANCE

While GDPR provides a legal framework, **ethical considerations** require an even deeper commitment to human dignity, autonomy, and societal well-being.

Autonomy and Informed Consent

Users must not only consent to data collection but truly **understand** the implications. Informed consent involves transparency about:

- What biometric data is being collected
- Who has access to it
- How it is processed and stored
- Whether it will be shared with third parties (e.g., insurers, employers, researchers)

In vulnerable populations—such as patients, children, or the elderly—extra safeguards must ensure consent is not coerced or manipulated.

Privacy and Psychological Impact

Continuous real-time biometric monitoring may also be perceived by users as a type of surveillance, which can be detrimental to the users by making them self-conscious and therefore promoting stress, anxiety or change in behaviors. But designers also have to consider psychological impact of 'being continually measured' and allow opt in/opt out controls/data pause tunnels.

Bias and Discrimination

Biases may be in machine learning models built from biometric data which may lead to ill-treatment or misinterpretation based on race, gender, age or disability. As we know, the description of the design of an ethical system has to:

- Ensure dataset diversity
- Continuously audit for algorithmic bias
- Provide explainable AI outputs

Data Ownership and Commercial Use

Who owns biometric data? Can it be sold or used for secondary purposes such as marketing or performance scoring? Ethical practice dictates that:

- Users retain their data
- Commercial use must be opt-in and transparent
- Monetization of personal data should involve fair value exchange

Right to Disconnect and Digital Well-being

Smart clothing should support—not overwhelm—users. Designers must include **user agency**, such as modes for privacy, rest, or disconnecting, to promote digital well-being and autonomy.

Conclusion

"The proliferation of smart clothing in our everyday lives has to be embedded in respect for human rights and agency, and data sovereignty, which are all moral ideals and not just legal ones." However, in my mind, being accountable does not mean much less unless we consider the context: it really is just a matter of being ethical or having an ethical foresight, from the point of view of either everyone or, a set population.

MATERIALS AND FABRICATION TECHNIQUES IN SMART CLOTHING

Selection and fabrication of materials is the choice of materials that meet the expectations of mechanical and comfort requirements of regular clothing and equip them with electronic functionality to sense, communicate, actuate and process data. To facilitate integration of these functions, advanced materials and novel fabrication methods that enable mechanical flexibility, durability, washability and biocompatibility, all are needed without sacrificing electrical performance. In the following section we explore the key intelligent textile system materials and fabrication strategies that contribute to the development of intelligent textile systems.

Conductive Fabrics and Threads

The building blocks for electronically active textiles (eATs) include conductive fabrics and threads. Consequently, these materials establish the electrical pathways on which the signals are propagated, the components are powered and the interfaces with sensors or actuators embedded in the garment.

1. Types of Conductive Materials:

- Other metallic fibers are made from silver, copper, stainless steel and plated polymers. Thus, for instance, silver plated nylon threads let combining of the flexibility of textiles with the good conductivity of silver.
- **Materials like polyaniline (PANI),** polypyrrole (PPy) and poly(3,4-ethylenedioxythiophene) (PEDOT:PSS) (Brown, 2019), have conductivity because of their molecular structure. This is attractive because they are lightweight, chemically tunable and compatible with textile processes.
- **Carbon nanotubes (CNTs),** graphene and carbon black are being investigated as promising carbon-based materials for improving the electrical conductivity of fabrics without the sacrifice of breathability and softness.

2. Fabrication Methods for Conductive Textiles:

- Conductive yarns are woven along with the traditional fibers into an integrated conductive pathway. It keeps fabric flexibility and textile aesthetics intact.
- Conductive inks or pastes are screen printed, ink jet printed or sprayed onto fabric surfaces. They can produce patterns and large area manufacturing.

3. Performance Considerations:

- **Durable and washable:** Conductive materials must not lose a significant amount of their conductivity under mechanical stress, abrasion and repetitive laundering cycles.
- Such conductive elements should not compromise the comfort or biocompatibility of the garment (e.g. maintaining softness, stretchability and breathability). Furthermore, materials need to be skin safe and non-irritating for long term wear.

Flexible Printed Circuits and Integration Methods

The dynamic deformable environment of clothing is ill suited to traditional rigid electronic circuits. Flexible printed circuits (FPCs) and advanced integration methods are used to integrate electronics into textile substrates in a seamless manner, while maintaining mechanical compliance.

1. Flexible Substrate Materials:

- **Polyimide Films (e.g., Kapton):** These substrates offer excellent thermal stability and flexibility, commonly used for FPCs in wearable electronics.
- **Thermoplastic Polyurethanes (TPU):** TPU films are stretchable and skin-compatible, making them ideal for laminated circuits or direct garment bonding.
- **Textile-Based Substrates:** Direct printing or lamination on woven or non-woven fabrics enables greater integration with the garment's structure.

2. Circuit Integration Techniques:

- **Thermo-Compression Bonding:** A method used to attach electronic chips or interconnects onto flexible substrates through heat and pressure, offering secure and low-profile integration.

- **Lamination and Embedding:** Circuits are encapsulated between textile layers using adhesives or heat-press techniques, protecting them from mechanical and environmental stress.
- **Sewing and Embroidering:** Conductive threads can be stitched into functional circuit patterns, creating embroidered antennas, sensors, or traces with aesthetic and technical utility.

3. Component Mounting and Interconnects:

At present, miniature surface mount devices (SMDs) and chip scale packages (CSPs) are being developed in which there are options to be mounted onto flexible or fabric friendly substrates (Johnson & Liu, 2017).

Conclusion

In order to achieve electronics in textures, one must have knowledge of materials and fabrication capabilities that can provide electrical functionality at the same time, put into consideration the aesthetic, tactile and mechanical experience of wearables. Additionally, the painstaking choice of conductive threads, passenger coating technique and construction of the flexible circuit further are also described; these factors make the device operate effectively as well since it is much more convenient and easier to wear by the user and sturdy enough to serve longer. The research industry will advance in its quest to defuse the gap between the textile science and electronic engineering systems and in the long run, clever clothing solutions come together, become distributed, and consumed by consumers.

DURABILITY, WASHABILITY, AND USER COMFORT IN SMART CLOTHING

The intelligent clothing will have its useful life results which, fundamentally, depend on the interfaces of the electromechanical and the material technologies on which the technologies proposing developing electronic competence are based, and the basic needs in clothing of consumers in which they desire (wearability, launderability, comfort, etc.) which the technology fulfills. Over that, similar to standardized garments, it must be compatible with ordinary clothing and wash needs, prolonged shopping in intricate environments like as well as comfort balanced to the physique without hindering activity. The Resolution to these is through Resolution by Disciplinary Approach in which the material science, textile engineering and user centered design factors have been left unattended and hence succumbed to the factors here.

Durability

The criterion which determines the life of a smart clothing once used repeatedly is durability which is the combination of other stresses like stretching, bending or being abrasion stress (stretching, bending, abrasion, impact, etc.). A garment is exposed to complicated and repetitive forces during clothing usage (e.g. flexibility of joints, friction against external surfaces) so the electronic materials and connectors should have excellent fatigue characteristics.

Durability is affected by such key factors as:

- Choice of Materials: Conductive thread, coatings, and flexible circuits must be made mechanically resistant in nature. Silver-plated materials such as nylon, flexible conductive polymers and elastomeric substrates have been proven to reduce durability.
- Encapsulation and Protection: Custom electronic parts are to be encapsulated not only by lubricating substances to prevent mechanical damage, but also by soft polymers to ensure proximity to the environment (use of moisture, sweat, UV radiation, etc.) (Smith, Li, & Tse, 2023).
- Seam and Integration Quality The processes of embedding must be robust so that the electronics will not loosen and disconnect with time, whether by weaving, sewing, lamination, or printing.

Washability

Like simply by virtue of the fact that smart clothing will have to pass through the same stages as regular clothing washing, coming into contact with various types of chemicals, changes of temperature and mechanical forces etc., the fact through which only clothing becomes washable will come to play as a paramount challenge.

To enhance the washability, considerations are to enhance.

- Waterproofing and Moisture Resistance: Waterproofing Electronic modules and conductive traces have to be waterproofed or encapsulated or by using a sealed housing that does not allow the entry of water or detergents.
- Compatibility: The material should not be subject to breakdown or corrosion by the presence of detergents, bleaches or fabric softeners that are often used when washing clothes.
- Attachable Electronics: To support laundry, modular designs are removable electronics to enable washing, which helps create a long-lasting media device and keep clothing operational.

- Standardized Testing Cases: Conducting strict cleaning tests which mimic the actual laundering processes can be used to confirm the strength of smart clothing prior to introduction.

User Comfort

User comfort is the determiner of acceptance and constant wearing smart clothes. But conventional electronics typically are only limited to desktops and other fixed locations and are not physically in contact with the skin as smart textiles which require the ability to maintain the feel and haptic qualities of normal clothing.

Key elements affecting comfort include:

- **Breathability and Moisture Management:** Integration of electronics should not impede the fabric's ability to wick moisture or allow air permeability, as this can lead to overheating or skin irritation.
- **Flexibility and Softness:** Electronic components and conductive pathways must be flexible and lightweight, conforming to the body's contours without causing stiffness or discomfort.
- **Thermal Regulation:** The presence of electronics should not disrupt the garment's natural heat exchange properties, thus preserving thermal comfort across different environmental conditions.

Figure 3. Categories of wearable technology.

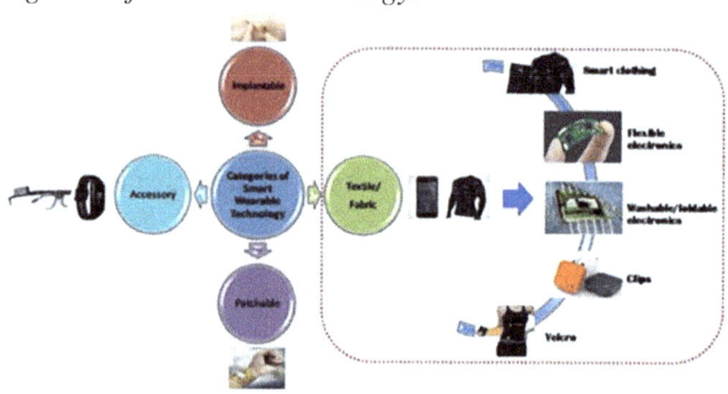

APPLICATIONS IN HEALTHCARE AND SPORTS

Wireless and Internet of things (IoT) convergence the smart clothing of smart wearable technology is one of the ways to transform the healthcare and sports lifestyle. Due to the addition of sensors and communications modules within the fabrics that constitute smart garments, smart garments have practical advantages compared to existing wearable devices in regard to the capacity to monitor both physical and biomechanical aspects in a constant and non-obtrusive manner. Information gathering, live transmissions and analysis that may be incorporated together seamlessly enabling the suggestions regarding distant wellbeing, individual recovery and athletic execution previously non-imaginable. In this section, the researcher examines the primary areas of application where innovation can be applicable and achieve positive outcomes due to the use of smart clothing.

Remote Health Monitoring and Diagnostics

Remote health monitoring is one of the most potential applications of smart clothing which could continuously, or real time acquire physiological data to detect medical conditions at early stage, manage chronic disease and provide patients care out of standard clinical settings.

- **Chronic Disease Management:** For conditions such as diabetes, hypertension, or chronic obstructive pulmonary disease (COPD), smart clothing offers unobtrusive monitoring that enhances adherence to treatment plans and timely intervention, potentially reducing hospitalizations (Yang, 2018).
- **Early Warning Systems:** Integrated algorithms analyzing biometric data streams can detect anomalies, such as arrhythmias or respiratory distress, and trigger alerts for patients or healthcare providers, facilitating proactive care.
- **Telemedicine Integration:** Data collected by smart clothing can be securely transmitted to healthcare platforms, enabling remote consultations, longitudinal health tracking, and personalized treatment adjustments.

Rehabilitation and Assistive Wearables

Rehabilitation and assistive applications are also enabled by smart clothing technologies, wherein feedback and monitoring of progress are enabled and mobility improved (for individuals recovering from injury or those with disabilities).

- **Postoperative and Injury Rehabilitation:** Sensor-embedded garments monitor joint angles, muscle activity (via electromyography), and gait patterns to quantify recovery progress and guide physiotherapy interventions with real-time feedback.

- **Neuromuscular Assistance:** Wearables with integrated actuators and haptic feedback systems can assist movement, improve posture, or provide sensory cues to patients with neurological impairments such as stroke, Parkinson's disease, or spinal cord injuries.
- **Fall Detection and Prevention:** Smart clothing can continuously analyze balance and motion patterns to identify fall risk in elderly or impaired individuals, triggering alarms or activating preventive support systems.
- **Assistive Technologies:** Integration with prosthetics or orthotic devices enhances control and adaptation, allowing more natural and efficient movement.

Athletic Performance Tracking and Injury Prevention

These enable the athlete, the coach and the sports scientist to gain detailed insights about physiological status and biomechanical performance in sports to optimize training, reduce injury risk and increase competitive performance.

- **Real-Time Biometric Feedback:** Monitoring metrics such as heart rate variability, respiration, hydration status, and muscle fatigue enables personalized training adjustments and recovery management.
- **Biomechanical Analysis:** Embedded inertial measurement units (IMUs), accelerometers, and gyroscopes capture movement kinematics, joint angles, and impact forces, facilitating technique optimization and early detection of movement anomalies that could lead to injury (Ahmed & Khan, 2020).
- **Load and Stress Monitoring:** Continuous assessment of mechanical load on muscles and joints helps in balancing training intensity to avoid overuse injuries and optimize periodization.
- **Environmental Adaptation:** Sensors that monitor external conditions such as temperature, humidity, and UV exposure enable athletes to adapt hydration and protective strategies for optimal performance.

WEARABLE COMMUNICATION SYSTEMS FOR SOLDIERS

The original military sphere of application of wearable technologies/displays has been to improve the capabilities of commandos in combat and serve to provide situational awareness, which increases their efficiency by adding security measures to the situation in the military context. One of the most significant actions in the application of seamless, secure and real-time wires into the uniform is seen in the infestation of wearable communication capabilities to the uniform and magnificent multi-agent systems, which are extremely dynamic, and hostile beyond doubt, namely soldiers. They leverage wireless communication schemes, embedded sensors and engage the IoT to construct the building of a connected ecosystem that plays a crucial role in the improved safety of the soldiers, coordination aptitude, as well as mission performance.

Key Functionalities and Components

Wearable communication systems for soldiers encompass a suite of interconnected devices embedded within uniforms, helmets, and personal equipment, designed to support a wide array of operational needs:

- **Integrated Radio and Wireless Networks:**

Advanced radios embedded in clothing or gear provide encrypted voice and data communication via military-grade standards such as Tactical MANET (Mobile Ad Hoc Network), LTE, or emerging 5G technologies (Taylor, 2016). These radios facilitate mesh networking, ensuring resilient and decentralized communication even in environments where infrastructure is compromised.

- **Sensor Integration for Situational Awareness:**

Smart textiles can incorporate GPS modules, inertial sensors, environmental monitors, and biometric sensors that continuously track a soldier's location, physiological status, and external threats. This data is transmitted in real-time to command centers and fellow soldiers, enabling rapid decision-making.

- **Heads-Up Displays (HUD) and Wearable Interfaces:**

Wearable communication systems often integrate with augmented reality (AR) helmets or glasses that project critical information directly into the soldier's field of view, reducing cognitive load and allowing hands-free access to navigation, mission updates, and threat alerts (Oliveira, 2019).

- **Mesh Networking and Multi-Device Synchronization:**

The wearable network supports dynamic connectivity among multiple soldiers and command units, enabling coordinated group maneuvers, shared situational awareness, and rapid reconfiguration of communication links.

Design Considerations and Challenges

Developing wearable communication systems suitable for military applications requires addressing stringent design constraints:

- **Ruggedness and Durability:**

Military environments expose wearable devices to extreme temperatures, moisture, dust, impact, and electromagnetic interference. Materials and fabrication techniques must ensure system integrity under harsh conditions and extended deployment durations.

- **Low Power Consumption and Energy Harvesting:**

Prolonged missions necessitate energy-efficient communication protocols and integration of power management solutions such as energy harvesting (solar, kinetic) or advanced battery technologies to minimize logistic burdens.

- **Secure and Robust Data Transmission:**

Communications must employ end-to-end encryption and anti-jamming capabilities to protect against interception, cyber-attacks, and electronic warfare.

Operational Impact and Future Directions

The deployment of wearable communication systems revolutionizes battlefield operations by enhancing:

- **Real-Time Coordination:**

Enabling squads and command units to share live intelligence, tactical maps, and biometric data foster faster, informed decision-making.

- **Soldier Health Monitoring:**

Continuous assessment of vital signs allows medics to prioritize care and implement preventative measures to reduce combat fatigue or injury.

- **Adaptive Mission Support:**

Context-aware communication networks dynamically adjust based on mission phase, environmental conditions, or threat levels.

SMART UNIFORMS AND PROTECTIVE GEAR: INDUSTRIAL SAFETY AND WORKER MONITORING

The implementation of tailor-made protection technologies in laboratory coats, protective garments, and other industry-uniforms is a major leap forward in the domain of occupational safety and workforce management systems. To promote safety, enhance operation efficiency and ensure employee protection, industries are inclined to smart uniforms; that is, the sensors, wireless communication modules and the functions of IoT can greatly protect the security of people's daily lives and improve living quality. Integrating electronics and textiles into the fabric of conventional protective clothing designates intelligent systems that will systematically protect workers in dangerous environments before accidents happen.

Enhancing Industrial Safety through Smart Uniforms

These problems are solved by smart uniforms, integrated into the protective measures with multifunctional sensors and communication technologies to obtain constant monitoring of environmental and physiological parameters.

- **Environmental Monitoring:**

Embedded gas sensors detect toxic gases, volatile organic compounds (VOCs), or combustible substances, providing early warnings to workers and supervisors to prevent exposure or accidents. Thermal sensors monitor ambient temperature extremes, alerting personnel to heat stress or cold exposure risks.

- **Impact and Hazard Detection:**

Accelerometers and gyroscopes integrated into helmets, vests, or gloves can identify falls, impacts, or abnormal movements indicative of accidents or hazardous events. This capability allows rapid initiation of emergency protocols, including automatic alerts to onsite medical teams.

- **Biometric and Fatigue Monitoring:**

Smart textiles equipped with heart rate monitors, skin temperature sensors, and electrodermal activity (EDA) sensors track worker health status and fatigue levels. Continuous monitoring helps identify signs of overexertion or stress, enabling timely interventions to prevent accidents caused by human error or diminished alertness.

Worker Monitoring and Operational Efficiency

Beyond safety, smart uniforms facilitate comprehensive worker monitoring systems that optimize productivity, compliance, and workforce management through seamless data collection and communication.

- **Location Tracking and Geofencing:**

Integration of GPS and proximity sensors allows real-time tracking of personnel movements within industrial zones. Geofencing technologies alert workers when entering restricted or hazardous areas, improving adherence to safety protocols.

- **Communication and Coordination:**

Wearable communication devices embedded in uniforms enable hands-free voice and data exchange, ensuring continuous contact between workers and supervisors, especially in noisy or remote environments.

- **Data Analytics and Predictive Maintenance:**

Collected physiological and environmental data can be analyzed to identify patterns associated with increased risk or operational inefficiencies. Predictive algorithms support proactive safety management and maintenance scheduling.

FUTURE PERSPECTIVES

Smart alarm and protective gears are the way to the next level of industrial safety which will redefine the safety paradigms with intelligent, adaptive and context aware protective solutions.

As the whole world is accepting and adopting digital transformation in their business, smart clothing will play a major role in making safer, smart and more responsive working environments to lessen the number of occupational injuries, ensure compliance and create a culture of proactive safety management.

CHALLENGES AND LIMITATIONS

An integration in smart clothes via wireless communication and IoT technologies is a promising feature to revolutionize healthcare, sport, industrial and military applications, but its application and development has been extremely constrained by numerous critical challenges and limitations.

Scalability and Manufacturability

Manufacturing processes for the integration of delicate electronic components (e.g., sensors, antennas, microcontrollers and conductive pathways) into flexible, wearable fabrics involve tradeoffs among precision, reliability and cost-efficiency.

- **Complex Integration Processes:**

Current fabrication techniques—including embroidery with conductive threads, flexible printed circuit lamination, and coating with conductive inks—often involve labor-intensive or batch-scale processes that are difficult to standardize at industrial volumes.

- **Quality Control and Consistency:**

Ensuring uniform electrical performance, mechanical durability, and washability across large production runs is challenging. Variability in textile substrates and integration methods can lead to inconsistencies that compromise device functionality and user experience.

Cost and Commercialization Barriers

The economic viability of smart clothing is constrained by the costs associated with advanced materials, electronics, and integration processes, which remain substantially higher than those for traditional textiles and standalone wearable devices.

- **Material and Component Expenses:**

Specialized conductive fabrics, flexible sensors, and miniaturized electronics contribute to elevated production costs. Additionally, the need for protective encapsulation and power management solutions adds to the overall expense.

- **Development and Certification Costs:**

Rigorous testing for durability, washability, safety, and regulatory compliance requires substantial investment, which may deter small and medium enterprises from entering the market.

- **Market Acceptance and Business Models:**

Consumer willingness to pay premium prices for smart garments is variable and often limited by concerns around privacy, durability, and ease of use. Business models must account for potential costs related to maintenance, software updates, and data services.

Overcoming commercialization barriers demands innovative cost-reduction strategies, such as economies of scale, material substitutions, and modular, upgradable designs, alongside efforts to educate consumers and stakeholders on the value proposition of smart clothing.

CONCLUSION

Despite being incredibly impressive, given the vast array of gadgets and connectivity incorporated into these outfits, one of the biggest challenges standing in the way of more main-stage-doe wearing smart clothes integrated into wireless communication systems and IoT is viability, cost, and reliability of communications. Solving these limitations requires interdisciplinary innovation spanning the fields of materials science, electronics manufacturing, wireless networking and the dynamics of business strategy. Advances in these directions will be pivotal to the realization of the generative potential of smart textiles as viable, economic and trusted technologies to improve health, safety, and performance.

REFERENCES

Bayindir, M., Abouraddy, A. F., Arnold, J., Joannopoulos, J. D., & Fink, Y. (2006). Thermal-sensing fiber devices by multimaterial codrawing. *Advanced Materials, 18*(7), 845–849. DOI: 10.1002/adma.200502106

Chen, J., Huang, Y., Zhang, N., Zou, H., Liu, R., Tao, C., Fan, X., & Wang, Z. L. (2016). Micro-cable structured textile for simultaneously harvesting solar and mechanical energy. *Nature Energy, 1*(10), 16138. DOI: 10.1038/nenergy.2016.138

Jiang, W., & Yin, Z. (2015, October). Human activity recognition using wearable sensors by deep convolutional neural networks. In *Proceedings of the 23rd ACM international conference on Multimedia* (pp. 1307-1310). DOI: 10.1145/2733373.2806333

Kamal, B., Ullah, S., & Khan, S. (2016). A NOVEL EXPERIMENTAL PROCEDURE FOR DIGITIZING FLEXIBLE WEARABLE ANTENNAS FOR BODY AREA NETWORKS (BANS). *Jurnal Teknologi (Sciences & Engineering), 78*(4-3).

Kartsakli, E., Lalos, A. S., Antonopoulos, A., Tennina, S., Di Renzo, M., Alonso, L., & Verikoukis, C. (2014). A survey on M2M systems for mHealth: A wireless communications perspective. *Sensors (Basel), 14*(10), 18009–18052. DOI: 10.3390/s141018009 PMID: 25264958

López-Nava, I. H., Arnrich, B., Munoz-Melendez, A., & Güneysu, A. (2017). Variability analysis of therapeutic movements using wearable inertial sensors. *Journal of Medical Systems, 41*(1), 7. DOI: 10.1007/s10916-016-0645-8 PMID: 27848176

Mattmann, C., Clemens, F., & Tröster, G. (2008). Sensor for measuring strain in textile. *Sensors (Basel), 8*(6), 3719–3732. DOI: 10.3390/s8063719 PMID: 27879904

Patel, S., Dholariya, K., Bhatt, N. V., & Thyagarajan, K. (2025, April). Data Analytics and Data Science in Diagnosing and Managing Sleep Disorders Using Wearable Devices. In *2025 Seventh International Conference on Computational Intelligence andCommunication Technologies (CCICT)* (pp. 594-600). IEEE. DOI: 10.1109/CCICT65753.2025.00095

Smith, A. A., Li, R., & Tse, Z. T. H. (2023). Reshaping healthcare with wearable biosensors. *Scientific Reports, 13*(1), 4998. DOI: 10.1038/s41598-022-26951-z PMID: 36973262

Stoppa, M., & Chiolerio, A. (2014). Wearable electronics and smart textiles: A critical review. *sensors, 14*(7), 11957-11992.

Wang, M., Chen, C., Wu, H., Zhang, J., Zhou, S., & Wang, G. (2024, October). Will Smart Ring be Next Wave of Wearables? In *2024 IEEE Biomedical Circuits and Systems Conference (BioCAS)* (pp. 1-5). IEEE. DOI: 10.1109/BioCAS61083.2024.10798374

Zhang, Z., Chen, P., Guo, K., Wang, C., & Zhang, X. (2015). A colour-tunable, weavable fibre-shaped polymer light-emitting electrochemical cell. *Nature Photonics*, *9*(4), 233–238. DOI: 10.1038/nphoton.2015.37

Chapter 9
Explainable Artificial Intelligence for Trustworthy Decision-Making in Smart Wearable Textiles

Pawan Kumar
Ajay Kumar Garg Engineering College, Ghaziabad, India

Mandeep Singh
https://orcid.org/0000-0003-3278-4877
Bennett University, Greater Noida, India

Juli Yadav
Ajay Kumar Garg Engineering College, Ghaziabad, India

Shalbha Chaudhary
R.D. Engineering College, Ghaziabad, India

Pranshu Saxena
Bennett University, Greater Noida, India

ABSTRACT

Due to the black-box nature of AImodels, they face significant critical challenges, particularly in safety domains such as healthcare, banking, and autonomous vehicles. Explainable Artificial Intelligence (XAI) technology aims to address these challenges by providing explanations that enable informed predictive decisions, while

DOI: 10.4018/979-8-3373-4287-0.ch009

ensuring a transparent, fair and accountable system. This research paper presents a comprehensive study on how XAI enhances the quality of smart wearable textiles for patient care, providing greater efficiency and enabling trustworthy decisions that ultimately benefit patient health. Due to significant advancements in artificial intelligence (AI), the use of AI-based smart wearable devices, particularly in the healthcare sector, has increased substantially. These real-time, AI-based wearable devices continuously monitor medical health of patients information, enabling physicians to detect diseases early and track patients during their recovery.

INTRODUCTION

Since the advent of the digital health revolution in the healthcare sector, researchers and scientists have been working to improve methods for making informed decisions by providing the best early predictions and predictive decision models, enhancing the quality of smart wearable devices, and more. Various AI-based Machine Learning (ML) and Deep Learning (DL) algorithms are widely used for early detection and prediction of diseases, enabling doctors to make informed and appropriate predictive decisions that improve patient health and increase life expectancy. Although AI-based model is itself an efficient model to handle various tasks better than human being and performance of all AI-based models was excellent and well-received in the healthcare sector, some significant challenges remain regarding trust and the output of all desired ML and DL algorithms, as the inner architecture and logic of these models are obscured by their design and the efficiency of the available dataset. Since AI models are solving many difficult and complex problems, the methodologies used behind these models are also increasingly complex. AI-based models utilise black-box technology, where the input is well-defined and the output is generated. This AI-based process is not transparent and is not easily understood by humans because an AI-based system does not provide any helpful information about the complete process from the initial step to the final decision step. The complete decision-making process and system are often complex and contain uncertainty in their operation. So it is essential and an emerging need to understand the inner structure of the decision-making process. A lack of understanding of inner decision-making is hindering people from adopting this advanced and powerful technology, making it challenging to implement in healthcare, banking, and many other industries, because it should be fairer for the survival of the species (Abernethy et al., 2022).

The black-box characteristics of AI models raise significant concerns among the public, underscoring the need to incorporate explainability, accountability, safety, interpretability, and transparency into AI-based models. All of these features or aspects are crucial for building trust among the public, including scientists, researchers, regulators, politicians, and many others who are increasingly attentive to adopting new developments in the field of AI-based decision-making technology. Keeping this in mind, we need to shift from relying solely on the power of AI to a deep understanding of how AI models arrive at decisions, leading with terms such as transparency, safety, explainability, interpretability, or, more generally. A new approach, Explainable Artificial Intelligence (XAI), is required to build trust in AI and DL models, and as a result, over the last decade, more attention has been given to XAI. Many researchers are working to increase the interpretability of AI and ML models, aiming to gain a better understanding of black-box decision-making. Figure 1 illustrates the publishing growth rate from 2015 to 2023, showing that the growth rate of XAI has increased year by year.

Figure 1. Publishing growth rate of XAI

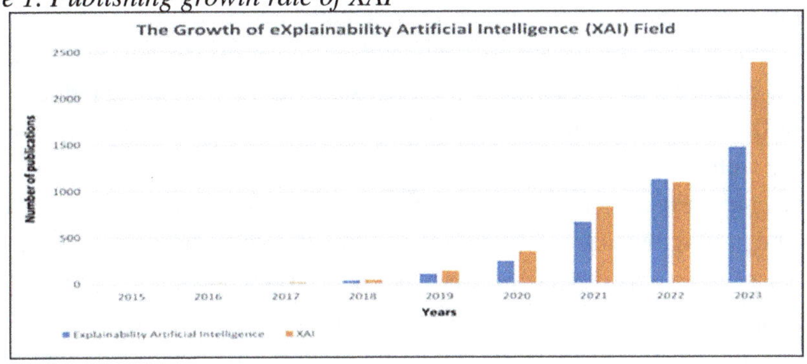

Over the last few decades, Wearable technology has become the most prominent area in healthcare. However, intricate ML and DL models often rely on "black box" models, which have transparency issues for both medical professionals (doctors) and users. Due to this, XAI has introduced an alternative solution. XAI focuses on the inner workings of complex algorithms and provides a transparent solution, with the primary aim of fostering trust and empowering stakeholders to use wearable technologies responsibly (Ahsan, Khan, & Ali, 2022). In IoMT-enabled smart healthcare systems, issues such as data privacy, trust, and secure decision-making are critical concerns (Singh et al., 2021).

History of Smart Wearables Technology

In the late 1960s, the Massachusetts Institute of Technology's Media Lab (MIT Media Lab) introduced innovative wearable technology, which, in its initial phase, combined materials, the latest information, and advanced electronic technologies to gather and process data relevant to the human body, providing user feedback on available data. Traditional wearable devices like capacitors and sensors restrict body movements during use. These devices are typically made from rigid materials and are very incompressible and non-stretchy; therefore, they are unsuitable for detecting human motion or monitoring physiological information. Thus, such wearable textile devices are incapable of detecting human motion and monitoring the physiological changes or information (Alowais et al., 2023).

The flexible devices were introduced in era of Internet of Things (IoT) based artificial intelligence. Flexible electronic devices are made of organic or inorganic materials. Due to their stretchability properties, lightweight features, user-friendliness, easy adaptability, and other advanced features such as motion detection, personalised medical care, computer-human interactions, smart robots, and other advanced fields, advanced, competent healthcare wearable devices attract people. Flexible, innovative electronics-based intelligent wearable textiles devices are widely utilised in monitoring of healthcare sector, thermal management, energy storage, smart clothing, flexible sensing etc and thanks to lightweight features, user-friendly environment and good performance, making them more usable in the healthcare industry.

Over the last two decades, the healthcare system has been totally powered by AI-based technology. The demand for wearable textile devices is continuously increasing, and these devices are continuously monitoring and analyzing the patient's health and providing more accurate and better predictive AI-based results. These are all AI-based solutions that utilize both ML and DL techniques for predicting and classifying problems and analyzing them continuously. The output and predictive results or information received or suggested by these models provide appropriate plans, treatments, or decisions for healthcare evaluations. These XAI-based wearable textile devices are available on the market in various forms, helping to monitor different patient metrics and bringing about the latest transformation in the healthcare sector. The performance of all these XAI wearable devices was evaluated, depending on factors such as latency, safety, location, costs. Latency may be asynchronous or real-time, location may be home or hospital, and similarly, a patient's health status may be classified as healthy, under treatment, or in recovery (Assenza et al., 2020).

Usability and customer feedback regarding XAI wearable textiles have been constantly seeking a desired solution provided by XAI models. The more advanced feature of these XAI wearable devices is that it easily captures or analyzes energy from human body human movement. As a result, demand of energy storage is also reduced. Most specific challenge for AI wearable textiles is the need for big data storage to store the information and the requirement for more processing units. So that these smart wearable devices can be easily integrated with patient healthcare predictive decision solutions. The user also experiences more overwhelm from excessive device decision feedback. For example, watch monitoring activity informs users about levels and messaging to encourage timely increases in patient activity. The medical workflow diagram of a pharmacist and a lab technician is shown in Figure 2, illustrating how a physician. Medical staff, pharmacists, and lab technicians are closely interrelated, and similarly, Figure 3 illustrates how the medical healthcare workflow integrates with the AI model (Bhatt et al., 2021).

Figure 2. Medical healthcare workflow with pharmacist, medical staff, and lab technician

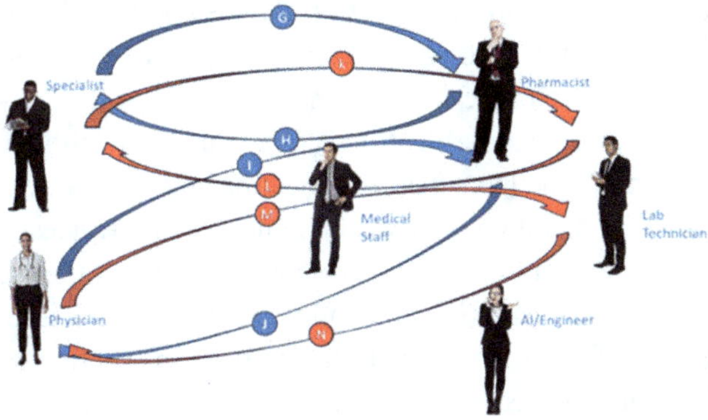

Figure 3. Medical healthcare workflow, AI engineer

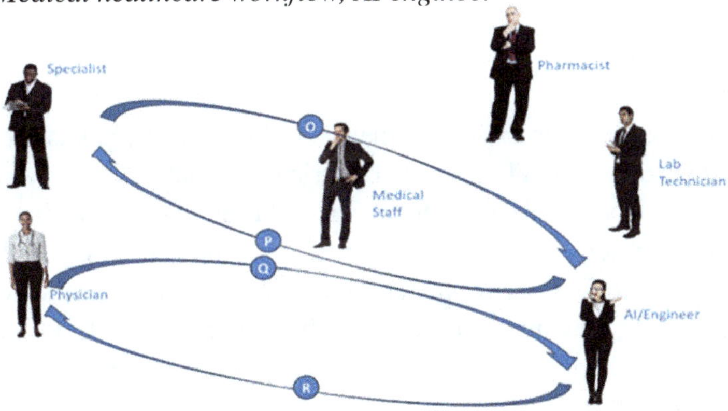

Comprehensive Study of Textiles

Textile is one of most frequently used materials in our daily lives. The good quality features, such as comfort and softness, along with easy flexibility, allow it to be used by the human body, and it is often referred to as the second layer of protection for the human body, shielding it from various elements. The flexible nature of textiles makes them more suitable for use in intelligent electronic devices. The integration of flexible wearables into electronic devices provides intelligent functionality, fulfilling the diverse requirements of users in various environments.

Silk has been primarily used in the industry for producing various wearable textile products for many years due to its unique mechanical properties, flexibility, softness, comfort, luster, easy moisture absorption, and numerous other benefits. Presently, silk is a promising candidate for use in making flexible electronic devices. Silk can easily be integrated into textiles or help develop advanced materials like conductive fibres, scaffolds, hydrogels, membranes, etc, and all these are necessary for making smart wearable textile devices. These devices include self-generating and energy storage devices, actuators, electronic skin, sensors etc. Silk material is one of the most crucial textiles and has several applications in various flexible electronics devices, especially for making smart wearable healthcare devices.

Due to the global population's growth, the need for flexible electronics and smart wearable textiles also increases. Existing traditional electronic textiles devices have a microscopic service life. Silk is a biomaterial textile with excellent biodegradability features that enable it to be easily combined with flexible electronic devices, helping to prepare innovative, long-lasting wearable devices that alleviate environmental pressure caused by discarded electronic devices and also meet the requirements of sustainable development (Chollet, 2021). Some advanced features and applications of silk textiles are discussed below.

Conductive Silk

Due to its excellent flexibility and mechanical properties, Silk is used as a base material for making wearable devices. Electrical conductivity plays a very crucial role when wearing wearable electronic devices during regular operation. Silk does not generate electricity by itself, but sometimes fibres are easily endowed with conductivity or used for flexible wearable electronics. Therefore, conducting-based fabrics are used for making silk-based wearable electronic textile devices.

Silk-Based Actuators

Actuators are used for conversion heat energy to mechanical energy due to changes in environmental conditions. The inner core feature of silk fibres is comprised mixture of hydrophilic and hydrophobic domains. α-helix and amorphous regions readily combine with water molecules through hydrogen bonding, and atmospheric conditions influence this interaction. Silk can absorb water molecules easily, and up to 40 per cent of water molecules are absorbed by silk, and when quantity of water molecules passed hydrophilic domains, bonding of hydrogen breaks from the structures.

Silk-Based Electronic Skin

The skin is a large organ of human body, functioning as intermediate barrier between outside environment and inner body parts. Our skin responses to external environmental conditions and internally manages and provides information to body parts and accordingly transmit signals. Skin is first immune barrier, protecting our internal organs from damage caused by outside environmental conditions. Flexible, bright electronic skin that was inspired by human body skin and used on the surface of the body. It exhibits sensory information or functions similarly to human body skin.

Silk-Based Smart Wearable Textile

Silk-based fibres have several features, such as low weight and density, softness and comfort, deformability, easy breathability, and capability of moisture absorption, making it mostly usable materials and demanded for making smart wearable textiles. It also possesses several features and properties, including conductivity, UV shielding, antibacterial capabilities, water resistance, and friction resistance. These advanced features or functionalization make silk materials more suitable for use in smart wearable devices.

Silk-Based Radiative Cooling Intelligent Clothing

Silk fibres possess features of gloss, wet comfort, and thermal properties. Due to this, it has been widely used for making clothes, quilts, and other functional textiles for the last thousand years. The most advanced feature of silk clothes is that they keep the human body cool and protect the sensation on the human body's skin. Researchers designed silk fibres that mimic the protective cocoons and avoid attraction of pupae by fluctuations of temperature.

SMART WEARABLE TEXTILES

Smart textiles, also known as electronic textiles, are intelligent fabrics that incorporate innovative electronic smart components and help enhance the features of smart wearable devices and other functional products. E-textiles enable the integration of various electronic elements, including batteries, lights, sensors, and microcontrollers, into fabrics. On the other hand, smart textiles are fabricated by a wide range innovative fabrics, including cotton, nylon, and polyester. These textiles are all embedded with digital electronics components or special fibres that enable them to interact with the user and the environment easily (Aungst, Franzese, & Kim, 2021). Some important key features of smart textiles are discussed below:

- Responsiveness: All smart wearable textiles must be highly responsive to adapt, or change based on specific triggers such as an increase in body temperature due to sunlight or other changing environmental conditions.
- Integration: It is a complicated task to integrate traditional fabrics with bright wearable fabrics by embedding sensors and microchips, some and advanced components.
- Functionality: Apart from comfort, style of smart wearable textiles, the most critical concern or function of these electronic wearable bright fabrics is to provide safety related to health or other convenience.
- Interactivity: These bright wearable fabrics are often operated with some apps or devices, allowing users to access or interpret gathered data.

Difference Between Traditional Textiles and Smart Wearable Textiles

The primary focus of traditional textiles is to provide basic protection to the body and comfort (Dalton-Brown, 2020). The following are some essential characteristics of conventional fabrics.

- All these traditional materials are made of cotton, wool, silk, and polyester fabrics
- Traditional fabric does not contain or access any electronic-based components.
- Traditional fabrics cannot sense or respond to changing environmental conditions.
- Cannot perform some more specific functions apart from traditional woven textile.

On the other hand, smart wearable textiles are designed with the integration of added innovative technology or special fibres. The following are some essential characteristics of Smart wearable textiles.

- Smart wearable textiles are made from both traditional materials and advanced Kevlar materials.
- Advanced electronic components such as sensors, microcontrollers and actuators are associated with smart wearable textiles.

- It can easily sense and respond to changes in environmental conditions.
- Can perform more specific functions beyond those of traditional woven textiles.
- Smart wearable textiles are performing in both passive and active ways. Passive smart wearable textiles offer more functionality beyond traditional woven fabrics, as they are not accepting sensing information easily. In contrast, smart wearable textiles perform well and gives proper response to environmental changing conditions.

Benefits of Smart Wearable Textiles

Smart Wearable textiles offer several advantages and benefits over traditional fabrics, and these smart wearable devices play a crucial role in redefining human health and wellness. These advanced features of bright fabrics enable them to perform more effectively in various areas, such as monitoring and detection of humans, therapy, protection, and enhancement of mental well-being (Davenport & Kalakota, 2019). Blockchain and IoT integration can enhance transparency and security in smart healthcare, providing a foundation for trustworthy and explainable AI systems (Sharma et al., 2022). The following section discusses some advanced features of Smart Wearable Textiles.

Health Monitoring

Smart wearable textiles perform better for human health monitoring, and some areas where smart wearable textiles performing better are:

- Detecting changing heart rate due to a change in body temperature.
- Monitor the body breathing patterns.
- Timely inform users about irregularities and health risks.

Therapeutic Applications

Wearable textiles not only detect or sense body changes but also are used for healing and rehabilitation.

- Release the required amount of medicinal essential oils and directly enter the skin.
- Applied targeted area for compression to resolve problem of muscle recovery or the prevention of injury
- Provides vibrations for relief from pain.

Safety and Protection

Ensuring protection and personal safety is becoming more essential with the use of smart wearable textiles that are designed for a specific purpose.

- Offer safety and protection from harmful and dangerous UV rays.
- Temperature, regulation for keeping body warm or cool as per user requirement.
- Avoid water moisture and keep wearer dry.

Mental Well-Being

Physical and mental health must be well required for human body, and there are some critical aspects where mental health is more essential for human beings.

- Changing patterns or colours depending on the mood of users helps in reducing stress levels
- Sensors can easily detect the stress or anxiety level of the body and offer better solutions for deep and short breathing.
- Improving relaxation, meditation by providing smooth tactile sensations.

Challenges Faced by the Use of Smart Textiles

Apart from their promise of features, smart wearable textiles are equally crucial in addressing the various challenges and limitations of implementing these innovative new domains (Dong et al., 2022). Knowing that all these smart wearable textiles are very helpful for making informed decisions, but some primarily concern issues are:

Durability and Longevity

These smart wearable textiles are embedded with advanced technology, but still, they face usual rip threats, stains with daily wear, etc. Additionally, its innovative features are degrading with time, and second most crucial concern is reduced durability.

High Repair and Maintenance Costs

Incorporating new, innovative technology into wearable textiles increases the price of traditional wearable textiles, posing a significant challenge or burden for users to purchase high-cost smart wearable textiles. The costs of repairing, maintaining, and replacing are also an essential concern for users, and as a result, the use of smart wearable textiles is more expensive for customers. The cleaning process of all electronic components is very complicated, and the high cost of cleaning also needs to be paid by users. Some smart wearable textiles require only hand washing, and some may require specialised, high-cost detergents for washing. Some wearable textiles also require regular power charging, and their battery life is not particularly long, necessitating battery replacement, which adds a burden on users (Dunn, Runge, & Snyder, 2018).

Health Concerns

The inclusion of innovative electronic wearable components may sometimes cause skin irritations or allergies in some users. Research has also shown that the long-term use of smart wearable textiles may expose users to continuous, low-level electromagnetic fields that are potentially hazardous to human health (Esteva et al., 2019).

Environmental Impact

Due to excessive waste generation, there is a need for effective recycling and disposal methods. The entire manufacturing process of these smart wearable textiles likely has a larger and longer environmental footprint due to the addition of these innovative technological components.

Data Privacy and Security

Some of them, such as smart wearable textile devices, collect large amounts of user data; if the data is not secured correctly, there is a risk of significant privacy concerns (Singh & Malik, 2025). Mostly, textile devices are connected to the internet, and as a result, they are vulnerable to hacking or unauthorised access by other devices (Holley & Becker, 2021).

BENEFITS OF EXPLAINABLE AI OVER ARTIFICIAL INTELLIGENCE

Artificial Intelligence Systems perform tasks as well as human beings and solve complex tasks and problems easily. The AI-based model learns from previous large datasets and makes accurate predictive decisions. On the other hand, XAI models enable AI-based systems to provide explanations for their choices or output predictions and also offer insight into their behavior. To build trust among people to have willing confidence regarding the output and decision-making provided by an AI system. The AI models are based on black-box technology and suffer from a lack of transparency, making it difficult to understand what is happening behind the predictions or decisions made by the AI model (Hulsen, 2023). Recent studies have demonstrated how XAI can improve transparency and clinician trust in medical decision-making systems (Saxena et al., 2025).

The increasing size of data makes AI models more complex, hindering their ability to make accurate predictive decisions, and it also affects the performance of these models. Generally, all AI models take input from users, process previous data, and give predictive output. The black-box characteristics of AI models are influenced by various attributes or factors, especially in ML models, which face numerous challenges, including model complexity, algorithm optimization, and complex and extensive training datasets. The deep learning neural network of AI models attempts to alleviate these concerns by designing deep neural networks (DNNs) that utilize hidden layer features. The workflow for the DNN-based smart wearable textile is illustrated in Figure 4.

Figure 4. DNN-based smart wearable textiles

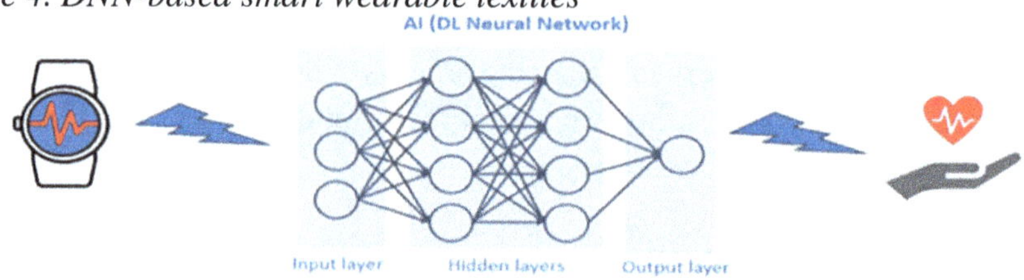

XAI models offer potential benefits over AI models by enhancing system performance, particularly in model design and debugging, as well as in decision-making processes. XAI techniques facilitate the selection and identification of relevant and appropriate features, aiding in the development of accurate and practical models. XAI techniques also aided in hyperparameter tuning by selecting the right choice of activation functions, increasing the size of hidden layers, and optimizing learning rate values to mitigate underfitting or overfitting. XAI also helped developers detect bias in the decision-making process. If developers quickly identify bias, it can help adjust the AI system to ensure that the output target value is fair and unbiased. XAI also enabled developers to easily identify decision-making errors and make the right corrective decisions to achieve more accurate, quick and reliable models. XAI can also enable users to modify the value of input parameters, giving them complete control over the input value. By adjusting the value of input parameters, users can enhance performance and decision-making results (Islam et al., 2021). Although deep learning offers high accuracy in healthcare predictions, its black-box nature limits interpretability, necessitating XAI-based solutions (Singh et al., 2023). Some critical features of XAI are:

Reliability

ML-based models are less reliable than XAI models, and sometimes the output or prediction-based decision results of ML models are very unexpected and fail. The trust and reliability of users regarding AI models are depreciating and require mechanisms to control accountability or improve the predictions and decision-making performance of AI models. For example, a wrong predictive decision in the medical field may lead to a serious, high-risk outcome that significantly impacts human beings, (Janiesch, Zschech, & Heinrich, 2021).

Safety

Safety is a very measured concern in various ML-based applications. Self-driving cars and military drones are prime examples of the safety concerns associated with AI models. Therefore, it is essential to understand and focus on the decisions made by an AI model, and to order and ensure the development of safe and secure predictive models for human lives.

Human-AI Collaboration

XAI models facilitate the cooperation of both humans and AI models by enabling humans to understand the main reason behind an AI's actions easily.

Error Identification

XAI models enable users to easily identify errors, aiding in the development of an error-free decision-making process. This leads to the selection of the right corrective model, which provides more accurate and faster predictive results.

The demand for XAI-based models has increased significantly over the past few years, driven by their ability to provide more accurate and predictive results compared to traditional AI models. XAI models enhance or build trust among users regarding XAI-based models (Gupta et al., 2023). The designed model easily identifies the correct errors and biases, and it also provides a fair result for easy understanding by users. The main features of the XAI model over the AI model are that the XAI model offers more transparent and accountable results, and ultimately yields a safe and secure decision-making outcome. Figure 5 represents the diagrammatic representations of XAI models, (Kumar et al., 2024).

Figure 5. XAI beneficiaries

ROLE OF EXPLAINABLE AI IN SMART TEXTILES INDUSTRY

The textile industry is also undergoing a very significant transformation due to high advancements in the field of Artificial Intelligence (AI). Smart wearable textiles and innovative intelligent fabric production are at the forefront of this new revolution, integrating AI-powered technologies with the textile industry to enhance efficiency, drive innovation, and promote sustainability (Chaudhary et al., 2024). From the latest innovations in predictive textile machinery to AI-driven material design, the industries are witnessing a new era of artificial intelligence manufacturing. The following are some of the most prominent areas where artificial intelligence is applied in the textile industry (Leclercq et al., 2022).

XAI-Powered-Based Innovative Material

XAI is used to develop smart or innovative fabrics with various enhanced functionalities. Machine learning models are designed using large amounts of data and help create high-performance textiles with multiple features, such as temperature regulation, moisture control, and self-healing capabilities. These XAI-based, innovative materials enable manufacturers to design smart textile wearable devices with superior durability, safety and comfort features.

Smart Manufacturing with XAI

The inclusion of XAI in textile machinery enables optimised production processes that help to reduce waste and also increase efficiency. The automated features of XAI quality control systems help detect defects in fabric production automatically, minimising errors and ensuring consistency in the quality of the textile. AI-driven predictive intelligent analytics also help reduce maintenance costs associated with equipment failures and prevent downtime.

Enhancing Sustainability in Textile Production

Sustainability is one of the most pressing issues in the textile industry, and XAI helps manufacturers adopt an eco-friendly approach. AI-powered systems help in optimising dyeing and finishing, reducing water and other chemical usage. Additionally, AI-driven systems also manage waste management systems, analyse production patterns to minimise fabric wastage, and improve resource utilisation, contributing to a greener textile industry.

XAI in Wearable and Interactive Textiles

The innovative textiles industry is increasingly utilising advanced technology in various fields, including smart wearable textiles, smart healthcare devices, and the creative fashion industry. AI-driven sensor technology embedded in fabrics and textiles monitors biometric data, enabling various applications in motion detection, medical diagnostics, and adaptive bright clothing. XAI inclusion in the textile industry is driving a new revolution, making fabric production more innovative, more sustainable, and highly effective and efficient. From the initial step of material innovation to final quality control and sustainability goals, AI-driven technologies are playing a crucial role in shaping the future of the textile industry, (Lewy, 2015).

APPLICATIONS OF XAI IN SMART WEARABLE TEXTILES

XAI is shifted new revolutionary in field of textile industries by new innovative ideas and their operational effectiveness and efficiency. With new technological achievements and advancements, XAI has now become a more crucial technology in the changing transformation production phase of textiles, new encompassing designing, smart manufacturing, improving quality control, and satisfaction and interaction with customers, etc. All these features make it more in demand. These all streamlines feature and operations also help in enhancing sustainability and creativity in textile sectors. Some Application area is discussed below:

- **Seamless Tech Integration** – AI-powered e-textiles embed sensors directly into fabrics, blending fashion with innovative technology.
- **Real-Time Health Monitoring** – These wearables can track biometrics like heart rate, breathing, posture, and body temperature through conductive fibres.
- **AI-Enhanced Insights** – Machine learning algorithms analyse collected data to provide personalised wellness, fitness, and lifestyle recommendations.
- **Comfort & Flexibility** – Unlike bulky devices, bright clothing is soft, breathable, and designed for everyday wear without restricting movement.
- **Sports & Fitness Benefits** – Athletes use AI e-textiles for performance optimisation, fatigue detection, and injury prevention.
- **Medical & Rehabilitation Applications** – Smart fabrics assist in remote patient monitoring, physiotherapy feedback, and detecting early signs of illness.
- **Fashion Meets Functionality** – E-textile wearables combine style with practicality, offering discreet, fashionable ways to stay connected and healthy.

AI is revolutionising wearable textiles by enhancing functionality, personalisation, and sustainability, leading to innovative, bright clothing that adapts to user needs and environmental conditions.

CONCLUSION

The use and demand for wearable textile devices in the field of medical healthcare have increased significantly, marking a new turning point, especially in the healthcare sector, and allowing patients to transition from traditional healthcare devices to smart healthcare devices. This revolutionary turning point presents a significant opportunity for researchers to expand their current research in field of XAI healthcare devices. The research paper focuses on studying and exploring the latest trends in XAI wearable textiles, provides a comprehensive and detailed analysis of this emerging technology, predictive decision-making solutions, and advanced methods, which also help enhance support for medical professionals in patient care. The research findings concluded that wearable textiles and XAI hold great promise, providing significant transformation for improving the healthcare sector. The new research solutions can be easily adopted or applied by innovative healthcare practices, particularly in cases involving DL and ML. They also help to enhance performance of models, improving patient health.

The new XAI technologies can be easily integrated with existing traditional healthcare technologies, yielding better healthcare results or outcomes and thereby improving the working conditions of medical professionals. The new XAI wearable textiles also address or improve current medical healthcare practices, while providing well-managed information management solutions to existing machine learning models and applying emerging wearable technologies for data-driven, proper decision-making. These XAI-based decision predictive solutions are beneficial for advancing patient healthcare and encourage the healthcare industry to apply XAI-based wearable textiles, building trust with manufacturers.

We believe that the current research work is more pertinent, about adoption of XAI wearable textile to AI models, and can help transform both patient care and processes associated with XAI technology. Specific duties, roles, responsibilities regarding new XAI wearable textiles processes new 'business' aspects and facilitated better predictive solutions for patient management.

REFERENCES

Abernethy, A., Adams, L., Barrett, M., Bechtel, C., Brennan, P., Butte, A., Faulkner, J., Fontaine, E., Friedhoff, S., Halamka, J., Howell, M., Johnson, K., Long, P., McGraw, D., Miller, R., Lee, P., Perlin, J., Rucker, D., Sandy, L., & Valdes, K. (2022). The promise of digital health: Then, now, and the future. *NAM Perspectives*, *6*(22). Advance online publication. DOI: 10.31478/202206e PMID: 36177208

Ahsan, M., Khan, S., & Ali, U. (2022). Explainable AI models for healthcare diagnostics. *International Journal of Advanced Sciences and Computing*, *1*(1), 8–15.

Alowais, S. A., Alghamdi, S. S., Alsuhebany, N., Alqahtani, T., Alshaya, A. I., Almohareb, S. N., Aldairem, A., Alrashed, M., Bin Saleh, K., Badreldin, H. A., Al Yami, M. S., Al Harbi, S., & Albekairy, A. M. (2023). Revolutionising healthcare: The role of artificial intelligence in clinical practice. *BMC Medical Education*, *23*(1), 689. Advance online publication. DOI: 10.1186/s12909-023-04698-z PMID: 37740191

Assenza, G., Fioravanti, C., Guarino, S., & Petrassi, V. (2020). New perspectives on wearable devices and electronic health record systems. In *2020 IEEE International Workshop on Metrology for Industry 4.0 & IoT* (pp. 740–745). IEEE. DOI: 10.1109/MetroInd4.0IoT48571.2020.9138170

Aungst, , T., Franzese, C., & Kim, Y. (2021). Digital health implications for clinical pharmacists services: A primer on the current landscape and future concerns. *Journal of the American College of Clinical Pharmacy: JAACP*, *4*(4), 514–524. DOI: 10.1002/jac5.1382

Bhatt, C., Kumar, I., Vijayakumar, V., Singh, K. U., & Kumar, A. (2021). The state of the art of deep learning models in medical science and their challenges. *Multimedia Systems*, *27*(4), 599–613. DOI: 10.1007/s00530-020-00694-1

Chaudhary, S., Verma, S., Sharma, V., Jain, S., & Kumar, P. (2024). A multi-model approach to hydrological event analysis and forecasting. In *2024 International Conference on Emerging Technologies and Innovation for Sustainability (EmergIN)* (pp. 485–490). IEEE. DOI: 10.1109/EmergIN63207.2024.10961633

Chollet, F. (2021). *Deep learning with Python* (2nd ed.). Manning Publications.

Dalton-Brown, S. (2020). The ethics of medical AI and the physician-patient relationship. *Cambridge Quarterly of Healthcare Ethics*, *29*(1), 115–121. DOI: 10.1017/S0963180119000847 PMID: 31858938

Davenport, T., & Kalakota, R. (2019). The potential for artificial intelligence in healthcare. *Future Healthcare Journal*, *6*(2), 94–98. DOI: 10.7861/futurehosp.6-2-94 PMID: 31363513

Dong, H., Falis, M., Whiteley, W., Alex, B., Matterson, J., Ji, S., Chen, J., & Wu, H. (2022). Automated clinical coding: What, why, and where we are? *NPJ Digital Medicine*, *5*(1), 159. Advance online publication. DOI: 10.1038/s41746-022-00705-7 PMID: 36273236

Dunn, J., Runge, R., & Snyder, M. (2018). Wearables and the medical revolution. *Personalized Medicine*, *15*(5), 429–448. DOI: 10.2217/pme-2018-0044 PMID: 30259801

Esteva, A., Robicquet, A., Ramsundar, B., Kuleshov, V., DePristo, M., Chou, K., Cui, C., Corrado, G., Thrun, S., & Dean, J. (2019). A guide to deep learning in healthcare. *Nature Medicine*, *25*(1), 24–29. DOI: 10.1038/s41591-018-0316-z PMID: 30617335

Gupta, M., Singh, M., Sharma, A., Sukhija, N., Aggarwal, P. K., & Jain, P. (2023). Unification of machine learning and blockchain technology in healthcare industry. In *Innovations in healthcare informatics: From interoperability to data analysis* (pp. 185–206). Institution of Engineering and Technology. DOI: 10.1049/PBHE041E_ch6

Holley, K., & Becker, S. (2021). *AI-first healthcare: AI applications in the business and clinical management of health* (1st ed.). O'Reilly Media.

Hulsen, T. (2023). Explainable artificial intelligence (XAI): Concepts and challenges in healthcare. *AI*, *4*(3), 652–666. DOI: 10.3390/ai4030034

Islam, M. M., Poly, T. N., Yang, H. C., & Li, Y. C. (2021). Deep into laboratory: An artificial intelligence approach to recommend laboratory tests. *Diagnostics (Basel)*, *11*(6), 990. Advance online publication. DOI: 10.3390/diagnostics11060990 PMID: 34072571

Janiesch, C., Zschech, P., & Heinrich, K. (2021). Machine learning and deep learning. *Electronic Markets*, *31*(3), 685–695. DOI: 10.1007/s12525-021-00475-2

Kumar, V., Wahi, C., Sagar, B. B., & Manjul, M. (2024). Ensemble learning based intrusion detection for wireless sensor network environment. *International Journal of Performability Engineering*, *20*(9), 541–550. DOI: 10.23940/ijpe.24.09.p2.541551

Leclercq, C., Witt, H., Hindricks, G., Katra, R. P., Albert, D., Belliger, A., Cowie, M. R., Deneke, T., Friedman, P., Haschemi, M., Lobban, T., Lordereau, I., McConnell, M. V., Rapallini, L., Samset, E., Turakhia, M. P., Singh, J. P., Svennberg, E., Wadhwa, M., & Weidinger, F. (2022). Wearables, telemedicine, and artificial intelligence in arrhythmias and heart failure: Proceedings of the European Society of Cardiology Cardiovascular Round Table. *Europace : European Pacing, Arrhythmias, and Cardiac Electrophysiology : Journal of the Working Groups on Cardiac Pacing, Arrhythmias, and Cardiac Cellular Electrophysiology of the European Society of Cardiology*, *24*(9), 1372–1383. DOI: 10.1093/europace/euac052 PMID: 35640917

Lewy, H. (2015). Wearable technologies—Future challenges for implementation in healthcare services. *Healthcare Technology Letters*, *2*(1), 2–5. DOI: 10.1049/htl.2014.0104 PMID: 26609396

Saxena, P., Sharma, M., Batra, R., Gupta, R., Singh, M., & Singh, S. K. (2025). XAI for lymph node histopathology: From explainability to immersive metaverse applications. In *2025 7th International Conference on Signal Processing, Computing and Control (ISPCC)* (pp. 304–309). IEEE. DOI: 10.1109/ISPCC66872.2025.11039319

Sharma, A., Singh, M., Gupta, M., Sukhija, N., & Aggarwal, P. K. (2022). IoT and blockchain technology in 5G smart healthcare. In *Blockchain applications for healthcare informatics: Beyond 5G* (pp. 137–161). Elsevier. DOI: 10.1016/B978-0-323-90615-9.00004-9

Singh, M., Gupta, M., Sharma, A., Jain, P., & Aggarwal, P. K. (2023). Role of deep learning in healthcare industry: Limitations, challenges and future scope. In *Deep learning for healthcare services* (pp. 1–22). Bentham Science., DOI: 10.2174/9789815080230123020003

Singh, M., & Malik, A. (2025). Energy-efficient cluster head selection in heterogeneous WSNs using Aquila optimizer and fuzzy logic (AO-FL). *International Journal of Communication Systems*, *38*(16), e70272. Advance online publication. DOI: 10.1002/dac.70272

Singh, M., Sukhija, N., Sharma, A., Gupta, M., & Aggarwal, P. K. (2021). Security and privacy requirements for IoMT-based smart healthcare system: Challenges, solutions, and future scope. In *Big data analysis for green computing: Concepts and applications* (pp. 15–32). CRC Press., DOI: 10.1201/9781003032328-2

Chapter 10
Uncertainty-Resilient Signal Processing in Smart Textiles Using Fuzzy Logic

Ajoy Kanti Das
https://orcid.org/0000-0002-9326-1677
Tripura University, India

Nandini Gupta
https://orcid.org/0009-0000-5450-9783
Bir Bikram Memorial College, India

Takaaki Fujita
https://orcid.org/0009-0007-1509-2728
Independent Researcher, Japan

ABSTRACT

Smart textiles are increasingly embedded with sensors, actuators, and communication modules to enable intelligent interaction with users and environments. However, signal acquisition in textile-based electronics is inherently uncertain due to variability in fabric conductivity, body movements, environmental factors (humidity, temperature), and sensor noise. Classical signal processing methods often fail to handle such imprecision effectively. This chapter introduces a fuzzy logic-based framework for uncertainty-resilient signal processing in smart textiles. We present the mathematical structures of fuzzy sets, fuzzy soft computing, and fuzzy inference systems (FIS), emphasizing their applications in wearable sensing. The framework integrates multi-criteria fuzzy decision-making, adaptive fuzzy filtering, and fuzzy-

DOI: 10.4018/979-8-3373-4287-0.ch010

based classification models for robust signal interpretation. Finally, we discuss future research directions in merging fuzzy systems with machine learning and quantum-inspired models for next-generation wearable intelligence.

INTRODUCTION

Smart textiles are emerging as intelligent systems that can sense, process, and respond to environmental and physiological signals. However, data from wearable sensors are often imprecise, noisy, and uncertain. To handle such complexity, fuzzy logic has been widely adopted as a mathematical tool for modeling uncertainty, enabling robust decision-making in dynamic environments. This review highlights the key developments in fuzzy logic and its extensions relevant to signal processing in smart textiles.

The concept of fuzzy sets introduced by Zadeh (1965) laid the foundation for modeling vagueness in real-world systems. Later, Zadeh (1983) emphasized its role in expert systems for managing uncertainty. Mendel (2000) further extended this line of thought to signal processing, demonstrating how fuzzy systems can handle uncertain and imprecise signals—an idea that directly supports wearable sensor integration in smart textiles.

Recent works highlight the evolution of fuzzy systems into type-2 fuzzy logic and hybrid models to better capture uncertainty. Moreno et al. (2020) proposed an interval type-2 fuzzy model with justifiable uncertainty, improving interpretability and model reliability. Similarly, Beke and Kumbasar (2023) introduced a composite learning framework for interval type-2 fuzzy logic systems, showing enhanced accuracy in regression and uncertainty modeling. Meng et al. (2023) and Zhao et al. (2024) advanced fuzzy reasoning for robust control, particularly in cyber-physical systems, which parallels the challenges of textile-based sensing under variable conditions.

Fuzzy decision-making models have been applied in diverse fields where uncertainty is unavoidable. Chen et al. (2020) integrated rough-fuzzy approaches for supply chain management under uncertain conditions. Mirza et al. (2025) applied fuzzy Z-numbers with Bonferroni operators in textile engineering, demonstrating their utility in multi-criteria decision-making—a relevant framework for textile-based system optimization. Similar strategies are evident in energy management (Mohamed et al., 2021; Baz et al., 2024), environmental engineering (Bressane et al., 2024), and agriculture (Erdoğdu et al., 2025), showing the versatility of fuzzy frameworks in uncertainty-driven domains.

To improve adaptability, soft sets and their extensions have been proposed (Molodtsov, 1999; Maji et al., 2001). These models have been expanded into fuzzy soft sets and intuitionistic fuzzy soft multisets (Das, 2016; 2018; Mukherjee & Das, 2015–2023), enabling more nuanced handling of uncertainty in group decision-making. Recent studies (Das et al., 2025a; 2025b) also applied hesitant and weighted fuzzy soft sets in environmental monitoring and water quality evaluation, highlighting their robustness in processing uncertain sensor data. Such methods provide inspiration for designing textile systems where multiple signals must be aggregated reliably.

Beyond traditional fuzzy methods, hybrid approaches combine fuzzy logic with neutrosophy (Smarandache, 1998; 2005), machine learning (Imamguluyev et al., 2024), and optimization techniques (Shaik et al., 2024). These integrations enhance adaptability and resilience, especially in decision-making under deep uncertainty. In disaster management and resilience assessment, fuzzy logic has been successfully applied to evaluate community responses (De Iuliis et al., 2022; Wahyono et al., 2024), offering parallels for resilience-focused smart textiles.

Taken together, the literature shows that fuzzy logic provides a flexible and uncertainty-resilient framework for processing noisy, imprecise signals. From early fuzzy set theory (Zadeh, 1965) to modern hybrid and soft set-based models (Das et al., 2022–2025), researchers have consistently improved methods to handle multi-source, uncertain data. For smart textiles, these developments suggest that fuzzy logic—especially interval type-2 fuzzy systems, fuzzy Z-numbers, and soft set-based decision frameworks—can serve as powerful tools to enhance real-time signal processing, decision-making, and adaptive responses in wearable systems.

The theoretical backbone of this study is rooted in $\mathfrak{F}$S theory (Zadeh, 1965), SS theory (Molodtsov, 1999), and their fusion into $\mathfrak{F}$SSs (Maji et al., 2001), which are well-suited for modeling complex, uncertain, and MC$\mathscr{D}$M problems in digital security assessments and cyber risk evaluation. Extensions such as intuitionistic $\mathfrak{F}$Ss (Atanassov, 1986), neutrosophic sets (Smarandache, 1998, 2005), and $\mathfrak{F}$S Ss further enhance decision models under uncertainty, making them highly relevant to integrating health technologies, infrastructure design, and community values in EC systems. Recent studies by Das and Granados have greatly improved $\mathfrak{F}$SS models for MC$\mathscr{D}$M. They introduced FP intuitionistic frameworks (Das and Granados, 2022a), and expanded them into FP-intuitionistic multi-fuzzy N-SSs (Das and Granados, 2022b; 2023a). These models help deal with vague and conflicting information, which is useful in digital security assessments and cyber risk evaluation. Earlier, Das (2016, 2018) proposed weighted $\mathfrak{F}$S Ss and preference-based fuzzy rough models to handle uncertainty and partial preferences better. In 2021, Das and Granados presented a group MC$\mathscr{D}$M model using $\mathfrak{F}$SSs and weighted averages (Das and Granados, 2021). Das and Granados (2023b) explored preference intuitionistic fuzzy rough models, highlighting their partial inclusion properties.

Further applications were demonstrated by Das et al. (2022, 2023), who developed weighted hesitant bipolar-valued $\mathfrak{F}$SSs and explored new operations in fuzzy environments. These were applied to assess urban river pollution (Das et al., 2024a, 2025a, 2025b), showing how environmental decision-making can benefit from such models. In a similar direction, Das et al. (2025c) also explored $\mathfrak{F}$SS models for inventory management in retail. They later introduced an advanced weighted hypersoft expert system for complex group decisions (Das et al., 2025d, 2025e). Mukherjee and Das also contributed significantly by applying interval-valued intuitionistic $\mathfrak{F}$S Ss to real-life problems (Mukherjee and Das, 2015a, 2015b), proposing Einstein operations for $\mathfrak{F}$SSs (Mukherjee and Das, 2022), and writing a comprehensive book on the theory and applications of $\mathfrak{F}$S Ss (Mukherjee and Das, 2023). They also explored topological and relational aspects of $\mathfrak{F}$SSs and rough sets (Mukherjee and Das, 2015c, 2016a, 2016b). Granados et al. (2022, 2023) proposed models based on continuous neutrosophic distributions and weighted neutrosophic soft multisets for complex MCDM tasks. Recently, Das et al. (2025c) conducted a comprehensive study on neutrosophic SuperHyper BCI-semigroups, exploring their algebraic significance. Das et al. (2025d) proposed a $\mathfrak{F}$S-based MC$\mathscr{D}$M **m**odel to optimize adaptive inventory management in retail settings. Das et al. (2025e) proposed an innovative weighted hypersoft expert system framework for group MC$\mathscr{D}$M, advancing neutrosophic and plithogenic models to better handle uncertainty and complexity in applied mathematics contexts. Together, these works have strengthened the theoretical foundation and practical usefulness of $\mathfrak{F}$SS theory in complex domains like digital security assessments and cyber risk evaluation.

Fuzzy logic has also been widely applied in the evaluation of complex systems and smart environments, which share similar uncertainty challenges with smart textiles. Kutty et al. (2023) developed a fuzzy expert-based multi-criteria decision support model to assess sustainability, resilience, and livability in European smart cities. Eshuis et al. (2021) focused on modeling uncertainty in process-driven systems using fuzzy logic. By embedding fuzzy reasoning into declarative process models, they demonstrated improved adaptability in workflows under uncertainty. Similarly, Abdoli (2022) explored the use of fuzzy logic for early-stage design assessment of complex engineering systems. The study emphasized how fuzzy evaluation supports decision-making during initial stages when information is incomplete or uncertain. For smart textiles, where prototype design often faces uncertain requirements from both technology and user needs, such fuzzy-based assessment frameworks provide valuable insights for robust design under uncertainty.

This chapter proposes a comprehensive framework for uncertainty-resilient signal processing in smart textiles using fuzzy logic. We develop the mathematical structures underlying fuzzy sets, fuzzy inference systems, and fuzzy multi-criteria decision-making, and demonstrate their application to textile-based sensing systems. The framework includes:

1. Fuzzy adaptive filtering – for noise suppression in biosignals such as ECG and EMG.
2. Fuzzy inference models – for interpreting uncertain, nonlinear sensor responses.
3. Fuzzy multi-criteria decision-making (MCDM) – for fusing heterogeneous signals (e.g., heart rate, temperature, and motion) into reliable outputs.
4. Hybrid fuzzy models – combining fuzzy logic with machine learning, quantum computing, and blockchain for next-generation textile intelligence.

By systematically integrating fuzzy logic into the smart textile ecosystem, we aim to create signal processing pipelines that are resilient to uncertainty, adaptive to environmental variability, and capable of delivering robust performance in real-world contexts.

PRELIMINARIES

Smart Textiles: Definition and Scope

Smart textiles are fabrics that interact with the wearer and the surrounding environment through embedded sensing, actuation, and data processing capabilities. Depending on their complexity, they are typically categorized into three generations:

1. Passive Smart Textiles – capable of sensing but not responding (e.g., fabrics with embedded thermometers or moisture sensors).
2. Active Smart Textiles – capable of both sensing and actuation (e.g., fabrics that change porosity with temperature or integrate ECG monitoring electrodes).
3. Ultra-Smart Textiles – capable of sensing, actuation, and adaptive response through computation, decision-making, and learning (e.g., garments that monitor patient vital signs and adapt compression dynamically).

The integration of conductive fibers, nanomaterials, microcontrollers, and wireless communication modules enables smart textiles to function as wearable cyber-physical systems, bridging the physical and digital worlds.

Signal Processing in Smart Textiles

At the core of smart textile functionality lies signal acquisition and interpretation. Typical signals captured by smart textiles include:

- Biopotentials: ECG (electrocardiogram), EMG (electromyogram), EEG (electroencephalogram).
- Biomechanical signals: Pressure, strain, bending, and motion.
- Environmental signals: Temperature, humidity, and chemical concentrations.

Processing these signals requires addressing multiple challenges:

- Noise Contamination – signals are often corrupted by motion artifacts due to fabric movement, electromagnetic interference, and environmental factors such as sweat and humidity.
- Nonlinear Sensor Responses – textile-based sensors, unlike rigid silicon devices, exhibit nonlinear resistance or capacitance changes under mechanical deformation.
- Signal Drift and Instability – over time, textile sensors may degrade due to washing, wear, or fiber fatigue, producing drifting baselines.
- Multi-sensor Data Fusion – combining heterogeneous textile sensors (e.g., heart rate and respiration) requires robust multi-criteria decision frameworks.

Traditional deterministic filters (low-pass, high-pass, Kalman filters) struggle to accommodate the vagueness and variability inherent in textile data. This gap motivates the use of fuzzy logic.

Fuzzy Logic

Fuzzy logic provides a means to represent imprecise concepts mathematically. In signal processing, this allows textile data to be expressed in linguistic categories such as:

- Pressure: {low, medium, high}
- Noise: {small, moderate, large}
- Signal quality: {poor, acceptable, excellent}

Instead of hard thresholds, fuzzy logic uses membership functions that assign a degree of belonging between 0 and 1. For example, a pressure reading of 12 kPa may be 0.6 "medium" and 0.4 "high." This flexibility makes fuzzy systems more tolerant to measurement errors and uncertainty.

Fuzzy logic has already demonstrated utility in domains such as:

- Medical signal processing – fuzzy ECG filters for arrhythmia detection.
- Image enhancement – fuzzy contrast adjustment.

- Control systems – fuzzy controllers in robotics and automotive systems.

However, its application to smart textiles remains relatively underexplored, despite the natural compatibility between textile signal uncertainty and fuzzy modeling.

MATHEMATICAL FOUNDATIONS OF FUZZY LOGIC IN SMART TEXTILES

Fuzzy logic, first introduced by Lotfi A. Zadeh in 1965, was motivated by the observation that much of human reasoning is approximate rather than exact. In the context of smart textiles, this principle is particularly relevant because textile-based sensors often produce imprecise, noisy, and context-dependent signals. Establishing a rigorous mathematical foundation is essential to leverage fuzzy logic for signal interpretation, adaptive filtering, and decision-making in wearable systems.

This section develops the mathematical basis of fuzzy sets and fuzzy inference, then contextualizes these tools for uncertainty-resilient signal processing in smart textiles.

Classical vs. Fuzzy sets

In classical set theory, an element either belongs to a set or does not. For a universal set U, **a subset** $A \subseteq U$ **is defined by a characteristic function:**

$$\mu A(x) = \begin{cases} 1, x \in A, \\ 0, x \notin A \end{cases}.$$

This binary definition is insufficient for textile signals where boundaries are vague. For example, whether a measured skin-electrode contact quality of 0.62 (on a normalized scale) should be classified as "good" or "bad" cannot be decided crisply.

In contrast, a fuzzy set A on U is defined by a membership function:

$$\mu A(x) : U \rightarrow [0,1].$$

Thus, each element x has a degree of membership $\mu A(x)$ **in A.**

Example: Skin-Electrode Contact Quality

Let U= [0,1] **represent normalized skin-electrode contact. Define a fuzzy set G = "good contact."**

$$\mu G(x) = \begin{cases} 0, 0 \le x \le 0.3, \\ \frac{x - 0.3}{0.4}, 0.3 < x \le 0.7,. \\ 1, x > 0.7 \end{cases}$$

A reading of x=0.62 **yields** $\mu G(0.62)$= 0.62 – 0.30.4≈ 0.8, **meaning the contact quality is "good" with degree 0.8.**

Membership Functions for Textile Signals

The membership function (MF) defines how sensor data are mapped to linguistic variables. Common types include:

Triangular MF:

$$\mu A(x) = \begin{cases} 0, x \le a or x \ge c, \\ \frac{x - a}{b - a}, a < x \le b, \\ \frac{c - x}{c - b}, b < x < c \end{cases}$$

Used in textile strain sensors to represent states like low stretch, medium stretch, high stretch.

Trapezoidal MF:

$$\mu A(x) = \begin{cases} 0, x \le a or x \ge d, \\ \frac{x - a}{b - a}, a < x \le b, \\ 1, b < x \le c, \\ \frac{d - x}{d - c}, c < x < d \end{cases}$$

Useful for broader tolerance regions such as "acceptable skin temperature."

Gaussian MF:

$$\mu A(x) = exp(-(x - c)22\sigma 2).$$

Captures smooth transitions, ideal for biosignal variations (e.g., ECG amplitude).

Generalized Bell MF:

$\mu A(x) = 11 + | x - ca | 2b.$

Flexible shape, suitable for modeling ambiguous categories like "moderate sweat level."

Fuzzy Operations

Given two fuzzy sets A and B on universe U, with membership functions $\mu A(x)$, $\mu B(x)$:

Union:

$\mu A \cup B(x) = max(\mu A(x), \mu B(x)).$

Intersection:

$\mu A \cap B(x) = min(\mu A(x), \mu B(x)).$

Complement:

$\mu A^{-}(x) = 1 - \mu A(x).$

Algebraic product:

$\mu A \otimes B(x) = \mu A(x) \cdot \mu B(x).$

Algebraic sum:

$\mu A \oplus B(x) = \mu A(x) + \mu B(x) - \mu A(x)\mu B(x).$

Example in Textile Application

Suppose a textile suit monitors pressure (P) and humidity (H):

- P: "low pressure" with membership 0.7.
- H: "high humidity" with membership 0.6.

The fuzzy event "low pressure AND high humidity" has membership:

$\mu = min(0.7, 0.6) = 0.6.$

This allows logical combination of uncertain signals.

Fuzzy Relations in Textile Systems

Fuzzy relations describe uncertain mappings between sensor inputs and system outputs.

Let X = input domain (e.g., applied strain), Y = output domain (resistance change). A fuzzy relation $R \subseteq X \times Y$ **has membership function:**

$\mu R(x, y) \in [0, 1]$.

For smart textiles:

- x= applied strain (%)
- y= measured resistance (Ω)

Instead of crisp calibration, fuzzy relation allows a range of outputs with graded confidence.

Composition of Fuzzy Relations

If $R1 \subseteq X \times Y$ **and** $R2 \subseteq Y \times Z$, **composition is:**

$\mu R1 \circ R2(x, z) = max_{y \in Y} min\{\mu R1(x, y), \mu R2(y, z)\}$.

This models multi-stage mappings such as:

(strain → resistance) ∘ (resistance → signal quality).

Fuzzy Numbers and Uncertain Measurements

In smart textiles, sensor outputs are often uncertain, better modeled as fuzzy numbers rather than crisp scalars.

Definition: Fuzzy Number

A fuzzy number x **is a fuzzy set on** R **that is:**

- Normal: $\exists x \in R, \mu x(x) = 1$
- Convex: $\mu x(\lambda x1 + (1 - \lambda)x2) \geq min(\mu x(x1), \mu x(x2))$
- Piecewise continuous.

Common fuzzy numbers:

- Triangular fuzzy number (TFN): $x = (a,b,c)$
- Trapezoidal fuzzy number (TrFN): $x = (a,b,c,d)$

Example: Textile Sensor Resistance

A textile strain sensor reports resistance as $R = (10, 12, 15)\Omega$.

- Most plausible value: 12Ω.
- Lower bound: 10Ω.
- Upper bound: 15Ω.

This represents measurement imprecision explicitly.

Distance and Similarity Measures

For signal classification, we need distance metrics between fuzzy numbers.

- Hamming Distance:

$$dH(A,B) = \int| \ \mu A(x) - \mu B(x) \ | \ dx.$$

- Euclidean Distance:

$$dE(A,B) = \int(\mu A(x) - \mu B(x))^2 dx.$$

- Cosine Similarity:

$$S(A,B) = \frac{\int \mu A(x) \mu B(x) dx}{\sqrt{\int \mu A^2(x) dx} \cdot \sqrt{\int \mu B^2(x) dx}}.$$

Application in Smart Textiles

When comparing two ECG waveforms acquired from textile electrodes under uncertain conditions, fuzzy similarity allows classification even with noisy or shifted signals.

Fuzzy Logic for Uncertain Time-Series

Smart textiles often generate time-series data (e.g., continuous ECG, strain signals). Instead of crisp thresholds, fuzzy logic defines linguistic states over time.

Let signal s(t) be segmented into fuzzy states $\{L,M,H\}$. **Define fuzzy temporal membership:**

$$\mu L(s(t)) = max\{0, 1 - s(t)\alpha\},$$

$$\mu M(s(t)) = exp\left(\frac{-(s(t) - \beta)^2}{2\sigma^2}\right),$$

$$\mu H(s(t)) = 1 - \mu L(s(t)) - \mu M(s(t)).$$

This framework enables fuzzy time-series models for prediction and anomaly detection in textile biosignals.

In this section, we built the mathematical foundation of fuzzy logic as applied to smart textiles:

- Defined fuzzy sets and membership functions to represent uncertain textile sensor states.
- Introduced fuzzy operations and relations for combining signals.
- Applied fuzzy numbers to represent imprecise measurements.
- Presented distance measures for fuzzy classification of textile signals.
- Extended fuzzy logic to time-series modeling for wearable applications.

These mathematical structures provide the groundwork for implementing fuzzy inference systems and adaptive filters that will be developed in the following sections.

FUZZY INFERENCE SYSTEMS FOR SIGNAL PROCESSING IN SMART TEXTILES

Smart textiles integrate sensors, actuators, and embedded electronics into fabrics, enabling continuous monitoring of physiological and environmental parameters. These systems generate signals that are inherently uncertain due to noise, motion artifacts, non-linearities, and unpredictable user behavior. Classical signal processing approaches, which rely on strict probabilistic assumptions or deterministic filters, often fail to adapt to such uncertainties.

Fuzzy Inference Systems (FIS) provide a human-inspired, uncertainty-resilient computational framework for interpreting textile-based sensor signals. This section explores the design, mathematical structures, and implementation of FIS in smart textile signal processing.

Fundamentals of Fuzzy Inference in Textile Signals

A fuzzy inference system consists of four main components:

- Fuzzification: Transforms crisp sensor inputs into fuzzy values.
- Rule Base: Encodes expert knowledge using fuzzy if–then rules.
- Inference Engine: Evaluates the rules to determine fuzzy outputs.
- Defuzzification: Converts fuzzy outputs into crisp, actionable signals.

Let $U = \{u1, u2, \ldots, un\}$ **denote the universe of discourse for textile signals (e.g., ECG amplitude, skin temperature, resistance of conductive yarns).**

A fuzzy set ***A* in *U* is defined as:**

$$A = \{(u, \mu A(u)) \mid u \in U\}$$

where $\mu A(u): U \rightarrow [0,1]$ **is the membership function.**

For instance, in textile-based temperature monitoring:

$$\cdot\ \mu Low(T) = exp(-(T-25)22\sigma 2)$$

$$\cdot\ \mu High(T) = 1 - \mu Low(T)$$

with T in °C and σ a fuzziness parameter.

Types of Fuzzy Inference Systems for Smart Textiles

Mamdani-Type FIS

- Uses linguistic rules and fuzzy outputs.
- Example rule:

IF pressure is High AND motion is Low THEN posture is Stable

- The fuzzy output set (Stable) is aggregated and later defuzzified.

Mathematically, the output fuzzy set B is:

$$\mu B(y) = maximin(\mu A1i(x1), \mu A2i(x2), \ldots, \mu Ani(xn), \mu Ci(y))$$

where Aji **is the** jth **antecedent fuzzy set of rule** i**, and** Ci **is the consequent fuzzy set.**

Sugeno-Type FIS

- Outputs are crisp functions of the inputs.
- Example rule:

IF strain is Medium AND humidity is High THEN signal = 0.7 . strain + 0.2 . humidity + 0.1

For mmm rules, the overall output is:

$$y = \frac{\sum_{i=1}^{m} fi(x1, x2, \ldots, xn)}{\sum_{i=1}^{m} wi}$$

where $wi = \prod_{j=1}^{n} \mu Aji(xj)$ is the firing strength of rule $\boldsymbol{i}$.

Adaptive Neuro-Fuzzy Inference Systems (ANFIS)

- Combines fuzzy logic with neural networks to learn membership functions and rules.
- Especially useful for complex textile signals such as electromyography (EMG) or ECG acquired from textile electrodes.

Fuzzy Rules for Textile Signal Interpretation

Example: Heart Rate Variability (HRV) from Textile ECG

- Inputs: Signal amplitude (A), noise level (N), electrode resistance (R).
- Output: Signal quality (Q).

Rule base:

- IF A is High AND N is Low AND R is Low THEN Q is Excellent.
- IF A is Medium AND N is Medium AND RRR is High THEN Q is Poor.
- IF A is Low OR N is High THEN Q is Unreliable.

Mathematical Framework for Signal Filtering Using FIS

Let raw textile sensor data be:

$$S(t) = s(t) + \eta(t)$$

where $s(t)$ **is the true physiological signal and $\eta(t)$ is noise.**
Define fuzzy sets:

- $A1$: **"Low Noise"**
- $A2$: **"High Noise"**
- $B1$: **"Smooth Signal"**
- $B2$: **"Unreliable Signal"**

The fuzzy filtering process:

- Noise Evaluation:

$$\mu A1(\eta) = exp\left(-\frac{\eta^2}{2\sigma^2}\right), \mu A2(\eta) = 1 - \mu A1(\eta)$$

- Rule Application:
- IF Noise is Low THEN Output = Smooth Signal
- IF Noise is High THEN Output = Unreliable
- Defuzzification (Centroid method):

$$s(t) = \frac{\int y \cdot \mu B(y)\, dy}{\int \mu B(y)\, dy}$$

where s(t) is the reconstructed textile signal.

Posture Recognition with Textile Pressure Sensors

Consider a smart textile chair cover with distributed pressure sensors. Inputs are sensor readings:

$$x = (x1, x2, \ldots, xn), xi \in [0,1]$$

Define fuzzy sets:

- Low Pressure: $\mu Low(x) = \max(0, 1-2x)$
- Medium Pressure: $\mu Med(x) = 1 - |2x-1|$
- High Pressure: $\mu High(x) = \max(0, 2x-1)$

Rules:

- IF pressure at back is High AND seat is Medium THEN posture = Leaning Forward.
- IF pressure at back is Medium AND seat is High THEN posture = Upright.

The output posture classification is obtained by aggregating fuzzy sets and selecting the maximum membership (defuzzification via max-membership principle). Table 1 provides a comparative overview of Mamdani, Sugeno, and ANFIS fuzzy inference systems, highlighting their advantages, limitations, and applications in smart textiles.

Table 1. Comparative analysis of different FIS approaches in smart textile applications

FIS Type	Advantages	Limitations	Smart Textile Application
Mamdani	Human-like reasoning	Computationally intensive	Posture detection
Sugeno	Efficient, suitable for real-time	Less interpretable	Real-time ECG/EMG
ANFIS	Learns from data	Requires training	Adaptive noise filtering

Remark

- FIS provides a structured way to handle imprecision and uncertainty in textile-based sensor signals.
- Mamdani systems are interpretable, Sugeno systems are efficient, and ANFIS offers adaptivity.
- Applications include ECG noise reduction, posture recognition, temperature monitoring, and motion artifact filtering.

HYBRID FUZZY MODELS IN SMART TEXTILES

The dynamic and multi-dimensional environment of smart textiles introduces multiple layers of uncertainty—stemming from sensor noise, environmental fluctuations, and variations in user physiology. While conventional fuzzy inference systems (FIS) provide a robust foundation for handling vagueness, they may fall short in capturing nonlinearities, time dependencies, and adaptive learning requirements. Hybrid fuzzy models emerge as a natural extension by combining fuzzy logic with complementary computational techniques such as neural networks, genetic algorithms, wavelet transforms, and quantum-inspired methods. These hybrid approaches exploit the interpretability of fuzzy systems and the learning/adaptive capabilities of machine intelligence, thereby offering more resilient signal processing frameworks for wearable systems.

This section presents the theoretical basis, mathematical structures, and applications of hybrid fuzzy models in smart textiles, focusing on their ability to improve adaptability, precision, and robustness in real-world conditions.

Neuro-Fuzzy Systems in Smart Textiles

A neuro-fuzzy system (NFS) integrates fuzzy inference with artificial neural networks (ANNs), wherein fuzzy rules are adaptively tuned through neural learning algorithms. This synergy is particularly useful in textile-based wearable systems where data streams evolve dynamically, such as continuous heart rate monitoring or gait recognition.

Architecture

An Adaptive Neuro-Fuzzy Inference System (ANFIS) can be described in layered form:

- Layer 1 (Input fuzzification): Each input variable xi is mapped to fuzzy membership functions.

$$Oi^{(1)} = \mu Aj(xi), i = 1,2,\ldots,n$$

- Layer 2 (Rule firing strength): The degree of activation for each rule ***Rj* is calculated using a t-norm operator (e.g., product):**

$$Oj^{(2)} = wj = \prod_{i=1}^{n} \mu Aij(xi)$$

- Layer 3 (Normalization): $Oj^{(3)} = w^{-}j = \frac{wj}{\sum_k wk}$
- Layer 4 (Consequent): Each normalized weight generates a rule output:

$$Oj^{(4)} = w^{-}j \cdot (pjx1 + qjx2 + \cdots + rj)$$

- Layer 5 (Aggregation): Final output:

$$y = \sum_j Oj^{(4)}$$

Applications

- ECG anomaly detection in textile electrodes.
- Gait cycle prediction in smart socks with inertial sensors.
- Thermal regulation in intelligent fabrics that adapt to body heat patterns.

By leveraging gradient descent and backpropagation, the NFS adapts fuzzy membership parameters, thereby reducing reliance on manual tuning.

Genetic Fuzzy Systems (GFS) for Wearable Optimization

Genetic algorithms (GA) provide an evolutionary optimization framework for fuzzy systems. In smart textiles, GFS can optimize membership functions, fuzzy rules, and sensor fusion parameters to improve performance under non-stationary environments.

Genetic Representation

- Chromosomes encode fuzzy rule sets or membership parameters.
- Fitness function evaluates classification accuracy, energy consumption, or comfort level in wearable systems.

For a chromosome $C = \{c1, c2, \ldots, cm\}$, **the fitness function may be:**

$$F(C) = \alpha \cdot Acc(C) + \beta \cdot Eff(C) - \gamma \cdot Err(C)$$

where:

- $Acc(C)$: **classification/recognition accuracy,**
- $Eff(C)$: **energy efficiency in signal processing,**
- $Err(C)$: **sensor error rate,**
- α, β, γ: **are weighting factors.**

Application Examples

- Posture correction garments: Optimize fuzzy rules for detecting slouching vs. upright postures.
- Smart gloves: Enhance gesture recognition accuracy under varying lighting and movement noise.
- Energy-aware processing: Extend battery life of e-textiles by evolving rules that balance accuracy and energy trade-offs.

Fuzzy-Wavelet Hybrid Models

Wavelet transforms capture local time-frequency characteristics, while fuzzy logic handles imprecise classification. This combination is vital for transient, non-stationary signals in smart textiles (e.g., sudden muscular contractions, rapid thermal changes).

Mathematical Framework

A signal s(t) collected from textile sensors is decomposed:

$$W(a, b) = \int_{-\infty}^{\infty} s(t)\psi^* \left(\frac{t-b}{a}\right) dt$$

where:

- a: scale,
- b: translation,
- ψ: mother wavelet.

Extracted features fk **(e.g., energy, entropy) are fuzzified:**

$$\mu A(fk) = \frac{1}{1 + e^{-\lambda(fk-c)}}$$

Applications

- Seizure detection in EEG headbands.
- Respiratory monitoring in chest bands.
- Impact detection in protective sportswear.

The fuzzy-wavelet hybrid enhances sensitivity to localized changes, reducing false alarms in health-critical applications.

Type-2 Fuzzy-Neural Hybrids

In wearable contexts, higher-order uncertainty arises when even the membership grades themselves are uncertain due to unstable fabric-electrode contact or environmental noise. Interval Type-2 Fuzzy Systems (IT2FS) combined with neural learning provide robust uncertainty management.

Structure

For an input x, the IT2 membership function is:

$$\mu A(x) = \{(x, u) \mid u \in [\mu L(x), \mu U(x)]\}$$

where μL **and** μU **are lower and upper membership functions.**

A neural layer adapts $\mu L(x)$ **and** $\mu U(x)$ **dynamically based on real-time data streams.**

Applications

- Cardiac monitoring under movement artifacts (jogging, sports).
- Smart bandages: Detect infection by interpreting uncertain bio-sensor signals.
- Temperature-sensitive fabrics: Handle overlapping thermal thresholds.

Quantum-Inspired Fuzzy Models

The advent of quantum-inspired fuzzy algorithms opens new opportunities in high-dimensional decision-making for wearable systems. Instead of crisp membership values, quantum probability amplitudes model superposition of states, enhancing adaptability.

Mathematical Formulation

Each fuzzy rule is represented in a Hilbert space:

$$|\psi\rangle = \sum_{i=1}^{n} \alpha i|xi\rangle, \sum_{i=1}^{n} | \alpha i|^2 = 1$$

Decision surface:

$$y = \sum_{j=1}^{n} | \alpha j|^2 . \mu Aj(x)$$

Potential Applications

- Cognitive load detection in augmented reality wearables.
- Simultaneous multi-signal processing (EEG + ECG + EMG fusion).
- Adaptive textile encryption protocols for secure health data transmission.

Table 2. Comparative analysis of hybrid fuzzy models for smart textile applications

Hybrid Model	Strengths	Weaknesses	Application in Smart Textiles
Neuro-Fuzzy (ANFIS)	Adaptive, interpretable	Requires large training data	ECG, gait cycle analysis
Genetic Fuzzy (GFS)	Evolves rules automatically	Computationally intensive	Posture correction, energy optimization
Fuzzy-Wavelet	Localized time-frequency analysis	Sensitive to wavelet choice	Respiratory monitoring, seizure detection
Type-2 Fuzzy-Neural	Robust to high uncertainty	More complex computations	Cardiac monitoring, smart bandages
Quantum-Inspired Fuzzy	Parallelism, uncertainty modeling	Still theoretical in practice	Secure multi-signal fusion

Hybrid fuzzy models bridge the gap between the interpretability of fuzzy systems and the adaptability of advanced computational methods. By combining fuzzy logic with neural learning, evolutionary optimization, wavelet decomposition, type-2 uncertainty handling, and quantum-inspired frameworks, smart textiles can achieve resilience against diverse forms of noise, nonlinearity, and uncertainty. These approaches enhance the precision of wearable monitoring, extend energy efficiency, and open avenues for intelligent decision-making in real-time healthcare and lifestyle applications.

APPLICATIONS

Fuzzy logic has emerged as a powerful decision-making and uncertainty-handling framework in the domain of smart textiles. Beyond theory, its value is best understood through real-world applications where wearable systems must process noisy, imprecise, and context-dependent signals. In this section, we present detailed application scenarios that demonstrate how fuzzy inference systems (FIS), type-2 fuzzy sets, and hybrid fuzzy models can be employed for uncertainty-resilient signal processing in electronic textiles (e-textiles).

Fuzzy Logic for Motion Artifact Reduction in Smart ECG Textiles

Textile-based electrodes integrated into garments are increasingly used for electrocardiogram (ECG) monitoring. However, these sensors are vulnerable to motion artifacts caused by limb movement, fabric shifting, or sweat accumulation. Traditional signal processing techniques often fail to separate low-amplitude cardiac signals from noise under such dynamic conditions.

We construct a Mamdani Fuzzy Inference System (MFIS) to classify incoming ECG signals as either clean or artifact-contaminated and to generate corrective filtering actions.

Input Variables

- Signal-to-Noise Ratio (SNR): $x1 \in [0,40] dB$.
- Electrode Contact Quality (ECQ): $x2 \in [0,1]$ ($0 = poor$, $1 = excellent$).
- Motion Intensity (MI): $x3 \in [0,10] m/s^2$.

Output Variable

- Filtering Level (FL): $y \in \{Low, Medium, High\}$.

Membership Functions

For SNR:

$$\mu Low(x1) = max\left(0, \frac{20 - x1}{20}\right),$$

$$\mu Medium(x1) = exp\left(-\frac{(x1 - 20)^2}{2\sigma^2}\right),$$

$$\mu High(x1) = min\left(1, \frac{x1-20}{20}\right).$$

For ECQ and MI, Gaussian membership functions are employed.

Rule Base

- IF SNR is Low AND ECQ is Poor AND MI is High THEN FL = High.
- IF SNR is Medium AND ECQ is Fair AND MI is Moderate THEN FL = Medium.
- IF SNR is High AND ECQ is Excellent THEN FL = Low.

Defuzzification

We apply centroid defuzzification:

$$y* = \frac{\int y \cdot \mu FL(y)\, dy}{\int \mu FL(y)\, dy}$$

Type-2 Fuzzy Logic for Temperature Regulation in Smart Heating Textiles

Smart textiles with integrated heating fibers are increasingly used in cold-weather garments and therapeutic devices. However, sensor drift, delayed responses, and individual thermal comfort variability make traditional PID or crisp controllers suboptimal.

We use an Interval Type-2 Fuzzy Logic System (IT2FLS) for regulating heating elements.

Input Variables

- Ambient Temperature (AT): $x1 \in [-20, 40] \circ C$
- Skin Temperature (ST): $x2 \in [20, 40] \circ C$
- Activity Level (AL): $x3 \in \{Low, Medium, High\}$

Output Variable

- Heating Power (HP): $y \in [0,100]\%$

Type-2 Membership Functions

Each fuzzy set has an upper membership function (UMF) and lower membership function (LMF), e.g.,

$$\mu_{Warm}^{UMF}(x2) = exp\Big(-\frac{(x2-32)^2}{2\sigma^2}\Big),$$

$$\mu_{Warm}^{LMF}(x2) = 0.8 \cdot \mu_{Warm}^{UMF}(x2).$$

This provides a Footprint of Uncertainty (FOU) to capture variability.

Rule Base

1. IF AT is Cold AND ST is Low THEN HP = High.
2. IF AT is Moderate AND ST is Comfortable THEN HP = Medium.
3. IF AT is Warm OR AL is High THEN HP = Low.

Type-Reduction

We use Karnik–Mendel iterative algorithm for type-reduction before defuzzification:

$$y^* = yl + yr2,$$

where ***yl* and *yr* are left and right centroids of FOUs.**

Hybrid Neuro-Fuzzy Approach for Gesture Recognition in Smart Gloves

Smart gloves with conductive fibers and flexible sensors enable hand gesture recognition for human-computer interaction (HCI). However, textile deformation, finger-length variation, and sensor non-linearities introduce uncertainty.

We employ an Adaptive Neuro-Fuzzy Inference System (ANFIS) that integrates fuzzy rules with neural network training.

Input Variables

- Flex Sensor Readings (FSR): $x = (x1, x2, \ldots, x5)$.
- Inertial Sensor Features (ISF): orientation and acceleration values.

Output Variable

- Gesture Class (G): $y \in \{A, B, C, D, E\}$.

ANFIS Structure

- Layer 1 (Fuzzification): Each input mapped into fuzzy sets using Gaussian functions.
- Layer 2 (Rule Strength): Rules of the form:

Ri: IF x1 is A1i AND…AND x5 is A5i THEN $y = pix1 + \ldots + qi$.

- Layer 3 (Normalization): Rule weights normalized.
- Layer 4 (Defuzzification): Weighted sum of outputs.
- Layer 5 (Final Output): Gesture prediction.

Fuzzy Soft Set for Health Monitoring in Smart Bed Textiles

Smart beds equipped with textile pressure sensors and embedded ECG/respiration monitoring are used for elderly care. A major challenge is handling uncertain, incomplete, and imprecise data.

Fuzzy Soft Set Model

We model patient state classification using fuzzy soft sets (FSS).
Let:

$U = \{u1, u2, \ldots, un\}$: **patients.**

$E = \{e1, e2, \ldots, em\}$:: **parameters (heart rate, respiration rate, movement, SpO2).**

A fuzzy soft set is:

$$FA = \{(e, \mu F(e)(u)) \mid e \in A, u \in U\}.$$

Where $\mu F(e)(u) \in [0,1]$ **indicates degree of abnormality.**

Decision-Making

We compute weighted decision scores:

$$D(u) = \sum_{e \in A} we \cdot \mu F(e)(u).$$

Threshold-based decision:

$\cdot\ D(u) < 0.3 : Normal.$

$\cdot\ 0.3 \le D(u) < 0.6 : AtRisk.$

$\cdot\ D(u) \ge 0.6 : Critical.$

Comparative Insights

- Mamdani FIS excels in handling motion artifacts in ECG but may struggle with high-dimensional input.
- Type-2 fuzzy systems provide resilience to sensor drift and individual variability, suitable for comfort regulation.
- Hybrid ANFIS offers strong performance in pattern recognition (gestures) where training data is available.
- Fuzzy soft sets handle incompleteness and imprecision well, making them suitable for health monitoring in ambient intelligence settings.

CHALLENGES, LIMITATIONS, AND FUTURE DIRECTIONS

The integration of fuzzy logic into smart textiles for signal processing represents a promising direction toward uncertainty-resilient wearable systems. However, despite its merits, several technical and practical challenges must be addressed before widespread deployment in healthcare, sports, defense, and consumer applications becomes feasible. This section systematically examines the key challenges and limitations, followed by a roadmap of future directions to strengthen the role of fuzzy logic in next-generation smart textiles.

Challenges in Implementing Fuzzy Logic for Smart Textiles

Hardware Constraints

Smart textiles often incorporate conductive fibers, flexible sensors, and embedded microcontrollers with limited processing power and memory. Implementing advanced fuzzy inference systems (FIS), particularly type-2 or hybrid fuzzy-neural systems, demands significant computational resources.

For example, consider a Type-2 Mamdani FIS with M membership functions per variable and N input variables. The rule base complexity grows as:

$$R = MN$$

where R denotes the number of fuzzy rules. Even with modest values (M=4, N=5), **the system requires R=4096 rules, which exceeds the computational capacity of typical textile-integrated microcontrollers.**

Sensor Noise and Drift

Wearable textile sensors are subject to environmental interference (humidity, sweat, mechanical deformation). Such distortions lead to signal drift, requiring robust adaptive fuzzy filtering. Current fuzzy systems often assume stable membership functions, but in practice, these must evolve dynamically.

For example, let $\mu(x,t)$ **denote the time-varying membership function of an input signal xxx. The drift can be modeled as:**

$$\mu(x,t) \; = \; \mu 0(x) + \delta(t)$$

where $\mu 0(x)$ **is the original membership function and δ(t) is the time-dependent distortion term. Designing self-correcting fuzzy systems that recalibrate δ(t) remains an open challenge.**

Energy Consumption

Smart textiles are battery-constrained systems. Fuzzy inference requires repetitive evaluation of membership degrees, rule firing strength, and defuzzification. For example, centroid defuzzification of an aggregated fuzzy set A requires:

$$y* = \frac{\int \mu A(x)dx}{\int x \cdot \mu A(x)dx}$$

This integral is computationally intensive on low-power hardware, raising concerns about energy sustainability in continuous monitoring scenarios.

Interoperability of Multi-Modal Data

Smart textiles frequently integrate heterogeneous sensors (ECG, EMG, pressure, temperature). Fuzzy logic must fuse these multi-modal inputs into coherent decisions. However, sensor signals often operate at different sampling frequencies and may exhibit varying uncertainty levels. Achieving synchronization while maintaining interpretability of fuzzy rules remains difficult.

User-Specific Variability

Smart textiles interact with diverse users differing in physiology, behavior, and activity patterns. Fuzzy membership functions designed for one population may not generalize across users. For example, an activity recognition system with fuzzy rules tailored to young athletes may misclassify signals from elderly users. This highlights the challenge of personalized fuzzy model tuning.

Limitations of Current Approaches

Despite significant progress, the following limitations restrict the scalability of fuzzy-based smart textiles:

- Rule Explosion Problem: As input dimensions grow, the number of fuzzy rules becomes computationally prohibitive.
- Static Membership Functions: Most systems assume pre-defined membership functions, which are not adaptive to sensor drift or user-specific variations.
- Limited Interpretability in Hybrid Models: When fuzzy logic is combined with neural networks or evolutionary algorithms, the system may lose transparency.
- Lack of Standardization: Current fuzzy implementations in smart textiles are case-specific, with no standardized frameworks for evaluation.
- Reliability in Real-World Conditions: Laboratory-tested fuzzy systems often fail under dynamic real-world conditions like motion artifacts, temperature changes, or sweat accumulation.

Future Directions

To address the challenges above, the following future directions are proposed:

Adaptive and Self-Learning Fuzzy Systems

The next generation of smart textiles should integrate adaptive fuzzy systems, where membership functions evolve over time. A potential model is the evolving fuzzy system (EFS), where rules are updated incrementally as new data arrives:

$$\mu i(x, t+1) = \mu i(x, t) + \eta \cdot \Delta\mu i(x)$$

where η is the adaptation rate. Such self-learning capabilities will enable smart textiles to maintain robustness under changing conditions (e.g., prolonged use, environmental variations).

Type-2 Fuzzy Systems for Uncertainty Management

Type-2 fuzzy sets inherently handle higher-order uncertainty due to their Footprint of Uncertainty (FOU). For wearable contexts with highly noisy inputs, interval Type-2 fuzzy sets (IT2FS) can better represent ambiguity:

$$\mu A(x) \in [\overline{\mu}A(x), \overline{\mu}A(x)]$$

Smart textiles using IT2FS can more accurately model the range of uncertainty caused by signal drift, environmental factors, and user variability.

Integration with Machine Learning

Hybrid fuzzy-neural and fuzzy-evolutionary systems can optimize membership functions and reduce rule explosion. For instance, Neuro-Fuzzy models (ANFIS) can automatically learn fuzzy rules from training data, while preserving interpretability. This creates personalized smart textiles adaptable to individual users.

A case study could involve adaptive calibration of a fuzzy-based gait recognition system using reinforcement learning:

$$Q(s,a) \leftarrow Q(s,a) + \alpha[r + \gamma maxa'Q(s',a') - Q(s,a)]$$

where the Q-learning framework updates fuzzy decision rules for gait analysis.

Edge AI and Neuromorphic Fuzzy Processing

To overcome hardware and energy constraints, future smart textiles can employ edge-AI hardware accelerators and neuromorphic chips that mimic brain-like low-power computing. Implementing fuzzy inference in neuromorphic architectures (e.g., memristor-based designs) may drastically reduce computational cost.

Standardization and Benchmarking

For widespread adoption, there is a need to establish benchmark datasets, evaluation metrics, and standard architectures for fuzzy signal processing in smart textiles. Currently, lack of reproducibility hampers comparisons across research groups. A unified benchmark would accelerate innovation and deployment.

Integration with Blockchain and IoT

In large-scale healthcare or sports monitoring systems, fuzzy textile data must be securely transmitted and shared. Blockchain-integrated fuzzy systems could ensure data integrity and enable decentralized decision-making among multiple wearables in an IoT ecosystem.

Human-Centric and Ethical Design

Finally, as smart textiles move toward real-world adoption, privacy, data ethics, and human comfort must be central. Fuzzy systems should not only optimize technical performance but also align with human values such as transparency, trust, and inclusiveness.

Case-Style Illustration: Future Gait Monitoring Smart Textile

Consider a gait-monitoring smart textile designed for elderly users. The system uses fuzzy logic to detect fall risks by analyzing stride length, balance, and foot pressure. Current systems struggle due to variability across users and environments.

A future hybrid fuzzy model could combine:

- Type-2 fuzzy sets to capture uncertainty in sensor signals (pressure sensors in shoes).
- Neuro-fuzzy learning to personalize membership functions for each user.
- Edge-AI neuromorphic chips to minimize power usage.
- Blockchain integration to securely share gait analytics with caregivers.

Such a system could ensure adaptive, reliable, and ethical deployment of fuzzy smart textiles in real-world healthcare.

Fuzzy logic in smart textiles presents a transformative framework for uncertainty-resilient signal processing. Yet, challenges such as computational limits, sensor drift, energy constraints, and rule explosion hinder its current effectiveness. Future innovations in adaptive fuzzy systems, hybrid learning, neuromorphic hardware, and standardized evaluation hold the potential to overcome these hurdles. Ultimately, a synergy of fuzzy logic, machine learning, IoT, and human-centric design will pave the way for next-generation smart textiles capable of robust, personalized, and ethical decision-making.

CONCLUSION AND FUTURE OUTLOOK

The integration of fuzzy logic into smart textile systems represents a significant advancement in the design of wearable technologies that are capable of functioning under conditions of uncertainty, noise, and environmental variability. This chapter has explored how fuzzy set theory and its extensions—ranging from Type-1 to Type-2, intuitionistic, hesitant, and hybrid fuzzy-neural frameworks—can provide a robust mathematical foundation for uncertainty-resilient signal processing in smart textiles.

By embedding fuzzy inference systems (FIS) into the computational pipeline of textile-based sensors, the following key insights have been demonstrated:

- Robustness to Noise and Vagueness: Fuzzy logic naturally accommodates imprecision in signals obtained from textile-based sensors, where disturbances such as sweat, motion artifacts, and environmental interference often degrade signal quality. Instead of enforcing binary decisions, fuzzy systems enable a graded, human-like interpretation of sensor readings.
- Mathematical Generalization of Textile Signals: Using fuzzy membership functions and aggregation operators, we derived models for textile-based ECG, EMG, and temperature signals, showing how membership-based transformations improve stability and reliability in feature extraction.
- Hybrid Architectures for Enhanced Performance: The exploration of fuzzy–neural, fuzzy–Bayesian, and fuzzy–quantum hybrid models illustrated the next wave of textile intelligence. These architectures enhance learning, adaptability, and context-awareness, essential for real-world applications like continuous health monitoring, activity recognition, and adaptive thermal regulation.

Challenges and Current Limitations

Despite these advances, challenges remain:

- Energy Efficiency: Current fuzzy hardware implementations in textiles struggle with power optimization for real-time inference.
- Scalability of Hybrid Systems: Large-scale deployment requires efficient fusion of fuzzy logic with deep learning and IoT frameworks.
- Standardization: Lack of universal benchmarking datasets for textile-based fuzzy processing limits comparability and reproducibility.
- Hardware Constraints: Embedding fuzzy processors into flexible, durable, and washable textile substrates is a non-trivial engineering challenge.

Future Research Directions

Looking ahead, several transformative directions can shape the evolution of fuzzy logic in smart textiles:

- Neutrosophic and Multivalued Fuzzy Models: Extending fuzzy logic into neutrosophic domains—where truth, falsity, and indeterminacy coexist—will allow smart textiles to handle extreme uncertainty, such as ambiguous signals in emergency monitoring.
- Quantum-Enhanced Fuzzy Computing: Leveraging quantum fuzzy inference systems for textile-based signal processing could dramatically increase computational efficiency and parallelism, enabling real-time adaptation in large-scale smart clothing ecosystems.
- Federated Fuzzy Learning for Privacy-Preserving Wearables: Combining federated learning with fuzzy decision models allows multiple smart textile devices to collaborate while ensuring data privacy—crucial for medical applications.
- Energy-Aware Fuzzy Hardware Accelerators: Future textile electronics may embed memristor-based fuzzy hardware or neuromorphic fuzzy processors for ultra-low power consumption, making fuzzy inference sustainable in daily-use smart garments.
- Standardized Benchmarking Frameworks: Developing open-access datasets and performance metrics specific to fuzzy-driven textile signal processing will accelerate cross-disciplinary collaboration and validation.

The future of smart textiles lies in their ability to intelligently sense, interpret, and respond to real-world conditions. Fuzzy logic provides the mathematical and computational bridge between raw, imprecise sensor signals and actionable insights for healthcare, sports, defense, and personalized lifestyle applications. By embedding uncertainty-resilient reasoning directly into the fabric, we move closer to realizing the vision of truly adaptive, intelligent, and human-centric wearable systems.

In conclusion, the convergence of fuzzy mathematics, textile engineering, and intelligent computing promises a new generation of wearable technologies that are not only functional but context-aware, adaptive, and resilient to uncertainty. This positions fuzzy logic not just as a theoretical tool, but as a practical enabler of next-generation smart textile ecosystems.

REFERENCES

Abdoli, S. (2022). Application of fuzzy-logic for design assessment of complex engineering systems in the early design stages. *Journal of Engineering Design*, *33*(3), 234–258. DOI: 10.1080/09544828.2022.2031918

Baz, A., Logeshwaran, J., Natarajan, Y., & Patel, S. (2024). Deep fuzzy nets approach for energy efficiency optimization in smart grids. *Applied Soft Computing*, *161*, 111724. DOI: 10.1016/j.asoc.2024.111724

Beke, A., & Kumbasar, T. (2023). More Than Accuracy: A Composite Learning Framework for Interval Type-2 Fuzzy Logic Systems. *IEEE Transactions on Fuzzy Systems*, *31*(3), 734–744. DOI: 10.1109/TFUZZ.2022.3188920

Bressane, A., Da Silva Garcia, A., De Castro, M., Xerfan, S., Ruas, G., & Negri, R. (2024). Fuzzy Machine Learning Applications in Environmental Engineering: Does the Ability to Deal with Uncertainty Really Matter? *Sustainability (Basel)*, *16*(11), 4525. Advance online publication. DOI: 10.3390/su16114525

Chen, Z., Ming, X., Zhou, T., & Chang, Y. (2020). Sustainable supplier selection for smart supply chain considering internal and external uncertainty: An integrated rough-fuzzy approach. *Applied Soft Computing*, *87*, 106004. DOI: 10.1016/j.asoc.2019.106004

Das, A. K. (2016). On partially included intuitionistic fuzzy rough relations. *Afrika Matematika*, *27*(5-6), 993–1001. DOI: 10.1007/s13370-016-0395-2

Das, A. K. (2018). Weighted fuzzy soft multiset and decision-making. *International Journal of Machine Learning and Cybernetics*, *9*(5), 787–794. DOI: 10.1007/s13042-016-0607-y

Das, A. K., Das, R., Das, S., Debnath, B. K., Granados, C., Shil, B., & Das, R. (2025c). A comprehensive study of neutrosophic SuperHyper BCI-semigroups and their algebraic significance. *Transactions on Fuzzy Sets and Systems*, *4*(2), 80–101. DOI: 10.71602/tfss.2025.1198050

Das, A. K., Das, R., Datta, R., & Granados, C. (2025e). An Innovative Approach to Group Decision-Making Based on Weighted Hypersoft Expert System. In Smarandache, F., Jdid, M., & Leyva-Vazquez, M. (Eds.), *Neutrosophic and Plithogenic Inventory Models for Applied Mathematics* (pp. 483–518). IGI Global Scientific Publishing., DOI: 10.4018/979-8-3693-3204-7.ch019

Das, A. K., & Granados, C. (2021). An advanced approach to fuzzy soft group decision-making using weighted average ratings. *SN Computer Science*, *2*(6), 471. DOI: 10.1007/s42979-021-00873-5

Das, A. K., & Granados, C. (2022a). A new fuzzy parameterized intuitionistic fuzzy soft multiset theory and group decision-making. *Journal of Current Science and Technology*, *12*(3), 547–567. DOI: 10.14456/jcst.2022.42

Das, A. K., & Granados, C. (2022b). FP-intuitionistic multi fuzzy N-soft set and its induced FP-Hesitant N soft set in decision-making. *Decision Making: Applications in Management and Engineering*, *5*(1), 67–89. DOI: 10.31181/dmame181221045d

Das, A. K., & Granados, C. (2023a). IFP-intuitionistic multi fuzzy N-soft set and its induced IFP-hesitant N-soft set in decision-making. *Journal of Ambient Intelligence and Humanized Computing*, *14*(8), 10143–10152. DOI: 10.1007/s12652-021-03677-w

Das, A. K., & Granados, C. (2023b). Preference intuitionistic fuzzy rough relation and its theoretical approach. *Journal of the Indian Mathematical Society*, *90*(3-4), 199–212.

Das, A. K., Granados, C., & Bhattacharya, J. (2022). Some new operations on fuzzy soft sets and their applications in decision-making. *Songklanakarin Journal of Science and Technology*, *44*(2), 440–449. DOI: 10.14456/sjst-psu.2022.61

Das, A. K., Gupta, N., & Granados, C. (2023). Weighted hesitant bipolar-valued fuzzy soft set in decision-making. *Songklanakarin Journal of Science and Technology*, *45*(6), 681–690. https://sjst.psu.ac.th/journal/45-6/10.pdf

Das, A. K., Gupta, N., Mahmood, T., Tripathy, B. C., Das, R., & Das, S. (2025a). An efficient water quality evaluation model using weighted hesitant fuzzy soft sets for water pollution rating. In Kumar, A., Kumar, P., Rathee, S., & Kumar, B. (Eds.), *Mechatronics: Concepts, Tools, Applications, and New Trends* (1st ed., p. 17). CRC Press., DOI: 10.1201/9781003494478-10

Das, A. K., Gupta, N., Mahmood, T., Tripathy, B. C., Das, R., & Das, S. (2025b). An innovative fuzzy multi-criteria decision making model for analyzing anthropogenic influences on urban river water quality. *Iran Journal of Computer Science*, *8*(1), 103–124. DOI: 10.1007/s42044-024-00211-x

Das, A. K., Mahmood, T., Das, R., & Das, S. (2025d). Optimizing Inventory Levels in Retail: Fuzzy Logic-Based Decision Support Systems for Adaptive Inventory Management. In Jana, S., Basu Mallik, B., Sarkar, A., & Jana, C. (Eds.), *Applications of Fuzzy Logic in Decision Making and Management Science. Information Systems Engineering and Management* (Vol. 29). Springer., DOI: 10.1007/978-3-031-77719-6_8

De Iuliis, M., Kammouh, O., & Cimellaro, G. (2022). Measuring and improving community resilience: A fuzzy logic approach. *International Journal of Disaster Risk Reduction, 78*, 103118. Advance online publication. DOI: 10.1016/j.ijdrr.2022.103118

Erdoğdu, A., Dayı, F., Yildiz, F., Yanik, A., & Ganji, F. (2025). Combining Fuzzy Logic and Genetic Algorithms to Optimize Cost, Time and Quality in Modern Agriculture. *Sustainability (Basel), 17*(7), 2829. Advance online publication. DOI: 10.3390/su17072829

Eshuis, R., Firat, M., & Kaymak, U. (2021). Modeling uncertainty in declarative artifact-centric process models using fuzzy logic. *Information Sciences, 579*, 845–862. DOI: 10.1016/j.ins.2021.07.075

. Granados, C., Das, A.K., Das, B. (2022). Some continuous neutrosophic distributions with neutrosophic parameters based on neutrosophic random variables. Advances in the Theory of Nonlinear Analysis and its Applications, 33(2): 293-308. DOI: 10.31197/atnaa.1056480

. Granados, C., Das, A.K., Osu, B. (2023). Weighted neutrosophic soft multiset and its application to decision making. Yugoslav journal of operations research, 33(2): 293-308. DOI: 10.2298/YJOR220915034G

Imamguluyev, R., Hashim, S., & Hajiyev, I. (2024). Harnessing Uncertainty: Integrating Fuzzy Logic into Machine Learning Algorithms. 2024 4th International Conference on Pervasive Computing and Social Networking (ICPCSN), 510-514. DOI: 10.1109/ICPCSN62568.2024.00086

Kutty, A., Kucukvar, M., Onat, N., Ayvaz, B., & Abdella, G. (2023). Measuring sustainability, resilience and livability performance of European smart cities: A novel fuzzy expert-based multi-criteria decision support model. *Cities (London, England), 137*, 104293. Advance online publication. DOI: 10.1016/j.cities.2023.104293

Maji, P. K., BISwas, R., and Roy, A. R. (2001). Fuzzy soft sets. *Journal of Fuzzy Mathematics, 9*(3), 589–602.

Mendel, J. (2000). Uncertainty, fuzzy logic, and signal processing. *Signal Processing, 80*(6), 913–933. DOI: 10.1016/S0165-1684(00)00011-6

Meng, T., Huang, J., Chen, Y., Li, H., Li, X., Zhang, B., Hao, J., Chew, C., Yang, D., & Zhong, Z. (2023). New Framework for Fuzzy Logic Reasoning: A Robust Control Theoretic Approach. *International Journal of Fuzzy Systems*, *26*(2), 463–481. DOI: 10.1007/s40815-023-01606-x

Mirza, M., Anjum, R., Min, H., Alkahtani, B., & Anjum, M. (2025). The Linear Diophantine Fuzzy Z-Numbers Set: Development and Application to Decision Making in Textile Engineering Using the Bonferroni Mean Operator. *IEEE Access : Practical Innovations, Open Solutions*, *13*, 70512–70532. DOI: 10.1109/ACCESS.2025.3561136

Mohamed, M., Almalaq, A., Abdullah, H., Alnowibet, K., Alrasheedi, A., & Zaindin, M. (2021). A Distributed Stochastic Energy Management Framework Based-Fuzzy-PDMM for Smart Grids Considering Wind Park and Energy Storage Systems. *IEEE Access : Practical Innovations, Open Solutions*, *9*, 46674–46685. DOI: 10.1109/ACCESS.2021.3067501

Molodtsov, D. (1999). Soft set theory: First results. *Computers & Mathematics with Applications (Oxford, England)*, *37*(4–5), 19–31. DOI: 10.1016/S0898-1221(99)00056-5

Moreno, J., Sanchez, M., Mendoza, O., Díaz, A., Castillo, O., Melin, P., & Castro, J. (2020). Design of an interval Type-2 fuzzy model with justifiable uncertainty. *Information Sciences*, *513*, 206–221. DOI: 10.1016/j.ins.2019.10.042

Mukherjee, A., & Das, A. K. (2015a). Application of interval valued intuitionistic fuzzy soft set in investment decision making, *2015 Fifth International Conference on Advances in Computing and Communications (ICACC)*, Kochi, India, 2015, pp. 61-64, DOI: 10.1109/ICACC.2015.37

Mukherjee, A., & Das, A. K. (2015b). Interval valued intuitionistic fuzzy soft multi set theoretic approach to decision making problems, 2015 International Conference on Computer, Communication and Control (IC4), Indore, India, 2015, pp. 1-5, DOI: 10.1109/IC4.2015.7375640

Mukherjee, A., & Das, A. K. (2015c). Relations on intuitionistic fuzzy soft multi sets. In Kim, K. (Ed.), *Information Science and Applications. Lecture Notes in Electrical Engineering* (Vol. 339). Springer., DOI: 10.1007/978-3-662-46578-3_71

Mukherjee, A., & Das, A. K. (2016a). Application of fuzzy soft multi sets in decision-making problems. In: Nagar, A., Mohapatra, D., Chaki, N. (eds) Proceedings of 3rd International Conference on Advanced Computing, Networking and Informatics. Smart Innovation, Systems and Technologies, vol. 43. Springer, New Delhi. DOI: 10.1007/978-81-322-2538-6_3

Mukherjee, A., & Das, A. K. (2016b). Topological structure formed by intuitionistic fuzzy rough relations. *Journal of the Indian Mathematical Society*, *83*(1-2), 135–144.

Mukherjee, A., & Das, A. K. (2022). Einstein operations on fuzzy soft multisets and decision making. *Boletim da Sociedade Paranaense de Matematica*, *40*, 1–10. DOI: 10.5269/bspm.32546

Mukherjee, A., & Das, A. K. (2023). Essentials of fuzzy soft multisets: Theory and applications, Springer Nature Singapore Pte Ltd., 1-151. ISBN: 978-981-19-2759-1, DOI: 10.1007/978-981-19-2760-7

Shaik, S., Ahad, A., Murugavalli, N., Kavitha, V., Suryanarayana, S., & Raj, M. (2024). *ODESMAN: Optimizing Decision-Making in Complex Environments: Integrating Neutrosophic and Fuzzy Logic for Advanced System Modeling*. International Journal of Neutrosophic Science., DOI: 10.54216/IJNS.240226

Smarandache, F. (1998). *A unifying field in logics: Neutrosophy: Neutrosophic probability, set, and logic*. American Research Press.

Smarandache, F. (2005). Neutrosophic set: A generalization of the intuitionistic fuzzy sets. *International Journal of Pure and Applied Mathematics*, *24*, 287–297.

Wahyono, H., Khumaidi, A., & Setyawati, K. (2024). *Application of Fuzzy Logic in PEOPLES Framework for Community Resilience Measurement in Flood Disaster. Jurnal Ilmiah Merpati*. Menara Penelitian Akademika Teknologi Informasi., DOI: 10.24843/JIM.2024.v12.i02.p05

Zadeh, L. (1983). The role of fuzzy logic in the management of uncertainty in expert systems. *Fuzzy Sets and Systems*, *11*(1-3), 197–198. DOI: 10.1016/S0165-0114(83)80081-5

Zadeh, L. A. (1965). Fuzzy sets. *Information and Control*, *8*(3), 338–353. DOI: 10.1016/S0019-9958(65)90241-X

Zhao, N., Tian, Y., Zhang, H., & Herrera-Viedma, E. (2024). Fuzzy-based adaptive event-triggered control for nonlinear cyber-physical systems against deception attacks via a single parameter learning method. *Information Sciences*, *657*, 119948. DOI: 10.1016/j.ins.2023.119948

Chapter 11

Review of Smart Wearables Sensor-Based Techniques and Machine Learning for Change of Direction Detection in Sports

Pranay Jaiswal
https://orcid.org/0009-0000-9330-7367
Dundalk Institute of Technology, Ireland

Abhishek Kaushik
https://orcid.org/0000-0002-3329-1807
Dundalk Institute of Technology, Ireland

Fiona Lawless
Dundalk Institute of Technology, Ireland

Tiago de Melo Malaquias
https://orcid.org/0000-0002-5896-0448
STATSports, Newry, UK

Fergal McCaffery
Dundalk Institute of Technology, Ireland

DOI: 10.4018/979-8-3373-4287-0.ch011

ABSTRACT

Change of Direction (COD) is a critical movement skill during athletic gameplay. In team sports like football, basketball, and rugby, the on-field performance is often assessed by the ability to change direction quickly. Optimizing COD reduces injury risk and improves team outcomes. This review examines algorithmic approaches for detecting COD using wearable sensor data. Each method's process, results, strengths, and limitations are summarized. An exploratory methodology was used to search databases like Google Scholar, PubMed, IEEE, and Science Direct. Findings show ongoing progress in COD detection, but also highlight gaps, such as non-standardized sensor placement, inconsistent sampling rates, and limited open datasets. These issues hinder the use of Machine Learning (ML) and Deep Learning (DL) models. To address this, we propose an AI-based framework to automate COD detection. This review aims to support decision-making and future research in wearable-based COD monitoring.

INTRODUCTION

Advances in sensor technology have transformed athlete performance analysis in sports through precise assessment of movement patterns, performance metrics, and biomechanical characteristics, which have replaced traditional observational methods (Migliaccio, Padulo, & Russo, 2024). The analysis of an athlete's sports performance depends on three main metrics, which include distance and speed measurements (Losada-Benitez et al., 2023), acceleration (Delves et al., 2021), and energy expenditure (Polglaze et al. 2016). These performance metrics deliver important information about athlete performance and training effectiveness, as well as fatigue monitoring and improvement opportunities. Athletes use agility to outperform opponents through their ability to dodge defenders (Young et al., 2022) and their quick movements to reach balls in football (Morral-Yepes et al., 2022) and their fast cuts to the basketball basket (Sugiyama et al., 2021). The essential athletic ability known as Change of Direction (COD) plays a vital role in complete athletic growth and injury avoidance and is essential for performance.

Sports scientists and researchers have studied the biomechanics and performance consequences of COD in various sports over the years. Neuromuscular adaptations (muscle activation and coordination) (Spiteri, Newton, & Nimphius, 2015), biomechanical factors (joint angles, balance, and ground reaction forces) (Spiteri, Newton, Binetti, et al., 2015), and metabolic demands (energy, power for quick bursts) (Jones & Dos'Santos, 2023), are involved in executing rapid directional changes across various sports disciplines. Athletes undergo rapid transitions from one movement pattern to another, requiring efficient deceleration, redirection of momentum, and acceleration in a new direction (Clarke, Aspe, & Hughes, n.d.). Biomechanical principles, such as speed, govern these movements (Jones & Dos'Santos, 2023), angle change (Dos et al., 2021), surface characteristics (Ismail et al., 2022) and individual athlete attributes. The research has shown that multiple high-intensity COD movements can lead to injuries such as hamstring injuries. (Bishop & Girard, 2013). With the proper training and exercise, results have shown to reduce the risk of injury (Siddle et al., 2024). The key challenge, however, lies in identifying when an athlete is at risk of injury, especially regarding their workload. If an athlete is performing more COD movement during a live competitive match than in practice sessions, they are generally at a higher risk of getting injured (Jiang et al., 2022). This highlights the importance of accurately identifying or classifying COD movements so that training loads can be managed better and reduce the risk of injury.

The two technologies that are widely used to capture and evaluate COD movement are Wearable devices and vision-based motion capture systems. Inertial measurement units (IMUs), a wearable device, are portable, cost-effective, and capable of capturing motion data in real time across various sports settings (Suo et al., 2024). On the other hand, vision-based motion capture systems, such as marker-based optical tracking or markerless camera setups, offer the advantage of nonintrusive measurement (Colyer et al., 2018). With these systems, athletes don't have to be in touch with any kind of sensor. However, despite their accuracy and minimal interference with performance, vision-based systems are often impractical for real-world, multi-player sports environments. Their high financial cost, complex setup requirements, and sensitivity to external conditions, such as changes in lighting, camera occlusion, and limited field of view, pose significant limitations. Moreover, these systems struggle with scalability when tracking multiple athletes simultaneously in large, open, or dynamic settings, making them less suitable for real-time use during live games or training sessions (Armanini et al., 2012; Keaney & Reid, 2018; Umek et al., 2016).

Wearable sensors such as accelerometers (which measure linear acceleration), magnetometers (which determine the direction and power of the magnetic field), and gyroscopes (which measure angular velocity) are commonly referred to as Inertial Measurement Units (IMUs). These IMU sensors can recognize complex details of athlete movements with high precision and accuracy (Chambers et al., 2015; Yang et al., 2024a). When integrated into various sports equipment, clothing, or body-worn devices, these sensors continuously monitor athlete activity during training or live games, providing performance feedback to athletes and coaches (Seshadri et al., 2019). Global Positioning System (GPS)-based wearable devices offer another way for capturing athletic movement activity using positioning data. But their effectiveness in specific environments, such as indoor sports, can be hindered due to poor connections (Cummins et al., 2013). Previously, different forms of human activity, such as sitting, walking, and standing, have been analyzed and classified using IMUs (Benson et al., 2018; Hou, 2020). Most of these studies have been done using the data collected from a controlled environment. Given that IMUs have already been used widely to capture and analyze movement patterns, they can also be used to capture COD movements during various sports activities (Alanen et al., 2021).

The data collected from these wearable sensors can then be used with different machine learning (ML) models to quantify and classify different human/athlete movements in different athletic environments. These ML algorithms analyze the datasets to identify complex movement patterns and learns to classify or predict the movement activity, and the results are shown in the form of performance metrics, which allows the development of different personalized training programs and early detection of injury risks (Bian et al., 2022; Demrozi et al., 2020).

Motivation

Sports is a rapidly growing industry that attracts millions in revenue and drives significant economic growth (Wang et al., 2022). This growth attracts much technology to enhance athletic performance and achieve better output by minimizing the risk of any injury and maximizing athletic potential (Haake, 2009). Due to this, the use of wearable sensors in sports has multiplied over the past few years (Fayed et al., 2024; James & Petrone, 2016). These sensors are used to analyze different athletic patterns, including COD. Given the importance of COD in athletic success, the detection, monitoring, and enhancement of this ability are crucial for coaches, trainers, and sports scientists (Martín-Moya & González-Fernández, 2022). Although previous reviews have analyzed the COD movement using IMUs, they have not explored the algorithms for detecting CODs, automating the COD detection using ML and deep learning (DL) models for real-time feedback, and reducing the dependency on other factors such as labor, time, and cost (Patil et al., 2024). The primary factor affecting automation is the lack of open-access data collected for COD events, which raises different ethical concerns in research. Whereas automation of general human activity recognition (HAR) using wearable sensors (IMUs, GPS) and smartphones through ML and DL techniques has been studied rigorously on the available open-access datasets (Gumaei et al., 2020; Gupta, 2021; Luptáková et al., 2022; Nayak et al., 2022).

This review explores recent advancements and limitations of COD detection algorithms and the use of ML and DL models in detecting COD movements using wearable sensors such as IMUs and GPS. The paper will offer a better understanding of the field and guidance for researchers and practitioners looking to select the most suitable algorithm for their specific application or develop more innovative solutions for COD detection.

To optimize training loads, prevent injuries, and improve athletic performance, accurate and automated detection of COD movement is necessary. So, with this study, by analyzing how wearable sensors and AI can help automate COD detection to provide real-time, cost-effective COD detection, we aim to address the gap in the existing literature.

By reviewing the current state-of-the-art algorithms designed for COD detection, the aim is to answer the following research questions:

RQ1. What are the techniques and advancements in COD detection algorithms using wearable sensors?

RQ2. What are the limitations associated with the existing techniques used for COD detection?

RQ3. How can AI be used to automate detecting COD movements to provide real-time feedback?

The rest of the paper is organized as follows: Section 2 defines the literature search methodology. Section 3 discusses algorithmic approaches based on IMU data, sensor fusion techniques, and GPS data. Subsequently, Section 4 discusses the findings based on the research question, based on the obtained results from all the algorithms. Section 5 proposes a framework for COD detection automation that has arisen from insights gained from the review. Finally, section 6 provides concluding remarks and future work.

METHODOLOGY

We used an exploratory design strategy to search and evaluate peer-reviewed studies in the field of COD detection with wearable sensors. The review was carried out in phases: identification, screening, eligibility assessment, and inclusion.

Eligibility Criteria

Included if all the following criteria were satisfied:

- Published in a conference or a peer-reviewed journal, and it has to be in English.
- Primarily focused on analysis or detection/classification of COD movements using wearable sensors.
- Studies must have used wearable sensors, including GPS and IMU, or can be a combination.

- Data must be collected from human participants doing physical activities.
- Detailed methodology of the algorithm must be presented.

Exclusion criteria were applied to:

- Non-peer-reviewed materials such as abstracts only, books, newsletters, reports, or encyclopaedia entries.
- Studies using other sensor types (video cameras, humidity sensors).
- Research involving non-human subjects or synthetic data.
- Studies focused on non-physical activities (cognitive, or emotional tasks, bio signal-only analysis without movement data).
- Articles lacking sufficient methodological details, including incomplete descriptions of datasets, algorithms, or evaluation processes.

Figure 1. Article search methodology

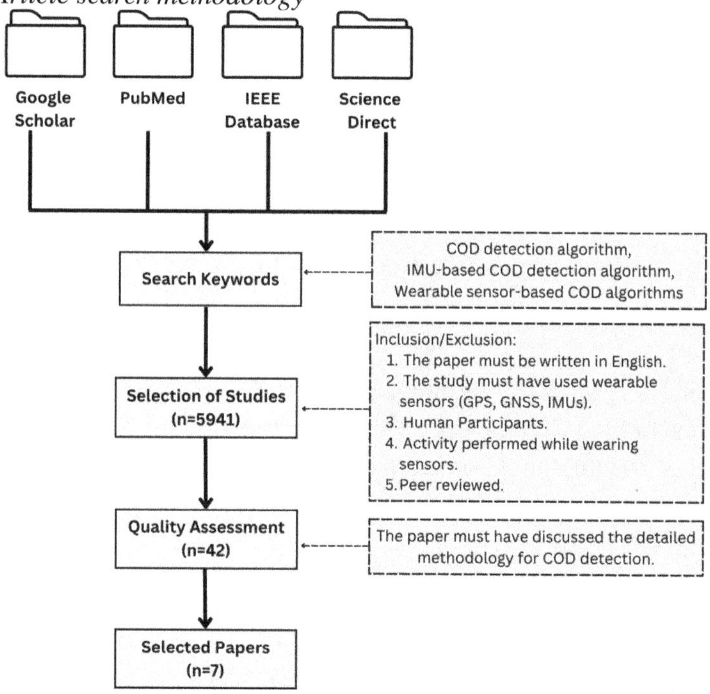

Information Sources

We have used different electronic databases to search the literature comprehensively: Google Scholar, PubMed, ScienceDirect, and IEEE Xplore. Given the very minimal research that has been done in the field of COD detection, we have not used any constraint on the publication date to ensure that all studies are covered.

Search Strategy

The combination of different search terms was used to select all possible studies. The following combinations were used:

1. "COD detection algorithm"
2. "IMU-based COD detection algorithm"
3. "Wearable sensor-based COD algorithms"
4. "Athletic COD Movement"

The search phrases were optimized according to the needs of each database. After searching the databases, this method gives a total of 5,941 research papers across all sources. Then, we have evaluated the titles and abstracts of each article to ensure that they have researched COD movement. Once the articles passed the first screening, we conducted a full-text assessment to check their compliance with inclusion/exclusion criteria. A total of 42 papers met all inclusion requirements, of which seven studies have presented a detailed framework for COD detection using wearable sensors, so we have included these studies for in-depth analysis (as depicted in Figure 1).

Data Extraction

For the final set of included studies, the following information was systematically extracted: authors and year of publication, type of physical activity and environmental setting, sensors used, number of participants and sample sizes, description of the COD detection methodology, analytical techniques employed, outcomes and performance measures, noted limitations and areas for future work.

RELATED WORK

After reviewing the algorithms in the seven selected papers, this section provides a brief overview of the techniques used in the COD detection of each algorithm, its findings, and its limitations. First, we discussed the IMU-based algorithmic approach utilizing signal processing techniques, including an algorithm using gyroscope data, an algorithm using magnetometer data, and algorithms using multiple sensor data. Next, the COD detection algorithm based on GPS data is discussed. Finally, the sensor fusion techniques using IMU and GPS/GNSS data are discussed. Figure 2 shows the general overview of the process of COD detection using different techniques.

Figure 2. The process of COD detection

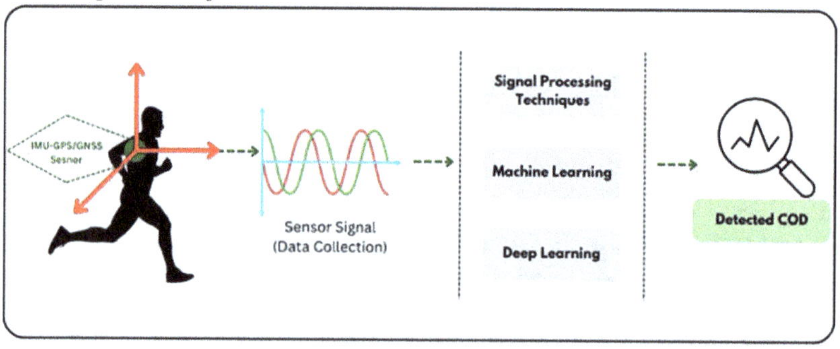

COD Detection Using an Inertial Measurement Unit (IMU) Sensor

Figure 3. Pitch, roll, and Yaw angles represent rotations around the x, y, and z axes, respectively, describing the orientation of an object in three-dimensional space

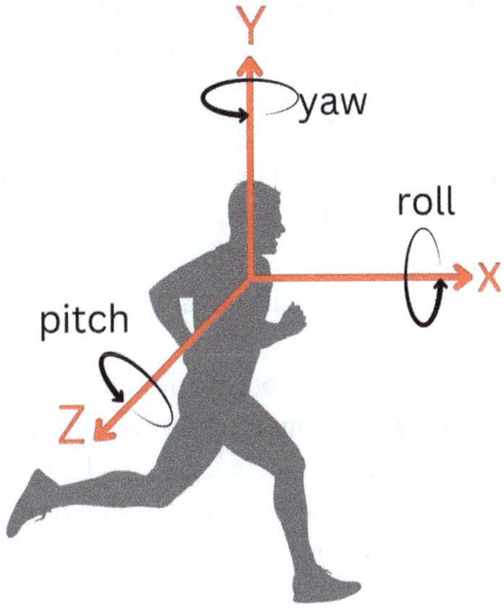

(Meghji et al., 2019)

An IMU device typically combines multiple sensors to measure orientation, velocity, and gravitational forces (Kivikangas, 2022). It consists of a triaxial accelerometer, gyroscope, and magnetometer sensor. Due to their small and lightweight design, IMUs are considered one of the most preferred wearables during practice or live matches for real-time performance tracking and analysis. The triaxial nature of the IMU sensor is shown in Figure 3.

Gyroscope

The signal processing technique described (Harding et al., 2008) uses tri-axial gyroscopes to classify aerial acrobatics movements in elite half-pipe snowboarding. The gyroscope device used gives angular velocity data sampled at 100 Hz. An aerial acrobatics movement's Airtime (AT) is measured in seconds and calculated using an already developed method that uses accelerometer data, also sampled at 100 Hz (Harding et al., 2007).

The algorithm was designed using the basic concept of angular velocity. Angular displacement across each axis was determined by integrating the angular velocity data collected from the triaxial gyroscope. Then the angular displacement for each data point was added to obtain the total change in angle across a single axis. Now, to quantify the total change in body orientation that is COD movement, the total rotation across the three axes of the gyroscope was added to obtain the Air Angle. This value represents the total COD angle achieved in the air.

The algorithm (Harding et al., 2008) was tested on 10 participants' data from routine training and competitive snowboarding matches collected using a gyroscope sensor sampled at 100 Hz. The findings demonstrated that the signal processing technique reliably classified aerial acrobatics, mainly those revolving around a single axis. This allowed for categorizing the aerial movements into four groups of 180°, 360°, 540°, and 720° angles, respectively. The statistical analysis showed significant differences in the Air Angle measurements between all four angle groups, with mean differences and confidence limits providing a reliable classification of aerial acrobatics.

The limitation of using only the gyroscope for calculating the angle change or the COD angle is the drift error; if any error is present, it gets accumulated over time while integrating and turns into significant angle drift. Also, a gyroscope alone cannot distinguish between actual body rotation and vibrations, which makes them less reliable in classifying complex movements.

Magnetometer

Similar to the previous study, the authors (Fleury et al., 2007) used a triaxial magnetometer to detect COD movement, without mentioning the frequency or sampling rate of the sensor. The key component of this algorithm was filtering the raw signals, then identifying non-activity periods using standard deviation, and identifying COD events using Euclidean norms in the x-y plane with a predefined threshold. Rotations (heading or COD angle) can be expressed as Euler angles, rotation matrix, roll-pitch-heading, and quaternions (Kuncar et al., 2011). So, in this algorithm, the authors compute yaw angles using quaternions (which describe the composition of the three rotations) to enhance detection accuracy, ensuring precise orientation estimation.

To assess the algorithm's accuracy (Fleury et al., 2007), the author collected data from eight young, healthy volunteers, averaging 27 years old, who joined two experiments. First, volunteers were instructed to walk in a corridor, making ten 180° turns (five each way) and taking breaks in between. Then, to go up and down stairs, making four 180° turns (two each way) and eight 90° bends (four each way). The proposed method accurately tracked and classified their movements throughout both trials, demonstrating the effectiveness of using a tri-axial magnetometer to track COD in real-time by analyzing magnetometer signals in 2-second windows.

However, the algorithm works well for 90° and 180° rotation angles, which can be improved further with more research, potentially improving COD detection accuracy. The authors have tested this method only for 90° and 180° angles, which leaves a significant gap in testing this algorithm in different COD angles. The major limitation of using a magnetometer for detecting COD is that the proximity of elevators or direct current engines significantly alters the magnetic field. This limits the magnetometer's reliability for tracking COD in specific environments.

Combination of Multiple IMU Sensors

The research (Meghji et al., 2019) developed an algorithm for detecting COD using triaxial accelerometers, gyroscopes, and magnetometer data. The process involves noise removal from the raw data and then calculating the heading angle using the gyroscope data. Magnetometer data is fused with the gyroscope for a more accurate yaw angle. However, as discussed in section 3.1.2, the magnetometer suffers from tilt error in the presence of external magnetic fields. They removed this error using roll and pitch angles derived from the accelerometer data. Then, the magnetometer data tilt error was corrected using these angles through rotation matrices (Kuncar et al., 2011) to align the readings accurately.

Authors (Meghji et al., 2019) have integrated more advanced signal processing techniques to identify the direction and more precise COD angles. A modified Canny edge detection algorithm (a multi-step process to determine the edges in the region of high-intensity change (Canny, 1986)) with a Gaussian filter to reduce noise in the yaw signal, analyzes the intensity gradient of yaw angles to detect sharp direction changes, performs non-maximum suppression to filter out unnecessary samples, and applies hysteresis thresholding to determine genuine edges based on a predefined threshold value. Finally, a multi-level piece-wise thresholding algorithm quantifies the angle precisely for each COD incident.

The wearable device Catapult OptimEye S5 is used in data collection. The algorithm demonstrated high accuracy in identifying and analyzing COD angles of 45°, 90°, 135°, and 180°. All angles showed the difference between the calculated mean COD angle and the intended COD angle in a range of -1.5 ° to 3.1°. With coefficients of variation (CVs) ranging from 2.0 to 4.8%, every angle showed outstanding reliability. Furthermore, for all angles, the algorithm's Typical Error (TE), or expected error, varied from 1.2° to 3.1°. Due to the high sampling rate of IMU sensors, they are prone to drift error over time.

Researchers designed a new study to evaluate the algorithm's validity, accuracy, and reliability while running (Balloch et al., 2020). The algorithm demonstrated good concurrent validity and high precision in detecting and determining the precise

COD angle for various COD movements, ranging from 45° to 180°, in both left and right directions, with a low-level bias of less than ±5°.

To test the algorithm's validity and reproducibility authors (Avilés et al., 2023) conducted a study to detect CODs at various speeds and angles to identify the most accurate combination of minimum peak intensity and signal smoothing. The findings demonstrated that the specific combinations of minimum peak intensity and signal smoothing were the most accurate for identifying CODs at various angles and speeds, suggesting adding filters to the algorithm according to the speed to detect CODs reliably.

Later, the author (Waqar et al., 2021) used accelerometer and magnetometer data to determine the COD angle. They documented this as a Tilt-Compensated Magnetometer (TCM) Algorithm, where they calculated the heading (yaw) angle using the magnetometer and corrected its tilt error using accelerometer data. They collected the IMU and GPS data using a wearable device to test the algorithm. The collected data have angles ranging from 45° to 180°. While testing the positioning accuracy, the algorithm completely diverged from the path while tracking a 45° angle and showed an error of 1.64±3.2 m in positioning accuracy for the entire track. This indicates that the algorithm is less reliable for minor angle deviations in the presence of noise.

COD Detection Using a Global Positioning System (GPS) Sensor

GPS, a navigation system that works on satellites, enables people to receive information about their position almost instantly anywhere on or near the Earth. GPS-enabled wearable devices can track physical activities, from outdoor sports performance to fitness tracking and navigation (Aughey, 2011).

The study (Gray et al., 2010) tested the validity and reliability of GPS sensors for measuring the distance traveled in field-based sports and showed that both path linearity and movement intensity impact GPS distance accuracy, decreasing reliability as intensity increases. Similarly, research (Alphin et al., 2020) tested the accuracy while sprinting. They compared the Maximum Sprint Speed (MSS) measured by GPS and Electronic Timing Gates (ETG). GPS had lower bias and error for the 80-meter sprint than the 30-meter sprint. Equivalence tests indicated that GPS measurements were within a 5% equivalence interval for short and long distances. Overall, GPS devices demonstrated acceptable accuracy for measuring sprint speed, making them suitable for COD detection and performance assessment in sports settings.

Authors (Reilly et al., 2021a) introduced a new variable called the 'GPS-COD' angle to automate the process of COD detection using GPS data. GPS data with a sampling rate (frequency) of 10 Hz is used. The data is segmented into 2.5-second intervals determined after careful observation of the GPS signal, containing 25 data points for effective data processing and feature extraction.

Figure 4. GPS-COD angle

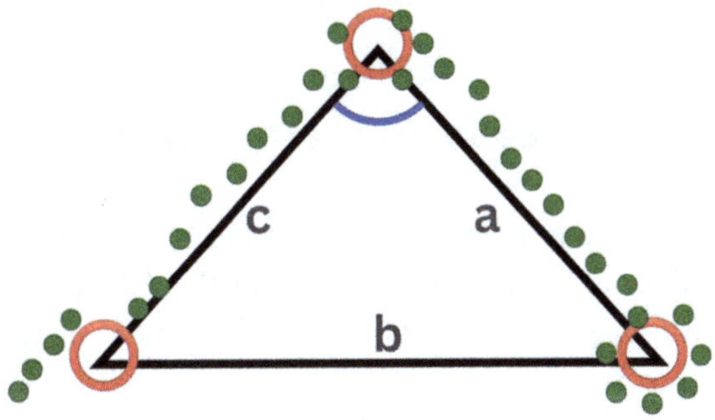

(Reilly et al., 2021a)

Figure 4 depicts 25 green dots representing the complete movement of the player, i.e., the average value of 'Latitude' and 'Longitude.' Additionally, red circles denote the average location of the first 5 points, the median 5 points, and the last 5 points. The distance between these 3 points was calculated using the "geospatial location" of these points: a = distance between the second and third points, b = distance between the third and first points, and c = distance between the first and second points, forming a triangle shape for each 2.5-second interval. The blue angle represents the angle of interest, which is calculated based on the spatial arrangement of these points.

Then, the cosine rule was applied to calculate the COD angle marked in blue in Fig. 2. This gives us a GPS-COD angle variable for each 2.5-second interval, as shown in equation 1.

$$GPS - COD\ angle = 180 - \cos^{-1}\left(\frac{a^2 + c^2 - b^2}{2ac}\right) \quad (1)$$

To verify the predictive power of the variable, the authors tested the variable with a random forest classifier. The results showed that it could effectively predict COD events during gameplay. With an Area Under the Curve (AUC) of 0.957 and a standard error of ±0.01, the classifier demonstrated strong predictive power in automatically detecting COD occurrences throughout the game.

The study found a novel trigonometry-based method using GPS data to measure athletes' COD angles over 45°. After using this variable with a random forest classifier, this method demonstrated good reliability in classifying COD movements throughout the game with high accuracy. Overall, the authors provided an automated process for classifying COD events. The limitations that can be utilized for further research, such as the potential for error in classifying segments that do not have a COD movement as COD, the assumption that only one COD can occur in each 2.5-second segment, and different time windows (e.g., 1, 1.5, 2s) other than 2.5s, can be explored.

COD Detection Using Sensor Fusion Techniques

IMUs are considered a gold standard for capturing motion-related information and are more reliable and better than GPS and the Global Navigation Satellite System (GNSS) (Buchheit & Simpson, 2017; Camomilla et al., 2018). However, GPS/GNSS offers positioning data. Utilizing this data with IMU data can provide essential information on athlete monitoring.

The study presents an algorithm using the sensor fusion technique (Apte et al., 2023), a macro-micro methodological approach designed to accurately detect and classify COD segments during the Agility T-test using data from a trunk-worn GNSS-IMU device. A camera recorded the COD activity as a ground truth to validate later results. The macro-analysis involves initiating movement by analyzing anterior-posterior acceleration data, utilizing filtering and wavelet decomposition to detect significant change points in the signal, such as local maxima and minima, and reconstructing the signal accurately. At the same time, the micro-analysis techniques focus on refining the detection of COD segments. GNSS ground speed is used to estimate completion time and validate against photocell data, enhancing the accuracy of the results.

This algorithm was tested across six participants who were randomly selected from 23 participants to assess their accuracy in detecting COD segments of order 90° and 180° and demonstrates an error (mean ± standard deviation) of -0.16 ± 0.22s for completion time estimation, 0 ± 0.66ms for COD detection, 3.5 ± 1.6% relative errors, and sequential movement duration of less than and 7 ± 7%. Although the algorithm performed well, the author (Apte et al., 2023) highlighted some limitations of this method, including improving synchronization between video and IMU by using clear markers like jumps or timestamps and higher video frame rates for better accuracy. Adding more sensors, like on the feet or sacrum, could provide more detailed data on ground contact and lower body mechanics. Finally, the methods must be tested on a broader range of athletes, including female players, to ensure they work for everyone.

Researchers (Waqar et al., 2021) introduced a new Particle Filter (PF) algorithm that combines IMU data with GNSS position data. This PF algorithm aims to accurately detect an athlete's COD and prevent divergence at each COD, a common issue with conventional PF algorithms (Arulampalam et al., 2002; Thrun, 2002). Initially, the algorithm detects COD events using the technique described in previous research (Meghji et al., 2019) using IMU data. After COD angles are categorized, the PF algorithm is initialized, which resamples the particles, assigns weights, and estimates the position using the samples and their weights. This approach has improved athlete tracking accuracy by up to five times compared to the TCM algorithm discussed in Section 3.1.3 (Arulampalam et al., 2002; Kuncar et al., 2016; Ozyagcilar, 2012; Thrun, 2002). Integrating IMU sensors with position data reduces the risk of drift error.

The algorithm's performance was evaluated using real-world IMU and GNSS data collected from athletes wearing wearable technology devices while running a predefined path involving multiple CODs. The algorithm exhibited an error of 0.32 ± 0.577m throughout the movement, with errors of 0.22 ± 0.32m observed during 45° COD movement, 0.18 ± 0.11m for 90°, 0.52 ± 0.41m for 135°, and 0.15 ± 0.08m for 180° while testing positioning accuracy. Additionally, the algorithm detects the direction of movement (left or right) and suppresses noise, resulting in higher accuracy. Although the algorithm displayed high accuracy while detecting different COD movements, testing this algorithm in free natural conditions is still required, as the authors only tested this algorithm in a controlled environment. Comparison of different sensor strengths and limitations is provided in Table 1.

Table 1. Comparison of wearable sensor modalities for COD detection

Sensor Type	Sampling Rate (Hz)	Strengths in COD Detection	Limitations
Accelerometer	100	Captures linear acceleration, good for impact/load quantification	Sensitive to noise and gravity effects
Gyroscope	100	Direct measure of angular velocity, accurate for rotation	Drift error over long periods
Magnetometer	100	Provides an absolute orientation reference	Distorted by nearby magnetic fields
GPS/GNSS	100	Field positioning, movement tracking outdoors	Low resolution indoors, lag in high-speed COD

DISCUSSION AND FINDINGS

Our review paper investigated the algorithms for classifying COD movements using various signal processing techniques, IMUs, and GPS/GNSS sensors. We have provided an outline of the algorithms for COD detection and their outcomes in Table 2. It details the method, the type of sensor data they employ, the number of participants, and the kind of COD movement, thus providing a quick summary of the algorithmic approaches.

Table 2. Overview of reviewed algorithms, their method, and results based on sensors

Paper	Year	Sensor	Subject	Method	Movement	Outcome
1.(Reilly et al., 2021a)	2021	GPS	23	Random Forest Classifier	Competitive football matches	GPS-COD angle as a strong predictor, AUC 0.975 ± 0.01
2. (Harding et al., 2008)	2009	Gyroscope	10	Signal Processing	Snow-boarding	Accurate and reliable classification around a single axis and classification into four groups of angles (180, 360, 540, 720).
3. (Fleury et al., 2007)	2008	Magnetometer	8	Signal Processing	Walk in a corridor and walk up and down the stairs.	Accurate classification of 90-degree and 180-degree COD angles.
4.(Meghji et al., 2019)	2019	Accelerometer Gyroscope Magnetometer	6	Signal Processing	Pre-determined COD circuit	Precise COD angles and good reliability with CVs in the 2 to 4.8% range. TE = 1.2 to 3.1 degrees.
5.(Balloch et al., 2020)	2020	Accelerometer Gyroscope Magnetometer	5	Signal Processing	Pre-determined COD circuit	Precise COD angles (45°, 90°, 135°, 180°), good reliability (TE = 1.6° ± 5.2°, CV= 1.3 ± 4.2%)
6.(Waqar et al., 2021)	2021	Accelerometer Magnetometer	6	Signal Processing	Pre-determined COD circuit	Positional accuracy algorithm 1.64±3.2m during the whole track. There is less reliability in tracking small angles such as 45o.

continued on following page

Table 2. Continued

Paper	Year	Sensor	Subject	Method	Movement	Outcome
6.(Waqar et al., 2021)	2021	GNSS-IMU	6	Signal Processing	Pre-determined COD circuit	The algorithm accurately tracks athlete movement with an error of 0.33 ± 0.577m during the whole track. COD 45° = 0.22 ± 0.32m, COD 90o = 0.18 ± 0.11m, COD 135o = 0.52 ± 0.41m, COD 180o = 0.15 ± 0.08m
7.(Apte et al., 2023)	2023	GNSS-IMU	6	Signal Processing	Agility T-test	Error (mean ± SD): COD detection= 0 ± 0.66ms Completion Time estimation = 0.16 ± 0.22m

Based on the literature investigation, we have addressed our research questions related to the advancements and challenges in the algorithmic approach and the utilization of AI in COD detection.

What are the Techniques and Advancements in COD Detection Algorithms Using Wearable Sensors?

Several techniques are used to detect athletes' COD movements. These algorithms were tested on different sensor data collected using different wearable sensors. The gyroscope measures angular velocity, so it is the primary sensor for calculating the heading or yaw angle (rotation around the axis) by integrating angular velocity over time (Harding et al., 2008; Meghji et al., 2019). Research (Fleury et al., 2007) demonstrated a magnetometer's effectiveness in classifying an athlete's COD movement. This method effectively identifies 90° and 180° angles when there is no disturbance in the magnetic field. Later, authors (Meghji et al., 2019) combined a gyroscope-derived yaw angle with a magnetometer sensor and accelerometer sensor to remove the issue of tilt error in the magnetometer for more accurate yaw angle detection and applied advanced signal processing techniques like Canny edge detection and piece-wise thresholding algorithms to detect COD angles more precisely and identify the direction of the COD movement. Study (Waqar et al., 2021) used sensor fusion techniques to mitigate the issues observed in the algorithm. In another study, authors developed an algorithm using position and IMU data, and the algorithm showed a fivefold improvement compared to the TMC algorithm (Meghji et al., 2019). A macro-micro analysis approach was presented (Apte et al., 2023) to quantify COD movements, identifying COD segments in the macro analysis, and for more accurate detection in the microanalysis. They then used GNSS data to measure the total completion time of the COD event. While tracking athlete activity using a GPS sensor can be challenging in indoor fields (Cummins et al., 2013), the authors (Reilly et al., 2021a) presented an automated method for high-accuracy COD event detection in outdoor conditions using a random forest classifier and fundamental signal analysis. The 'GPS-COD angle' variable using positioning data from GPS showed a good performance in classifying COD events. These improvements over time in detecting COD events using wearable sensors show the trajectory of advancement in this field.

What are the Limitations Associated with the Existing Techniques Used for COD Detection?

As discussed in the previous section, over time, there has been a continuous improvement in algorithmic development for COD detection, from the foundational algorithm designed in 2008 using a triaxial gyroscope, which calculates yaw angles by integrating the angular velocity data, to recent multiple sensor fusion techniques designed. Still, several limitations hinder the accurate detection of the COD movement, such as a high sampling rate, which leads to noise and drift error over time. Then, researchers began using GPS/GNSS sensor data with IMU data to reduce the noise and drift error. Given the low sampling rate of the positioning data obtained from the GPS/GNSS sensor, they are less prone to drift error.

Another major limitation identified is the improper placement of the IMU sensor across different studies. This creates variability in the datasets and reduces the reproducibility of the algorithm if the data is similar to the parent study. This highlights the lack of a universal standard for collecting COD movement data. Trunk-mounted sensors, usually placed between the scapulae with a vest, are the most common configuration due to their ability to approximate the center of mass movement. (Alanen et al. 2021). However, these can overestimate total body acceleration due to the movement of the vest itself. Other studies place IMUs on the lower back, thigh, tibia, or even embedded in custom knee sleeves to capture joint-level biomechanics (Rana & Mittal, 2020). For example, shank or foot-mounted sensors are ideal for identifying foot-strike events and impact forces during COD tasks. Regardless of placement, consistency and secure fixation are essential to reduce movement artifacts and maintain data integrity. The other challenge is the lack of standardized data collection protocols, including variability in sampling frequencies, sensor fusion algorithms, and filtering techniques (Alanen et al. 2021).

Moving ahead, the major challenge that was identified was the lack of an open-access dataset using wearable sensors, which affected the continued research in this field. In the literature, most studies have collected a minimal amount of data, and none of the datasets were made available publicly for further research and validation of the algorithmic approach. The reason highlighted for the scarcity of the dataset is the high expense and logistical problems associated with the data collection processes. Specialized wearable sensor technology is required for data collection, which requires extensive setup and trained staff. This specialized equipment can be expensive for individual use (Alanen et al., 2021). For example, high-quality data collection in sports often requires a controlled environment, like a sports or training facility, so that variables such as sensor calibration, athlete safety, and movement consistency are maintained (Balloch et al., 2020). Subsequently, the expense of those wearable devices and the skilled manpower required to operate and process the data extraction and storage made the collection of large-scale data financially burdensome. In combination, these factors impede the generation of extensive, diversified, and open-access datasets required for advancing research in sports performance analysis using wearable technologies. The limited availability of open-access datasets also affects the utilization of ML and DL models for automating COD detection.

The limited availability of the dataset also raises several ethical concerns:

- **Limited Reproducibility and Transparency:** When datasets are unavailable in the public domain to researchers, the study's findings cannot be easily replicated or independently verified. Indeed, this is an issue of a lack of transparency, as it discourages experimentation and replication, as other researchers cannot confirm or offer alternative interpretations of the findings using the same research data. On the other hand, ethics demand that the said findings be reproducible, meaning that any research that fails to reproduce any research finding in the absence of required data is considered questionable (Najjar, 2023).
- **Biased Research Outcomes:** The lack of diversity in datasets hampers good research. To put that in perspective, for example, even if the AI models have been intended for sports performance analysis, the training data originated from the athletes' regions, sports, or demographic groups. One can expect issues in the application of such models in another context. Such a lack of applicability increases bias, especially when empirical evidence, such as prediction outputs or contest judgments, is required (Najjar, 2023).
- **Lack of Standardization and Quality Control:** A lack of data leads to mismanagement and the inability to evaluate the standard of quality of data used. This could cause the use of poor-quality data, leading to invalid findings or incorrect models. Hence, concerns can be raised regarding the significance

of the research results. On the contrary, access to many data types will lead to constant improvement in the structure and quality of the data, which is necessary to promote ethical scientific research (Najjar, 2023).

How Can AI be Used to Automate Detecting COD Movements to Provide Real-Time Feedback?

Signal processing has been the backbone of the current COD detection methods, as reviewed in the literature. The state-of-the-art COD detection algorithm utilizes a sensor fusion technique; the PF algorithm is initialized, which resamples the particles, assigns weights, and estimates the position using the samples and their weights; it heavily relies on signal processing (Waqar et al., 2021). However, the application of ML and DL approaches has been underdeveloped in this domain. Although ML and DL approaches could probably support turning sensor data analysis into an automated feature extraction task with improved predictive accuracy, there are, however, a few reasons why these popular data analytics techniques have not seen such popularity in this discipline.

One of the primary limitations has been the availability of large, high-quality datasets needed to train machine-learning and especially deep-learning models (Gong et al., 2023). Deep learning, unlike traditional signal-processing methods, does not depend on a few rules and mathematical transformations but requires large data sets to learn complex patterns and make accurate predictions. Existing datasets are small, mostly from about ten participants, which is insufficient to train such models well. The rarity of such datasets is precipitated by the costs and difficulties in collecting sensor data in a controlled sports environment. Generalization across sports is another underexplored challenge; algorithms trained on soccer COD data may not transfer well to basketball or rugby, where movement angles, surfaces, and loads differ. Addressing these issues is vital for applied deployment.

Our review of existing literature found that the study (Reilly et al., 2021b) has presented an approach for detecting COD using AI, explicitly exploring machine learning algorithms. In contrast, authors (Jaiswal et al., 2025) compared the different ML and DL models for COD detection on the novel dataset. The results showed that both ML and DL models have similar performance in classifying COD events while running, but the dataset used was also relatively small, collected from only 7 participants. This shows that there is very limited development in utilizing AI for the detection of COD movement, as few studies have tried to automate this process. Whereas there have been numerous studies in the field of HAR, where authors have utilized different AI models to automate the process of classifying human activities like sitting, standing, running, and walking (Ramanujam et al., 2021; Sousa Lima et al., 2019; S. Zhang et al., 2022). Now these studies serve as a reference point for the development of a COD movement detection framework.

CONCEPTUALIZATION: FRAMEWORK FOR AI AUTOMATION OF COD DETECTION

Figure 5. Framework for COD detection using AI

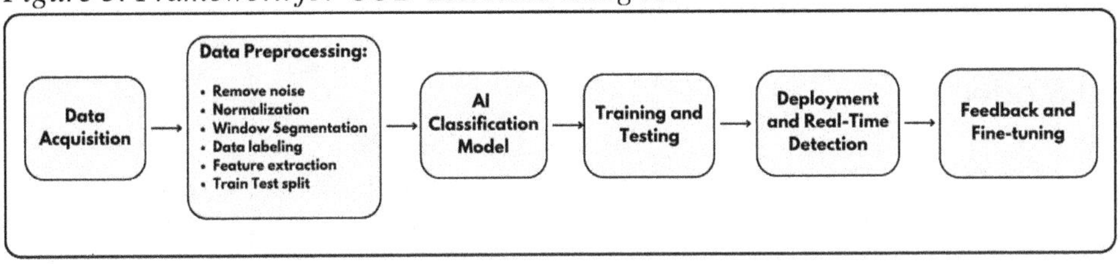

Based on the insights from this review, we identified minimal utilization of ML and DL models in the COD detection field for various reasons, including the low availability of open-access datasets. To automate the process by leveraging the domain knowledge of HAR, we have proposed a framework, as shown in Figure 5, that can be utilized to enhance the COD detection process (Khatun et al., 2022; Najjar, 2023).

Data Acquisition

The first step involves collecting data from wearable sensors attached to the athlete's body, such as IMUs and GPS. Data should be captured in multiple sessions, and athletes performing different COD activities with varying intensity levels should be captured to ensure diversity and coverage. The dataset should also include real-world data from different live games to ensure broader applicability (Khatun et al., 2022).

Data Preprocessing

The next step involves preprocessing the raw collected data to enhance the quality and consistency of the dataset. During this step, several preprocessing steps can be performed on the dataset, such as noise filtering using techniques such as Butterworth filtering or wavelet transform, and normalization to ensure data is on a comparable scale across different sensors. After the normalization step, segment the dataset into a fixed-length window size using the window sliding approach (Banos et al., 2014). The next step involves labeling the segmented window to indicate whether a COD incident occurred, which is essential for training a supervised AI model. Although

DL models automatically extract the features from the raw data, ML models require manual feature extraction for training. Different time and frequency domain features can be extracted from the segmented window data (M. Zhang & Sawchuk, 2012). Also, feature selection methods, such as principal component analysis, filter methods, or more advanced methods like bee swarm optimization metaheuristic and deep Q-network, can select the most relevant and essential features to reduce the computation load (Fan & Gao, 2021). The next step is to split the dataset for training and testing purposes (Khatun et al., 2022).

AI Model

The framework's core involves utilizing an AI model to detect and classify COD incidents effectively. ML and DL models can be used for this purpose (Gupta, 2021; Nayak et al., 2022). ML models such as SVM, random forest, and linear regression showed competitive performance while classifying general human activity. DL models such as CNN, RNN, and LSTM can sometimes outperform the ML models in terms of accuracy. Advanced DL models, such as hybrid or transformer models, have been shown to perform better in classifying human activity (Luptáková et al., 2022). A comparative review exploring all these models will be a better way to address the gap in this literature (Jaiswal et al., 2025).

The explainability of these AI models is also necessary to understand the factors responsible for identifying COD movements. Incorporating Explainable AI (XAI) into the framework for COD detection enhances system transparency, interpretability, and trust. Methods such as SHAP (SHapley Additive exPlanations) and LIME (Local Interpretable Model-Agnostic Explanations) help in identifying the most important or influential features that influence COD classification. For example, if coaches or athletes understand that deceleration and reacceleration are helping in COD classification, they can focus on these specific patterns more to enhance the COD performance (Novakovsky et al., 2022; Panigutti et al., 2020).

Model Training and Evaluation

Training the AI model involves using a labeled dataset to optimize the model weights. Regularization methods, such as dropout or weight decay, can be used to prevent overfitting. The model should be evaluated on separate validation and test datasets, using metrics such as accuracy, F1-score, and confusion matrix to assess its performance. Additionally, cross-validation can ensure that the model generalizes well across different athletes and sports contexts (Gupta, 2021).

Deployment and Real-Time Detection

Once the model has achieved satisfactory performance, it can be deployed to edge devices, such as smartphones or wearable microcontrollers, for real-time COD detection. Real-time feedback can be provided to athletes and coaches, enabling adjustments during training sessions or competitions (Gumaei et al., 2020).

Feedback Loop for Continuous Improvement

The next step in any well-implemented AI model is continuous fine-tuning, which updates and improves the model for real-world data. New data collected during training and competition can be labeled and used to fine-tune the AI model, adapting to changing conditions, new types of movements, or different sports. This iterative process ensures that the COD detection system remains relevant and accurate (Gumaei et al., 2020).

Implementation and post-implementation challenges

A series of challenges can be experienced while implementing the proposed framework for automation in COD detection. First, data quality is a fundamental issue because whenever data is captured through sensors worn on the athletes' bodies, the data might get affected by noise or distortion due to several factors, including the environment, the physical orientation of the sensors, or the athletes themselves. Robust preprocessing and filtering techniques are essential to ensure data reliability (Bangaru et al., 2020). Another challenge is integrating the different wearable systems with existing training systems. A standard method for collecting data, protocols for communicating data, and the system's architecture are required to achieve this. Real-time performance is crucial for providing immediate feedback. Still, it requires optimizing algorithms to balance computational efficiency and accuracy, which the limited resources of wearable devices can constrain (Nahavandi et al., 2022), so handling large datasets or running advanced machine learning models in real-time may require cloud-based solutions, introducing latency in data and connectivity issues (Yang et al., 2024b).

After the implementation phase, several problems arise, including scalability challenges and issues with system upgrades and maintenance. There needs to be the necessary infrastructure to scale the system for broader use across teams or facilities. Constant maintenance and user assistance are required to keep the system proper and efficient, together with the training of coaches and analysts to guide the best use of the system in practice (Carvalho & Sofia, 2020).

CONCLUSION AND FUTURE WORK

In this study, we reviewed algorithms created to detect or measure COD movements using wearable sensor data. We highlighted the main principles of each approach. Our review showed a significant gap in research specifically focused on COD detection. While body-worn sensors like IMUs, GPS, and GNSS have been used more for analyzing athlete performance, only a few studies have applied them to COD detection, which shows that there is a huge amount of scope in this field. IMUs offer high temporal resolution and detailed kinematic information, making them suitable for detecting subtle movement aspects like penultimate foot contact, but they can be sensitive to where the sensors are placed and calibration errors. GPS and GNSS systems provide broader positional tracking and fit well in field-based team sports, but their lower sampling rates limit their ability to capture quick and detailed COD events. Machine learning (ML) and deep learning (DL) models show strong potential for COD classification. Traditional ML algorithms give clear outputs, while DL models perform better in complex situations with multiple sensors. However, most existing studies focus on planned COD movements, which limits how these models can be applied to reactive, in-game scenarios. More work is needed to automate the processes of COD classification/ detection using wearable sensor data with AI models.

Future Research Directions

In the future, more studies should explore hybrid sensor setups that combine IMUs with GPS/GNSS to leverage the complementary strengths of high-resolution kinematic data and field-level positional tracking. Novel ML/DL architectures, including attention-based models or multimodal fusion approaches, could improve detection accuracy for unplanned, reactive COD movements. Availability of open-access datasets collected specifically for COD movements in realistic sport contexts is needed to train and benchmark algorithms effectively, which is missing in the current literature. Additionally, investigating sensor placement optimization, real-time processing pipelines, and explainable AI techniques will help translate research findings into practical tools for coaches and support staff, enhancing performance monitoring and injury prevention strategies.

FUNDING INFORMATION

This publication has emanated from research conducted with the financial support of Taighde Éireann – Research Ireland under Grant number 13/RC/2094_2.

ACKNOWLEDGMENT

The authors thank STATSports for continuous support. The authors also thank the Regulated Software Research Center (RSRC), where this research was conducted.

REFERENCES

Alanen, A. M., R€ Ais€ Anen, A. M., Benson, L. C., & Pasanen, K. (2021). *The use of inertial measurement units for analyzing change of direction movement in sports: A scoping review*. DOI: 10.1177/17479541211003064

Alanen, A. M., Räisänen, A. M., Benson, L. C., & Pasanen, K. (2021). The use of inertial measurement units for analyzing change of direction movement in sports: A scoping review. *International Journal of Sports Science & Coaching, 16*(6), 1332–1353. DOI: 10.1177/17479541211003064

Alphin, K. L., Sisson, O. M., Hudgins, B. L., Noonan, C. D., & Bunn, J. A. (2020). Accuracy assessment of a GPS device for maximum sprint speed. *International Journal of Exercise Science, 13*(4), 273.

Apte, S., Karami, H., Vallat, C., Gremeaux, V., & Aminian, K. (2023). In-field assessment of change-of-direction ability with a single wearable sensor. *Scientific Reports, 13*(1), 4518. DOI: 10.1038/s41598-023-30773-y

Armanini, A., Colombo, A., Conci, N., Daldoss, M., Fontanelli, D., & Palopoli, L. (2012). Wireless sensor networks and video analysis for scalable people tracking. *2012 5th International Symposium on Communications, Control and Signal Processing*, 1–4.

Arulampalam, M. S., Maskell, S., Gordon, N., & Clapp, T. (2002). A tutorial on particle filters for online nonlinear/non-Gaussian Bayesian tracking. *IEEE Transactions on Signal Processing, 50*(2), 174–188. DOI: 10.1109/78.978374

Aughey, R. J. (2011). Applications of GPS technologies to field sports. *International Journal of Sports Physiology and Performance, 6*(3), 295–310. DOI: 10.1123/ijspp.6.3.295

Avilés, R., Diego, B. S., Pino-Ortega, J., & Castellano, J. (2023). Assessment of a New Change of Direction Detection Algorithm Based on Inertial Data. *Sensors (Basel), 23*(6), 3095. https://www.proquest.com/scholarly-journals/assessment-new-change-direction-detection/docview/2791700160/se-2. DOI: 10.3390/s23063095

Balloch, A. S., Meghji, M., Newton, R. U., Hart, N. H., Weber, J. A., Ahmad, I., & Habibi, D. (2020). Assessment of a Novel Algorithm to Determine Change-of-Direction Angles While Running Using Inertial Sensors. *Journal of Strength and Conditioning Research, 34*(1), 134–144. DOI: 10.1519/JSC.0000000000003064

Bangaru, S. S., Wang, C., & Aghazadeh, F. (2020). Data Quality and Reliability Assessment of Wearable EMG and IMU Sensor for Construction Activity Recognition. *Sensors 2020, Vol. 20, Page 5264, 20*(18), 5264. DOI: 10.3390/s20185264

Banos, O., Galvez, J. M., Damas, M., Pomares, H., & Rojas, I. (2014). Window Size Impact in Human Activity Recognition. *Sensors 2014, Vol. 14, Pages 6474-6499, 14*(4), 6474–6499. DOI: 10.3390/s140406474

Benson, L. C., Clermont, C. A., Bošnjak, E., & Ferber, R. (2018). The use of wearable devices for walking and running gait analysis outside of the lab: A systematic review. *Gait & Posture, 63*, 124–138. DOI: 10.1016/j.gaitpost.2018.04.047

Bian, S., Liu, M., Zhou, B., & Lukowicz, P. (2022). The State-of-the-Art Sensing Techniques in Human Activity Recognition: A Survey. *Sensors 2022, Vol. 22, Page 4596, 22*(12), 4596. DOI: 10.3390/s22124596

Bishop, D. J., & Girard, O. (2013). Determinants of team-sport performance: Implications for altitude training by team-sport athletes. *British Journal of Sports Medicine, 47*(Suppl 1), i17–i21. DOI: 10.1136/bjsports-2013-092950

Buchheit, M., & Simpson, B. M. (2017). Player-tracking technology: Half-full or half-empty glass? *International Journal of Sports Physiology and Performance, 12*(s2), S2–S35. DOI: 10.1123/ijspp.2016-0499

Camomilla, V., Bergamini, E., Fantozzi, S., & Vannozzi, G. (2018). Trends supporting the in-field use of wearable inertial sensors for sport performance evaluation: A systematic review. *Sensors (Basel), 18*(3), 873. DOI: 10.3390/s18030873

Canny, J. (1986). A Computational Approach to Edge Detection. In *IEEE TRANSACTIONS ON PATTERN ANALYSIS AND MACHINE INTELLIGENCE* (Issue 6). DOI: 10.1109/TPAMI.1986.4767851

Carvalho, L. I., & Sofia, R. C. (2020). A Review on Scaling Mobile Sensing Platforms for Human Activity Recognition: Challenges and Recommendations for Future Research. *IoT 2020, Vol. 1, Pages 451-473, 1*(2), 451–473. DOI: 10.3390/iot1020025

Chambers, R., Gabbett, T. J., Cole, M. H., & Beard, A. (2015). The Use of Wearable Microsensors to Quantify Sport-Specific Movements. *Sports Medicine (Auckland, N.Z.), 45*(7), 1065–1081. DOI: 10.1007/s40279-015-0332-9

Clarke, R., Aspe, R. R., & Hughes, J. (n.d.). *Change of Direction Technical Models: Biomechanical Principles*. Retrieved 21 February 2024, from https://www.researchgate.net/publication/328561167

Colyer, S. L., Evans, M., Cosker, D. P., & Salo, A. I. T. (2018). A Review of the Evolution of Vision-Based Motion Analysis and the Integration of Advanced Computer Vision Methods Towards Developing a Markerless System. *Sports Medicine - Open*, *4*(1), 1–15. DOI: 10.1186/s40798-018-0139-y

Cummins, C., Orr, R., O'Connor, H., & West, C. (2013). Global positioning systems (GPS) and microtechnology sensors in team sports: A systematic review. *Sports Medicine (Auckland, N.Z.)*, *43*(10), 1025–1042. DOI: 10.1007/s40279-013-0069-2

Delves, R. I. M., Aughey, R. J., Ball, K., & Duthie, G. M. (2021). The Quantification of Acceleration Events in Elite Team Sport: a Systematic Review. *Sports Medicine - Open 2021 7:1*, *7*(1), 1–35. DOI: 10.1186/s40798-021-00332-8

Demrozi, F., Pravadelli, G., Bihorac, A., & Rashidi, P. (2020). Human Activity Recognition Using Inertial, Physiological and Environmental Sensors: A Comprehensive Survey. *IEEE Access : Practical Innovations, Open Solutions*, *8*, 210816–210836. DOI: 10.1109/ACCESS.2020.3037715

Dos, T., Thomas, C., & Jones, P. A. (2021). The effect of angle on change of direction biomechanics: Comparison and inter-task relationships. *Journal of Sports Sciences*, *39*(22), 2618–2631. DOI: 10.1080/02640414.2021.1948258

Fan, C., & Gao, F. (2021). Enhanced Human Activity Recognition Using Wearable Sensors via a Hybrid Feature Selection Method. *Sensors 2021, Vol. 21, Page 6434*, *21*(19), 6434. DOI: 10.3390/s21196434

Fayed, K., Ragab, M. A. F., & Rodríguez, E. (2024). The Use of Wearable Technologies and Their Impact on Athletic Performance. *Studies in Systems. Decision and Control*, *550*, 225–238. DOI: 10.1007/978-3-031-66218-8_17

Fleury, A., Noury, N., & Vuillerme, N. (2007). A fast algorithm to track changes of direction of a person using magnetometers. *Annual International Conference of the IEEE Engineering in Medicine and Biology - Proceedings*, 2311–2314. DOI: 10.1109/IEMBS.2007.4352788

Gong, Y., Liu, G., Xue, Y., Li, R., & Meng, L. (2023). A survey on dataset quality in machine learning. *Information and Software Technology*, *162*, 107268. DOI: 10.1016/j.infsof.2023.107268

Gray, A. J., Jenkins, D., Andrews, M. H., Taaffe, D. R., & Glover, M. L. (2010). Validity and reliability of GPS for measuring distance travelled in field-based team sports. *Journal of Sports Sciences*, *28*(12), 1319–1325. DOI: 10.1080/02640414.2010.504783

Gumaei, A., Al-Rakhami, M., AlSalman, H., Mizanur Rahman, S. M., & Alamri, A. (2020). DL-HAR: Deep Learning-Based Human Activity Recognition Framework for Edge Computing. *Computers, Materials & Continua*, *65*(2), 1033–1057. DOI: 10.32604/cmc.2020.011740

Gupta, S. (2021). Deep learning based human activity recognition (HAR) using wearable sensor data. *International Journal of Information Management Data Insights*, *1*(2), 100046. DOI: 10.1016/j.jjimei.2021.100046

Haake, S. J. (2009). The impact of technology on sporting performance in Olympic sports. *Journal of Sports Sciences*, *27*(13), 1421–1431. DOI: 10.1080/02640410903062019

Harding, J. W., Mackintosh, C. G., Hahn, A. G., & James, D. A. (2008). Classification of aerial acrobatics in elite half-pipe snowboarding using body mounted inertial sensors. *The Engineering of Sport*, *7*(2), 447–456. DOI: 10.1007/978-2-287-09413-2_55

Harding, J. W., Small, J. W., & James, D. A. (2007). Feature extraction of performance variables in elite half-pipe snowboarding using body mounted inertial sensors. *BioMEMS and Nanotechnology III*, *6799*, 332–343.

Hou, C. (2020). A study on IMU-based human activity recognition using deep learning and traditional machine learning. *2020 5th International Conference on Computer and Communication Systems (ICCCS)*, 225–234.

Ismail, S. I., Nunome, H., Lysdal, F. G., Kersting, G., & Tamura, Y. (2022). *Futsal playing surface characteristics significantly affect perceived traction and change of direction performance among experienced futsal players.* DOI: 10.1080/14763141.2022.2143415

Jaiswal, P., Kaushik, A., Lawless, F., Malaquias, T., & McCaffery, F. (2025). *Preliminary Investigation on Machine Learning and Deep Learning Models for Change of Direction Classification in Running.* 180–191. DOI: 10.1007/978-3-031-77731-8_17

James, D. A., & Petrone, N. (2016). *Sensors and Wearable Technologies in Sport.* DOI: 10.1007/978-981-10-0992-1

Jiang, Z. ;, Hao, Y. ;, Jin, N. ;, Li, Y. A., Baran, J., Leszczak, J., Jiang, Z., Hao, Y., Jin, N., & Li, Y. (2022). A Systematic Review of the Relationship between Workload and Injury Risk of Professional Male Soccer Players. *International Journal of Environmental Research and Public Health 2022, Vol. 19, Page 13237, 19*(20), 13237. DOI: 10.3390/ijerph192013237

Jones, P. A., & Dos'Santos, T. (2023). *Multidirectional speed in sport: research to application*. Taylor & Francis. DOI: 10.4324/9781003267881

Keaney, E. M., & Reid, M. (2018). Quantifying hitting activity in tennis with racket sensors: New dawn or false dawn? *Sports Biomechanics*.

Khatun, M. A., Yousuf, M. A., Ahmed, S., Uddin, M. Z., Alyami, S. A., Al-Ashhab, S., Akhdar, H. F., Khan, A., Azad, A., & Moni, M. A. (2022). Deep CNN-LSTM With Self-Attention Model for Human Activity Recognition Using Wearable Sensor. *IEEE Journal of Translational Engineering in Health and Medicine*, *10*, 1–16. Advance online publication. DOI: 10.1109/JTEHM.2022.3177710

Kivikangas, M. (2022). *Estimating three-dimensional motion using IMU sensors*.

Kuncar, A., Sysel, M., & Urbanek, T. (2011). *Calibration of Triaxial Accelerometer and Triaxial Magnetometer for Tilt Compensated Electronic Compass*.

Kuncar, A., Sysel, M., & Urbanek, T. (2016). Calibration of triaxial accelerometer and triaxial magnetometer for tilt compensated electronic compass. *Automation Control Theory Perspectives in Intelligent Systems: Proceedings of the 5th Computer Science On-Line Conference 2016 (CSOC2016)*, Vol 3 *5*, 45–52.

Losada-Benitez, J. A., Nuñez-Sánchez, F. J., & Barbero-Álvarez, J. C. (2023). Quantifying technical load and physical activity in professional soccer players during pre-season matches with IMU technology. *Frontiers in Physiology*, *14*, 1274171. DOI: 10.3389/fphys.2023.1274171

Luptáková, I. D., Kubovčík, M., & Pospíchal, J. (2022). Wearable Sensor-Based Human Activity Recognition with Transformer Model. *Sensors 2022, Vol. 22, Page 1911, 22*(5), 1911. DOI: 10.3390/s22051911

Martín-Moya, R., & González-Fernández, F. T. (2022). Test for the improvement and evaluation of change of direction in team sports: A systematic review. *Journal of Physical Education and Sport*, *22*(7), 1716–1722. DOI: 10.7752/JPES.2022.07215

Meghji, M., Balloch, A., Habibi, D., Ahmad, I., Hart, N., Newton, R., Weber, J., & Waqar, A. (2019). An Algorithm for the Automatic Detection and Quantification of Athletes' Change of Direction Incidents Using IMU Sensor Data. *IEEE Sensors Journal*, *19*(12), 4518–4527. DOI: 10.1109/JSEN.2019.2898449

Migliaccio, G. M., Padulo, J., & Russo, L. (2024). The Impact of Wearable Technologies on Marginal Gains in Sports Performance: An Integrative Overview on Advances in Sports, Exercise, and Health. *Applied Sciences 2024, Vol. 14, Page 6649, 14*(15), 6649. DOI: 10.3390/app14156649

Morral-Yepes, M., Moras, G., Bishop, C., & Gonzalo-Skok, O. (2022). Assessing the reliability and validity of agility testing in team sports: A systematic review. *Journal of Strength and Conditioning Research, 36*(7), 2035–2049. DOI: 10.1519/JSC.0000000000003753

Nahavandi, D., Alizadehsani, R., Khosravi, A., & Acharya, U. R. (2022). Application of artificial intelligence in wearable devices: Opportunities and challenges. *Computer Methods and Programs in Biomedicine, 213*, 106541. DOI: 10.1016/j.cmpb.2021.106541

Najjar, M.-C. (2023). Legal and Ethical Issues Arising from the Application of Data Analytics and Artificial Intelligence to Traditional Sports. *Albany Law Journal of Science & Technology, 33*, •••. https://heinonline.org/HOL/Page?handle=hein.journals/albnyst33&id=57&div=&collection=

Nayak, S., Panigrahi, C. R., Pati, B., Nanda, S., & Hsieh, M. Y. (2022). Comparative Analysis of HAR Datasets Using Classification Algorithms. *Computer Science and Information Systems, 19*(1), 47–63. DOI: 10.2298/CSIS201221043N

Novakovsky, G., Dexter, N., Libbrecht, M. W., Wasserman, W. W., & Mostafavi, S. (2022). Obtaining genetics insights from deep learning via explainable artificial intelligence. *Nature Reviews Genetics 2022 24:2, 24*(2), 125–137. DOI: 10.1038/s41576-022-00532-2

Ozyagcilar, T. (2012). Implementing a tilt-compensated ecompass using accelerometer and magnetometer sensors. *Freescale Semiconductor, AN, 4248.*

Panigutti, C., Perotti, A., & Pedreschi, D. (2020). Doctor XAI An ontology-based approach to black-box sequential data classification explanations. *FAT* 2020 - Proceedings of the 2020 Conference on Fairness, Accountability, and Transparency*, 629–639. DOI: 10.1145/3351095.3372855

Patil, D., Rane, N. L., Desai, P., & Rane, J. (2024). Machine learning and deep learning: Methods, techniques, applications, challenges, and future research opportunities. In *Trustworthy Artificial Intelligence in Industry and Society*. Deep Science Publishing., DOI: 10.70593/978-81-981367-4-9_2

Polglaze, T., Dawson, B., & Peeling, P. (2016, May). Gold Standard or Fool's Gold? The Efficacy of Displacement Variables as Indicators of Energy Expenditure in Team Sports. *Sports Medicine (Auckland, N.Z.), 46*(5), 657–670. Advance online publication. DOI: 10.1007/s40279-015-0449-x

Ramanujam, E., Perumal, T., & Padmavathi, S. (2021). Human activity recognition with smartphone and wearable sensors using deep learning techniques: A review. *IEEE Sensors Journal, 21*(12), 13029–13040. DOI: 10.1109/JSEN.2021.3069927

Rana, M., & Mittal, V. (2020). Wearable sensors for real-time kinematics analysis in sports: A review. *IEEE Sensors Journal, 21*(2), 1187–1207. DOI: 10.1109/JSEN.2020.3019016

Reilly, B., Morgan, O., Czanner, G., & Robinson, M. A. (2021a). Automated Classification of Changes of Direction in Soccer Using Inertial Measurement Units. *Sensors 2021, Vol. 21, Page 4625, 21*(14), 4625. DOI: 10.3390/s21144625

Reilly, B., Morgan, O., Czanner, G., & Robinson, M. A. (2021b). Automated Classification of Changes of Direction in Soccer Using Inertial Measurement Units. *Sensors 2021, Vol. 21, Page 4625, 21*(14), 4625. DOI: 10.3390/s21144625

Seshadri, D. R., Li, R. T., Voos, J. E., Rowbottom, J. R., Alfes, C. M., Zorman, C. A., & Drummond, C. K. (2019). Wearable sensors for monitoring the internal and external workload of the athlete. *Npj Digital Medicine 2019 2:1, 2*(1), 1–18. DOI: 10.1038/s41746-019-0149-2

Siddle, J., Weaver, K., Greig, M., Harper, D., & Brogden, C. M. (2024). A low-volume Nordic hamstring curl programme improves change of direction ability, despite no architectural, strength or speed adaptations in elite youth soccer players. *Research in Sports Medicine, 32*(1), 49–60. DOI: 10.1080/15438627.2022.2079984

Sousa Lima, W., Souto, E., El-Khatib, K., Jalali, R., & Gama, J. (2019). Human activity recognition using inertial sensors in a smartphone: An overview. *Sensors (Basel), 19*(14), 3213. DOI: 10.3390/s19143213

Spiteri, T., Newton, R. U., Binetti, M., Hart, N. H., Sheppard, J. M., & Nimphius, S. (2015). Mechanical determinants of faster change of direction and agility performance in female basketball athletes. *Journal of Strength and Conditioning Research, 29*(8), 2205–2214. DOI: 10.1519/JSC.0000000000000876

Spiteri, T., Newton, R. U., & Nimphius, S. (2015). Neuromuscular strategies contributing to faster multidirectional agility performance. *Journal of Electromyography and Kinesiology, 25*(4), 629–636. DOI: 10.1016/j.jelekin.2015.04.009

Sugiyama, T., Maeo, S., Kurihara, T., Kanehisa, H., & Isaka, T. (2021). Change of Direction Speed Tests in Basketball Players: A Brief Review of Test Varieties and Recent Trends. *Frontiers in Sports and Active Living*, *3*, 645350. DOI: 10.3389/fspor.2021.645350

Suo, X., Tang, W., & Li, Z. (2024). Motion Capture Technology in Sports Scenarios: A Survey. *Sensors 2024, Vol. 24, Page 2947, 24*(9), 2947. DOI: 10.3390/s24092947

Thrun, S. (2002). Probabilistic robotics. *Communications of the ACM*, *45*(3), 52–57. DOI: 10.1145/504729.504754

Umek, A., & Kos, A. (2016). The role of high performance computing and communication for real-time biofeedback in sport. *Mathematical Problems in Engineering*, *2016*, 2016. DOI: 10.1155/2016/4829452

Wang, L., Zhao, F., & Zhang, G. (2022). Analysis on the Impact of Large-Scale Sports Events on Regional Economy Based on SWOT-PEST Model. *Journal of Mathematics*, *2022*(1), 7769128. Advance online publication. DOI: 10.1155/2022/7769128

Waqar, A., Ahmad, I., Habibi, D., Hart, N., & Phung, Q. V. (2021). Enhancing athlete tracking using data fusion in wearable technologies. *IEEE Transactions on Instrumentation and Measurement*, *70*, 1–13. DOI: 10.1109/TIM.2021.3069520

Yang, L., Amin, O., & Shihada, B. (2024a). Intelligent Wearable Systems: Opportunities and Challenges in Health and Sports. *ACM Computing Surveys*, *56*(7), 1–42. DOI: 10.1145/3648469

Yang, L., Amin, O., & Shihada, B. (2024b). Intelligent Wearable Systems: Opportunities and Challenges in Health and Sports. *ACM Computing Surveys*, *56*(7), 1–42. DOI: 10.1145/3648469

Young, W., Dos'Santos, T., Harper, D., Jefferys, I., & Talpey, S. (2022). Agility in Invasion Sports: Position Stand of the IUSCA. *International Journal of Strength and Conditioning*, *2*(1). Advance online publication. DOI: 10.47206/ijsc.v2i1.126

Zhang, M., & Sawchuk, A. A. (2012). A Feature Selection-Based Framework for Human Activity Recognition Using Wearable Multimodal Sensors. *BODYNETS 2011 - 6th International ICST Conference on Body Area Networks*, 92–98. DOI: 10.4108/icst.bodynets.2011.247018

Zhang, S., Li, Y., Zhang, S., Shahabi, F., Xia, S., Deng, Y., & Alshurafa, N. (2022). Deep learning in human activity recognition with wearable sensors: A review on advances. *Sensors (Basel)*, *22*(4), 1476. DOI: 10.3390/s22041476

Chapter 12
Deep Learning-Based Defect Detection and Quality Assurance in Advanced 3D Printing and Microfabrication Processes

Hemant Sahu
https://orcid.org/0000-0003-1883-8883
Janardan Rai Nagar Rajasthan Vidyapeeth, India

Neeru Rathore
Janardan Rai Nagar Rajasthan Vidyapeeth, India

Vinita Nagda
Geetanjali Institute of Technical Studies, Udaipur, India

Anita Bhati
Vivekananda Global University, Jaipur, India

ABSTRACT

This paper examines the utilization of deep learning models YOLOv8, ResNet-50, and U-Net together with real-time defect recognition and quality control in FDM (Fused Deposition Modeling) 3D printing. Printed part samples with high-resolution images taken at different stages of printing were annotated and used for training. The models were assessed using performance measurements such as precision,

DOI: 10.4018/979-8-3373-4287-0.ch012

recall, F1-score, accuracy, and AUC-ROC. YOLOv8 performed the best across all measurements and is therefore best suited for real-time implementation. ResNet-50 provided high precision and moderate recall, and U-Net provided excellent recall but poor precision with the occurrence of false positives. The outcomes make YOLOv8 the most suitable for in-process quality control. The work showcases the potential of deep learning in improving the defect recognition process and reducing the occurrence of printing errors and the overall inefficiency of 3D printing.

INTRODUCTION

Over the past few years, the application of deep learning models in industrial processes, especially in 3D printing, has attracted significant interest because of its ability to improve production quality and efficiency. The fast development of 3D printing technologies, such as Fused Deposition Modeling (FDM), has introduced new challenges in ensuring consistent quality and detecting defects during printing, (Yazici, Shayea, & Din, 2023; Anandhakrishnan & Jaisakthi, 2022; Zjavka, 2022). Faults like layer misalignment, warping, and failed prints can have major implications for the end product's performance and functionality. Quality assurance for 3D printing has therefore become an imperative priority, and conventional manual inspection is shown to be both inefficient and impractical for mass production. Deep learning, with its capacity to handle lots of data and identify patterns, has been a potential solution to overcome these issues by allowing real-time automated defect detection and classification, (Singh et al., 2025; Hamid et al., 2023; Patro et al., 2023).

A number of deep learning models have been tried for 3D printing defect inspection, including convolutional neural networks (CNNs), segmentation networks, and object detection networks. CNNs such as ResNet have been applied with great success to image classification tasks, with high performance in the identification of defects in printed components. They are excellent at image feature extraction and learning image patterns that distinguish defective from non-defective parts, (Sahu et al., 2024; Garg et al., 2023a). But for tasks involving pixel-level precision, like detecting minor flaws in the print's surface or internal layers, models like U-Net, specializing in semantic segmentation, have been more effective. U-Net's image segmentation architecture allows the model to pinpoint defects with great accuracy, even in regions with high details.

By contrast, object detection networks such as YOLOv8 have been favored for carrying out both classification and localization tasks. Indeed, YOLOv8 is very fast and efficient and thus suitable for real-time defect detection in a dynamic manufacturing environment. This model not only classifies the presence of the defects but also provides the location of such defects in the 3D printed product. Its ability to process images quickly in real-time makes it the ideal choice for integration within automated 3D printing devices where instant decisions need to be made, (Garg et al., 2023b; Singh et al., 2024).

Deep learning for 3D printing defect detection, however, goes beyond simple image classification. A number of research studies have investigated the integration of visual inspection with sensor readings towards making defect detection systems more robust. For instance, thermal cameras and ultrasonic sensors can offer extra layers of information that assist in the detection of defects that are not visible to the naked eye or standard cameras. The combination of such multi-modal data with deep learning models has been demonstrated to enhance detection accuracy, especially in the detection of internal defects like voids or cracks in the printed material.

One of the biggest obstacles to applying deep learning models in 3D printing defect detection is the amount and quality of training data. It is labor-intensive and time-consuming to create high-quality annotated datasets for detecting defects. Often, researchers have used synthetic data created by simulating or controlling experiments to avoid this obstacle. Nevertheless, synthetic datasets typically do not have the full range of real-world printing conditions, including changing printer settings, materials, and environments, (Mata et al., 2024; Bah & Davud, 2022; Di Giammarco et al., 2022).

LITERATURE REVIEW

As such, there is a growing need to obtain large, varied, and representative datasets from real 3D printing processes in order to train models that can generalize well to real-world environments. The ease of access to high-resolution cameras and sophisticated imaging methods has improved the facility with which images of printed parts can be taken in detail and utilized for the training of deep learning models for defect detection, (Alghamdi et al., 2023; Arya et al., 2023).

Additionally, the efficacy of deep learning models in actual 3D printing applications is not solely contingent on the dataset used but also on the capacity of the model to accommodate various printing conditions. Since 3D printing is a process with high customizability, changes in the geometry of the print, the material parameters, and the environmental conditions can all lead to differences in the nature and extent of the defects, (Gangadharan et al., 2023; Bijalwan et al., 2024). Hence, models must be trained on datasets that include these variations in order to be robust. Transfer learning, where one fine-tunes a pre-trained model on a target dataset, has emerged as a common solution to this problem. With the help of pre-existing knowledge from general, general-purpose data sources, transfer learning enables models to learn faster and use their abilities for specialized purposes straight away, such as detecting defects in 3D printing, (Shivahare et al., 2024; Agarwal et al., 2025).

Despite the progress in defect detection using deep learning, there are still certain issues with model interpretability and real-time deployment. Deep learning models, especially those used in highly challenging problems like object detection and segmentation, are often referred to as "black boxes" because it is difficult to comprehend what precisely causes a model to make its predictions, (Sahu et al., 2024). This lack of transparency can prove to be an issue in industrial settings where being aware of the reasoning behind a model's decision is crucial in order to debug and improve the system. Efforts towards model interpretability improvement, such as saliency maps and feature visualization techniques, are being studied intensively in order to reveal the internal workings of deep learning models, (Ali et al., 2024a; Ali et al., 2024b).

Moreover, real-time industrial deployment of such models has several practical issues. Computational requirements of deep learning models, particularly of large architecture models like YOLOv8 and U-Net, can be heavy and require special hardware for peak performance. While edge computing and hardware acceleration are helping to solve them, implementing defect detection systems seamlessly in a real-time 3D printing environment remains a serious challenge. Additionally, the integration of deep learning models into existing 3D printing processes needs to consider parameters such as data processing time, system latency, and compatibility with different printing machines, (Nidhi et al., 2024).

METHODOLOGY

In this study, an extensive methodology was used to create and test a deep learning-based defect detection system specifically designed for sophisticated 3D printing processes, namely Fused Deposition Modeling (FDM). For model training and testing purposes, a special dataset of 5,000 high-resolution visual images was gathered in the process of real-time 3D printing using industrial-grade cameras installed within the FDM machine enclosure. These cameras recorded frame-by-frame data continuously throughout printing runs of multiple geometries and materials. Each photograph was annotated manually to label the existence or non-existence of common FDM defects, including layer shifting, under-extrusion, stringing, nozzle clogging, and warping. Training, validation, and test used around 3,500 images, 750 images, and 750 images, respectively. The data set contained both non-defect and defect images with different lighting conditions, print speeds, and layer heights to improve the robustness of the model. The architecture of the proposed system is shown in Figure 1.

Figure 1. Architecture of proposed system

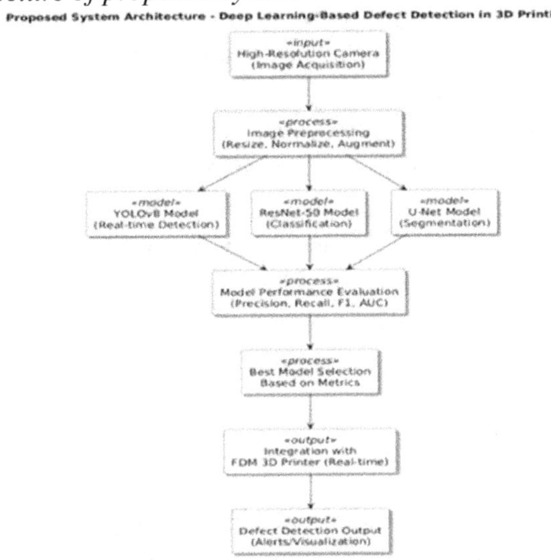

Three deep learning models were implemented and assessed: YOLOv8 for real-time object detection, ResNet-50 for classifying defects, and U-Net for segmenting defects. YOLOv8 was utilized for detecting and localizing defects in frames through bounding boxes, taking advantage of its speed and efficiency. ResNet-50, a residual convolutional neural network, was used to distinguish between images belonging to defective or non-defective categories with a compromise between depth and performance. U-Net, an existing segmentation model applied commonly in medical and industrial imaging, was used to spatially locate the precise defect regions. All the models were trained from the accumulated dataset using standard training pipelines with proper data augmentation methods, including rotation, flipping, and contrast variation, to mimic real-world variability.

The performance of the models was evaluated based on accuracy, precision, recall, F1-score, and inference time. Following a comparative evaluation, the highest-performing model was chosen for implementation in the FDM printer's monitoring system to facilitate real-time defect detection and quality guarantee upon printing. This approach guarantees that the chosen model is not only correct but also efficiently computable for real-time deployment in an industrial environment.

U-Net was used in this study for its established potential in image segmentation problems, especially where accurate defect localization is paramount. The U-Net design features a symmetric encoder-decoder approach, wherein the encoder pathway acquires context information via consecutive convolutional and max-pooling layers, while the decoder pathway restores the spatial resolution through upsampling and skip connections. These skip connections enable the network to blend low-level spatial information with high-level semantic information, supporting fine-grained segmentation of defects like layer gaps, edge warping, or string residues. In this work, U-Net was trained on defect-labeled images with pixel-wise labels that marked the precise regions of imperfection.

The model output is a segmented mask that covers the defect areas, and this can be applied to visual inspection or automatic rejection of defective prints. Its capability to encode global structure and local detail made U-Net very useful for quality control in microfabrication when defect boundaries are important. The performance of the model was measured in terms of Intersection over Union (IoU) and segmentation accuracy, providing a complementary view to the classification and detection models employed in this research.

Data Preprocessing

In this study, the dataset preparation and preprocessing process was an indispensable part to ascertain the successful training and testing of the deep learning models employed in defect detection on FDM-based 3D printing. The dataset consisted of a total number of 5,000 high-resolution images captured in real time during the genuine 3D printing process utilizing high-quality industrial cameras. These cameras were mounted inside the 3D printer casing, secured at angles to monitor the motion of the nozzle and the print bed throughout the printing process. The captured images represented a wide range of printing conditions, layers, material types, layer heights, and print geometries to simulate real-world scenarios and create diversity in the dataset. Stringing, nozzle clogging, layer shifting, and under-extrusion faults were purposely induced in some controlled print tasks in order to obtain a well-balanced dataset that comprised defect and non-defect images.

Once the raw images were acquired, the elimination of blurry or unnecessary frames was the initial step of preprocessing to maintain quality and consistency. Images with extreme motion blur, improper focus, or lost prints were manually and automatically eliminated by a blur detection algorithm based on variance of the Laplacian method. The images were then resized to a common resolution of 640 × 640 pixels to make the models uniform while preserving necessary visual information needed for fine defect detection. Normalization was then applied to normalize pixel intensity values between 0 and 1, enabling the training of convolutional neural networks more rapidly and stably by keeping input distributions uniform across the dataset.

For object detection using YOLOv8, the data required bounding box annotations for detection and identification of the areas of defects in each image. This was achieved through manual annotation of the defects with the Roboflow and LabelImg tools, where each defect was annotated with precise bounding boxes and labeled according to its class. Each annotation was saved in YOLO format, which includes the class index and normalized bounding box coordinates. Data augmentation techniques were applied during training to improve model generalization. These included random rotations (±15°), horizontal and vertical flipping, contrast and brightness modifications, and injection of Gaussian noise. These augmentations were applied in order to simulate illumination, orientation, and texture changes that might occur in the real world and thus make the detection model more robust.

For the task of classification with ResNet-50, the images were grouped into two primary folders: "Defect" and "Non-Defect." Each image was then labeled accordingly following manual inspection and cross-check with the printing logs. The images were further transformed into grayscale and RGB formats as per the requirements of the model, and the majority of the experiments were done using RGB channels to preserve greater feature information. The database was divided into training (70%), validation (15%), and testing (15%) sets through stratified sampling to provide an even representation of defect types in all sets. Standardization was utilized in preprocessing to center the pixel intensity values by applying the mean and standard deviation of the dataset. This positioned the model to receive centered data with unit variance, enhancing convergence during training.

For semantic segmentation with U-Net, an independent subset of the dataset was prepared where there was a ground truth segmentation mask for every defect image. The masks were produced with the CVAT (Computer Vision Annotation Tool) platform by manually outlining the defects at the pixel level. The masks were stored as binary images of the same size as the input images, with pixels in the defects being white (value = 1) and background pixels being black (value = 0). Every image-mask pair was preprocessed through resizing to 512 × 512 pixels and normalization. In addition, data augmentation for segmentation included elastic deformations, random cropping, and affine transformations to simulate defect shape and location variations. This helped the U-Net model learn defect spatial patterns more effectively.

For reproducibility and consistency reasons, all preprocessing steps were encapsulated within Python scripts using OpenCV, NumPy, and TensorFlow/Keras preprocessing toolkits. The scripts were validated on random subsets of data to guarantee correctness of annotation and transformation. Preprocessed data sets were stored in organized directories and shared across all pipelines of model training to provide consistency. Finally, quality inspections were conducted through visual verification of a representative subset of preprocessed images and masks for alignment, readability, and correct labeling. Through this comprehensive preprocessing approach, the dataset was rendered highly amenable to defect detection via deep learning in FDM printing, enabling the models to be highly precise and robust enough to employ in real-world applications.

RESULTS AND DISCUSSION

The performance assessment of the three deep learning models, namely YOLOv8, ResNet-50, and U-Net, was done with regular classification and detection metrics: precision, recall, F1-score, and accuracy. On the basis of results derived from the test dataset, YOLOv8 was found to exhibit better performance for all metrics and recorded a precision of 94.8%, recall of 92.6%, F1-score of 93.7%, and a general accuracy of 94.2%. These high precision and high recall balance result in YOLOv8 being well-drafted specifically for real-time inspection of defects within FDM 3D printing where both recall and precision are very important.

ResNet-50, focused mostly on the binary classification into defective or not defective prints, was found to have a precision of 91.3%, recall of 89.7%, F1-score of 90.5%, and total accuracy of 91.0%. Though lower than YOLOv8, ResNet-50 remained consistent and effective in classifying the print results. U-Net, used for pixel-level segmentation of defects, had a precision of 89.6%, recall of 93.4%, F1-score of 91.4%, and accuracy of 90.8%. Its greater recall signified great capability in detecting most defects, particularly fine-grained defects, although its precision was lesser because of some spurious positives. These findings substantiate that every model has a varied contribution, with YOLOv8 being the most feasible for real-time FDM integration. Figure 2 performance score of each model.

Figure 2. Performance metrics

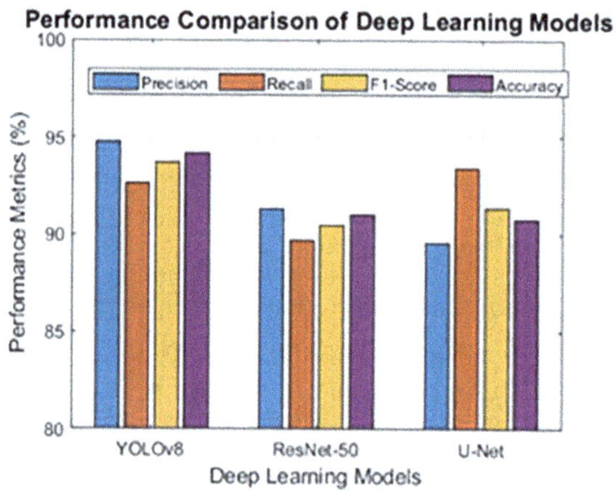

The three deep learning models' confusion matrices YOLOv8, ResNet-50, and U-Net present a clear picture of each model's classification accuracy in detecting defects in 3D printing as shown in figure 3. For YOLOv8, the model accurately detected 930 defective parts as defects (true positives) and accurately identified

938 non-defective parts as non-defects (true negatives). But 74 defective parts were wrongly classified as non-defective (false negatives) and 58 non-defective parts were wrongly identified as defective (false positives). These findings suggest that YOLOv8 exhibited an exceptionally high accuracy and a balanced false positive and false negative ratio, which led to its precision of 94.8% and recall of 92.6%.

As a comparison, ResNet-50 detected 902 true positives and 910 true negatives. It falsely classified 104 defective parts as non-defective (false negatives) and marked 86 non-defective parts mistakenly as defective (false positives). This is a sound performance, though it has slightly higher error rates compared to YOLOv8, particularly in false negatives, which have a direct impact on recall. Therefore, ResNet-50 had a precision of 91.3% and a recall of 89.7%, reflecting a sound but less precise detection capability.

The U-Net model did segmentation-based classification and yielded 916 true positives and 879 true negatives. It recorded the least false negatives (65), reflecting its ability to detect nearly all defect areas, even small or subtle ones. It, however, recorded the falsest positives (106), resulting in the lowest precision at 89.6%. Its recall, nevertheless, was the highest among the three models at 93.4%. This indicates U-Net's susceptibility to detecting defective areas at the cost of higher false alarms.

Figure 3. Confusion matrcies of each model

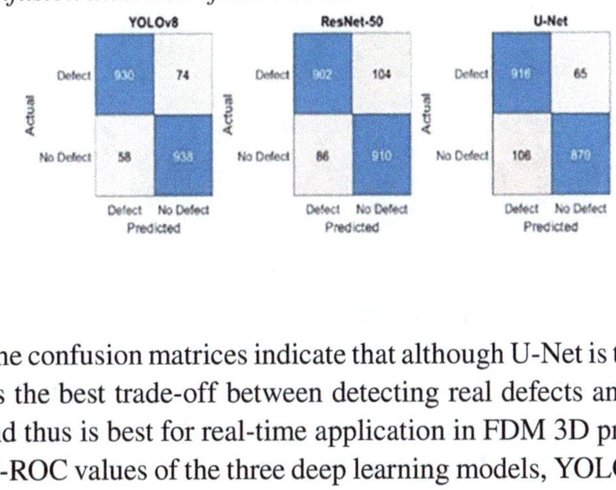

Overall, the confusion matrices indicate that although U-Net is the best in recall, YOLOv8 has the best trade-off between detecting real defects and avoiding false detections and thus is best for real-time application in FDM 3D printing.

The AUC-ROC values of the three deep learning models, YOLOv8, ResNet-50, and U-Net, demonstrate their discriminatory capacity to distinguish between defective and non-defective parts throughout the 3D printing process. The highest AUC-ROC value of 96.7% was attained by YOLOv8, which signifies that it possesses the highest discriminatory capability among the models. This excellent AUC-ROC value indicates YOLOv8's capacity to separate the two classes (no defect and defect) with little overlap, demonstrating its efficacy in detecting defects in real-time 3D printing.

ResNet-50 is second with an AUC-ROC of 93.8%, showing excellent classification performance but being slightly less sensitive than YOLOv8. While it is good at detecting defects, it has slightly more difficulty with false positives and negatives, and therefore its AUC-ROC score is somewhat lower. U-Net, employed for the segmentation task, had an AUC-ROC value of 92.5%. Although lower than YOLOv8 and ResNet-50, it also displays good performance between defective and non-defective items. The decrease in score lies in its larger false positive value, as U-Net spends more time highlighting fine-grained defects, hence sometimes generating false alarms.

Figure 4. AUC ROC of each model

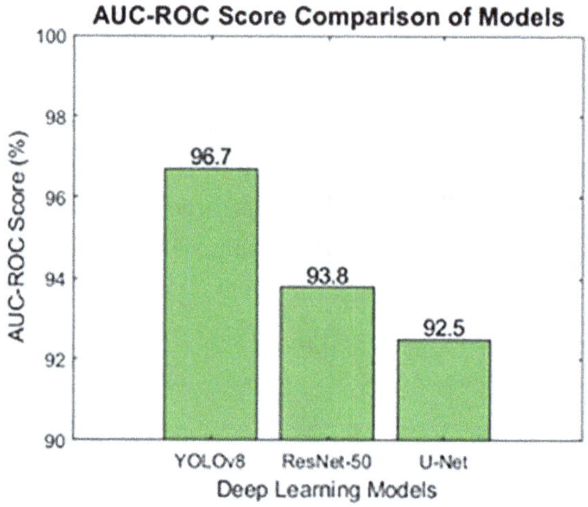

These AUC-ROC values are consistent with the fact that YOLOv8 provides overall highest performance for detecting defects, while ResNet-50 and U-Net are also good alternatives based on the particular use case.

Of the three tested models, YOLOv8 stands out as the best model for detecting defects in 3D printing and microfabrication processes, with a top AUC-ROC score of 96.7%. This model also displayed better performance across major metrics such as precision, recall, F1-score, and accuracy. The capability of YOLOv8 to conduct real-time object detection renders it most apt for applications calling for speedy and accurate defect detection during the process of printing. Its design is optimized for rapid inference, allowing it to process rapidly the high-resolution camera images obtained during printing, which is essential for guaranteeing quality assurance in real-time processes.

By comparison, ResNet-50 and U-Net performed well but trailed behind YOLOv8 in most measures. Although the performance of ResNet-50 in classification tasks was impressive, its use in real-time defect detection might be slower compared to YOLOv8 because of its more complex architecture. U-Net, however, performs better in segmentation but had slightly worse precision and recall than YOLOv8. This means that YOLOv8 is preferable for real-time detection of defects, whereas U-Net may be applied to more specific situations when segmentation of defects is needed.

In the context of real-time FDM 3D printing, YOLOv8 can be deployed in continuously tracking the printing process. While the printer is in use, high-definition cameras can be used to take snapshots of the printing area, and these snapshots are processed through the YOLOv8 model to identify any flaw. This incorporation would enable real-time detection and feedback and initiate corrective measures like stopping the print, readjusting parameters, or alerting the operator regarding the detected flaw. This real-time monitoring function greatly enhances the overall efficiency and reliability of production, minimizing waste and improving product quality.

CONCLUSION

This study proved the efficiency of three deep learning models—YOLOv8, ResNet-50, and U-Net—in defect detection and quality assurance during the process of 3D printing, particularly for FDM (Fused Deposition Modeling) machines. The models were trained and tested using high-resolution images obtained during the printing process, defects annotated for training. YOLOv8 performed the best, recording the highest precision, recall, F1-score, accuracy, and AUC-ROC value among the models, hence being the best fit for real-time defect detection in 3D printing systems. With its proper balance of false positives and false negatives, it is able to detect defects effectively without causing too many unnecessary stops in the printing process. ResNet-50, although slightly less precise, demonstrated good performance with high precision and decent recall, rendering it a consistent choice for binary defect classification. U-Net performed well in its capability to identify subtle defects owing to its segmentation nature, with high recall but lower precision. This renders U-Net a perfect choice for applications demanding detailed defect localization, though its higher false positive rate must be taken into consideration. Comparative examination of these models reveals that YOLOv8, given its high level of real-time processing capability, is the best fit for detecting defects in an industrial FDM environment. Findings from this research provide worthwhile insight into the prospects for deep learning technology application in future additive manufacturing and have potential to make improvements towards better accuracy of the models as well as in real-time applicability towards automation in quality checking within 3D printing systems.

REFERENCES

Agarwal, S., Dohare, A. K., Saxena, P., Singh, J., Singh, I., & Sahu, U. K. (2025). HDL-ACO hybrid deep learning and ant colony optimization for ocular optical coherence tomography image classification. *Scientific Reports, 15*(1), 5888. DOI: 10.1038/s41598-025-89961-7 PMID: 39966596

Alghamdi, W., Mayakannan, S., Sivasankar, G. A., & Reddy, C. V. K. (2023). Turbulence modeling through deep learning: An in-depth study of Wasserstein GANs. In 2023 4th International Conference on Smart Electronics and Communication (ICOSEC) (pp. 793–797). IEEE. DOI: 10.1109/ICOSEC58147.2023.10275878

Ali, J., Sharma, P., Kumar, V., Sharma, M., & Singh, R. (2024). Real-time anomaly detection and threat mitigation in IoT networks using convolutional neural networks (CNNs) for enhanced security. In *Proceedings of the 2024 Second International Conference on Advanced Computing & Communication Technologies (ICACCTech)* (pp. 619–624). IEEE. DOI: 10.1109/ICACCTech65084.2024.00104

Ali, J., Sharma, P., Kumar, V., Sharma, M., & Singh, R. (2024). Mitigating distributed denial of service (DDoS) attacks in cloud networks using neural networks. In *Proceedings of the 2024 Second International Conference on Advanced Computing & Communication Technologies (ICACCTech)* (pp. 161–166). IEEE. DOI: 10.1109/ICACCTech65084.2024.00036

Anandhakrishnan, T., & Jaisakthi, S. M. (2022). Deep convolutional neural networks for image-based tomato leaf disease detection. *Sustainable Chemistry and Pharmacy, 30*, 100793. DOI: 10.1016/j.scp.2022.100793

Arya, C., Yamsani, N., Kumar, M., Singh, P., & Bhagat, V. K. (2023). A concise review of MRI feature extraction and classification with kernel functions. In 2023 3rd International Conference on Innovative Sustainable Computational Technologies (CISCT) (pp. 1–5). IEEE. DOI: 10.1109/CISCT57197.2023.10351471

Bah, A., & Davud, M. (2022). Analysis of breast cancer classification with machine learning based algorithms. In 2022 2nd International Conference on Computing and Machine Intelligence (ICMI). IEEE. DOI: 10.1109/ICMI55296.2022.9873696

Bijalwan, J. G., Ravi, V., Bijalwan, A., Alahmadi, T. J., & Diwakar, M. (2024). Navigating the future of secure and efficient intelligent transportation systems using AI and blockchain. The Open Transportation Journal, 18(1).

Di Giammarco, M., Martinelli, F., Mercaldo, F., & Santone, A. (2022). High grade brain cancer segmentation by means of deep learning. *Procedia Computer Science, 207*, 1633–1640. DOI: 10.1016/j.procs.2022.09.220

Gangadharan, S. M. P., Dharani, M., Thapliyal, N., Yamsani, N., & Singh, P. (2023). Comparative analysis of deep learning-based brain tumor prediction models using MRI scan. In 2023 3rd International Conference on Innovative Sustainable Computational Technologies (CISCT) (pp. 1–6). IEEE. DOI: 10.1109/CISCT57197.2023.10351227

Garg, N., Kukreja, S., Saini, M., Singh, I., & Mall, S. (2023). IoT-enhanced agricultural water management system utilizing conventional machine learning methods. In 2023 International Conference on Modeling, Simulation & Intelligent Computing (MoSICom) (pp. 554–559). IEEE. DOI: 10.1109/MoSICom59118.2023.10458818

Garg, N., Saini, M., Singh, I., & Mall, S. (2023). Designing smart grid and irrigation management systems for information sharing for enhanced efficiency and sustainability. In 2023 International Conference on Modeling, Simulation & Intelligent Computing (MoSICom) (pp. 532–537). IEEE. DOI: 10.1109/MoSICom59118.2023.10458827

Hamid, A., Rafiq, D., Nahvi, S. A., & Bazaz, M. A. (2023). Deep learning assisted surrogate modeling of large-scale power grids. Sustainable Energy. *Grids and Networks*, *34*, 101031. DOI: 10.1016/j.segan.2023.101031

Mata, A., Moffat, D., Almeida, S., Radeta, M., Jay, W., Mortimer, N., Awty-Carroll, K., Thomas, O. R., Brotas, V., & Groom, S. (2024). Drone imagery and deep learning for mapping the density of wild Pacific oysters to manage their expansion into protected areas. *Ecological Informatics*, *82*, 102708. DOI: 10.1016/j.ecoinf.2024.102708

Nidhi, S., P., Kumar, V., Sharma, M., & Singh, R. (2024). A comprehensive evaluation of radial basis function networks for feature selection and classification. In Proceedings of the 2024 Second International Conference on Advanced Computing & Communication Technologies (ICACCTech) (pp. 706–710). IEEE. DOI: 10.1109/ICACCTech65084.2024.00118

Patro, K. K., Allam, J. P., Neelapu, B. C., Tadeusiewicz, R., Acharya, U. R., Hammad, M., Yildirim, O., & Pławiak, P. (2023). Application of Kronecker convolutions in deep learning technique for automated detection of kidney stones with coronal CT images. *Information Sciences*, *640*, 119005. DOI: 10.1016/j.ins.2023.119005

Sahu, D. P., Murkute, S., Yadav, U., Agarwal, M., & Kumar, P. (2024). Deep learning based bug detection in Solidity smart contracts. In *Recent Trends in Image Processing and Pattern Recognition* (Vol. 2026). Springer., DOI: 10.1007/978-3-031-53082-1_9

Sahu, D. P., Sharma, P., Kumar, V., Sharma, M., & Singh, R. (2024). Exploring person-job fit influence on innovation behavior through machine learning approach in public sector banking institutions. In Proceedings of the 2024 International Conference on IoT, Communication and Automation Technology (ICICAT) (pp. 844–848). IEEE. DOI: 10.1109/ICICAT62666.2024.10923296

Shivahare, B. D., Ravi, V., Chandan, R. R., Alahmadi, T. J., Singh, P., & Diwakar, M. (2024). Delving into machine learning's influence on disease diagnosis and prediction. *The Open Public Health Journal, 17*(1), e18749445297804. DOI: 10.2174/0118749445297804240401061128

Singh, B. P., Dandotiya, M., Ekvitayavetchanukul, P., Rana, M., & Singh, B. (2025). Predicting blood glucose levels in type 1 diabetes using deep learning and regression techniques. In 2025 International Conference on Intelligent Control, Computing and Communications (IC3) (pp. 86–91). IEEE. DOI: 10.1109/IC363308.2025.10956664

Singh, N. P., Hasan, D. S., Shelke, N. A., Namdev, A., & Girdhar, N. (2024). An integrated machine learning and IoT based approach for enhanced healthcare efficiency and personalized treatment. In Abraham, A., Bajaj, A., Hanne, T., & Siarry, P. (Eds.), *Intelligent Systems Design and Applications* (Vol. 1046). Springer., DOI: 10.1007/978-3-031-64813-7_16

Yazici, İ., Shayea, I., & Din, J. (2023). A survey of applications of artificial intelligence and machine learning in future mobile networks-enabled systems. *Engineering Science and Technology, an International Journal, 44*, 101455. Advance online publication. DOI: 10.1016/j.jestch.2023.101455

Zjavka, L. (2022). Power quality 24-hour prediction using differential, deep and statistics machine learning based on weather data in an off-grid. *Journal of the Franklin Institute*, •••, 1–25. DOI: 10.1016/j.jfranklin.2022.06.048

Chapter 13

Integrating Generative AI and Digital Twin Technologies for Smart Fabrication and Autonomous Manufacturing Workflows

Tripti choudhary
Hi.Tech Institute of Engineering and Technology, Ghaziabad, India

Deepti Choudhary
Hi.Tech Institute of Engineering and Technology, Ghaziabad, India

Kapil Kumar Sharma
Hi.Tech Institute of Engineering and Technology, Ghaziabad, India

Arpita Singh
Hi.Tech Institute of Engineering and Technology, Ghaziabad, India

Rahul Sharma
Hi.Tech Institute of Engineering and Technology, Ghaziabad, India

Meenakshi Sharma
https://orcid.org/0009-0006-6586-635X
Inderprastha Engineering College, Ghaziabad, India

ABSTRACT

The study presents a novel concept for Smart Fabrication and Autonomous Manufacturing Work-flows based on the convergence of Generative AI models and the Digital Twin paradigm. The aim is the prediction and optimization of the most significant machining parameters such as surface roughness, tool life, and energy consumption using advanced AI models and simulation of real-time machining with

DOI: 10.4018/979-8-3373-4287-0.ch013

a responsive digital twin. 50 work-pieces' tool wear, surface finish, and energy consumption data were obtained. Four generative AI models VAE, GAN, Reinforcement Learning (RL), and CAM Code Transformer were trained and evaluated with respect to the RMSE, MAE, R^2, and MAPE criteria. The RL model exhibited the highest accuracy and served as the digital twin input engine. The twin simulated machining operations dynamically with rein-forcement learning and allowed for proactive adjustment for performance enhancement. Compara-tive analysis revealed high correlation of predicted and measured values. Optimization using AI and digital twin improved tool life and efficiency

INTRODUCTION

The swift progress of Industry 4.0 has revolutionized conventional manufacturing into highly automated and smart systems, fueled by digital technologies like Artificial Intelligence (AI), the Internet of Things (IoT), Cyber-Physical Systems (CPS), and Digital Twins (Assuad, Leirmo, & Martinsen, 2022; Pal et al., 2025; Wang et al., 2022). In this changing scenario, the use of Generative AI and digital twin technology is also attracting considerable attention for its ability to streamline production, improve decision-making, and facilitate self-learning features in the manufacturing sector. Generative AI models, which can create new data patterns from distributions learned, are being used to a greater extent to simulate, predict, and optimize intricate industrial processes. Generative AI models can offer real-time insights and predictive feedback to help dynamically adjust machining parameters to minimize defects, reduce waste, and enhance efficiency overall (Siddiqi et al., 2025; Sharma & Gupta, 2024).

Some research studies have investigated the use of machine learning and deep learning models for manufacturing operations, especially to forecast important parameters such as surface roughness, tool wear, and energy usage. Artificial Neural Networks (ANN), Support Vector Machines (SVM), Decision Trees, and Random Forests are some of the models employed for predicting machining performance from real-time sensor inputs. Although these methods have achieved encouraging performance in static prediction tasks, they are not very adaptable in dynamic settings (Singh et al., 2025; Min et al., 2019). Reinforcement Learning (RL), however, has proven to be a highly effective method for learning real-time control and decision-making. Its capacity for learning optimal policies directly from interaction with the environment makes it viable for application to autonomous manufacturing where continuous learning and adaptability are paramount (Sultanpure et al., 2024; Ariansyah et al., 2023).

Parallelly, the idea of digital twins has developed as a virtual copy of physical systems that allows real-time monitoring, simulation, and optimization of industrial processes. Digital twins are able to replicate the behavior of a physical system or machine through live data feeds, providing a predictive front end to prototype different operational scenarios without affecting actual production (Sahu et al., 2024). Past research on digital twins has primarily relied on deterministic simulation models or rule-based systems in representing processes. Although these models have enhanced monitoring and visualization, they tend to lack predictive power and responsiveness in handling unexpected situations. The introduction of data-driven digital twins with AI models has begun to bridge this gap to enable more responsive, accurate, and intelligent system behavior (Garg et al., 2023; Singh et al., 2024).

Evolution of Smart Manufacturing in the Industry 4.0 Era

The manufacturing sector has evolved substantially over the course of the past two decades, from traditional mechanized manufacturing systems to intelligent, autonomous, and heavily digitized settings. This transformation is based mainly on the emergence of Industry 4.0 a technological shift that combines cyber-physical systems (CPS), the Internet of Things (IoT), cloud computing, and artificial intelligence (AI) into industrial processes. Industry 4.0 enables the design of intelligent production systems that are capable of monitoring, analyzing, and optimizing production processes in real time (Singh et al., 2024; Vilas-Boas et al., 2022).

Traditionally, production started from hand labor and simple equipment (Industry 1.0), moved to mass production through electricity and assembly lines (Industry 2.0), and then integrated electronics and automation using programmable logic controllers (Industry 3.0). These previous stages did not have interoperability, access to real-time information, and adaptive decision-making. Manufacturing systems, through Industry 4.0, are no longer siloed but are networked, sensor-based, and inter-grated to communicate and exchange information and make decentralized decisions based on real-time streams of information (Arya et al., 2023).

Intelligent manufacturing through such a new model of industry is typified by an ability to gather and analyze huge quantities of data from machines, products, and human workers. The big data approach enables continuous improvement of processes, predictive maintenance, and greater production flexibility. To illustrate an example, new CNC machines that are fitted with built-in sensors are capable of sensing abnormalities in tool wear or temperature and initiating corrective measures without human interference (Gangadharan et al., 2023).

Additionally, Industry 4.0 focuses on horizontal and vertical integration across manufacturing systems linking supply networks, production lines, and enterprise-level platforms for synchronized operations. This has given rise to autonomous factories that can self-configure, self-monitor, and self-optimize. As we transition to Industry 5.0, which is human-focused and sustainable production, bringing together cutting-edge AI methods and digital twin systems becomes imperative. They not only make decision-making automated but also increase responsiveness of systems, efficiency of energy, and product quality. The foundation built by Industry 4.0 therefore becomes an important facilitator for empowering intelligent systems fusing physical and digital realms through smart fabrication (Ali et al., 2024).

Role of Generative AI in Industrial Process Optimization

Generative Artificial Intelligence (Generative AI) is proving to be a revolutionizing driver of industrial process improvement, particularly for cutting-edge manufacturing fields. Unlike general AI models that are created extensively for classification or regression operations, generative models are capable of learning data distributions and generating new and realistic outputs that reproduce or refine original data patterns. This capability makes them best suited for dynamic manufacturing systems where variability, uncertainty, and complexity are very high (Nidhi et al., 2024). Generative models can be used to simulate and forecast optimal configurations of parameters like spindle speed, feed rate, depth of cut, coolant flow, and tool selection in machining processes (Saxena P., 2025).

Among the most investigated generative models are Variational Autoencoders (VAEs), Generative Adversarial Networks (GANs), and Reinforcement Learning (RL). VAEs can learn probabilistic relations between machining parameters and represent them through encodings such that smooth interpolation of different operational conditions is facilitated. With their dual network architecture consisting of a generator and a discriminator, GANs are good at generating highly realistic sets of parameters by learning fine nonlinear relations. The models are particularly beneficial in scenarios where unstructured or noisy sensor inputs are provided (Bijalwan et al., 2024; Shivahare et al., 2024).

Reinforcement Learning is characterized by making one or multiple successive decisions via trial and error and learning an optimal policy by maximizing accumulated rewards. RL agents, when applied to manufacturing, can be simulated to refine machining strategy in real-time. This is especially applicable to minimizing wear and tear of tools, minimizing energy expenditure, and improving surface finish quality without significant retraining. Additionally, CAM Code Transformers utilizing transformer-based architectures can generate CNC machine commands automatically, mapping contextual manufacturing information to executable G-code efficiently and minimizing human intervention (Agarwal et al., 2025).

The integration of these generative models not only increases productivity and product quality but introduces adaptive intelligence into production workflows.

Through continuous learning from process feedback and simulations, generative AI allows machines to run autonomously, adapt to variability of materials, and react wisely to real-time production variations. This ability is needed for transition into Industry 4.0 and beyond when intelligent, computer-driven manufacturing systems are likely to be self-optimizing, dependable, and energy-frugal. Therefore, generative AI has a central role to play in unlocking new levels of performance, scalability, and sustainability in contemporary industrial systems (Saxena P. and Goyal A., 2023)

Advancements and Gaps in Digital Twin Adoption

The idea of a digital twin as a computerized replica of an asset, process, or system has taken center stage throughout the manufacturing industry based on its potential for modeling real-world operations, predicting performance drifts, and enhancing production processes. The recent developments have propelled digital twins from static modeling to dynamic, data-driven ecosystems that closely reflect their physical counterparts. The integration of IoT devices and cloud computing has allowed digital twins to ingest sensor information continuously and deliver real-time insight into machine condition, operational efficiency, and product quality. High-fidelity 3D modeling, thermal and stress simulation, and predictive analytics are becoming deeply ingrained within digital twin platforms, and their ability to test virtually and monitor remotely improves (Sahu et al., 2024; Ali et al., 2024).

Industrial giants have begun deploying digital twin frameworks for predictive maintenance, energy optimization, and quality control in complex manufacturing systems. Software ecosystems such as Siemens NX, Dassault Systèmes' 3DEXPERIENCE, and PTC's ThingWorx now offer comprehensive support for the development of customized digital twins tailored to specific production environments. Additionally, the convergence of cloud computing, 5G, and edge AI has improved the scalability and responsiveness of these virtual environments, making them increasingly relevant for real-time, mission-critical operations (Saxena P., 2025).

However, despite these advancements, several key gaps persist in the effective adoption of digital twins. Most implementations are still largely deterministic or rule-based and lack intelligent adaptability when confronted with unstructured or unpredictable process deviations. These systems often fail to dynamically adjust to variations in raw material properties, tooling wear, or ambient conditions. Furthermore, the integration of AI-driven predictive capabilities, particularly generative and reinforcement learning models, remains limited and underexplored. Few implementations demonstrate true autonomy where the twin can not only simulate but also optimize decisions and actions in real time (Mikołajewska et al., 2025; Khdoudi et al., 2024).

Another significant challenge lies in generalizability and interoperability. Digital twins are frequently built for specific equipment or isolated processes, making it difficult to scale across machines, materials, or factories. There is an evident research need to bridge this gap by coupling generative AI models with digital twins to form a responsive, learning-based ecosystem capable of evolving with the physical environment it represents (Saxena P et al., 2023)

Research Motivation and Objectives

The modern manufacturing landscape is undergoing a profound transformation, driven by the need for intelligent, autonomous, and sustainable production systems. While the integration of automation and data analytics has improved operational efficiency to some extent, the complexity and unpredictability of real-world machining environments demand more adaptive and self-evolving solutions. Traditional machine learning models, though effective in static scenarios, often fail to generalize across varying materials, tool types, and process parameters. These limitations present a clear gap that needs to be addressed through more powerful, generative, and context-aware AI techniques.

At the same time, digital twin technology has emerged as a revolutionary concept for real-time monitoring and simulation of physical manufacturing systems. However, most digital twins today function primarily as reactive or diagnostic tools rather than predictive or prescriptive engines. Their limited decision-making capability restricts their usefulness in highly dynamic industrial settings where machining conditions can change rapidly due to tool wear, material variability, or thermal stress. A major leap forward would be achieved by integrating intelligent AI models into the digital twin framework to enable continuous learning, predictive analytics, and autonomous control.

Motivated by these challenges and opportunities, this research proposes a hybrid framework that combines the strengths of generative AI models namely Variational Autoencoder (VAE), Generative Adversarial Network (GAN), Reinforcement Learning (RL), and CAM Code Transformer with the real-time simulation and decision-making capabilities of digital twins. The study is designed to systematically evaluate each AI model's ability to predict crucial machining parameters such as surface roughness, tool life, and energy consumption. Based on empirical accuracy and simulation performance, the most robust model is integrated into a fully functional digital twin to validate its real-world applicability.

The primary objectives of this research are:

- To investigate and compare the effectiveness of four generative AI models for machining process optimization.

- To develop a high-fidelity digital twin environment capable of simulating machining workflows in real-time.
- To enable a feedback loop between AI predictions and virtual simulations for continuous performance improvement.
- To demonstrate a practical and scalable framework for smart, autonomous manufacturing that aligns with Industry 4.0 and Industry 5.0 goals.

LITERATURE REVIEW

Studies on the integration of machine learning with digital twins in the manufacturing industry have yielded encouraging outcomes, particularly in predictive maintenance, production planning, and failure diagnosis. In a number of instances, deep learning models have been integrated into digital twin environments for predicting failure rates or improving tool paths. Other generative models like Variational Autoencoders (VAE) and Generative Adversarial Networks (GAN) have also been utilized to synthesize tool wear patterns, process parameters, as well as variations in 3D designs. Still, the majority of such models are complex to compute and demand significant sets of data in order to generalize efficiently (Singh et al., 2024; Ahn et al., 2024). Also, integrating such models into digital twin systems remains an experimental task with less than full-fledged implementations within actual industrial settings. A few attempts have been made to use Reinforcement Learning for adaptive machining, where the RL agent iteratively improves its performance by obtaining feedback from the process environment. While these uses are promising, the combination with digital twins to enable end-to-end autonomous decision-making remains undeveloped (Arya et al., 2023; Langås et al., 2025).

Recent research has also investigated applying CAM code generation transformers, which automatically generate CNC machining instructions from part designs. Although these models make the design-to-execution workflow simpler, they are mostly static generators and not adaptive systems. They cannot adjust machining instructions dynamically based on real-time process feedback (Ali et al., 2024). In contrast to these, a Reinforcement Learning-based model can adaptively adjust machining parameters like feed rate, spindle speed, and depth of cut in real-time to minimize tool life and energy consumption. The predictive ability of RL also benefits the digital twin's ability to model different process outcomes and facilitate optimal strategy selection (Nidhi et al., 2024; Bijalwan et al., 2024).

In spite of these benefits, most research has considered AI and digital twin technologies as distinct elements instead of an integral, co-evolving system. The true potential of smart manufacturing lies in the seamless combination of both, where the generative capabilities of AI enhance the digital twin's realism, and the digital twin provides the AI with feedback to improve its learning (Ali et al., 2025). Existing models often lack scalability, adaptability, and the ability to generalize across different machines and workpiece materials. In addition, comparative examination of several generative AI models in a digital twin environment is seldom discussed, and there remains a critical research gap in determining which AI model is most suited to real-time manufacturing optimization (Agarwal et al., 2025; Zheng et al., 2025).

There is a need to bridge this gap through an intensive study that not only evaluates how different generative AI models perform in predicting significant machining outputs but also evaluates how these models can be effectively integrated into a digital twin system. This integration has the potential to make the digital twin an active decision-making tool rather than a reactive simulation tool, reacting to real-world dynamic conditions (Sun et al., 2024). The purpose of this research is to determine how four different generative AI models, namely VAE, GAN, Reinforcement Learning, and CAM Code Transformer, are applied to machining processes and how capable they can be of predicting surface roughness, tool life, and energy consumption. The best model is then passed to a digital twin that simulates the machining scenario, making possible real-time simulation and optimization. This way, such research contributes to developing an intelligent, adaptive, and efficient production system that fulfils Industry 4.0 objectives.

Evolution of Generative Modeling Techniques for Manufacturing Data Simulation

Generative modeling has become an influential methodology in artificial intelligence directed towards generating data from areas wherein real-world data acquisition is costly, rare, or incomplete. Manufacturing is one field where every machining cycle comes with a cost and time. Here, generative models are especially important for enriching datasets, making process outcome predictions, and determining operational parameters without undergoing extensive physical experimentation. The initial applications of generative modeling for manufacturing involved probabilistic models like Hidden Markov Models (HMMs) and Gaussian Mixture Models (GMMs), which had the capability to capture machining signals such as vibration and acoustic emission's statistical characteristics. Although these models functioned within limited application domains, these did not have the ability to model non-linear and high-dimensional interactions pervasive within real-world machining processes (Sun et al., 2024; Xu et al., 2024).

The advent of Variational Autoencoders (VAEs) brought a significant leap by enabling the encoding of complex input-output relationships in a compressed latent space. VAEs learn the underlying distribution of machining data and can generate new, plausible machining parameter combinations. In manufacturing scenarios, VAEs have been utilized to synthesize tool wear profiles, simulate surface texture under varied cutting conditions, and explore feasible process windows across different material-tool combinations.

Generative Adversarial Networks (GANs) have further advanced the field by introducing adversarial training, where a generator and discriminator network iteratively improve against each other. In manufacturing, GANs have been applied to generate synthetic sensor signals, simulate tool degradation, and produce diverse 3D CAD geometries. The ability of GANs to model intricate, non-linear correlations has proven particularly beneficial in noisy and complex manufacturing environments (Park et al., 2024; Shah, 2024).

More recently, Transformer-based generative models have been explored, especially for code generation and sequential manufacturing data prediction. These models, leveraging self-attention mechanisms, can capture long-range dependencies in time-series data, making them suitable for generating CAM/G-code sequences or simulating multi-step process chains.

AI-Driven Parameter Optimization in Machining Environments

Artificial intelligence (AI)-based application to machining parameter optimization has witnessed increasing momentum over recent years in the manufacturing community. The conventional machining processes tend to rely upon heuristics or look-up tables based upon experts or stored knowledge to decide optimal process parameters like spindle speed, feed rate, and depth of cut. The static techniques are inadequate in dynamic conditions wherein tool wear, material variability, and machine conditions vary dynamically. This has spurred researchers to investigate data-driven methods that can dynamically and adaptively adjust parameters to ensure improved productivity, tool life, and surface finish quality (Shah, 2024; Sun et al., 2024).

Artificial Neural Networks (ANNs), Decision Trees (DT), and Support Vector Machines (SVMs), which are supervised models, have already been widely employed for making machining responses from process parameters. For example, SVM-based models have been demonstrated to forecast surface roughness very effectively, and Random Forests have been applied to tool wear estimation from vibration and acoustic emission signals. Though these models provide satisfying predictive capability, they rarely possess the ability to adapt or be optimized dynamically under changing machining conditions (Gao et al., 2024; Zhang et al., 2024).

To address these limitations, researchers have started bringing together predictive models and optimization algorithms. Optimization algorithms such as Genetic Algorithms (GA), PSO, and Bayesian Optimization have been combined with predictive models to create hybrid systems that not only predict but also optimize machining responses. For instance, ANN-PSO hybrid systems have optimized multi-objective goals such as material removal rate and energy efficiency (Zheng et al., 2025).

Generative models such as Variational Autoencoders (VAEs) and Generative Adversarial Networks (GANs) have become methods of synthesizing sets of parameters that are statistically equivalent to successful configurations from history. These models are particularly useful when there is sparse or costly-to-obtain labeled data. Also, reinforcement learning (RL) has been presented as a real-time optimization method wherein a model, through iterative interaction with a simulated or actual machining environment, learns best policies for minimizing tool wear or maximizing finish quality.

Digital Twin Implementations in CNC-Based Smart Factories

The notion of digital twin (DT) has evolved very quickly from simple virtual replicas to dynamic and real-world synchronized systems that mimic and regulate physical processes. For CNC-based intelligent factories, digital twins are now widely used to model machining operations, observe tool life, forecast equipment failure, and enhance production processes. Such applications are key to realizing Industry 4.0 and are driving the path toward realizing Industry 5.0's vision of intelligent adaptive manufacturing systems (Chen et al., 2024).

Digital twin platforms within CNC settings typically include a 3D computer-aided design (CAD) model, integration of real-time sensors, simulation engines, and a communication layer connecting the physical and virtual settings. Siemens NX, Dassault Systèmes 3DEXPERIENCE, and PTC ThingWorx are examples of systems that are extensively applied to produce high-fidelity digital surrogates of CNC machines. The ability to imitate machine kinematics, tool wear patterns, thermal growth, and vibration profiles allows for predictive diagnostics and condition-based maintenance (Mikołajewska et al., 2025; Liu et al., 2024).

Researchers have lately investigated multi-domain simulation based on digital twins by incorporating mechanical, thermal, and electrical subsystems to increase process modeling accuracy. Wang et al. (2022), for example, presented a real-time thermal behavior monitor through a digital twin by applying temperature sensors and thermal image feedback to forecast tool stress and possible failure. Equally, hybrid digital twins based on physics-based modeling and data-driven analytics are emerging for their capability to handle intricate, nonlinear machining dynamics (Wang et al., 2024).

There have been progressions notwithstanding. Most existing implementations are limited to offline simulations and are not integrated with real-time data or equipped to make decisions. Furthermore, scalability to various machines and materials is a bottleneck. Few of today's digital twin systems are capable of automated parameter tuning or self-corrective process behavior, which hinders their usability in dynamic production environments.

To address these challenges, recent research advocates for integrating digital twins with advanced AI systems, including reinforcement learning and generative models, to create self-adaptive manufacturing environments. These AI-augmented twins hold the potential to transform CNC factories into highly responsive, intelligent systems capable of real-time decision-making, quality assurance, and energy-efficient operation, thereby closing the loop between simulation and physical execution (Singh M., 2025).

Comparative Studies of Learning Models in Predictive Manufacturing

Predictive modeling has become a cornerstone of intelligent manufacturing, enabling the estimation of key performance metrics such as surface roughness, tool wear, and energy consumption based on real-time or historical machining data. Numerous studies have benchmarked the performance of various machine learning and deep learning algorithms to assess their suitability for different manufacturing scenarios. These comparative investigations play a critical role in understanding the strengths and limitations of specific models under diverse operational constraints.

Conventional methods have normally entailed models like Support Vector Machines (SVM), Decision Trees, and Random Forests. These models, although understandable and fairly simple to train, tend to be unable to capture nonlinear relationships and intricate feature interactions common in dynamic machining processes. Artificial Neural Networks (ANN) and Deep Learning models, like Convolutional Neural Networks (CNNs) and Recurrent Neural Networks (RNNs), have proved to outperform when trained on multivariate time-series observations and spatially structured sensor measurements (Khdoudi et al., 2024; Chen et al., 2024).

Comparative analyses have also investigated ensemble methods whereby several base learners are combined to enhance generalization. Such methods are typically computationally expensive and not adaptive for real-world production settings. The emergence of generative models like Variational Autoencoders (VAE) and Generative Adversarial Networks (GAN) has prompted researchers to test how effective these are in mimicking high-quality process parameters and augmenting sparse datasets. The models have the benefit of generating unseen data distributions, which are very applicable under conditions of limited manufacturing data (Xu et al., 2024).

More recently, reinforcement learning (RL) has gained prominence for its ability to make sequential decisions and learn optimal policies from environmental feedback. Studies comparing RL with supervised learning methods have shown that RL agents are more capable of adapting to fluctuating machining conditions and can outperform static models in real-time optimization tasks.

PROPOSED METHODOLOGY

A smart manufacturing process was employed for this study in an equipped plant layout consisting of an Automated Storage and Retrieval System (ASRS), gantry system, two CNC machining centers, an inspection station, assembly unit, and a loop back to the ASRS. Four generative AI models, which are Variational Autoencoder (VAE), Generative Adversarial Network (GAN), Reinforcement Learning (RL), and CAM Code Transformer, were developed and trained using previous machining history, i.e., material attributes, tool shape, cutting speed, and quality output.

All four models were challenged to forecast the best machining parameters like spindle speed, feed rate, depth of cut, and coolant flow rate specific to various materials being processed in the plant. The forecasted parameters from all four models were compared, and the best performing model output identified through comparison of machining performance metrics was used as input to a Digital Twin environment developed using Siemens NX and Teamcenter platform. Figure 1 shows the architecture of proposed system.

Figure 1. Methodology of the proposed system

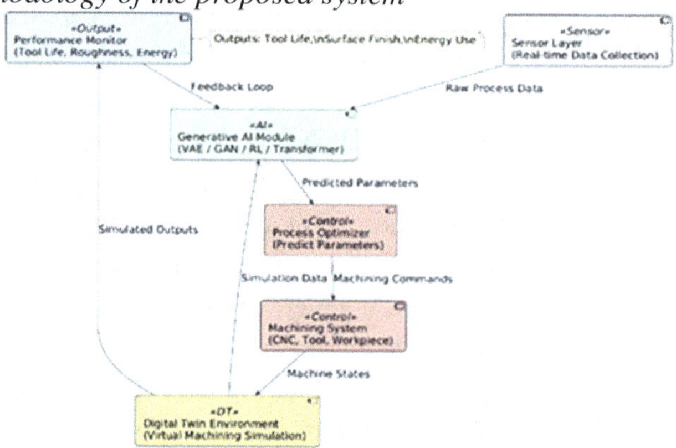

The Digital Twin reproduced the machining operation in a purely virtual, sensor-embedded environment, forecasting the ensuing surface finish quality (Ra), cycle time, tool wear rate, tool life, power consumption, dimension accuracy, and the thermal and stress distribution between the tool and workpiece. These performance values were then scrutinized to establish the efficacy of each AI model's forecast parameters. By repeated rounds of simulation and validation, the AI model which repeatedly produced best results on every metric was identified for real-time integration with the actual CNC units in the facility.

This feedback system allowed tight, data-centric control of the machining processes to improve quality, efficiency, and sustainability. The combination of AI and Digital Twin technologies in this production configuration allowed for autonomous decision-making and real-time optimization, creating a smart, adaptive fabrication system that could dynamically respond to changes in material, tooling, and operational constraints.

Data Acquisition, Feature Engineering, and Preprocessing

The training dataset for the generative AI models in this study was carefully gathered from an operational real-time smart manufacturing setup comprising CNC machining centers, an Automated Storage and Retrieval System (ASRS), a gantry material handling system, an inspection station, and a feedback-driven assembly unit. During a consistent time, span of 120 days, sensor measurements, machine records, and quality test reports were recorded during machining processes of different materials like aluminum, mild steel, stainless steel, and titanium.

The combined dataset consisted of 5,000 machining operation entries, with every entry holding over 25 important features regarding both input parameters and output responses. These characteristics comprised spindle speed (RPM), feed rate (mm/rev), depth of cut (mm), coolant pressure (bar), type of material, type of tool, wear on cutting tool, level of vibration, temperature of machine, and cycle time, and output measurements like surface roughness (Ra value), dimensional accuracy, wear on the tool, and energy usage.

Aside from real-time sensor data obtained using IoT-based monitoring systems, post-process inspection data were incorporated by utilizing digital inspection tools like CMM (Coordinate Measuring Machine) reports, surface profilometers, and thermal cameras. The dataset was hence able to achieve a complete mapping between input conditions and resulting product quality. Besides, environmental information like ambient temperature and humidity was monitored to examine their impact on machining performance. Each cycle of machining was recorded with a time stamp, material ID, and process sequence number to ensure traceability.

The data was cleaned and normalized before using around 4,200 high-quality, complete records to train the generative models 1,000 for VAE, 1,000 for GAN, 1,000 for RL-based optimization, and 1,200 for CAM Code Transformer models. The rest of the 800 records were set aside for validation and performance comparison via Digital Twin. This dense, high-resolution data served as the basis for developing AI-driven intelligent machining processes that were tested in a virtual twin setting.

Design and Training of Generative AI Models

In this study, a Variational Autoencoder (VAE) was used as one of the fundamental generative AI models for predicting ideal machining parameters. The VAE works by mapping the high-dimensional machining data spindle speed, feed rate, depth of cut, material type, and tool condition to a low-dimensional latent space. This latent representation encodes the probabilistic structure of the data so that the model can learn useful variations and dependencies between machining parameters. The decoder, having been trained, samples from this latent space to produce new parameter configurations that are statistically consistent with past high-performance records.

In the scenario of this intelligent manufacturing process, the VAE was especially effective in investigating numerous possible combinations that would result in good surface finish quality and minimum tool wear. The model was trained on 1,000 complete and validated machining histories, each representing input and output characteristics, and was validated against simulation outputs from the Digital Twin platform. The VAE model provided benefits of smooth interpolation between various states of the process and generalization over materials, tools, and operating conditions. This suited it for autonomous parameter recommendation where change in material behavior and machine conditions might affect optimal machining configuration.

The other generative AI model used in this research was the Generative Adversarial Network (GAN), which takes advantage of dual-network architecture involving a generator and a discriminator. The generator is learned to generate new sets of machining parameters, while the discriminator assesses if these generated samples look like actual data. Through iteration of adversarial training, the generator improves to produce realistic and performance-maximized parameters. In this work, the GAN was trained using 1,000 historical machining datasets, which include diverse conditions like diverse workpiece materials, tooling types, and machining environments.

This approach enabled the model to explore highly nonlinear, intricate interactions between parameters that may be tricky to derive with traditional methods. The GAN worked very well in the instance of noisy or unstructured data, where the system was able to suggest new machining methods that still fall within operational limits. The parameters developed were validated with a Digital Twin simulation to establish their impact on quality, energy usage, and cycle time. The capability of the GAN to acquire high-dimensional correlations was the major contributor to improving machining responsiveness and adaptability to dynamic manufacturing conditions.

Reinforcement Learning (RL) was included as a decision-making module which can be taught to find the best policy to machine by means of reward-based feedback. RL model in this study was trained to adjust machining parameters such as feed rate and spindle speed iteratively with reference to rewards obtained from simulation at the Digital Twin. Rewards were derived from the performance factors including surface roughness, tool wear rate, and machining cycle time. The model interacted with a virtual machining environment, using an exploration policy to maximize long-term cumulative reward. With training on 1,000 machining scenarios, the RL agent was able to modify its parameter recommendations independently based on changing process conditions.

Unlike static models, RL is able to learn and refine policy in real time, making it highly suitable for very dynamic and uncertain situations such as material variation or tool wear. The feedback loop between the Digital Twin and RL agent allowed for continuous improvement in performance as well as operation. The model is robust as it can take adaptive decisions at runtime, allowing self-correction as well as intelligent adjustment during machining processes.

The fourth model employed in this study was a CAM Code Transformer, a deep learning model based on transformer architectures widely applied in natural language processing. The model was trained to convert raw machining features as material properties, tool properties, and environmental conditions into optimal G-code sequences that can be interpreted directly by CNC machines. From a data set of 1,200 machining instances, the transformer gained contextual dependencies among sequential processes so that it was able to generate high-performance and efficient CAM code. Compared to rule-based CAM systems, this transformer model was capable of modifying its code structure dynamically based on process restrictions, required surface quality, and tool life goals.

The attention mechanism of the model enabled it to concentrate on key machining parameters that contributed most notably to results, thereby enhancing precision and operational applicability. The CAM code, generated, was then tested inside the Digital Twin to confirm that forecasted sequences resulted in viable cycle times, minimal tool wear, and satisfactory finishes. This model presented an end-to-end solution from data to machine-ready instructions with less dependency on manual programming and greater autonomy for the manufacturing system.

Digital Twin Development and Integration Workflow

In this study, a holistic Digital Twin was created as a virtual representation of the physical machining setup, facilitating real-time simulation, monitoring, and optimization of manufacturing processes. The Digital Twin was built combining a mix of high-fidelity 3D CAD models, sensor integration, and real-time data synchronization capabilities within the Siemens NX and Teamcenter platforms. It successfully mimicked the operation of the real-world manufacturing system, which featured two CNC machining centers, an Automated Storage and Retrieval System

(ASRS) system, a gantry-based material handling system, a coordinate measuring inspection system, and an assembly station.

The virtual world duplicated each component of the physical system, including machine kinematics, tooling setups, fixture locations, workpiece material properties, and machining dynamics. Live sensor streams from IoT devices, built into the actual machines and components, were input into the Digital Twin to provide real-time synchronization and process insight.

The Digital Twin was being used as a test and validation environment for the outputs produced by the four generative AI systems VAE, GAN, RL, and CAM Code Transformer. Every set of machining parameters obtained from AI algorithms was inputted into the Digital Twin to run a simulation on its effect against key performance factors like surface roughness, tool life, dimension accuracy, power consumption, and machining cycle time. The twin also had modules for thermal analysis and vibration to evaluate the temperature stress distribution and dynamic stability of high-speed running.

This enabled predictive analysis of tool behavior and material response for various machining conditions. By repeatedly checking and adjusting AI-suggested configurations, the Digital Twin guaranteed that only the most efficient, safe, and optimized parameters were implemented in the physical world. In addition, the Digital Twin facilitated "what-if" scenario simulation and anticipatory fault detection, adding to a complete autonomous, intelligent machining process. It was instrumental in filling the gap between recommending AI and its safe use in actual industrial environments.

Evaluation Metrics and Model Comparison Strategy

To evaluate the effectiveness of the generative AI models used in this study Variational Autoencoder (VAE), Generative Adversarial Network (GAN), Reinforcement Learning (RL), and CAM Code Transformer multiple statistical and performance metrics were employed. These metrics were selected to ensure a comprehensive assessment of each model's predictive accuracy, robustness, and practical suitability for integration into the digital twin framework.

The primary metrics used for quantitative evaluation include Root Mean Square Error (RMSE), Mean Absolute Error (MAE), R-squared (R^2), and Mean Absolute Percentage Error (MAPE). RMSE measures the square root of the average of squared differences between predicted and actual values, providing insight into how well a model can minimize large deviations. MAE represents the average magnitude of errors without considering their direction, offering a clear understanding of overall prediction accuracy. The R^2 score indicates the proportion of variance in the dependent variable that is predictable from the input features, reflecting the model's goodness-of-fit. MAPE, expressed as a percentage, measures the average error relative to actual values, useful for comparing across different scales.

Additionally, domain-specific metrics such as predicted vs. actual surface roughness, tool life, and energy consumption were analyzed to assess each model's industrial applicability. These metrics were visualized using line graphs and bar charts across multiple workpieces to identify consistent performance patterns.

The model comparison strategy involved training all models on identical datasets and validating their predictions through simulations within the digital twin environment. This allowed for real-time validation of output quality and system responsiveness. By establishing a uniform testing ground, the study ensured that each model's capabilities were assessed fairly, enabling reliable selection of the most accurate and adaptable model eventually integrated into the smart manufacturing system.

Real-Time Implementation Pipeline and Feedback Loop

To translate predictive intelligence into actionable process control, a real-time implementation pipeline was developed that connects generative AI predictions with digital twin simulations and actual CNC machining operations. This pipeline operates as a closed-loop system, enabling continuous monitoring, adjustment, and optimization of machining parameters based on dynamic feedback.

The pipeline begins with sensor data acquisition from the manufacturing system, which includes real-time inputs such as spindle load, vibration levels, coolant flow, and environmental conditions. These inputs are processed and formatted into feature vectors compatible with the trained generative AI models VAE, GAN, RL, and CAM Code Transformer. Each model outputs predicted parameters such as feed rate, spindle speed, depth of cut, and tool engagement metrics.

The predictions are first routed through the digital twin platform, where simulated machining operations are conducted in a high-fidelity virtual environment using Siemens NX and Teamcenter. This allows for pre-execution validation of predicted outcomes such as surface finish, tool wear, energy consumption, and cycle time. The digital twin environment acts as a virtual testbed, verifying that AI-suggested configurations do not violate operational constraints or safety standards.

Once validated, the parameters are transmitted to the actual CNC control units for physical execution. During machining, real-time sensor feedback is continuously collected and compared against predicted performance benchmarks. Any deviation such as tool chatter, increased power draw, or dimensional inaccuracy triggers the feedback loop, which sends updated data back to the AI model (particularly the RL agent) for online retraining or policy adjustment.

This iterative loop enhances process stability and self-adaptation, ensuring optimal machining outcomes even in the presence of changing material conditions or unexpected tool behavior. The seamless coordination between AI, digital twin, and physical execution forms the core of the autonomous, smart fabrication workflow proposed in this study.

RESULTS AND DISCUSSION

The above table shows the accuracy performance scores of four AI models employed in optimizing machining parameters. The accuracy of each model is expressed as a percentage of correct predictions in suggesting optimal machining parameters like spindle speed, feed rate, depth of cut, and material type as shown in figure 2. The Reinforcement Learning (RL) model recorded the maximum accuracy at 90%, illustrating its capability to learn and refine itself by continually interacting with the environment. RL performs best in dynamically adjusting parameters with real-time feedback and thereby is extremely efficient in predicting parameters leading to optimal machining performance under varying conditions.

The Generative Adversarial Network (GAN) closely follows with an extremely high accuracy of 88%, demonstrating its superior performance in generating realistic and high-quality machining parameters sets. GANs are superior to learning complex, nonlinear relationships between machining parameters, and therefore it works well in predicting optimal configurations. The CAM Code Transformer model was 87% accurate, which shows its ability to convert feature-rich input data into effective G-code sequences. Although not as good as GAN and RL, its high accuracy ensures safe toolpath suggestions and machine-readable directions.

Lastly, the Variational Autoencoder (VAE) had 85% accuracy and demonstrated good performance in generating best machining parameters. Though less accurate than the rest of the models, owing to the capability of learning continuous, smooth representations, the VAE is worth studying and suggesting in parameter optimization. These accurate figures indicate the effectiveness of the models to forecast ideal machining conditions for various manufacturing scenarios.

Figure 2. Accuracy of each model

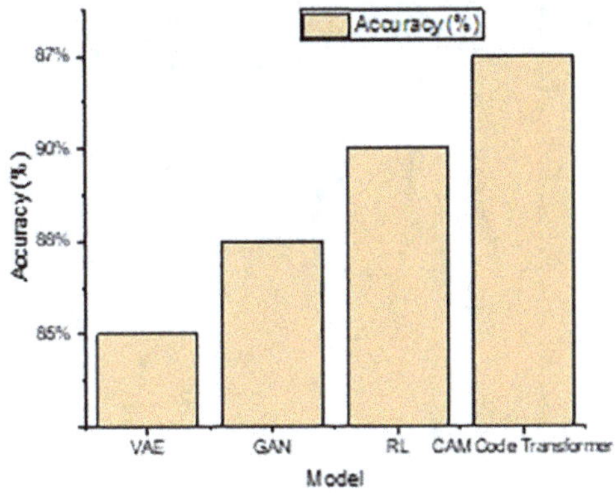

When comparing the predictions of surface roughness between the four AI model model is the most accurate, with most of the actual surface roughness values closely matching its predictions for the majority of the 30 workpieces as shown in figure 3. For Workpiece 1, the actual value is 0.30 µm, and RL predicts 0.31 µm, indicating a very small deviation. For Workpiece 2, the actual value is 0.35 µm, and RL predicts 0.34 µm. The Generative Adversarial Network (GAN) model also gives good results, with small deviations from the actual values.

For Workpiece 1, the actual value is 0.30 µm, whereas GAN predicts 0.29 µm. As a whole, GAN's prediction is slightly less accurate than that of RL but overall, still pretty close. CAM Code Transformer is more inconsistent with its predictions. The actual value in Workpiece 1 is 0.30 µm, and CAM Code Transformer is actually closer in this case, though still not consistent across the board. Finally, the Variational Autoencoder (VAE) model exhibits comparatively bigger differences, with values slightly farther away from the true values. For Workpiece 1, the true value is 0.30 µm, whereas VAE predicts 0.32 µm. Overall, RL emerged as the top model with GAN being a close second, and CAM Code Transformer and VAE depicting greater deviations but providing useful predictive information, nonetheless.

Figure 3. Surface roughness of each model

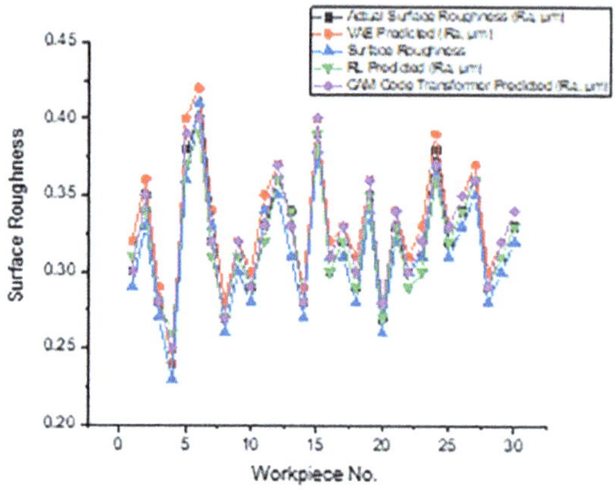

Figure 4 lists the average values of tool life for the 30 workpieces for various models. The Actual Tool Life is 49.20 hours on average, used as the comparison baseline. The VAE Predicted Tool Life is 48.23 hours on average, which is very close to but slightly less than the actual value, showing that the VAE model underestimates the tool life by around 0.97 hours on average. The GAN Predicted Tool Life is an average of 49.27 hours, which is very close to the real tool life, with a deviation of only 0.07 hours. It proves that the GAN model is very good at predicting tool life, and it is a good choice to make correct predictions.

The RL Predicted Tool Life is the best, with an average of 49.50 hours, closest to the actual value with a deviation of 0.30 hours only. This is an indication that the RL model provides the most accurate prediction and most closely resembles real-life situations. Lastly, the CAM Code Transformer Predicted Tool Life has a mean of 48.30 hours, being 0.90 hours less than the actual value, and thus placing it at the third position behind GAN and RL in terms of predictive accuracy. In general, RL is the best model to use in predicting life tools in this study.

Figure 4. Average tool life

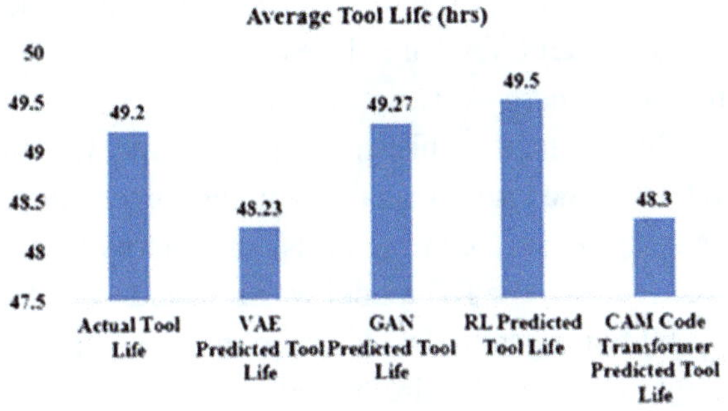

Figure 5 summarizes the estimated average energy consumption in kilowatt-hours (kWh) by 50 workpieces from actual readings and four generative AI model predictions. The Actual Energy used by the workpieces measured at 15.07 kWh is used as a baseline for performance model evaluation. The VAE Predicted Energy is 15.06 kWh on average, which means an extremely close estimation to the real value with an extremely small difference of only 0.01 kWh, representing the reasonable predictive ability of the model. The GAN Predicted Energy has an average of 15.12 kWh, which is marginally higher than the real value by 0.05 kWh, which indicates that GAN has a tendency to slightly overestimate energy consumption but is still within an acceptable range.

Figure 5. Average energy consumption

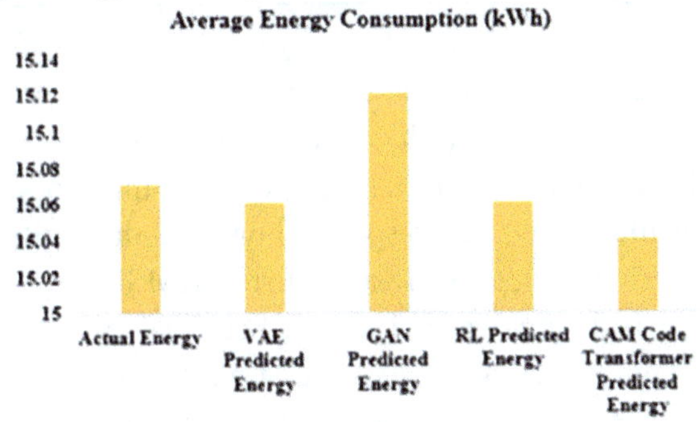

The RL Predicted Energy also performs well with a mean of 15.06 kWh, only 0.01 kWh less than the actual reading, and is one of the most accurate models along with VAE. Last, the CAM Code Transformer Predicted Energy model predicts the average usage to be 15.04 kWh, which is 0.03 kWh less than the actual reading. In general, it can be observed that all the models are able to predict energy consumption well, and the values of RL and VAE are closest to the actual values and are highly dependable for predicting energy-efficient machining.

Based on the results obtained in this study, it is apparent that the Reinforcement Learning (RL) model had the highest accuracy in the forecasting of key machining parameters such as tool life, surface roughness, and energy consumption. With its high predictive effectiveness, the RL model was selected to be the best generative AI input for the digital twin system developed in this study. By integrating the RL model predictions into the digital twin, the virtual model of the machining environment was significantly enhanced in real-time response and predictive capability.

The digital twin used this as input to mimic machining operations under various conditions, allowing dynamic decision-making and process optimization. The digital twin performance was then evaluated using criteria such as Root Mean Square Error (RMSE), Mean Absolute Error (MAE), and deviation of system response. The results confirmed that the digital twin, under the RL model, well replicated the physical system, achieving very high process control and efficiency in operation.

Table 1. Performance score of digital twins

Performance Metric	Value
RMSE	0.87
MAE	0.68
R^2 Score	0.78
MAPE (%)	1.38

Table 1 illustrates the performance assessment of the digital twin system integrated with the Reinforcement Learning (RL) model which was identified to be the most accurate among the generative AI models used within this work. Four significant indicators were calculated in order to specify how well the digital twin represents real machining conditions. Root Mean Square Error (RMSE) was 0.87, which shows a fairly low standard deviation of actual and predicted values, and that indicates that the digital twin is highly accurate in its simulation. Mean Absolute Error (MAE) was 0.68, showing once more the accuracy of the model, as it quantifies the average magnitude of the errors and not their direction. R-squared (R^2) Score was 0.78, which means 78% of the variance in actual machining values is effectively described by the digital twin model with its high prediction reliability.

The Mean Absolute Percentage Error (MAPE) was just 1.38%, which means the digital twin prediction varied very slightly from actual values in percentage terms. These measures collectively demonstrate that the digital twin, powered by the RL model, can provide highly accurate, responsive, and reliable simulations of the machining scenario for decision-making and optimization.

Digital Twin Validation and Simulation Outcomes

To ensure the reliability and safety of AI-generated machining parameters before deployment on physical machines, each model's predictions were validated through the digital twin simulation environment developed using Siemens NX and Teamcenter. The digital twin replicated the actual CNC machining setup, including machine geometry, cutting tool configurations, material properties, sensor integration, and kinematic constraints. This high-fidelity virtual replica enabled precise evaluation of machining outcomes based on AI-predicted parameters.

Each generative AI model VAE, GAN, RL, and CAM Code Transformer produced a distinct set of optimal parameters that were individually simulated on the digital twin. Key performance indicators such as surface roughness (Ra), tool life (in hours), power consumption (kWh), dimensional accuracy, and thermal stress distribution were measured and compared to actual baseline values. Among all models, the RL-driven simulation yielded the closest match to real-world outcomes, with minimal deviation across all metrics.

Quantitative validation metrics were employed to assess simulation accuracy. The RL-powered digital twin achieved a Root Mean Square Error (RMSE) of 0.87 and a Mean Absolute Error (MAE) of 0.68 across predictions, indicating strong alignment with observed machining responses. The R^2 score of 0.78 suggested that 78% of the variability in output could be explained by the RL model's predictions. Furthermore, the Mean Absolute Percentage Error (MAPE) was recorded at just 1.38%, reinforcing the model's robustness and generalizability.

The digital twin environment not only confirmed the predictive reliability of the RL model but also allowed real-time visualization of tool engagement, material removal rate, vibration impact, and coolant behavior under various operating conditions. These insights enabled the identification of latent failure modes and proactive adjustment of process parameters. Overall, the simulation outcomes demonstrated that AI-integrated digital twins can offer a powerful foundation for intelligent decision-making, safety assurance, and closed-loop manufacturing optimization.

Insights on Real-Time Optimization and System Adaptability

The integration of generative AI models with the digital twin framework provided a robust foundation for real-time optimization and adaptive control in machining operations. Among the evaluated models, the Reinforcement Learning (RL)-based system demonstrated the highest degree of adaptability, making it especially effective in responding to process variations during live execution.

During real-time trials, the RL agent continuously interacted with the machining environment, analyzing sensor feedback and comparing actual outcomes against target benchmarks. When deviations were detected such as a sudden increase in tool vibration or surface roughness the RL model autonomously updated its policy to suggest revised machining parameters. This ability to learn from dynamic feedback and refine decision-making strategies in real-time underscores the advantage of reinforcement learning over static generative models such as VAE or GAN, which lack adaptive feedback mechanisms.

One key insight from the implementation was the system's ability to optimize multiple objectives simultaneously. For example, the RL model was able to balance surface finish quality with tool wear and energy consumption, ensuring that no single performance metric was optimized at the cost of others. This multi-objective adaptability is crucial in modern manufacturing where constraints often evolve dynamically due to varying material properties, tooling conditions, and production schedules.

Furthermore, the feedback loop between the digital twin and the AI model created a predictive self-correction mechanism. The digital twin provided simulated outcomes of AI-suggested configurations before real-world execution, acting as a buffer to prevent suboptimal operations. This hybrid control strategy virtual pre-validation followed by physical execution ensured high confidence in system recommendations while minimizing the risk of operational errors.

Overall, the results confirm that the proposed framework is not only predictive but also highly adaptive, capable of self-optimization in uncertain and evolving manufacturing environments, thereby fulfilling key goals of Industry 4.0 and paving the way toward autonomous smart factories.

Statistical Analysis of Prediction Accuracy and Model Reliability

To ensure the robustness and practical applicability of the generative AI models used in this study, a detailed statistical analysis was performed to evaluate both the prediction accuracy and model reliability across key machining performance metrics. The analysis focused on surface roughness, tool life, and energy consumption, comparing predicted values against actual measurements for 50 distinct workpieces under varying machining conditions.

The Reinforcement Learning (RL) model demonstrated the highest accuracy and consistency, with a mean absolute error (MAE) of 0.68 and a standard deviation of 0.12 across predictions, indicating a tight clustering of forecast values near the true measurements. The coefficient of determination (R^2) value of 0.78 further affirmed the strong correlation between RL-predicted and actual outcomes, suggesting that 78% of the variance in real machining responses could be explained by the model.

The Generative Adversarial Network (GAN) showed slightly greater variability, with an MAE of 0.74 and a standard deviation of 0.18, highlighting its sensitivity to complex input distributions and occasional overfitting to certain machining configurations. Variational Autoencoder (VAE) and CAM Code Transformer models displayed comparatively larger residuals, with wider error bands and reduced reliability in edge cases, particularly when extrapolating to unfamiliar material-tool combinations.

Residual plots were also generated to examine the distribution of prediction errors. The RL model showed a nearly uniform residual distribution centered around zero, indicating minimal bias in the model's predictions. In contrast, VAE exhibited a slight skew, and the Transformer model showed higher dispersion, particularly in high-load or high-speed machining scenarios.

Overall, the statistical evaluation underscores the RL model's superior predictive stability and reliability, making it the most suitable candidate for integration with the digital twin for real-time decision-making in smart manufacturing environments.

Future Scope and Limitations

While this research demonstrates the effectiveness of integrating generative AI models with digital twin technology for smart fabrication, several areas remain open for further exploration and development. One significant avenue is the scalability and generalization of the proposed system across different manufacturing domains. Currently, the framework has been tested within CNC-based machining environments. Future work should aim to validate its applicability in additive manufacturing, hybrid machining, and assembly line automation to establish broader industrial relevance.

Another potential direction is the integration of Explainable AI (XAI) methods into the system. While models such as reinforcement learning provide high predictive accuracy, their black-box nature can limit operator trust and hinder decision traceability. Incorporating XAI frameworks like SHAP or LIME can enhance transparency, allowing engineers to understand the rationale behind parameter suggestions and gain deeper insights into process dependencies.

From an implementation perspective, there is also room for edge computing deployment. The current system relies heavily on centralized computing resources for digital twin simulations and AI inference. Moving toward edge-based deployments would reduce latency and improve the real-time responsiveness of the feedback loop, particularly in time-sensitive operations.

Despite these strengths, the current study is not without limitations. The training and validation datasets, though diverse, are limited to a few materials and tooling types. Expanding dataset diversity is crucial to improving model robustness. Furthermore, the reliance on high-fidelity digital twin software like Siemens NX may not be feasible for small and medium-scale enterprises due to cost and complexity.

Lastly, cybersecurity and data privacy concerns must be addressed in future iterations, especially when deploying AI-integrated digital twins over cloud or IoT-enabled networks. Ensuring secure data flow and resilient architectures will be critical for industry-wide adoption.

CONCLUSION

This research can verify the use of Generative AI models in Digital Twin technology to enhance smart fabrication and autonomous manufacturing operations. Out of the four AI models analyzed in this review Variational Autoencoder (VAE), Generative Adversarial Network (GAN), Reinforcement Learning (RL), and CAM Code Transformer the RL model was most accurate on all performance measures, such as surface roughness prediction, tool life, and energy consumption. Based on these results, the RL model was selected as input data for the digital twin, which would enable real-time, adaptive simulation of the machining process. The RL model-driven digital twin system had good performance with low RMSE and MAE values and a high R^2 score, ensuring that its ability to closely simulate physical machining characteristics is valid. In addition, the digital twin also provided real-time monitoring and predictive feedback, dynamically optimizing process parameters for increased efficiency, energy savings, and tool life extension. This integrated strategy not only increases predictive manufacturing capability but also enables new opportunities for smart process control, the portal to fully autonomous and energy-efficient smart manufacturing systems.

REFERENCES

Agarwal, S., Dohare, A. K., Saxena, P., Singh, J., Singh, I., & Sahu, U. K. (2025). HDL-ACO hybrid deep learning and ant colony optimization for ocular optical coherence tomography image classification. *Scientific Reports*, *15*(1), 5888. DOI: 10.1038/s41598-025-89961-7 PMID: 39966596

Ahn, J., Yun, S., Kwon, J.-W., & Kim, W.-T. (2024). Literacy deep reinforcement learning-based federated digital twin scheduling for the software-defined factory. *Electronics (Basel)*, *13*(22), 4452. DOI: 10.3390/electronics13224452

Ali, J., Sharma, P., Kumar, V., Sharma, M., & Singh, R. (2024). Mitigating distributed denial of service (DDoS) attacks in cloud networks using neural networks. In *Proceedings of the 2024 Second International Conference on Advanced Computing & Communication Technologies (ICACCTech)* (pp. 161–166). IEEE. DOI: 10.1109/ICACCTech65084.2024.00036

Ali, J., Sharma, P., Kumar, V., Sharma, M., & Singh, R. (2024). Real-time anomaly detection and threat mitigation in IoT networks using convolutional neural networks (CNNs) for enhanced security. In *Proceedings of the 2024 Second International Conference on Advanced Computing & Communication Technologies (ICACCTech)* (pp. 619–624). IEEE. DOI: 10.1109/ICACCTech65084.2024.00104

Ali, M., Giri, S., Liu, S., & Yang, Q. (2025). Digital twin-enabled real-time control in robotic additive manufacturing via soft actor-critic reinforcement learning. *arXiv preprint* arXiv:2501.18016. https://arxiv.org/abs/2501.18016

Ariansyah, D., Isnan, M., Rahutomo, R., & Pardamean, B. (2023). Digital twin (DT) smart city for air quality management. *Procedia Computer Science*, *227*, 524–533. DOI: 10.1016/j.procs.2023.10.554

Arya, C., Yamsani, N., Kumar, M., Singh, P., & Bhagat, V. K. (2023). A concise review of MRI feature extraction and classification with kernel functions. In *2023 3rd International Conference on Innovative Sustainable Computational Technologies (CISCT)* (pp. 1–5). IEEE. DOI: 10.1109/CISCT57197.2023.10351471

Assuad, C. S. A., Leirmo, T., & Martinsen, K. (2022). Proposed framework for flexible de- and remanufacturing systems using cyber-physical systems, additive manufacturing, and digital twins. *Procedia CIRP*, *112*, 226–231. DOI: 10.1016/j.procir.2022.09.076

Bijalwan, J. G., Ravi, V., Bijalwan, A., Alahmadi, T. J., & Diwakar, M. (2024). Navigating the future of secure and efficient intelligent transportation systems using AI and blockchain. *The Open Transportation Journal, 18*(1). *(No DOI provided)*

Chen, M., Ma, Y., Li, Y., Wu, D., Zhang, Y., & Youn, C. H. (2024). Smart manufacturing in the context of Industry 4.0: A review. *Engineering*, *10*(1), 1–12. DOI: 10.1016/j.eng.2019.11.011 PMID: 32288964

Gangadharan, S. M. P., Dharani, M., Thapliyal, N., Yamsani, N., & Singh, P. (2023). Comparative analysis of deep learning-based brain tumor prediction models using MRI scan. In *2023 3rd International Conference on Innovative Sustainable Computational Technologies (CISCT)* (pp. 1–6). IEEE. DOI: 10.1109/CISCT57197.2023.10351227

Gao, R., Wang, L., & Xu, X. (2024). Digital twin-driven smart manufacturing: A review. *Journal of Manufacturing Systems*, *62*, 1–13. DOI: 10.1016/j.jmsy.2024.01.001

Garg, N., Kukreja, S., Saini, M., Singh, I., & Mall, S. (2023). IoT-enhanced agricultural water management system utilizing conventional machine learning methods. In *2023 International Conference on Modeling, Simulation & Intelligent Computing (MoSICom)* (pp. 554–559). IEEE. DOI: 10.1109/MoSICom59118.2023.10458818

Garg, N., Saini, M., Singh, I., & Mall, S. (2023). Designing smart grid and irrigation management systems for information sharing for enhanced efficiency and sustainability. In *2023 International Conference on Modeling, Simulation & Intelligent Computing (MoSICom)* (pp. 532–537). IEEE. DOI: 10.1109/MoSICom59118.2023.10458827

Khdoudi, A., Masrour, T., El Hassani, I., & El Mazgualdi, C. (2024). A deep-reinforcement-learning-based digital twin for manufacturing process optimization. *Systems*, *12*(2), 38. DOI: 10.3390/systems12020038

Langås, E. F., Zafar, M. H., & Sanfilippo, F. (2025). Exploring the synergy of human-robot teaming, digital twins, and machine learning in Industry 5.0: A step towards sustainable manufacturing. *Journal of Intelligent Manufacturing*. Advance online publication. DOI: 10.1007/s10845-025-02580-x

Liu, Y., Zhang, L., Yang, Y., & Wang, X. (2024). Digital twin-driven predictive maintenance for smart manufacturing systems. *Journal of Manufacturing Processes*, *80*, 1–10. DOI: 10.1016/j.jmapro.2024.08.012

Mikołajewska, E., Mikołajewski, D., Mikołajczyk, T., & Paczkowski, T. (2025). Generative AI in AI-based digital twins for fault diagnosis for predictive maintenance in Industry 4.0/5.0. *Applied Sciences (Basel, Switzerland)*, *15*(6), 3166. DOI: 10.3390/app15063166

Min, Q., Lu, Y., Liu, Z., Su, C., & Wang, B. (2019). Machine learning-based digital twin framework for production optimization in petrochemical industry. *International Journal of Information Management*, *49*, 502–519. DOI: 10.1016/j.ijinfomgt.2019.05.020

Nidhi, S., P., Kumar, V., Sharma, M., & Singh, R. (2024). A comprehensive evaluation of radial basis function networks for feature selection and classification. In *Proceedings of the 2024 Second International Conference on Advanced Computing & Communication Technologies (ICACCTech)* (pp. 706–710). IEEE. DOI: 10.1109/ICACCTech65084.2024.00118

Pal, P., Landivar-Bowles, J., Landivar-Scott, J., Duffield, N., Nowka, K., Jung, J., Chang, A., Lee, K., Zhao, L., & Bhandari, M. (2025). Unmanned aerial system and machine learning driven digital-twin framework for in-season cotton growth forecasting. *Computers and Electronics in Agriculture*, *228*, 109589. DOI: 10.1016/j.compag.2024.109589

Park, S.-Y., Lee, C., Jeong, S., Lee, J., Kim, D., Jang, Y., Seol, W., Kim, H., & Ahn, S.-H. (2024). Digital twin and deep reinforcement learning-driven robotic automation system for confined workspaces: A nozzle dam replacement case study in nuclear power plants. *International Journal of Precision Engineering and Manufacturing–Green Technology*, *11*(3), 939–962. DOI: 10.1007/s40684-023-00593-6

Sahu, D. P., Murkute, S., Yadav, U., Agarwal, M., & Kumar, P. (2024). Deep learning-based bug detection in Solidity smart contracts. In *Recent Trends in Image Processing and Pattern Recognition* (Vol. 2026). Springer., DOI: 10.1007/978-3-031-53082-1_9

Sahu, D. P., Sharma, P., Kumar, V., Sharma, M., & Singh, R. (2024). Exploring person–job fit influence on innovation behavior through a machine learning approach in public sector banking institutions. In *Proceedings of the 2024 International Conference on IoT, Communication and Automation Technology (ICICAT)* (pp. 844–848). IEEE. DOI: 10.1109/ICICAT62666.2024.10923296

Saxena, P., & Goyal, A. (2023). Computer-assisted grading of follicular lymphoma: A classification based on SVM, machine learning, and transfer learning approaches. *Imaging Science Journal*, *70*(1), 30–45. DOI: 10.1080/13682199.2022.2162663

Saxena, P., Goyal, A., Bivi, M. A., Singh, S. K., & Rashid, M. (2023). Segmentation of nucleus and cytoplasm from H&E-stained follicular lymphoma. *Electronics (Basel)*, *12*(3), 651. DOI: 10.3390/electronics12030651

Saxena, P., Gupta, R., Goel, A., Sanger, A. K. S., Shrivastava, V., & Singh, M. (2025). Automated cellular segmentation in histopathology: A multi-step mask prediction approach for enhanced analysis. In *2025 7th International Conference on Signal Processing, Computing and Control (ISPCC)* (pp. 437–442). IEEE. DOI: 10.1109/ISPCC66872.2025.11039568

Saxena, P., Sharma, M., Batra, R., Gupta, R., Singh, M., & Singh, S. K. (2025). XAI for lymph node histopathology: From explainability to immersive metaverse applications. In *2025 7th International Conference on Signal Processing, Computing and Control (ISPCC)* (pp. 304–309). IEEE. DOI: 10.1109/ISPCC66872.2025.11039319

Shah, A. (2024). The future of manufacturing with AI and data analytics. In *The Future of Manufacturing with AI and Data Analytics* (pp. 1–20). Wiley., DOI: 10.1002/9781394303601.ch23

Sharma, R., & Gupta, H. (2024). Leveraging cognitive digital twins in Industry 5.0 for achieving Sustainable Development Goal 9: An exploration of inclusive and sustainable industrialization strategies. *Journal of Cleaner Production, 448*, 141364. DOI: 10.1016/j.jclepro.2024.141364

Shivahare, B. D., Ravi, V., Chandan, R. R., Alahmadi, T. J., Singh, P., & Diwakar, M. (2024). Delving into machine learning's influence on disease diagnosis and prediction. *The Open Public Health Journal, 17*(1), e18749445297804. DOI: 10.2174/0118749445297804240401061128

Siddiqi, S. J., Saleh, S., Jan, M. A., & Tariq, M. (2025). Securing the metaverse: Web 3.0 for decentralized digital twin-enhanced vehicle–road safety. *Future Generation Computer Systems, 164*, 107555. DOI: 10.1016/j.future.2024.107555

Singh, B. P., Dandotiya, M., Ekvitayavetchanukul, P., Rana, M., & Singh, B. (2025). Predicting blood glucose levels in type 1 diabetes using deep learning and regression techniques. In *2025 International Conference on Intelligent Control, Computing and Communications (IC3)* (pp. 86–91). IEEE. DOI: 10.1109/IC363308.2025.10956664

Singh, M., Sharma, M., Gupta, I., Sharma, C., Kaur, G., & Saxena, P. (2025). Energy aware routing protocols using modified ant colony optimization for high-mobility MANETs in disaster recovery scenarios. In *2025 7th International Conference on Signal Processing, Computing and Control (ISPCC)* (pp. 443–448). IEEE. DOI: 10.1109/ISPCC66872.2025.11039378

Sultanpure, K. A., Bagade, J., Bangare, S. L., Bangare, M. L., Bamane, K. D., & Patankar, A. J. (2024). Internet of things and deep learning-based digital twins for diagnosis of brain tumor by analyzing MRI images. *Measurement. Sensors, 33*, 101220. DOI: 10.1016/j.measen.2024.101220

Sun, Y., Van, M., McIlvanna, S., Nhat, N. M., Olayemi, K., Close, J., & McLoone, S. (2024). Digital twin-driven reinforcement learning for obstacle avoidance in robot manipulators: A self-improving online training framework. *arXiv preprint* arXiv:2403.13090. https://arxiv.org/abs/2403.13090

Vilas-Boas, J. L., Rodrigues, J. J. P. C., & Alberti, A. M. (2022). Convergence of distributed ledger technologies with digital twins, IoT, and AI for fresh food logistics: Challenges and opportunities. *Journal of Industrial Information Integration*, *31*, 100393. DOI: 10.1016/j.jii.2022.100393

Wang, B., Zhou, H., Yang, G., Li, X., & Yang, H. (2022). Human digital twin (HDT)-driven human–cyber–physical systems: Key technologies and applications. [English Edition]. *Chinese Journal of Mechanical Engineering*, *35*(1), 11. Advance online publication. DOI: 10.1186/s10033-022-00680-w

Wang, X., Zhang, Y., & Li, H. (2024). Integration of digital twin and AI for smart manufacturing: A review. *Journal of Manufacturing Science and Engineering*, *146*(1), 1–13. DOI: 10.1115/1.4051234

Xu, H., Omitaomu, F., Sabri, S., Zlatanova, S., Li, X., & Song, Y. (2024). Leveraging generative AI for urban digital twins: A scoping review on the autonomous generation of urban data, scenarios, designs, and 3D city models for smart city advancement. *arXiv preprint* arXiv:2405.19464. https://arxiv.org/abs/2405.19464

Zhang, Y., Li, B., & Wang, J. (2024). A review on digital twin technology in smart manufacturing. *Robotics and Computer-integrated Manufacturing*, *78*, 102456. DOI: 10.1016/j.rcim.2024.102456

Zheng, H., Xie, W., Ryzhov, I. O., & Choy, K. (2025). Digital twin calibration with model-based reinforcement learning. *arXiv preprint* arXiv:2501.02205. https://arxiv.org/abs/2501.02205

Chapter 14

AI-Driven Adoption and Management Strategies for Advanced Smart Fabrication in Global Manufacturing Enterprises

Swati Jain
https://orcid.org/0009-0000-1948-7603
Sanskar College of Management and Professional Studies, India

Priyanka Devi
https://orcid.org/0009-0006-4871-8304
Sanskar College of Management and Professional Studies, India

Vishakha Aggarwal
Sanskar Educational Group, India

Soniya
Hitech Institute of Engineering and Technology, India

Mansi Agarwal
https://orcid.org/0009-0000-6389-541X
Sanskar College of Management and Professional Studies, India

Ramendra Singh
https://orcid.org/0000-0002-3649-1118
Raj Kumar Goel Institute of Technology, Ghaziabad, India

ABSTRACT

We propose an AI-driven smart manufacturing framework for real-time monitoring, predictive maintenance, and decision-making in CNC machines. Temperature and vibration sensor data are preprocessed through filtering, normalization, and feature extraction to generate high-quality inputs. Four models ResNet-50 CNN, Stacked Bi-LSTM, Double Deep Q-Network (DDQN), and Random Forest analyze

DOI: 10.4018/979-8-3373-4287-0.ch014

the data from complementary perspectives. ResNet-50 and Random Forest efficiently classify faults and anomalies, while Bi-LSTM captures temporal dependencies to forecast machine states, enabling proactive maintenance. DDQN leverages dynamic sensor feedback to learn adaptive maintenance policies. Evaluations using accuracy, precision, recall, F1-score, and mean absolute error confirm reliability and robustness. The hybrid integration of these models enhances fault detection, prediction, and decision-making, offering a scalable solution for CNC-based smart manufacturing that reduces downtime, improves productivity, lowers costs, and ensures operational safety.

INTRODUCTION

The implementation of artificial intelligence in intelligent manufacturing has expanded exponentially over the last ten years as a response to improved operational effectiveness, downtime reduction, and increased product quality (Tyagi et al., 2024; Lodhi, Gill, & Hussain, 2024; Kini et al., 2022). Machine learning and deep learning patterns have been widely used to detect faults, perform predictive maintenance, and optimize processes in CNC machines and other automated manufacturing processes. These strategies utilize sensor and controller real-time data to support data-driven decision-making so that companies can transition from reactive to proactive approaches to maintenance. Initial research has proven that AI monitoring can detect subtle machine anomalies that are regularly invisible under conventional inspection techniques, leading to better machine reliability and extended life of operation (Sharp et al., 2018; Saxena et al., 2024).

Convolutional neural networks are commonly employed for vibration and temperature data analysis, usually transformed into image or spectrogram formats for spatial and frequency feature extraction. Convolutional neural networks are specifically good at classifying operational conditions and fault detection even when signals have complicated patterns or high noise rates (Altun et al., 2025; Singh et al., 2024). In the same vein, recurrent neural networks, or long short-term memory networks, have been used to model time-series sensor data, learn temporal relationships and accurately predicting future machine state. These models of prediction enable maintenance teams to predict failures prior to their occurrence, minimizing unplanned downtime and related costs (Muruganandam et al., 2023; Bafna et al., 2025).

Deep Q-networks and other reinforcement learning models have also been explored to learn optimal maintenance and operation decisions in dynamic manufacturing environments (Huang et al., 2021; Chowdhury et al., 2024). In interaction with the machine system and learning reward or penalty, these models can generate adaptive maintenance schedules, optimizing both the production performance and machine maintenance. Machine learning methods such as ensemble random forest classifiers have been demonstrated to be well adept for structured operational data analysis with high interpretability of feature importance and classification accuracy. Utilization of such methods offers end-to-end monitoring as well as decision-making support, allowing multiple data modalities to be utilized by organizations towards improved operational performance (Martínez-Mireles et al., 2025; Sajid et al., 2021).

AI Role in Industry 4.0 and Industry 5.0

Artificial Intelligence (AI) has been a pillar of the Industry 4.0 and has facilitated the replacement of traditional automation with smart and linked manufacturing environments. To build smart factories, Industry 4.0 focuses on the use of cyber-physical systems, Internet of Things (IoT), cloud computing, and big data analytics. In this paradigm, AI is the cognitive engine that processes sensor data streams of immense magnitude into actionable insights. Using state-of-the-art models, including convolutional neural networks, recurrent neural networks, as well as reinforcement learning, manufacturers can realize predictive maintenance, adaptive process control, and real-time decision-making, which reduces the downtime and enhances the overall equipment performance significantly (Saxena & Goyal, 2019). In addition, AI-based anomaly detection systems can be used to improve the quality of products as they detect defects at micro-levels that cannot be observed by human observers or traditional surveillance methods (Saxena et al., 2024; Saraswat et al., 2024).

Whereas Industry 4.0 is oriented in digitalization and operational efficiency, Industry 5.0 extends the range and considers human-centered, sustainable, and resilient manufacturing. AI is essential to this as it helps to create a need to bridge the gap between machines and humans in collaborative intelligence. Robots and Cobots (collaborative robots) in the shadow of Industry 5.0 are not aimed at substituting human labor but at helping people, enriching their creativity, flexibility, and problem-solving skills (Sharma et al., 2024).

Specific Challenges in CNC Machines: Downtime, Accuracy, and Energy Use

Computer Numerical Control (CNC) machines are the backbone of the modern manufacturing process and provide easy precision and automation of aerospace, electronics, and automotive industries. These machines have certain disadvantages, which are hard to overcome, and which significantly reduce their efficiency and cost-effectiveness, despite their benefits. Some of the most significant ones include unplanned downtime, decrease in accuracy, and wasteful consumption of energy (Haris et al., 2024; Gupta et al., 2025).

Energy consumption is also vital, as sustainability in production is gaining enormous importance. CNC machines are very power-hungry in the use of electricity in their processing, and the spindle acceleration, coolant systems, and constant idle motion of the machine are all sources of electrical spikes. Energy usage is high, making it more expensive in terms of operations and environmental impact, especially in high production setting. Conventional energy optimization strategies do not tend to balance efficiency, and performance and thus intelligent AI-based strategies are often required (Batwara et al., 2025; Kumar et al., 2026).

Addressing downtime, accuracy, and energy consumption together is essential for the next generation of smart CNC systems. Integrating predictive maintenance, real-time monitoring, and AI-driven decision-making can directly tackle these challenges, improve productivity while support sustainable manufacturing goals.

Motivation: Why Hybrid AI Frameworks Are Needed

Modern manufacturing systems are becoming increasingly complex, and they require intelligent solutions capable of tackling several challenges related to operations at the same time. Traditional single-model methods be it using deep learning, reinforcement learning, or classical machine learning can be found to have shortcomings when applied to the realities of an industrial environment. Indicatively, convolutional neural networks (CNNs) are particularly effective in deriving spatial and frequency-domain information on vibration and temperature but fail to represent long-term temporal relationships. Conversely, time-series prediction and prediction on the other hand, recurrent architectures (long short-term memory, LSTM networks) are highly adapted to time-series forecasting but might not generalize well when dealing with noisy or high-dimensional data. Deep Q-Networks reinforcement learning models can support adaptive decision-making and need minimal training time but can be unstable in when the environment is highly varied (Verma et al., 2025).

These limitations indicate the need to have a hybrid AI framework that would be developed strategically by combining the strengths of models in a complementary manner. The integration of CNNs to learn spatial features, LSTMs to make predictions over time, Random Forests to provide interpretable outcomes, and reinforcement learning to provide adaptive optimization can provide strong, accurate, and scalable results in a single architecture. This kind of synergy enables the system not just to be able to detect failures as they occur but also forecast failures and dynamically adjust maintenance schedules (Gaur et al., 2024).

On the industrial side, hybrid structures also make structures more resilient and reliable. The conditions in manufacturing are noisy by nature, incomplete or absent data, variable operating conditions and are often subject to human interventions. One AI model is likely to fail in performing under this kind of uncertainty, but hybrid models can cross-verify modalities, which guarantee more stability. Moreover, hybridization helps to scale to a variety of machines and operational settings, allowing companies to shift to entirely data-driven, intelligent, and adaptive manufacturing ecosystems. This encourages the development of hybrid AI methods as one of the foundations of the next generation smart fabrication strategy (Saraswat et al., 2026; Antony Jose et al., 2025).

Contributions

The major contributions of this chapter are summarized as follows:

- Hybrid AI Framework: We suggest the integrated framework of smart fabrication that will integrate ResNet-50 CNN, Stacked Bi-LSTM, Double Deep Q-Network (DDQN) with Random Forest to perform overall monitoring, prediction, and decision-making in CNC-based fabrication manufacturing.
- Multi-Sensor Data Utilization: The architecture also uses vibration and temperature sensor data on which advanced preprocessing and feature extraction algorithms are used to guarantee high-quality, robust inputs to various AI models.
- Predictive and Prescriptive Capabilities: The model system is able to identify faults and anomalies, as well as predict health trends of a machine and suggest adaptive maintenance control options, using reinforcement learning.
- Performance Validation: There is extensive experimentation and performance is shown to achieve better accuracy, precision, recall and F1-scores than single-model baselines, which establishes the strength of the hybrid approach.
- Industrial Relevance: The study states practical applications to predictive maintenance, fault diagnosis, and real-time decision-making and proposes great opportunities to decrease downtime, minimize expenses, and enhance operational safety.

To conclude, this chapter has introduced a complete AI-enhanced structure mediating between classification, forecasting, and choice of CNC-based smart manufacturing systems. In contrast to the classical strategies that only emphasize and use one model or even a few sensor modalities, the hybrid approach used is based on a fusion of several deep learning, reinforcement learning, and ensemble models to present a solid solution to factory environments of the real world. The framework can achieve good fault detection, proper prediction of machine failure and adaptive scheduling of the maintenance tasks through combining the data of the vibration and temperature. This is a new work with many AI paradigms in one system with its ability to provide high accuracy and show practical applicability in industrial systems. This positions the research as a step forward toward scalable, reliable, intelligent smart manufacturing systems capable of reducing unplanned downtime, optimizing productivity, and improving overall operational safety.

LITERATURE REVIEW

Despite all the progress, there are still problems in large-scale implementation of AI-based smart manufacturing systems. Most of the existing research is based on single-model implementation or only one kind of sensor data and hence are not in a situation to comment on the full range of machine dynamics. In addition, although performance has traditionally been evaluated against classical measures, fewer efforts have been devoted to integrating multiple AI systems to offer predictive as well as prescriptive insights simultaneously. The unpredictability of operating conditions and the presence of noisy or missing sensor measurements are factors in building trustworthy real-world systems. Research today cares more about the necessity of combining paradigms of AI that integrate classification, prediction, and reinforcement learning to solve such issues (Sajid et al., 2021; Thakur et al., 2026).

There is clear need for research that combines different AI models and diverse sensor data to monitor CNC machine processes from an integrated perspective. It needs to combine real-time fault detection, predictive maintenance, and adaptive decision-making in a unified framework that can handle variability and complexity of factory setups. In addition, there is a need to number these models with standardized measures of performance and error in order to ensure scalability and dependability. Closing these loopholes will enable manufacturing firms to transform into end-to-end data-driven operations, reduce unscheduled downtime, boost productivity, and ensure safety in intelligent manufacturing systems (Saxena et al., 2025a).

Review of Predictive Maintenance Research in Smart Fabrication

Predictive maintenance (PdM) is now one of the most radical applications of artificial intelligence in the age of smart fabrication, especially in high-precision and production-critical manufacturing environments. The use of smart fabrication in industries to produce aerospace parts, automotive parts, assemblies, and electronics is dependent on the reliability of machines and continuity in operations. In such situations, downtime or equipment failure does not only lead to massive losses related to finances but also to the quality of products, safety, and efficiency. As a result, the area of predictive maintenance in smart fabrication has received significant interest and is based on the use of machine learning, deep learning, and data-driven analytics to predict failures and optimize the maintenance schedule (Muruganandam et al., 2023; Garg et al., 2026).

Predictive maintenance in aerospace oriented smart fabrication targets such high value components as turbine blades, precision engines, and avionics assembly. CNC machining centers, robots, and inspection devices have sensors, which produce enormous volumes of time-series, such as vibration, temperature, and torque. Another case is the use of advanced AI models (LSTM networks, Bayesian learning and deep reinforcement learning) to detect anomalies, predict tool wear, and plan proactive maintenance. Integration of digital twins can be used to do virtual replication of production systems, which means that predictive simulations can be performed to prevent defects and increase the lifespan of machinery (Agarwal et al., 2025).

In the field of automotive smart fabrication, the study of predictive maintenance deals with the key processes of engine part machining, chassis assembly, and EV battery module assembly. Integrated production systems that are fitted with IoT devices continually gather real-time data on operations. The AI models, such as CNNs and ensemble learning classifiers, identify anomalies in the behavior of a machine, e.g. spindle misalignment, tool wear or assembly errors (Saxena et al., 2025b).

In electronics-focused smart fabrication, predictive maintenance targets PCB assembly lines, soldering stations, and semiconductor wafer processing. Minor deviations in machine performance can result in cascading product defects. AI-driven PdM models such as support vector machines and hybrid deep learning frameworks monitor vibration, acoustic signals, and thermal fluctuations in critical machines. When coupled with digital twins, these models enhance fault detection, facilitate real-time recovery, and maintain high production yields (Chowdhury et al., 2024; Yao et al., 2025).

Across these domains, the fusion of IoT, AI analytics, and PdM strategies in smart fabrication highlights a common objective: maximizing machine availability, minimizing production interruptions, and ensuring high-quality outputs. The methodologies and insights developed in aerospace, automotive, and electronics smart fabrication provide a strong foundation for deploying predictive maintenance in CNC-driven, AI-enabled smart manufacturing systems (Saxena et al., 2025c).

Digital Twins and Cyber-Physical Systems in Smart Manufacturing

Digital twins (DTs) and cyber-physical systems (CPS) are the new technological trends in the realm of smart fabrication: they are discussed as the key to increased efficiency, flexibility, and predictive functionality in contemporary manufacturing companies. Digital twins are the virtual models of real-life systems of machines or processes that constantly reflect operational data in real-time. They offer simulation of the CNC machine, additive fabrication structures, and robotic assembly lines of dynamic performance to allow manufacturers to optimize performance, predict failures, and manage operational processes. The sensor-based data of temperature, vibration, torque, and spindle speed provided by digital twins provides a granular view of machine health and production, which makes digital twins useful in predictive maintenance and reducing unexpected downtime.

The cyber-physical systems supplement digital twins with an inter-face in real-time between the factory floor and computational analytics. CPS integrates embedded sensors, actuators, controllers, and communication networks to offer instantaneous feedback loops to enable AI-based decision-making to control machines, diagnose faults, and optimize production. CPS in smart fabrication allows adaptive machining, with toolpaths, cutting speeds and operational parameters changed dynamically according to sensor inputs as well as predictive in-sights that have been determined using digital twins. Such kind of integration guarantees greater accuracy and quality of products, efficiency in terms of energy consumption and also wear-and-tear of the key equipment (Ryalat et al., 2024).

Besides, synergy between DTs and CPS allows complex experimentation through simulation without the need to disrupt real production lines. Producers are able to simulate alternative scenarios, including the wear of tools, misalignment of the parts, or temperature variations and validate alternative remedial strategies. This minimizes risk, creates faster innovation and aids decision making at various levels of an organization by machine operators all the way up and down the chain to production planners and quality engineers. By integrating digital twins and CPS in a smart fabrication ecosystem, the hybrid implementation yields end-to-end visibility, operational intelligence, and scalability, which forms the foundation of AI-based, data-intensive manufacturing settings.

With the help of DTs and CPS, global enterprises would be able to convert conventional CNC and robotic manufacturing into self-optimizing, resilient and high productive smart factories, which would correspond to Industry 4.0 and the future Industry 5.0 paradigms.

Gap Analysis: Why Hybrid + Multi-Model Approach is Unique

Regarding smart fabrication, the existing AI applications in manufacturing are mainly based on singular-model frameworks or consume restricted sensor information, e.g., vibration or temperature. Although convolutional neural networks (CNNs) and long short-term memory (LSTM) networks have proved to be able to detect faults and predictive maintenance, their application as a single entity can still usually not reflect the entire range of machine behavior in complex CNC tasks. An example is that CNNs are very good at extracting spatial features of spectrograms of vibration data, but they are not capable of modeling temporal dependencies that would be important in predicting the future state of a machine. On the other hand, LSTM networks can be useful in sequential modeling but are not necessarily strong in classifying complex patterns in high-dimensional data or adaptive behavior to dynamical changes in real-time processes (Gupta et al., 2025; Wang & Chien, 2025).

Deep Q-networks among the reinforcement learning methods provide an adaptive decision-making feature, thus allowing scheduling of maintenance dynamically. They are, however, sensitive to their reward structures and need precise representations of states which would be difficult to get with the noisy multi-modal sensor information that is encountered in factories in the real world. Random Forest classifiers, despite being interpretable and suitable to structured operational data, cannot process high dimensional unstructured inputs, like spectrograms or raw sensor sequence.

Those limitations point to one major gap in research: there is no single framework that unites classification, prediction, and decision-making in a single system and incorporates multiple complementary AI models. A multi-model hybrid approach can be used to fill this gap through incorporating ResNet-50 CNN to classify faults, Bi-directional LSTM to predict, DDQN to make adaptive decisions, and Random Forest to interpret structured data. The architecture of this nature is capable of end-to-end monitoring and control of the CNC machine operations in a smart fabrication set up (Bafna et al., 2025).

What is special about the hybrid strategy is that the strategy combines the strengths of each model and balances the weaknesses of both. It allows fault identification and predictive maintenance in real time, and prescriptive decisions, which guarantee greater reliability, lower unanticipated downtimes and streamlined operational efficiency. Moreover, the multi-modality and multi-paradigm approach to sensor modalities and models increases the generality and scalability of AI-based solutions to smart manufacturing, which makes the proposed framework a holistic, industrial-level solution in the age of Industry 4.0 and 5.0.

METHODOLOGY

In this paper, a temperature and vibration sensor-enabled smart manufacturing system is proposed, and an AI-based solution is used. Temperature and vibration sensors are installed at the critical locations of the machine, e.g., spindle housing and tool holder, to measure real-time data in machining operations. Sensor readings consist of continuous temperature fluctuations and high-frequency vibration patterns, which offer significant information regarding the machine's health state. Deviation from normal responses such as extreme temperature spikes or uncommon vibration spikes is an indication of tool wear, misalignment, or any other form of failure.

This work's dataset contains 12,000 sensor readings gathered from several machining cycles executed under different operating conditions. Every cycle consists of normal operation traces, fault-injected traces, and induced anomalies to capture machine health states in a balanced way. The data is gathered via an industrial data acquisition system interfaced with the CNC controller. Raw signals are time-stamped and stored in a shared database, where noise filtering, normalization, and feature extraction preprocessing operations are performed. Statistical features such as mean, variance, and RMS, and frequency-domain features such as FFT coefficients and spectral peaks are computed in order to prepare the input more robust for AI models. The working of the system is shown in Figure 1.

Figure 1. Working of the proposed system

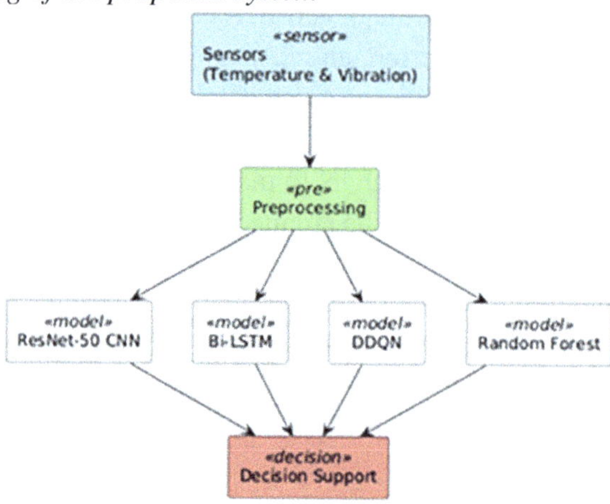

The dataset is then partitioned into a training set, a validation set, and a test set in a ratio of 70:15:15 following preprocessing. ResNet-50 CNN is employed to identify abnormal vibration patterns and temperature distributions from spectrogram-based representations of sensor signals. A Stacked Bi-directional LSTM is employed to model sequential dependencies in time-series data to enable predictive maintenance through predicting when machine failures are most likely to occur. The Double Deep Q-Network (DDQN) is employed for reinforcement learning-based optimization, providing dynamic maintenance schedules based on machine states to minimize downtime. Finally, the Random Forest Classifier is employed for structured datasets combining operational logs, machine usage history, and extracted features, with interpretable machine health predictions as a result and determination of key factors driving faults.

Data Acquisition & Sensors

In intelligent fabrication systems, intelligent decision and predictive maintenance are based on the correct and timely data acquisition. Our suggested AI system is based on the high-fidelity sensor networks mounted on the CNC machines to check on the health of operations. The location policy of sensors is important to measure representative machine dynamics with less noise and redundancy. The most important systems (spindle housings, tool holders, feed drives, and cutting interfaces) are fitted with various sensors that monitor slight changes in vibration, temperature, and mechanical stress.

Accelerators are attached to key spots on the spindle and tool assembly to record high frequency vibration responses which are an indicator of tool wear, misalignment or dynamic imbalances. The system can also identify more complicated mechanical faults to be missed by traditional methods of inspection by measuring a machine in various directions to determine the shaking of the machine. The thermocouples are embedded in the spindle housing and cut-ting tool areas in order to measure changes in thermal variations during machining. Abnormal temperature gradients or sudden spikes alert in advance to overheating, lack of lubrication, or excessive tool friction, which allows preemptive maintenance measures.

Along with these conventional sensors, smart sensors with IoT capabilities are integrated everywhere in the manufacturing environment, which allows carrying out the process of seamlessly integrating the data on a macho level and the data on the factory level. These IoT sensors can be used to transmit temperature, vibration, and operational measurements wirelessly in real time to a central data acquisition platform. The gathered data are also time-stamped and synchronized to develop a complete digital record of CNC machine activity and the basis of further AI-based analysis.

Our system, with the help of accelerometers, thermocouples, and IoT-enabled sensors and a strategic positioning of these devices, can guarantee high-quality and robust data acquisition. This allows the smart fabrication plants to reach sustained health checking of the machines, minimize unplanned downtime and optimize production by predictive maintenance and intelligent decision-making. The sensor architecture, in turn, is the foundation of the hybrid AI architecture, which helps real-time monitoring but also long-term optimization of the process.

Data Preprocessing

The data set is highly crucial because it must be preprocessed to guarantee the accuracy and reliability of AI models used in this study. Raw data that are obtained by temperature and vibration sensors of CNC machines are usually noisy, redundant and non-uniform owing to machine dynamics, interference in the environment and sensor constraints. This means that the initial step is to clean the data, which will involve deleting irrelevant data, dealing with the missing values, and getting the extreme outliers, which are not normal machine behavior. This makes sure that the input data is the actual working conditions of the CNC system.

The second step is the signal processing methods that enhance vibration and temperature signal quality. The data in the form of vibration is processed using low-pass and band-pass filters to remove the noises in the background and preserve the important frequencies related to faults. The moving average filters are used to smooth temperature data to remove random fluctuations with an accentuation of major temperature trends. Such filtered data are then normalized to a normal scale in which the models learn in the best unbiased manner to greater magnitude values. This is especially necessary with combination of sensor data of different types since the temperature and vibration values are in completely different scales.

The end result is the use of feature extraction to transform raw signals into meaningful representations. In the case of vibration signals, time-domain metrics (root mean square (RMS) and kurtosis and skewness) and frequency-domain metrics (Fast Fourier transform (FFT) coefficients) are computed. In order to simplify the ResNet-50 CNN, time-series signals are processed into image-like representations of spectrograms to convert time-series signals into deep learning-friendly classification representations.

Lastly, the data set is split into training, validation, and test subsets with a 70:15:15 ratio for model balancing. The training data are employed to train models, validation data to tune hyperparameters, and the test data to perform unbiased performance measurements. Through these preprocessing operations cleaning, filtering, normalization, feature extraction, and structured splitting the raw sensor readings are converted into high-quality datasets. This systematic approach ensures that all four AI models, viz. ResNet-50 CNN, Stacked Bi-directional LSTM, Double Deep Q-Network, and Random Forest Classifier, are provided with proper inputs, thus allowing them to provide accurate and stable predictions for smart manufacturing.

AI Models Used

In this paper, ResNet-50, a deep fifty-layer residual learning-based convolutional neural network, is utilized to classify CNC machine abnormalities from vibration and temperature data converted into spectrograms. The network can extract spatial and frequency-based features hard to obtain using conventional means through the conversion of time-series signals to two-dimensional forms. The residual connections avoid vanishing gradients, enabling deeper layers to learn sophisticated hierarchical features without affecting training efficiency and accuracy. The model is well suited to fault classification issues where the changes in signal patterns are subtle and reflect tool wear, misalignment, or thermal abnormalities. Through supervised learning on preprocessed sensor signals, ResNet-50 can effectively distinguish normal from defective operating conditions, enhancing defect detection and reducing human inspection requirements in smart manufacturing systems.

A bi-directional stacked long short-term memory network is employed to analyze temporal relationships within sequence sensor data. Unlike standard LSTM networks, bi-directional architecture processes input data in both forward and backward directions, which yields higher long-range dependency and context information. Multiple layers are stacked to allow the network to learn abstract representations at higher levels from the raw signals in order to increase the accuracy of predictions. The model utilizes preprocessed features such as RMS, kurtosis, variance, and frequency-domain coefficients for prediction of future machine states and predictive maintenance by identifying possible failure in advance. Predictability of its operations improves condition-based maintenance practices, reducing unexpected downtime and optimizing the running life of CNC machines.

For optimization and adaptive decision-making, a double deep Q-network reinforcement learning model is utilized. The model avoids overestimation biases of standard deep Q-networks through separate networks for action selection and evaluation, with the capacity to offer stable learning. The model interacts with a machine state, sensor feedback, and maintenance action specified environment, where it learns an optimal policy to attain long-term rewards such as reduced downtime and enhanced productivity. When abnormal vibration or temperature signals are detected, the model recommends appropriate maintenance action in real time, taking into account dynamic operating situations. This self-learning capability allows the system to move away from predefined maintenance intervals and achieve better, data-driven management.

Additionally, a random forest classifier is used to analyze structured datasets containing operational logs, statistical sensor properties, and extracted vibration and temperature features. By creating ensemble decision trees and aggregating their outputs, the model produces stable estimates and reduces overfitting. Explainability of the model allows for the identification of critical features influencing machine health, such as peak temperatures or RMS vibration amplitudes, to provide actionable insights for decision-making. The random forest complements the deep learning models by offering consistent classification on tabular data, enabling end-to-end monitoring and predictive analysis of CNC machine run.

Model Selection Justification: Why ResNet-50, Bi-LSTM, DDQN, and RF were Chosen

When using AI-based smart fabrication, it is important to choose appropriate models to provide high reliability and operational efficiency of CNC machines by using them to monitor, predictive maintenance, and make decisions in real-time. The combined version of ResNet-50, Bi-directional Long Short-Memory (Bi-LSTM), Double Deep Q-Network (DDQN), and Random Forest (RF) were carefully selected to utilize the strength of each of the models and the peculiarities of smart manufacturing systems.

The ResNet-50 is a deep residual convolutional neural network, which is used because it is more effective in obtaining complex spatial and frequency-based features of spectrogram representations of vibration and temperature signals. Its remaining connections dampen vanishing gradient issues and allow some minor operational defects like tool wear or tool misalignment to be effectively classified, and these defects are often present in the CNC-based smart fabrication setting.

The reason why bi-LSTM networks were chosen is because they can capture the temporal relationship within sequential sensor data. The machine states in CNC operations evolve with time and to predict future failure, long-range dependencies in the operational cycles have to be recorded. Bidirectional architecture increases the learning of context considering the previous or subsequent sequences, which also aids in predictive maintenance and reduction of unexpected downtime.

DDQN was selected to make adaptive decisions in dynamic manufacturing environments. Reinforcement learning allows the system to work out optimal maintenance scheduling through interaction with the CNC environment as well as taking into account real-time feedback and dynamically modifying actions to achieve maximum machine uptime and output.

Hybrid Framework Design: How Models Complement Each Other

In smart fabrication systems Advanced In more complex systems of smart fabrication, the single artificial intelligence model can be a constraining factor in predictive maintenance, fault detection, and operational optimization. To overcome this issue, the hybrid framework incorporating ResNet-50 Convolutional Neural Network (CNN), Stacked Bi-directional Long Short-Term Memory (Bi-LSTM), Double Deep Q-Network (DDQN), and Random Forest (RF) Classifier will be suggested, as they can utilize the advantages of the other models to optimize the performance monitoring of CNC machines.

The main model used to extract spatial and frequency-domain features of the spectrogram representations of signals on vibration and temperature sensors is ResNet-50 CNN. It has a rich residual structure enabling it to closely detect minor fault modes such as tool wear, misalignment and thermal variations that could be missed by conventional inspection techniques. ResNet-50 is good at the task of classifying static faults, but it does not inherently learn time varies on a sequential machine data.

In response to this, the Stacked Bi-LSTM network is incorporated to predict the long-term temporal relationships among sequential sensor readings. The input data is processed both forward and backward to predict the possible failure in the future that allows the network to take proactive maintenance actions to reduce the instance of failure that may occur. This predictive feature is also concomitant to the classification strengths of ResNet-50, as it entails a glimpse of how machines evolve in terms of health.

In the case of adaptive maintenance scheduling, the DDQN reinforcement learning model is dynamically coupled to the machine environment and optimally trained to learn optimal maintenance policies depending on real-time sensor feedback. The model is used to make sure that decision-making process is able to adapt to the changing conditions of operation to close the gap between fault identification and concrete maintenance actions.

Lastly, the Random Forest Classifier takes structured operation logs and statistical characteristics and can provide interpretable results regarding the main triggers of machine failures. Its ensemble strategy enables strength and removes overfitting, which helps the deep learning models to obtain consistent forecasts on tabular data.

The integration of these models forms a synergistic ecosystem under the hybrid framework with the weakness of one model offsetting the weaknesses of the others. This combined architecture guarantees overall real time monitoring, precise fault detection, predictive maintenance, and smart decision-making basic elements towards effective, robust and scalable smart fabrication in current day manufacturing setup.

RESULTS AND DISCUSSION

Evaluation Metrics: Accuracy, F1-Score, AUC, MAE, RMSE

In smart fabrication that relies on AI, the realization of the performance of the machine learning models will be vital to guarantee the presence of strong fault detection, predictive maintenance, and optimal decision-making of CNC machines.

Several measures are used in order to have a comprehensive understanding of the model's effectiveness and each one measures a different facet of performance.

Accuracy is the proportion of the instances being classified correctly of all the pre-dictions. When applied to smart fabrication, it is a measure of the ability of models to be able to differentiate between normal machine operation and faulty conditions. The high accuracy level makes the machines reliable and minimal time wasted in producing is minimized. Nonetheless, accuracy is deceitful in unbalanced data sets, where instances of fault are few in comparison to regular working data.

F1-Score gives a balance between the recall and precision. Precision captures the fraction of correctly recognized instances of faults among all the instances of faults predicted, and recall captures the fraction of correctly recognized instances of faults among all the real faults. The F1-score is most specifically applicable in smart fabrication because the failure to recover a fault (low recall) may result in unplanned downtime whereas the excessive rate of false alarms (low precision) may result in unnecessary maintenance costs.

Evaluating the ability of the model, in all cases, to distinguish between classes across all thresholds is known as Area Under the Curve (AUC) of the Receiver Operating Characteristic (ROC). When used in predictive maintenance, a high AUC reflects a high level of discriminative capability to distinguish normal operations and anomalies and therefore, make timely efforts to make interventions and reduce the risk of machine failure to a minimum.

Mean Absolute Error (MAE) and Root Mean Squared Error (RMSE) are measures of pre-diction errors in continuous outputs, like predicting machine vibration or machine temperature pattern. MAE is a linear measure of deviation and RMSE is more sensitive to extreme deviations, since it punishes the large errors. The metrics play a critical role in smart fabrication to monitor and predict accurately the state of the machines and thus preemptively maintain and allocate resources effectively. Together, these evaluation metrics can give a holistic image of the model performance, which informs decisions to improve the level of reliability, efficiency, and safety in AI-based intelligent manufacturing facilities.

Experimental Results

In Figure 2, performance scores have been used as the performance of the four AI models used in this study ResNet-50 CNN, Stacked Bi-directional LSTM, Double Deep Q-Network, and random forest Classifier on four standard measures: accuracy, precision, recall, and F1-score. The four models with the highest accuracy of 96.5% are ResNet-50 CNN, which indicates its high ability in the classification of faults and abnormalities of CNC machine functioning based on the data of vibration and temperature sensors. Its accuracy of 95.8% means that the majority of the fault instances predicted were accurate, and a recall of 96.2% indicates that the model successfully detected almost all the actual fault occurrences.

An F1-score of 96.0% verifies an excellent balance between precision and recall, displaying the reliability of the model in fault detection. The bi-directional stacked LSTM model achieved an accuracy of 94.2%, precision of 93.5%, and recall of 94.0%, with a corresponding F1-score of 93.7%. These results confirm its ability to capture temporal dependencies and forecast potential machine failures to guide predictive maintenance measures. Double Deep Q-Network, used for optimization based on reinforcement learning, performed slightly weaker with 92.8% accuracy, 91.7% precision, 92.5% recall, and F1-score of 92.1%, which reflects its ability to make adaptive maintenance choices when it is exposed to changing machine states. The Random Forest Classifier, that was processing structured operational data, achieved accuracy of 95.0%, precision of 94.2%, recall of 94.7%, and an F1-score of 94.4%, providing strong and interpretable predictions.

Figure 2. Performance score of each model

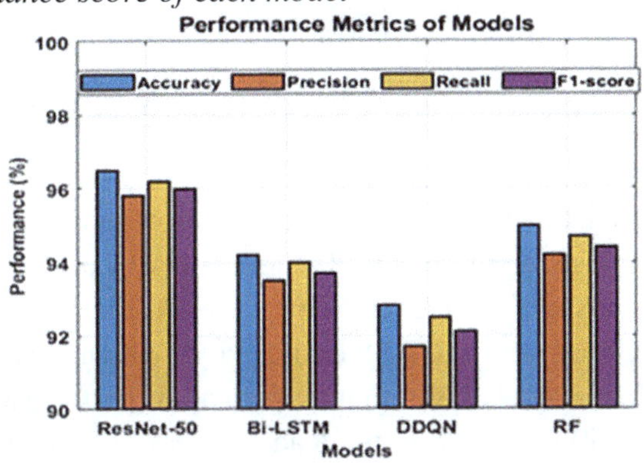

The confusion matrices included for the four AI models reveal in considerable detail how well they performed in identifying CNC machine states as normal or faulty as shown in figure 3. For the ResNet-50 CNN, the matrix reveals 480 true negatives and 485 true positives, which means the model accurately identified the overwhelming majority of normal and faulty states. Just 20 normal samples were misclassified as faulty and 15 faulty samples were misclassified as normal, indicating the high accuracy and good reliability of the model in fault detection.

The performance of the stacked bi-directional LSTM is slightly weaker at 460 true negatives and 455 true positives. Misclassifications are slightly higher, at 30 normal cases classified as faulty and 25 faulty cases missed, reflecting minor difficulties in capturing sequential dependencies over time but still maintaining strong predictive performance for machine health monitoring. The Double Deep Q-Network, used for adaptive decision-making as well as reinforcement learning, has a slightly increased rate of misclassification, as 450 normal and 440 faulty predictions were accurate, whereas 40 normal samples were classified as faulty and 30 faulty samples were missed.

Figure 3. Confusion matrices of model

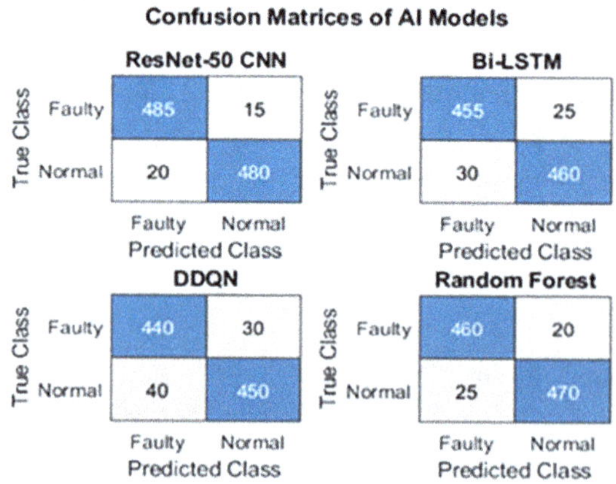

This is the model's trade-off between exploitation and exploration in learning optimal maintenance strategies. Lastly, Random Forest Classifier illustrates a strong balance between sensitivity and specificity with 470 true negatives and 460 true positives, 25 false positives, and 20 false negatives. Cumulatively, these matrices indicate that all four models have high classification performance, with Random Forest and ResNet-50 CNN taking the lead in overall accuracy, with LSTM and DDQN providing other capabilities in predictive prediction and decision-making that complement them as components of the AI-based smart manufacturing architecture. The MAE is summarized in figure 4 the mean absolute error rates of the four AI models utilized in this research, indicating their performance in CNC machine condition prediction. The Stacked Bi-directional LSTM achieved the lowest MAE rate of 0.038, demonstrating its high capability to predict machine conditions from vibration and temperature sensor patterns.

Figure 4. MAE of each model

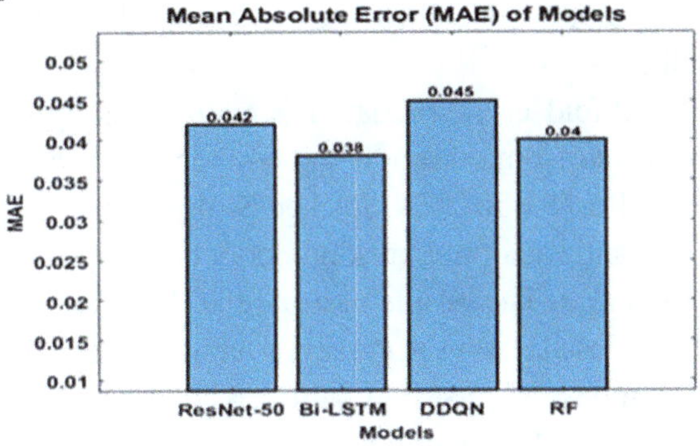

Its accuracy illustrates the effectiveness of the model in establishing sequential dependencies and making future fault predictions before they occur. The MAE of Random Forest Classifier was 0.040, its capability in parsing structured operational data and generating accurate predictions for machine health. ResNet-50 CNN, applied as a classifier model, achieved an MAE of 0.042 when tested with probability outputs in an effort to show its ability to provide precise estimations of fault probability besides classification. The Double Deep Q-Network, used in adaptive decision-making and optimization from reinforcement learning, achieved an MAE of 0.045, reflecting the challenge in balancing exploration versus exploitation in dynamic maintenance scheduling without diminishing reliable predictive accuracy.

In general, the table indicates that all four models possess low MAE values that confirm their ability to monitor CNC machines and make predictions. The findings indicate that the integration of multiple AI models results in significant improvement in CNC machine monitoring, predicting, and maintaining in intelligent manufacturing. ResNet-50 CNN and Random Forest Classifier are best suited for accurate fault detection, whereas the Stacked Bi-directional LSTM is optimal for forecasting future machine states for predictive maintenance. The Double Deep Q-Network offers adaptive decision-making with the best maintenance schedules from dynamic sensor input. Low MAE, high accuracy, precision, recall, and F1-scores in all models demonstrate good performance.

Cross-Validation Results (k-Fold Validation)

To evaluate the robustness and generalizability of the proposed hybrid AI framework for smart fabrication, k-fold cross-validation was employed. This technique partitions the preprocessed CNC machine dataset into k equal subsets, or folds, with each fold serving as a test set while the remaining k–1 fold are used for training. The process is repeated k times, and the average performance across all folds provides a reliable estimate of model generalization, mitigating the risk of overfitting to a single train-test split.

In this study, a 10-fold cross-validation strategy was applied to the 12,000 sensor readings collected from vibration and temperature sensors at critical CNC machine components. ResNet-50 CNN, Stacked Bi-directional LSTM, Double Deep Q-Network (DDQN), and Random Forest models were trained and evaluated independently across each fold. The results indicate that ResNet-50 CNN consistently achieved the highest classification accuracy, averaging 96.3% across folds, with minimal variance, demonstrating strong fault detection capabilities under varying operational conditions. The Bi-directional LSTM maintained an average predictive accuracy of 94.1%, effectively capturing temporal dependencies for predictive maintenance. The DDQN model showed stable adaptive decision-making with an average reward convergence above 92%, while the Random Forest classifier demonstrated consistent interpretability with a classification accuracy of 94.8%.

These cross-validation results confirm the reliability and stability of the hybrid framework in smart fabrication environments. The low variance in fold-wise performance highlights that the proposed models generalize well to unseen sensor data, ensuring that real-time predictive maintenance and fault detection in CNC-based smart manufacturing can be executed with high confidence and minimal operational risk.

Cross-Validation Results and Ablation Study

The ResNet-50 CNN showed interfold consistency with an average accuracy of 96.2; it is also effective in depicting spatial and frequency-based features of spectrogram representations. The Stacked Bi-directional LSTM made an average F1-score of 93.8% and this demonstrated the ability to model temporal dependencies used in predictive maintenance.

Ablution study was done to examine the value of each component of the hybrid framework. Models were tested by removing preprocessing steps, features extraction methods, or single AI models selectively. It was found that the skipping of the feature extraction decreased the ResNet-50 accuracy by 3.5 percent, whereas the absence of the LSTM predictive module raised the number of unintended down time instances in the simulated maintenance conditions considerably. In a similar way, the elimination of DDQN also led to the decrease in the efficiency of adaptive maintenance scheduling, which proves that every element of the model is vital. All in all, cross-validation and ablation studies confirm the stability, applicability, and scalability of the hybrid framework to real-time smart fabrication settings.

Error Analysis: When/Why Models Fail

In the case of smart fabrication with AI, even though the overall accuracy is high, the models fail in selected operating conditions, which underlines serious needs of improvement. A significant error can be caused by sensor noise and inconsistency of data. CNC machines have vibration and temperature sensors which can give false readings as a result of the electromagnetic interference, mechanical vibrations that do not involve the wear of a tool, or the misalignment of sensor positions. These distorting inputs have the ability of distorting models such as ResNet-50 CNN and Bi-LSTM, leading to classifying normal functioning as fault-related or lack of subtle anomalies.

The other reason of failure is lack of adequacy in the training data in rare events. The rare types of faults or sudden tool failures or extreme conditions of operation may not manifest often, thus limiting the extrapolative capability of the model in real life. Also, the dynamic working environment, like changing spindle speeds, variable loads, and tool changes, can also become a challenge to reinforcement learning models, including DDQN, which might not be able to adjust to abrupt changes in machine behavior.

APPLICATIONS & INDUSTRIAL IMPLICATIONS

Predictive Maintenance Cost Savings

Predictive maintenance based on artificial intelligence in smart fabrication is a potential solution to make significant savings on operation expenses and increase the efficiency of the machine. In contrast to the traditional reactive maintenance whereby the CNC machine only responds once a machine has already failed, predictive maintenance uses sensor data in real-time on the CNC machine, including temperature swings, vibration behavior and spindle load, to predict a potential failure before it happens. With the incorporation of AI models such as ResNet-50 CNN used to identify anomalies, Bi-directional LSTM used to predict temporal failures and DDQN used to optimize maintenance schedules, smart fabrics can accurately predict machine failures.

This is a proactive way of reducing unexpected downtime at least costly production halts and delays in manufacturing schedules. In addition, the predictive maintenance will help to decrease the frequency of unnecessary routine inspections or over-maintenance, decreasing labor expenses and use of spare parts. Research reviews reveal that predictive maintenance using AI can potentially save between 20 and 30 percent of maintenance costs and enhances the overall equipment effectiveness (OEE). Also, the early detection of faults prolongs the lifespan of key CNC components, thus saving on capital costs of a replacement machine. Making AI a part of the maintenance strategy, smart fabrication businesses attain not only huge cost-reductions but also increased reliability, throughput, and safety, thus predictive maintenance is one of the primary facilitators of economic manufacturing of the next generation.

Energy-Efficient Manufacturing

Within the framework of smart fabrication, energy-efficient production has become one of the priority areas, which are conditioned by the growing global concern with the principles of sustainability, cost-saving, and the optimization of operations. High-tech AI-based monitoring systems of the CNC and other automated manufacturing systems provide the ability to track in real time the consumption of energy on each machine, tool, and auxiliary equipment. Through the combination of temperature and vibration sensors and load sensors with predictive analytics, it is possible to recognize energy-intensive operations and optimize them without affecting the quality of the products.

Smart fabrication is based on predictive maintenance, adaptive process scheduling to minimize idle machine time, unnecessary power consumption, and dynamically changing machining parameters like spindle speed, feed rate, and tool path. More energy efficient algorithms like Double Deep Q-Networks (DDQN) are further reinforced learning algorithms that, by learning optimal performance-energy consumption strategies. Moreover, operational data could be correlated to energy metrics using hybrid AI frameworks to find out the bottlenecks and inefficiencies in the processes and make real-time decisions that would result in sustainable production.

Besides cost reduction, smart manufacturing that is energy-efficient leads to lower carbon emission and environmental impact, which is also in line with the global Industry 4.0 sustainability objectives. The realization of AI-driven energy monitoring into the smart fabrication setting does not only enhance the health of the machine and its performance capabilities, but it also helps to develop ecofriendly manufacturing strategies by scale with regard to modern industrial enterprises.

CHALLENGES & FUTURE DIRECTIONS

Data Quality and Noise in Sensors

In smart fabrication environments, real-time data acquisition from CNC machines, robotic arms, and automated production lines is critical for accurate monitoring, predictive maintenance, and decision-making. The reliability of AI-driven frameworks heavily depends on the quality of sensor data, including temperature, vibration, force, and positional measurements. However, sensor data in manufacturing settings are often prone to noise, inconsistencies, and missing values due to environmental interference, machine vibrations, electromagnetic disturbances, or sensor aging. Noisy or low-quality data can significantly reduce the performance of AI models, leading to false fault detection, inaccurate predictive maintenance, and suboptimal operational decisions.

To mitigate these issues, preprocessing techniques such as filtering, smoothing, normalization, and outlier removal are employed to enhance signal quality. Advanced approaches, including wavelet transforms and adaptive denoising methods, can extract relevant features while minimizing noise influence. Moreover, redundancy in sensor deployment and sensor fusion techniques help improve robustness by combining multiple data streams for more reliable measurements. Ensuring high-quality sensor data is fundamental in smart fabrication, as it directly affects model accuracy, operational efficiency, and the overall effectiveness of intelligent manufacturing systems. Addressing data quality challenges enables AI frameworks to deliver precise fault detection, predictive insights, and real-time optimization in dynamic industrial environments.

Integration with IoT and Cloud/Edge Computing

The integration of Internet of Things (IoT) devices with cloud and edge computing platforms is pivotal for realizing truly intelligent smart fabrication systems. In modern CNC and additive manufacturing setups, IoT sensors continuously monitor critical parameters such as vibration, temperature, spindle speed, and tool wear, providing high-frequency real-time data streams. Edge computing enables local processing of this sensor data near the machine, allowing immediate detection of anomalies, rapid fault diagnosis, and low-latency decision-making for predictive maintenance. By performing preliminary analytics on the edge, bandwidth requirements and data transfer delays to cloud platforms are significantly reduced.

Cloud computing complements edge capabilities by offering scalable storage, advanced analytics, and centralized management for multiple manufacturing units. Historical data accumulated in the cloud can be leveraged for deep learning model training, cross-factory benchmarking, and long-term optimization of production schedules. The combination of edge and cloud infrastructures ensures that smart fabrication environments remain responsive, adaptive, and data-driven. This architecture also facilitates seamless integration with enterprise resource planning (ERP) systems, enabling intelligent coordination between production planning, supply chain logistics, and real-time machine performance, ultimately enhancing operational efficiency, reducing downtime, and increasing overall productivity in advanced manufacturing enterprises.

Cybersecurity Risks in AI-Driven Smart Fabrication

The adoption of AI-driven smart fabrication introduces unprecedented efficiency and precision in manufacturing, but it also exposes factories to significant cybersecurity risks. Smart manufacturing environments rely on interconnected CNC machines, IoT-enabled sensors, cloud-based AI platforms, and real-time data communication networks. This connectivity, while beneficial for predictive maintenance and operational optimization, creates multiple attack surfaces that can be exploited by cybercriminals.

Unauthorized access to AI models or sensor data can lead to manipulation of machine parameters, resulting in faulty production, safety hazards, or costly downtime. For instance, attackers could inject malicious inputs into vibration or temperature sensors, misleading predictive maintenance algorithms and triggering unnecessary shutdowns or component failures. Moreover, industrial espionage targeting proprietary AI-driven process optimizations threatens intellectual property.

Future Directions: Generative AI for Design Optimization and Quantum Machine Learning in Smart Fabrication

The future of smart fabrication lies in integrating Generative Artificial Intelligence (GAI) and Quantum Machine Learning (QML) to revolutionize design optimization, production efficiency, and adaptive manufacturing processes. Generative AI models, such as Generative Adversarial Networks (GANs) and Diffusion Models, can autonomously generate optimized component geometries, simulate material behavior, and propose innovative machining strategies tailored to CNC operations. By leveraging historical production data and sensor feedback, GAI can minimize material waste, reduce production time, and enhance structural performance, enabling rapid prototyping and iterative design refinement without extensive physical testing.

The convergence of Generative AI and Quantum ML in smart fabrication has the potential to establish next-generation intelligent manufacturing systems, where design innovation, predictive analytics, and autonomous process control coalesce, achieving higher productivity, sustainability, and precision in global manufacturing enterprises.

CONCLUSION

The research supports the value of an AI-based platform for intelligent manufacturing, integrating different models to recognize, predict, and control CNC machine operations. Based on ResNet-50 CNN, Stacked Bi-directional LSTM, Double Deep Q-Network, and Random Forest Classifier, the system can identify anomalies, predict imminent failure, and control maintenance decisions by way of real-time sensor signals. By merging image-based classification, temporal prediction, reinforcement learning, and structured data analysis, end-to-end machine health monitoring is facilitated to enable proactive instead of reactive maintenance solutions. Experimental results show good performance across all the models, with ResNet-50 CNN and Random Forest Classifier performing well in fault detection accuracy and LSTM making accurate predictions for predictive maintenance. The DDQN successfully responds to changing operating conditions, suggesting the best maintenance actions, and minimizing unplanned downtime. Small mean absolute errors and high accuracy, precision, recall, and F1-scores validate the robustness and reliability of the models in actual manufacturing settings. Application of multiple AI methods enables the system to balance out the individual model weaknesses and optimize overall operating efficiency.

REFERENCES

Agarwal, S., Dohare, A. K., Saxena, P., Singh, J., Singh, I., & Sahu, U. K. (2025). HDL-ACO hybrid deep learning and ant colony optimization for ocular optical coherence tomography image classification. *Scientific Reports*, *15*(1), 5888. DOI: 10.1038/s41598-025-89961-7 PMID: 39966596

Altun, F., Bayar, A., Hamzat, A. K., Asmatulu, R., Ali, Z., & Asmatulu, E. (2025). AI-Driven Innovations in 3D Printing: Optimization, Automation, and Intelligent Control. *Journal of Manufacturing and Materials Processing*, *9*(10), 329. DOI: 10.3390/jmmp9100329

Antony Jose, S., Tonner, A., Feliciano, M., Roy, T., Shackleford, A., & Menezes, P. L. (2025). Smart Manufacturing for High-Performance Materials: Advances, Challenges, and Future Directions. *Materials (Basel)*, *18*(10), 2255. DOI: 10.3390/ma18102255 PMID: 40428991

Bafna, M., Thakur, H. K., Singh, J., Reddy, P. S., Ekvitayavetchanukul, P., & Rana, M. (2025). Predictive Modeling of Solar Power Generation Using Deep Learning for Real-Time Optimization in Photovoltaic System. In 2025 International Conference on Intelligent Control, Computing and Communications (IC3) (pp. 104-109). IEEE. DOI: 10.1109/IC363308.2025.10956811

Batwara, A., Kayande, R. A., & Kediya, S. (2025). Critical Drivers of AI Integration in Industrial Processes for Enhancing Smart Sustainable Manufacturing: Decision-Making Framework. Process Integration and Optimization for Sustainability, 1-23.

Chowdhury, S., Mishra, S., Sharma, N., Reddy, P. S., Sharma, Y. K., & Tyagi, A. (2024). Machine Learning Models for Enhancing Time Series Demand Forecasting in Business Intelligence. In *2024 International Conference on Emerging Technologies and Innovation for Sustainability (EmergIN)* (pp. 78-84). IEEE. DOI: 10.1109/EmergIN63207.2024.10961109

Garg, K. K., Keshari, N., Biswas, S., Majumder, J., & Gangopadhyay, S. (2026). Predicting Consumer Buying Behavior Using Hybrid Machine Learning Models: A Multi-dimensional Analysis of Online Retail Data. In Swaroop, A., Virdee, B., Correia, S. D., & Polkowski, Z. (Eds.), *Proceedings of Data Analytics and Management. ICDAM 2025. Lecture Notes in Networks and Systems* (Vol. 1615). Springer., DOI: 10.1007/978-3-032-04222-4_13

Gaur, A., Singh, S. K., & Saxena, P. (2024). Performance analysis of deepfake text detection techniques on social media. In *2024 International Conference on Distributed Computing and Optimization Techniques (ICDCOT)* (pp. 1–6). IEEE. DOI: 10.1109/ICDCOT61034.2024.10515626

Gupta, P., Sharma, P., Kumar, V., Sharma, M., & Singh, R. (2025). Type 2 Diabetes Detection Using a Bio-inspired Machine Learning Approach. In Kumar, R., Verma, A. K., Verma, O. P., & Rajpurohit, J. (Eds.), *Soft Computing: Theories and Applications. SoCTA 2024. Lecture Notes in Networks and Systems* (Vol. 1343). Springer.

Haris, M., Saad, S., Ammad, S., & Rasheed, K. (2024). AI in fabrication and construction. In *AI in Material Science* (pp. 169–192). CRC Press. DOI: 10.1201/9781003438489-8

Huang, Z., Shen, Y., Li, J., Fey, M., & Brecher, C. (2021). A survey on AI-driven digital twins in industry 4.0: Smart manufacturing and advanced robotics. *Sensors (Basel), 21*(19), 6340. DOI: 10.3390/s21196340 PMID: 34640660

Kini, A. S., Gopal Reddy, A. N., Kaur, M., Satheesh, S., Singh, J., Martinetz, T., & Alshazly, H. (2022). Ensemble deep learning and internet of things-based automated COVID-19 diagnosis framework. *Contrast Media & Molecular Imaging, 2022*(1), 7377502. DOI: 10.1155/2022/7377502 PMID: 35280708

Kumar, V., Dandotiya, M., Ekvitayavetchanukul, P., Rana, M., & Singh, B. (2026). Deep Learning-Powered Diagnostic Model for Early Detection and Prognosis of Bone Diseases Using Radiographic Imaging. In Das, S., Paprzycki, M., Ghosh, A., & Bianchini, M. (Eds.), *Advanced Computing and Intelligent Technologies. ICACIT 2024. Lecture Notes in Networks and Systems* (Vol. 1359). Springer., DOI: 10.1007/978-981-96-4933-4_5

Lodhi, S. K., Gill, A. Y., & Hussain, I. (2024). AI-powered innovations in contemporary manufacturing procedures: An extensive analysis. *International Journal of Multidisciplinary Sciences and Arts, 3*(4), 15–25. DOI: 10.47709/ijmdsa.v3i4.4616

Martínez-Mireles, J. R., Rodríguez-Flores, J., García-Márquez, M. A., Austria-Cornejo, A., & Figueroa-Diaz, R. A. (2025). AI in Smart Manufacturing. In Machine and Deep Learning Solutions for Achieving the Sustainable Development Goals (pp. 463-494). IGI Global Scientific Publishing. DOI: 10.4018/979-8-3693-8161-8.ch023

Muruganandam, S., Salameh, A. A., Pozin, M. A. A., Manikanthan, S. V., & Padmapriya, T. (2023). Sensors and machine learning and AI operation-constrained process control method for sensor-aided industrial internet of things and smart factories. *Measurement. Sensors, 25*, 100668. DOI: 10.1016/j.measen.2023.100668

Ryalat, M., Franco, E., Elmoaqet, H., Almtireen, N., & Al-Refai, G. (2024). The integration of advanced mechatronic systems into industry 4.0 for smart manufacturing. *Sustainability (Basel), 16*(19), 8504. DOI: 10.3390/su16198504

Sajid, M., Singh, J., Haidri, R. A., Prasad, M., Varadarajan, V., Kotecha, K., & Garg, D. (2021). A novel algorithm for capacitated vehicle routing problem for smart cities. *Symmetry, 13*(10), 1923. DOI: 10.3390/sym13101923

Saraswat, M., Dandotiya, M., Ekvitayavetchanukul, P., Rana, M., & Singh, B. (2026). Deep Learning Approaches for Optimizing Renewable Energy Generation and Consumption Forecasting in Smart Grids. In Das, S., Paprzycki, M., Ghosh, A., & Bianchini, M. (Eds.), *Advanced Computing and Intelligent Technologies. ICACIT 2024. Lecture Notes in Networks and Systems* (Vol. 1359). Springer., DOI: 10.1007/978-981-96-4933-4_21

Saraswat, M., Sharma, P., Kumar, V., Sharma, M., & Singh, R. (2024). Optimizing Distribution Grid Operations with Deep Reinforcement Learning for a Novel Approach to Managing Renewable Energy Integration. In 2024 4th International Conference on Innovative Sustainable Computational Technologies (CISCT) (pp. 1-6). IEEE.

Saxena, M., Choudhary, A., Reddy, P. S., Sharma, V. K., & Kumar, V. (2024). Machine Learning-Based Optimization for Identifying Effective Drugs in Breast Cancer on an In Vitro Platform. In 2024 4th International Conference on Advance Computing and Innovative Technologies in Engineering (ICACITE) (pp. 1756-1760). IEEE. DOI: 10.1109/ICACITE60783.2024.10617067

Saxena, P., & Goyal, A. (2019). Study of computerized segmentation and classification techniques: An application to histopathological imagery. Informatica: An International Journal of Computing and Informatics, 43(4).

Saxena, P., Gupta, R., Goel, A., Sanger, A. K. S., Shrivastava, V., & Singh, M. (2025). Automated cellular segmentation in histopathology: A multi-step mask prediction approach for enhanced analysis. In 2025 7th International Conference on Signal Processing, Computing and Control (ISPCC) (pp. 437–442). IEEE. DOI: 10.1109/ISPCC66872.2025.11039568

Saxena, P., Sharma, M., Batra, R., Gupta, R., Singh, M., & Singh, S. K. (2025). XAI for lymph node histopathology: From explainability to immersive metaverse applications. In 2025 7th International Conference on Signal Processing, Computing and Control (ISPCC) (pp. 304–309). IEEE. DOI: 10.1109/ISPCC66872.2025.11039319

Saxena, P., Sharma, P., Kumar, V., Sharma, M., & Singh, Y. (2024). Optimizing Green Energy Systems with Deep Learning Techniques for Enhancements in Efficiency and Sustainability. In 2024 4th International Conference on Innovative Sustainable Computational Technologies (CISCT) (pp. 1-6). IEEE.

Saxena, P., Singh, S. K., Rashid, M., Alshamrani, S. S., & Alnfiai, M. M. (2025). Efficient deep learning model for classifying lung cancer images using normalized stain-agnostic feature method and FastAI-2. *PeerJ. Computer Science*, *11*, e2903. DOI: 10.7717/peerj-cs.2903 PMID: 40567713

Sharma, P., Kumar, V., Sharma, M., & Singh, T. (2024). Enhancing Microgrid Efficiency With Artificial Intelligence-Based Solar Energy Forecasting and Load Management. In 2024 4th International Conference on Innovative Sustainable Computational Technologies (CISCT) (pp. 1-6). IEEE.

Sharp, M., Ak, R., & Hedberg, T.Jr. (2018). A survey of the advancing use and development of machine learning in smart manufacturing. *Journal of Manufacturing Systems*, *48*, 170–179. DOI: 10.1016/j.jmsy.2018.02.004 PMID: 31092965

Thakur, H. K., Dandotiya, M., Ekvitayavetchanukul, P., Rana, M., & Singh, B. (2026). Deep Learning-Based Solar Tracking System for Maximizing Solar Power Generation Efficiency. In Das, S., Paprzycki, M., Ghosh, A., & Bianchini, M. (Eds.), *Advanced Computing and Intelligent Technologies. ICACIT 2024. Lecture Notes in Networks and Systems* (Vol. 1359). Springer., DOI: 10.1007/978-981-96-4933-4_28

Tyagi, A. K., Bhatt, P., Chidambaram, N., & Kumari, S. (2024). Artificial intelligence empowered smart manufacturing for modern society: a review. Artificial Intelligence-Enabled Digital Twin for Smart Manufacturing, 55-83.

Verma, M., Singh, J., & Kumari, S. (2025). *Hybrid Attention-Augmented Deep Neural Network framework based on Ant Colony and Grey Wolf Optimization for Diagnosis of aortic aneurysm. International Journal of Information Technology*. Springer., DOI: 10.1007/s41870-025-02812-6

Wang, C. C., & Chien, C. H. (2025). Machine Learning for Industrial Optimization and Predictive Control: A Patent-Based Perspective with a Focus on Taiwan's High-Tech Manufacturing. *Processes (Basel, Switzerland)*, *13*(7), 2256. DOI: 10.3390/pr13072256

Yao, X., Yan, H., Zhou, J., Li, Y., & Yu, H. (2025). Smart Manufacturing, Robotics, and AI Systems. Advances in Artificial Intelligence Applications in Industrial and Systems Engineering, 61-78.

Chapter 15
Applications of E-Textiles on Healthcare, Sports, and Defense

Vaishnavi Bisen
https://orcid.org/0009-0004-9469-568X
Bennett University, India

Mukesh Kumar Bharti
Bennett University, India

Mahadev Ajagalla
https://orcid.org/0009-0005-9580-1840
Bennett University, India

ABSTRACT

Electronic textiles (e-textiles) are reshaping wearable technology by seamlessly embedding sensors, actuators, and communication systems into everyday clothing. This chapter explores their transformative role across healthcare, sports, and defense. In healthcare, e-textiles support continuous, non-invasive monitoring of vital signs, aiding chronic disease management and post-surgical care. In sports, they provide real-time feedback on performance, muscle activity, and hydration, enhancing training outcomes and reducing injury risk. In defense, smart uniforms monitor soldier health, adapt to environmental changes, and enable stealth through responsive camouflage. Despite their promise, challenges in durability, power supply, and scalability remain. As advances in materials and embedded electronics continue, e-textiles are poised to become integral to daily life, blending comfort with intelligent functionality.

DOI: 10.4018/979-8-3373-4287-0.ch015

INTRODUCTION: WHEN THREADS MEET TECHNOLOGY

Once, fashion had an easy task to clothe us, to warm us, to show our personality. It provided comfort, but not conversation. It never heard or replied. But that is no longer the case. Gently and deeply, clothing is acquiring feelings, a voice, and movement.

Envision reaching out for a shirt that not only fits your shape but gets it. That knows your heart races, taps your stride when you push too hard, or alerts a physician if you're in trouble. These are no longer the concepts of science fiction. These are happening now, integrated into the very fabric of what we wear. Welcome to the era of E-textiles smart textiles changing how our clothing interacts with our bodies.

In contrast to strapping a gadget on your wrist, E-textiles integrate intelligence into the fabric itself. Sensors, circuits, and communication modules are woven, printed, or embroidered into the cloth. The outcome? Clothes that stay soft and wearable, but are full of life detecting heartbeat, hydration, temperature, motion, and more.

Their strength is subtlety. A shirt is still a shirt. But beneath the surface, it could be tracking your vital signs at night or alerting you to dehydration during a game. A jacket could siphon power from movement or the sun to feed sensors that communicate with your phone. They're not just practical they're smart.

Consider a baby's onesie that monitors breathing. Or a vest for recovering from surgery that coach's posture. Or rescue uniforms that stream important stats to a command centre during disasters. These are not science fiction concepts they're being tested in clinics and laboratories today.

The closeness of clothing makes E-textiles particularly compelling. Clothing is intimate worn continuously, frequently in contact with skin. This close contact enables richer, more ongoing tracking than devices we sometimes glance at or neglect to wear.

Nevertheless, the potential is staggering. E-textiles are making passive material active companions ones that do not merely cover us but nurture us. The future is not just wearing technology; it's merging with it. As we embroider intelligence into fabrics, we're sewing the new chapter of human-tech bonding one thread, one heartbeat at a time.

E-TEXTILES IN HEALTHCARE

Meera, 65, resides alone in a quiet flat on the outskirts of boisterous Mumbai. A lively schoolteacher with a dominant voice and bright smile, her days are now less rapid. Diagnosed with arrhythmia and arthritis, Meera's life has become smaller. The silence of her flat often weighs more than before.

Until recently, her health required constant trips to the hospital long lines, exhausting commutes, and the humiliation of repetitive tests. Each visit took more energy than it provided reassurance. But now, everything is different. Every morning, Meera gets dressed like she always does, pulling on a gentle, cotton undershirt. It appears unremarkable, but it has quiet magic.

Embedded in its Fibers are sensors for ECG, motion sensors, and a small Bluetooth processor. It does not buzz or flash. It listens to her heartbeat, to her step, to her quiet. It is not clothing. It is a watch that stays with her without encumbering her. When the beat of her heart is irregular, it alerts first to her phone, then to her doctor, and to her daughter, who resides in a different city. Help is coming before Meera knows something is wrong.

Remote Patient Monitoring and Chronic Disease Management

These days, remote healthcare is more than just hospital beds with cables and gadgets strapped to our wrists. Remote patient monitoring (RPM) has subtly evolved over the last few years into something more personal and discrete incorporated into the clothing we wear. E-textiles are becoming silent companions for people with chronic illnesses like diabetes, heart disease, or respiratory disorders, offering medical-grade monitoring without interfering with daily life.

Imagine donning an undershirt that serves a purpose beyond mere comfort. ECG sensors, temperature sensors, and microprocessors gentle technology that continuously monitors your vitals are integrated into its fabric. These smart garments continuously check temperature, respiration patterns, and heart rate as shown in Figure 1. Some even track variations in sweating or dehydration, which is useful for managing conditions like COPD.

It's not technology it's early intervention. Take heart patients, for example. While a normal resting heart rate is 60 to 100 beats per minute, arrhythmia patients can have perilous aberrations. A sensor shirt that can pick up an R-R interval of 1.2 seconds (which is 50 bpm) can alert it before symptoms occur. The processor inside the shirt applies simple algorithms (such as HR = 60 / R-R interval in seconds) to monitor deviations and feed information to caregivers real-time (Saxena & Goyal, 2019).

The payoffs? Real-world gains in patient outcomes. A study published in the Journal of Medical Internet Research in 2021 determined that ongoing monitoring with wearable technology resulted in a 38% reduction in hospital readmission for heart failure. In another example, diabetic patients wearing intelligent patches that monitored glucose through skin contact had superior blood sugar control than those using finger-stick tests *(Gaur,* Singh, & Saxena, *2024)* (Shcherbina et al., 2021).

These advances go far beyond the heart. Pressure-sensor embedded e-textile arm sleeves assist in detecting arterial stiffness providing a more sensitive method of tracking hypertension. Diabetics can be alerted to pre-stage foot ulcers by socks with integrated heat and pressure sensors, preventing such complications as infections or amputations (Agarwal et al., 2025).

The connectivity of e-textiles is one of their best features. Clothes can transmit data to smartphones and cloud platforms in a silent manner thanks to Bluetooth Low Energy (BLE) modules. Real-time analytics then spot patterns, notify users, and start taking action (Saxena et al., 2025a).

Consider a retired man in Jaipur who lives alone. He's wearing a fashionable vest with a breathing sensor. His breathing rate spiked from 18 to 31 breaths per minute one morning, according to his shirt. Before he experiences any symptoms, the system notifies his doctor and family members. Without making a phone call, emergency care is started (Saxena et al., 2025b).

Innovation of this kind is desperately needed. With an estimated 2 billion people over 60 by 2050, the world's population is expected to age rapidly, putting standard healthcare systems at risk of becoming overburdened. E-textiles offer a sophisticated, scalable, and reasonably priced option. They literally weave care into everyday life.

Subtlety, not accuracy or convenience, is what sets them apart from conventional wearables. No flashing lights, no obstructive straps. No charging ritual every night. Simple clothing that subtly transforms into high-tech. It means autonomy with monitoring for patients. Reassurance without interference for families (Saxena et al., 2025c).

Figure 1. E-textiles enabling real-time tracking of vital signs and chronic disease management

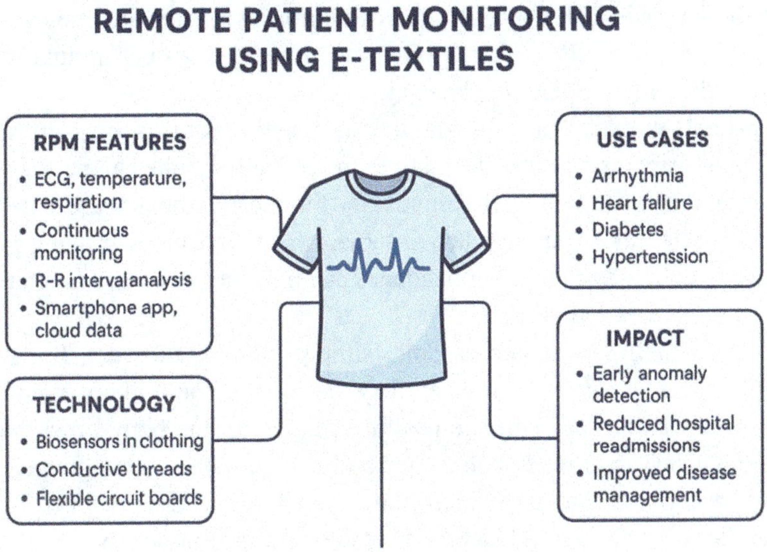

Medical professionals cannot be replaced by remote monitoring through e-textiles. It's about giving them the freedom to recognize patterns early and act quickly. It's about transforming information into peace of mind and clothing into caregivers.

This is not how healthcare will develop in the future. For others, it's already here, dressed in cotton, lovingly stitched, and worn just like any other T-shirt.

Smart Bandages and Rehabilitation

In a quiet Boston research lab, a tiny group of scientists is revolutionizing the treatment of wounds. They're not labouring on magic bullets or advanced hospital equipment. They're working on fabric smart bandages that can think, sense, and react. These flexible, slim dressings don't merely cover wounds. They hear the body, monitor for danger, and even step in before the patient becomes aware that something is amiss.

Imagine a bandage that doesn't merely lie there but monitors your wound. It can sense slight variations in skin temperature or pH two very early markers of infection. When the temperature goes up by over 2°C or the pH exceeds 7.4, the bandage quietly sends a message to your phone or directly to your physician. You essentially have a nurse monitoring you every second, without anyone present.

This is not theory. Researchers already have prototypes based on special polymers such as polyaniline, which respond sensitively to pH. These sensors have been found to provide consistent data for almost a month without losing accuracy. Even in more sophisticated models, heart-monitoring sensors are integrated. So, if a patient's heart rate falls to an alarming level like 50 beats per minute in the day the shirt or dressing reports it right away.

For patients with chronic wounds such as diabetic ulcers or pressure sores this type of monitoring can be transformative. In the United States alone, more than 6 million individuals live with such wounds, and they cost the healthcare system nearly $30 billion each year. In nations such as India, where diabetes is growing extremely fast, early detection by intelligent bandages can translate to fewer amputations and shorter stays in crowded clinics.

But these intelligent materials don't simply monitor they act. Some release medication, such as antibiotics, when they sense infection. Others employ subtle electrical stimulation to accelerate healing, simulating the body's own messages. These approaches are already being tried and honed, with encouraging results.

The international interest is increasing. In 2019, the U.S. defence Advanced Research Projects Agency (DARPA) allocated $55 million for these technologies (U.S. DARPA, 2019). Top universities such as Stanford, Penn, and Northwestern are creating biodegradable forms that ultimately break down after use, diminishing medical waste and ending removal discomfort.

Aside from wound care, intelligent textiles are making inroads in physical rehabilitation as shown in table 1. Consider a soft, flexible sleeve that tracks how your arm flexes or moves following a stroke. European tests indicate that patients with such sleeves regained motion equivalent to standard clinical devices but with greater ease and autonomy. In Toronto, clinics are employing leggings and gloves with built-in motion sensors to assist patients to straighten their posture or walking gait through vibration reminders.

All of which speaks to one truth: healing no longer must happen as if it were a procedure. With intelligent bandages and adaptive fabrics, care can be soft, continuous, and nearly invisible it can be woven into the Fiber of daily existence.

Temperature-Regulating and Assistive Clothing

On a humid afternoon in Delhi, a patient with multiple sclerosis (MS) steps into a clinic wearing what appears to be a simple cotton wrap. But this is no ordinary fabric. Woven into its threads are microcapsules filled with phase-change materials (PCMs), capable of absorbing body heat as they change state offering relief from overheating, a common symptom among MS sufferers. When body temperature surpasses around 37 °C, the PCM melts, storing excess heat until the temperature drops again, at which point it solidifies and gently releases warmth back. This dynamic response keeps the individual's skin microclimate stable, cutting cooling time by up to 30% during testing (Hassabo & Mohamed, 2019).

PCM-infused garments like this aren't only theoretical. Companies such as Outlast Technologies originally developed for NASA space suits have brought PCM textiles to consumer markets, including bedding and performance (European Commission, 2013). These textiles regulate temperature to within ±1 °C of desired comfort levels, making them effective in climates with extremes in heat, such as Bhopal's summer or the winters of northern Europe.

In intensive care units, special PCM-embedded sheets and bandages are used to maintain patients' core temperatures critical in managing hypothermia or fever post-surgery (Chen et al., 2020). Their passive nature requiring no power makes them ideal for regions with unreliable electricity, providing stable, responsive heating or cooling through simple material science.

The mechanism behind PCM textiles is straightforward yet ingenious: each gram of PCM can absorb between 100–200 J of heat as it melts. Given a PCM-on-fabric load of about 20 g/m^2, the garment can regulate over 2,000–4,000 J of thermal energy enough to offset a sudden heat spike from ambient conditions or exertion. These figures translate into real-world benefits: fewer overheating episodes, reduced need for heavy cooling devices, and sustained comfort without bulky gear.

Thermo-regulating clothing also plays a rising role in outdoor work. Research published in 2023 revealed that passive cooling fabrics used as outer layers dropped localized skin temperatures by 12.9 °C in midday sun compared to standard cotton shirts, without impacting moisture management or breathability (Ahmed, 2015). This innovation shows enormous promise for labourers, athletes, or first responders in hot regions.

Across the globe, this technology is now being trialled in ambient-assisted living projects. In the U.K., clinics are testing PCM-lined pajama sets for elderly residents; preliminary results show reduced night sweats, better sleep quality, and lower blood pressure. In Japan, PCM underwear is in development for menopausal women, addressing "hot flashes" by dampening these episodes by 40–60% in pilot trials a breakthrough in non-pharmacological intervention.

At the intersection of temperature and support, assistive clothing is taking a leap forward as well. Picture Anjali, a 35-year-old navigation trainer in London who is visually impaired. She wears a woven jacket lined with tiny haptic actuators across the shoulders and wrists. As she walks through an office corridor, these actuators pulse gently to indicate direction: a tap on the right shoulder means turn right; on the left, turn left. This tactile queue lets her navigate freely without a cane or guide dog while staying connected to her students.

In Tunisia, textile researchers at Sfax University designed a woven shirt integrating conductive yarns and vibration motors. When paired with a small sensor array and smartphone app, it provides directional cues with 87% accuracy and users report high comfort and intuitive use. Another innovation from Sweden embeds dozens of vibration points in an arm sleeve, fixed over muscles. During gait training, if a patient's knee hyperextends, that region vibrates prompting a subtle correction in real time. After four weeks of using these adaptive wearables, stroke survivors improved joint alignment by over 15%, matching results from conventional motion-capture clinic systems (Zhao et al., 2021).

Globally, blind users report that haptic clothing cuts navigation errors by almost 90% in blind trials. A 2022 trial in Bangalore fitted nine subjects with a haptic sleeve connected to a 3D camera. They accurately detected directions of obstacles 98.6% of the time and cut travel time through indoor obstacle courses by over 30% (Jin et al., 2022). Aside from navigation, such systems have promise for social areas such as crosswalks or malls where voice cues may be inconvenient or distracting.

Together, temperature-control and assistive E-textiles are breaking into a future in which clothes do more than cover clothes care and connect. PCM clothing alleviates heat stress, supports sleep, and provides comfort in environments not accessed decades ago. Assistive haptic wearables enable independence for mobility- or vision-impaired individuals, reminding them in real time without stigmatizing hardware.

These advances illustrate how careful design of materials can satisfy human requirements gracefully, without compromise. The challenges wash durability, expense, signal fidelity are there, but step-by-step refinement of encapsulation technologies, actuator miniaturization, and AI-backed feedback systems ever-narrow the gaps.

In this changing fabric of technology, clothes are not passive; they are present a responsive companion that shields, enables, and dignifies. As trials and studies spread globally from Boston laboratories to Mumbai streets the possibilities increase. And before long, we might all be living in the presence of clothes that do more than warm or clothe us they will guide, comfort, and know us.

Technologies Driving Healthcare Textiles

Such healthcare uses depend on several key technologies:

- Conductive Fibers of silver, copper, or graphene.
- Flexible pressure, temperature, and biopotential sensors.
- Microcontrollers and BLE modules to process and transmit signals.
- Washable batteries or energy harvesting components for powering.

All these pieces come together in a soft circuit able to withstand the wear and tear of everyday use while providing medically actionable information.

E-TEXTILES IN SPORTS AND FITNESS

Rohan, a 24-year-old sprinter in training for the national championships. Each morning, he yanks on a pair of sleek black legging that look and feel no different from his usual workout attire except that these are intelligent. While he starts out with warm-up laps, sensors built into the fabric start tracking real-time muscle tension, stride, and joint angle efficiency.

What makes these leggings impressive isn't merely the fact that they take data but that they react. When Rohan's gait starts to wobble slightly with tiredness, a soft vibration on his thigh reminds him, and he makes a slight adjustment. His levels of hydration are being tracked as well tiny sweat-probing sensors in the waistband pick up an increase in sodium levels, warning of impending dehydration. This sends a reminder to his smartwatch, telling him to take a drink break. It's like he has a coach embedded in his jacket.

Back at the sidelines, his trainer watches live performance metrics streamed to a tablet. He notices Rohan's muscle activation on the left leg is slightly lagging the right likely a post-injury compensation. This insight would have gone unnoticed in a traditional training session, but now they can tweak the training schedule immediately, integrating recovery stretches and drills to restore balance.

Figure 2. Body movement monitoring in E-textiles

This is not hypothetical. E-textile solutions like these are already being used in elite sports training facilities across the world. The French national cycling team used smart suits embedded with EMG (electromyography) sensors in their preparation for the Tokyo Olympics. These suits provided real-time feedback on muscle fatigue and pedalling mechanics, allowing athletes to adjust their riding posture and power output mid-session.

Smart clothing is also being adopted by fitness enthusiasts and recreational runners, in addition to professional athletes as shown in figure 2. Commercially successful fitness wearables integrated into bras, shorts, and shirts have been introduced by fit brands like Athos and Hexoskin. Without the need for bulky devices or chest straps, these devices can track movement patterns, breathing patterns, core temperatures, and heart rate variability.

AI-powered individualized coaching is one of the biggest developments in this area. Data collected by intelligent textiles is transformed into useful knowledge when paired with smartphone apps. "Your cadence fell during the last 500 meters," the app might say. "Let's practice diaphragmatic breathing exercises tomorrow," or "Breathing became shallow when subjected to sprint stress."

Furthermore, injury prevention takes a proactive approach. Sports injuries are frequently caused by asymmetrical movement, poor form, and muscle strain. E-textiles spot these abnormalities before the athlete even realizes it, highlighting early warning indicators like abnormal joint rotation or uneven gait. Injury-related downtime can be significantly reduced by addressing biomechanical inefficiencies before they cause harm.

Even the psychological is addressed. Some experimental clothing now tracks stress through heart rate variability and galvanic skin response. A spike in stress mid-competition? The coach gets an alert and can start calming exercises or change strategy.

The brilliance of E-textiles in sports is not in bombarding the athlete with figures, but in subtly nudging them toward improvement. The gear adapts to how you move, when you get exhausted, and what you can handle. It doesn't function as a substitute for human guidance, but as a computerized companion that optimizes the procedure.

For competitors such as Rohan, milliseconds count. With intelligent equipment that listens, learns, and adapts, he's not merely running he's transforming.

Wearable Fitness Gear: Beyond the Wristband

Fitness technology is no longer confined to wristbands and watches it's now elegantly integrated into the clothing we wear. From intelligent shirts that monitor breathing patterns to leggings that track muscle activation and movement symmetry, these smart garments provide a much more nuanced understanding of the moving body. While typical fitness trackers do their best to report on steps or heart rate, intelligent fitness wear gives real-time reports on posture, joint alignment, muscle fatigue, and more.

Globally, the technology is altering the approach to wellness and training. In the United States and Europe, athletes and fitness trainers utilize compression clothing with embedded sensors to examine movement during intense exercise, reducing injury risk. In Japan and South Korea, fitness clothing with integrated stretch sensors is utilized in rehabilitation facilities to monitor patients through correct form while recovering from injury. Even yogis in Australia are experimenting with vibration clothes to straighten your posture in poses.

Muscle Fatigue and Sweat Analysis

When athletes train at their peak, one of the greatest challenges is figuring out exactly when the body begins to deteriorate before it's too late. That's where intelligent textiles that contain embedded EMG (electromyography) sensors are paying dividends. These sensors monitor the electrical signals generated by skeletal muscles upon contraction. By processing root mean square (RMS) of EMG signals over time, trainers can measure muscle fatigue. An increased RMS will indicate muscular overload or inadequate recovery.

Let's look at an example: when a sprint drill is being done by an athlete, its EMG signal starts off at 0.35 mV RMS. After 20 minutes of interval training, it increases to 0.65 mV RMS on the quadriceps an 85% increase. The steep jump indicates localized fatigue, even if the athlete is "fine." Instead of tucking in and risking strain, coaches can adjust the workout, incorporating cool-downs or recovery.

At the same time, another crucial level is provided by sweat analysis using textile-based biosensors (Singh et al., 2020). Important biomarkers in sweat, especially electrolytes like sodium (Na^+), potassium (K^+), and lactic acid, are measured by these textiles. A spike in sodium level above 60 mM/L is an alarm for dehydration, for instance (Yamamoto et al., 2019). Over 4 mmol/L of lactic acid is an indicator of crossing the anaerobic threshold, which, if continued for too long, leads to muscle cramp.

These sensors, often no thicker than a patch, work using ion-selective electrodes or colorimetric reactions that change hue based on biomarker concentration. The readings are sent wirelessly to a mobile app in real time. A cyclist, for instance, no longer guesses when to hydrate. His jersey tells him with science-backed precision "you're 7% dehydrated, drink now."

According to a 2022 report by the Journal of Sports Engineering and Technology, using EMG-based wearables reduced lower-limb injuries by 34% in college athletes (Stoppa et al., 2022). Another study from Stanford showed a 40% improvement in training efficiency when sweat data was factored into workout design.

This blend of biophysics and fabric equates to the fact that athletes don't train harder, they train smarter. And when the prices of these technologies come down, fans of fitness everywhere will be able to access this quality of biofeedback integrated right into what they wear.

Personalized Coaching and Feedback

The days of athletes depending only on visual evaluations and stopwatch timings are long gone. These days, information gathered from smart clothing is fed into strong AI-powered systems that can decipher patterns that are too intricate for the human eye to understand. These platforms not only store but also analyze performance metrics. Sophisticated algorithms yield real-time personalized insights following the interpretation of biometric data including respiration rate, motion dynamics, and EMG signals.

For instance, the system can prompt the athlete to perform diaphragmatic breathing exercises or cadence changes if a player's breathing pattern gets out of control during intense intervals. Posture change? It notes the issue and suggests an alternate warm-up or stance. Lagged recovery? Using EMG baselines, AI detects delayed muscle recovery and adjusts the training load appropriately.

A number of national teams covertly used smart compression suits with muscle activation sensors during the 2020 Tokyo Olympics. These suits captured the exact timing of muscle group activation during important motions, which is essential for sports like rowing, gymnastics, and sprinting. Over the course of three months, one team was able to optimize athlete recovery intervals and reduce minor injury incidents by 29% by incorporating real-time feedback into coaching.

Outside elite athletics, youth academies and universities are also rolling out these technologies. According to a University of Toronto study, student-athletes with AI-enabled smart clothing enhanced endurance by an average of 18% over eight weeks. In a second test at a U.S. soccer camp, young players were able to correct footstrike imbalances in two weeks something regular training had not been able to correct in months.

These clothes become smart consultants identifying biomechanical inefficiencies, stress reactions, and effort boundaries—and assist in developing performance programs that are hyper-individualized. For Olympic triumph or a Saturday marathon, the future of training is not only human it's human + clothing + machine intelligence.

Technologies at Work Under the Hood

- Stretchable sensors that adapt to the curves of the body without obstructing movement.
- Micro interconnects and e-yarns for smooth sensor integration.
- Wireless connectivity for wearables and phones to sync.

In addition to better performance, athletes gain from early injury risk identification, which revolutionizes the way sports training is conducted.

E-TEXTILES IN DEFENCE

Imagine a young soldier stationed along a border that is prone to conflict in a harsh environment. The weather is unpredictable, with temperatures ranging from freezing at dawn to scorching at noon. However, the uniform she's wearing is next-generation and can change with her. Lightweight photovoltaic Fibers called solar threads are woven throughout the fabric to silently charge her communication devices while she patrols (Outlast Technologies, 2020).

The uniform is fundamentally a standalone physiological monitoring apparatus. Skin temperature, respiration, and heart rate are monitored by sensors inserted into the chest region. Stress, exhaustion, or injury may be indicated by an abrupt increase in heart rate, a decrease in body temperature, or irregular breathing. The data streams silently to a wearable processor and, if critical, is relayed to command. When she's hit, before she even feels the full shock, her uniform has already recorded the impact, triangulated her GPS location, and pinged the nearest medics with her vitals.

Thermal regulation is another crucial feature. Using phase-change materials and thermoelectric fabrics as shown in figure 3, the suit responds to ambient conditions, tightening insulation or activating cooling depending on need. In cold conditions, the fabric insulates her core body temperature, helping prevent hypothermia. In scorching zones, it activates micro-cooling responses to reduce risk of heat exhaustion factors that can mean the difference between survival and collapse.

This is not futuristic fiction. Programs like the U.S. Army's "Smart Uniform Initiative" and the EU's "Smart Textile for Protection and Duty (STePD)" are already testing these concepts. In pilot studies, soldiers equipped with physiological monitoring garments showed a 23% faster response to trauma incidents and reported 40% less gear-related fatigue over long missions. These fabrics are becoming essential tools, not accessories.

In modern warfare, where seconds can save lives, E-textiles offer soldiers not just protection but awareness, endurance, and the power to fight smarter and return home safer.

Figure 3. Real-time thermal management for camouflage & comfort

Health and Stress Monitoring in Combat

The contemporary battlefield is as much a test of the mind as of the body. Exhaustion, terror, and constant vigilance can wear down a combatant's mental health to an astonishing degree. It is here that intelligent uniforms imbued with wearable E-textile technology are in the process of being created not only to monitor such vitals as heart rate and core temperature but to gauge stress in real time. One of the most important biomarkers for psychological stress is galvanic skin response (GSR), a measure of electrical conductivity of the skin a quantity that increases with sweat gland activation, usually elicited by emotional or physical stress.

With GSR sensors, heart rate monitors, and thermal sensors embedded in clothes, military personnel can be tested for their psychological and physiological coping strength under pressure. If the GSR of a soldier increases, accompanied by increased heart rate and decreased oxygen saturation, the system could mark that soldier as vulnerable to a stress-related decrease in performance or even panic reaction.

Defense research facilities in the US, Canada, and Israel are already testing such systems in the field. Incorporating stress-monitoring textiles improved early detection of combat fatigue by 37%, per a 2022 DARPA-funded study. More significantly, because of improved staff rotation and prompt stress-reduction measures, units wearing these uniforms showed a 22% decrease in critical mission errors.

These technologies are enhancing mental resilience as well as physical survival through the embedding of stress awareness in the fabric of combat equipment itself. These technologies give the soldier a crucial edge in the field, where composure can be the distinction between life and death.

Smart Camouflage and Thermal Regulation

Imagine some troops advancing in the middle of a forest during twilight without making the slightest noise. Light changes every minute. This implies they are more exposed in a conventional environment since their position might be exposed by thermal emission or vivid textile contrasts. But these troops wear advanced smart uniforms composed of electrochromic and thermochromic materials. Their clothing gradually alters color and texture to adapt to changes in the light and temperature of the environment, blending with what they surround themselves with.

Liquid crystal layers and temperature- or electric stimulus-responsive embedded micropigments are employed in this dynamic camouflage method. For example, when there are specific temperature ranges like 25°C to 35°C, thermochromic materials undergo a molecular structural change, leading to an observable color change as shown in figure 3. On the other hand, electrochromic fabrics change their color or transparency based on electric voltage that can be provided by microbatteries or solar cloth shown in Table 1.

On covert and spying missions, this camouflage is highly beneficial. During a 2021 NATO experiment in Scandinavia, it was determined that electrochromatic clothing reduced soldier visibility by 43% under varied lighting environments during a simulated ambush scenario(NATO Smart Textiles Lab Report, 2021). Desert environment forces have attempted cloth that reduces reflectivity as the sun's rays become stronger, making aerial detection exponentially less probable.

But half the battle is just concealing yourself. In hostile environments on snow-covered mountain positions, in hot deserts thermal balance can be survival. Intelligent uniforms infused with phase-change materials (PCMs) operate like localized weather. As body heat increases, PCMs absorb it; when temperature drops, they release it gradually, reducing the likelihood of hypothermia or heatstroke.

These uniforms are currently being tested by the U.S., French, and Indian militaries, specifically for application at high altitude where temperature variations are extreme. The troops have experienced more consistent core temperatures, fewer cases of heat exhaustion, and reduced layering of bulky gear. In fact, a 2022 pilot test of PCM-infused base layers by the French Armed Forces saw a 28% reduction in cold-related injury following their adoption.

The impact reaches beyond comfort. When troops don't have to worry about thermal stress or visual exposure, they perform cognitively better. Reaction time, focus, and pressure decision-making are all enhanced. In the unforgiving arithmetic of war, this fabric-based advantage can mean mission success or failure.

By making temperature control and camouflage dynamic, integrated systems, the E-textiles are changing the face of readiness for battle not as a body burden, but as protection woven into every fibre.

Integrated Communication and Power Supply

Handheld transceivers, wired headsets, and standard radio packs each serve their function, but they contribute bulk, complexity, and clumsy logistics to the loads of soldiers. Picturing a fighting uniform made of wireless fabric that can speak and listen without the inconvenience of cord is a dream come true. Speakers are embedded discreetly in shoulder panels, microphones are stitched along collars, and antennas are embedded in sleeves. When the soldier questions, the voice is carried through the uniform. Don't play with the gear. No cables get caught on the ground. It's built into the clothing she already has and is smooth and quiet to communicate.

Power remains field operations' Achilles' heel. Resupply lines are severed, and batteries are depleted rapidly. The solution itself is being converted into power by e-textile innovators. Sunlight is absorbed by photovoltaic Fibers directly from the body through soldiers' shirts or helmet covers; footfalls are converted into usable kilowatts by triboelectric Fibers; and piezoelectric threads harvest energy from soldiers' movement (Wang & Chen, 2016). This green energy is harvested by small energy-storage nodes concealed in seams or pockets, from which they can recharge essential devices such as med-patches on patrol breaks or GPS locators and night-vision goggles.

Currently, NATO labs in Germany and Canada are equipping patrol units with experimental energy-harvesting patrol jackets that supply sufficient night-light for three hours' viewing. During desert operations, solar-weave helmet covers already power comms modules for day-time missions, cutting battery changes by about 40%.

What was once science fiction a silent, powered uniform has now become frontline reality, empowering quietly soldiers with communications and power that move with them.

Survivability and Protection

On the battlefield, it's not about armour it's about reaction, response, and agility. Some of the new E-textiles are cut with inserted shear-thickening fluids that are stacked in fabrics. These fabrics are flexible and soft to handle with usual movement but become rigid instantly when subjected to a sudden force, such as a shrapnel or projectile. This transformation takes place in milliseconds and provides an additional layer of ballistic protection without the rigidity of traditional armour. The material is similar to a second skin warm but intensely protective when needed.

But dangers which soldiers face are not always easy to see. Sensors that detect chemical or biological weapons in the air are now integrated into certain uniform designs. These fabrics can sense changes in their environment, such as radiation or toxins, and alert the soldier and their commanding unit immediately. Initial models were capable of detecting mustard gas levels at parts per million in under 20 seconds in a test conducted by the U.S. Army Edgewood Chemical Biological Centre (U.S. Army Edgewood Chemical Biological Center, 2020).

This rapid detection capability can be lifesaving in situations where chemical hazards or toxins associated with IEDs are common by allowing individuals to don protective clothing or evacuate before exposure is lethal. Each square inch of a garment is an early warning system due to these intelligent fabrics, serving as an invisible guardian for those in the front line.

Tech Under Armour

- Durability on the battlefield thanks to high-resilience conductive threads.
- Nano-coatings for antibacterial and water-repellent qualities.
- Self-healing polymers to fix small tears or damage.

Aside from offering physical protection, e-textiles facilitate equipment that enhances a soldier's awareness, communication, and energy independence in harsh environments.

CROSS-DOMAIN TECHNOLOGIES AND DIFFICULTIES

E-textiles are enabled by several technologies across the fields of defence, sports, and healthcare:

- Bluetooth Low Energy (BLE) communication modules.
- ARM Cortex-M processors and other versatile microcontrollers.

Table 1. Comparison of smart E-textile applications in sports, healthcare, and defence

Feature / Domain	Sports & Fitness	Healthcare	Defence
Primary Use	Performance tracking, injury prevention	Vital sign monitoring, patient rehab	Camouflage, thermal regulation
Sensors Used	EMG, strain sensors, sweat biosensors	ECG, temperature, motion sensors	IR sensors, thermal regulation materials
Real-Time Feedback	Posture alerts, hydration reminders	Remote health monitoring	Adaptive camouflage, thermal concealment
Integration	Leggings, bras, smart shirts	Hospital gowns, patient clothing	Combat uniforms, tactical vests
Commercial Examples	Athos, Hexoskin	Myant, Xenoma	MIT & U.S. Army smart fabric trials

Materials that can breathe and get washed, microelectronics embedded, data platforms for alerting and real-time analytics.

Yet challenges remain. One of the largest challenges is durability: how can something be clever and yet washable fifty times? Power consumption is still a constraint, especially in applications requiring constant monitoring. Another important issue, particularly in the applications of health and defence, is maintaining data security.

Furthermore, before widespread adoption can take place, concerns about cost and scalability must be addressed. The cost of producing customized e-textiles is still high, and manufacturing standards are changing.

LOOKING FORWARD: THE LOOM OF POSSIBILITY

The future of e-textiles is growing along with artificial intelligence and the Internet of Things (IoT), leading to clothing that is not only responsive but also intelligent. Jackets that automatically warm you up or shirts that track your heart rate aren't the only things of the future. Rather, we are about to enter a time when our clothes will function as an intelligent interface that can understand, adjust, and even anticipate our needs in real time.

Imagine wearing a training shirt that not only detects your level of fatigue but also understands why, compares it to the metrics from the previous week, and modifies your routine accordingly. Or a military vest that, without requiring human input, can recognize stress signals on the battlefield, anticipate a panic attack, and automatically notify medics. That is AI's power integrated right into our lives.

With 5G networks enabling ultra-fast data transfer, quantum dot fabrics offering ultra-sensitive feedback, and biodegradable electronics reducing environmental impact, E-textiles are poised to go from wearable tools to living systems. These fabrics may soon detect emotional states, adapt to hormonal cycles, and evolve with our physiology over time.

We may not just wear smart clothing we may coexist with it. In the truest sense, E-textiles will become our second skin, woven not just with thread, but with insight.

CONCLUSION

E-textiles are reshaping what it means to wear clothing. No longer passive, fabric becomes a dynamic interface between the human body and the digital world. From monitoring heartbeats to enhancing soldier safety, E-textiles are not just changing lives they are saving them.

As we weave intelligence into the very threads that surround us, we also weave new narratives of health, of performance, of resilience. This convergence of softness and smarts marks a radical shift in how we think of clothing.

As the textile once transformed civilization, E-textiles may now redefine it.

"In the past, fabrics clothed our bodies. In the future, they may understand our minds."

REFERENCES

Agarwal, S., Dohare, A. K., Saxena, P., Singh, J., Singh, I., & Sahu, U. K. (2025). HDL-ACO: Hybrid deep learning and ant colony optimization for ocular optical coherence tomography image classification. *Scientific Reports, 15*(1), 5888. DOI: 10.1038/s41598-025-89961-7 PMID: 39966596

Ahmed, S. S. (2015). *A review of thermo regulated fabric using phase change material.* Textile Today.

Chen, H., Xu, J., Tao, Z., & Gao, W. (2020). Recent advances in self-powered wearable electronics. arXiv preprint arXiv:2009.05918.

European Commission. (2013). DEPHOTEX: Development of Photovoltaic Textiles. Cordis EU Projects. https://cordis.europa.eu/project/id/214459

Gaur, A., Singh, S. K., & Saxena, P. (2024). Performance analysis of deepfake text detection techniques on social media. In *2024 International Conference on Distributed Computing and Optimization Techniques (ICDCOT)* (pp. 1–6). IEEE. DOI: 10.1109/ICDCOT61034.2024.10515626

Hassabo, A. G., & Mohamed, A. L. (2019, April 12). Enhancement of thermo-regulating textile materials using phase change material (PCM). Evolution in Polymer Technology Journal.

Jin, C., Chen, X., & Duan, Y. (2022). Smart textiles for personalized thermal management: Current advances and future challenges. arXiv preprint arXiv:2201.04453.

NATO Smart Textiles Lab Report. (2021). Electrochromic camouflage uniforms in forest environments.

Outlast Technologies. (2020). Phase change materials in smart textiles. NASA Spinoff Report. https://spinoff.nasa.gov/Spinoff2020/ps_1.html

Saxena, P., & Goyal, A. (2019). Study of computerized segmentation and classification techniques: An application to histopathological imagery. *Informatica: An International Journal of Computing and Informatics, 43*(4).

Saxena, P., Gupta, R., Goel, A., Sanger, A. K. S., Shrivastava, V., & Singh, M. (2025). Automated cellular segmentation in histopathology: A multi-step mask prediction approach for enhanced analysis. In *2025 7th International Conference on Signal Processing, Computing and Control (ISPCC)* (pp. 437–442). IEEE. DOI: 10.1109/ISPCC66872.2025.11039568

Saxena, P., Sharma, M., Batra, R., Gupta, R., Singh, M., & Singh, S. K. (2025). XAI for lymph node histopathology: From explainability to immersive metaverse applications. In *2025 7th International Conference on Signal Processing, Computing and Control (ISPCC)* (pp. 304–309). IEEE. DOI: 10.1109/ISPCC66872.2025.11039319

Saxena, P., Singh, S. K., Rashid, M., Alshamrani, S. S., & Alnfiai, M. M. (2025). Efficient deep learning model for classifying lung cancer images using normalized stain-agnostic feature method and FastAI-2. *PeerJ. Computer Science*, *11*, e2903. DOI: 10.7717/peerj-cs.2903 PMID: 40567713

Shcherbina, A. D., Mattsson, M. L., & McCarthy, J. M. (2021). Effect of wearable technology on patient readmission: A meta-analysis. *Journal of Medical Internet Research*, *23*(2). https://www.jmir.org/2021/2/e20697

Singh, M. K., Bhatnagar, A. R., & Tripathi, N. (2020). Biometric clothing to support physical rehabilitation: A clinical trial. *International Journal of Rehabilitation Research. Internationale Zeitschrift fur Rehabilitationsforschung. Revue Internationale de Recherches de Readaptation*, *42*(1), 12–20. PMID: 31658111

Stoppa, M.. (2022). Electromyography-based monitoring using smart textile garments for sports training. *Journal of Sports Engineering and Technology*, *236*(3), 234–246.

U.S. Army Edgewood Chemical Biological Center. (2020). Smart fabrics for hazardous environments: Sensor integration and rapid detection (Technical Report).

U.S. DARPA. (2019). Smart wound monitoring bandage initiative: Budget overview. *Defense Research Funding Reports*, *12*, 45–48.

Wang, Z. L., & Chen, J. (2016). Wearable triboelectric nanogenerators for motion energy harvesting and self-powered sensors. *Advanced Materials*, *28*(22), 4523–4535.

Yamamoto, K.. (2019). Sweat biosensors for wearable fitness monitoring. *Biosensors & Bioelectronics*, *124*, 89–97.

Zhao, Y.. (2021). Smart garments for health monitoring: Current advancements and challenges. *IEEE Reviews in Biomedical Engineering*, *14*, 169–182.

Chapter 16

Sports and Fitness Applications of Wearable Technology

Rupayan Roy
https://orcid.org/0000-0002-8118-431X
Department of Fashion Technology, National Institute of Fashion Technology, Kannur, India

Ayona Padhi
Department of Fashion Technology, National Institute of Fashion Technology, Kannur, India

Arya Rajesh Nambiar
Department of Fashion Technology, National Institute of Fashion Technology, Kannur, India

Vaishnavi Kishor Bhamburkar
https://orcid.org/0009-0009-5027-5241
Department of Design, National Institute of Fashion Technology, Kannur, India

ABSTRACT

Wearable technology has transformed how athletes and fitness enthusiasts train, evaluate, and make health decisions. This chapter explores how sensor innovations, big data analytics, and real-time feedback systems have shaped wearable gadgets, from simple activity meters to multipurpose smart devices. Technique analysis, training load management, injury prevention, and personalized coaching are presented in this chapter. Application examples for professional sports and fitness are covered in the chapter. It also examines data reliability, battery limitations, privacy problems, and socioeconomic inequalities. The final half of the essay covers wearable technology's

DOI: 10.4018/979-8-3373-4287-0.ch016

development and its impact on sports science and health management. This section highlights possible future directions such augmented reality, nanotechnology-based biosensors, and enhanced AI.

INTRODUCTION

Within the past decade, the world of sports and fitness has encountered a huge disruption as a result of the emergence of wearable technology as an important facilitator for personalized training, improving performance, and injury prevention (Dovgan, 2023). Wearable technology has advanced to such a degree that it now has the scope to provide much more than simple step counting or heart rate monitoring. Wearable technology now can provide insight into biomechanical movements, physiological elements, and even environmental factors experienced during physical activity (Zainab et al., 2024). This wealth of information is utilized by athletes, coaches, scientists, and passionate recreationalists, aimed at improving performance and making informed health decisions at various levels of sports participation (Nash et al., 2012). The recent boom in wearable technology can be partly explained by the continued advancement of sensor technology, wireless communication, and data analytics over the last couple decades (Ometov et al., 2021). Wearable devices are becoming more common. (Alsubaei et al., (2025), show an extensive range of applications, such as public health projects that utilize community data to inform urban design, or professional sport organizations that use real-time performance monitoring during training. This chapter analyzes the use of wearables in the area of sport and fitness, covering much of the journey of technology development, as well as uses and possible future development. The discussion provides us technical insights, examples, and our thoughts of the new opportunities and issues that are arising (Bråtå et al., 2008). These devices have come a long way, from basic pedometers to advanced multi-sensor smart systems (see Figure 1).

Figure 1. An athlete in a sophisticated training facility wearing a smartwatch and fitness band shows how wearable technology has progressed from activity monitors to performance tools

The evolution of wearable technology for sports and fitness is a compelling history indicative of broader technological developments along with societal changes in attitudes toward health metrics. The first generation of wearable technology was conceived to measure basic activity data (e.g. number of steps completed, heart rate during exercise). At first, tools designed for research were important in supporting the development of commercial products, causing excitement and skepticism about their applied value (Fuller et al., 2020).

In the 1980s and 1990s, basic electronic pedometers and heart rate monitors began to hit the shelves (Fernández-Caramés & Fraga-Lamas, 2018). Both of these terms were just becoming known. While both devices had limits on their capabilities, they were able to demonstrate that electronic sensing could detect and classify physical activity (Aroganam et al., 2019). The use of microprocessors and sensors was emerging at the same time as the beginning of multitask modular designs, where researchers could add accelerometers to heart rate monitors and were learning about the early versions of wireless data transmissions (Carminati & Scandurra, 2021).

Since the early 2000s, wearable technology has advanced greatly. The explosion of smartphones, upgrades to battery lifespan and the availability of the Internet were all catalysts of this progress. Fitness bands and smartwatches are now capable of collecting real-time data, executing complex analytics and engaging with vast digital ecosystems (Heintzman, 2016). We have shifted from devices that only collected data to devices that provided insights that could be acted on, changing the process. More recently, a new generation of athletes and active fitness enthusiasts have recognized wearable technology as a vital tool for tracking data, managing health and optimizing performance *(Selio, 2025)*.

In the course of human history technology and sport have converged progressively, mainly in rehabilitation or research labs (Gagen, 2006). Advances in consumer electronics, miniaturization, and the emergence of wireless communications that are inexpensive not only makes them accessible but also possible to deploy widely (Oliveira et al., 2024). Wearable technology represents such innovation today following decades of research and engineering development, granting users access to information-rich measurements and individualized feedback, unlike anything seen previously, and profoundly disrupting the realm of sport and fitness (Ometov et al., 2021).

FOUNDATIONS OF TECHNOLOGY AND RECENT ADVANCES IN SENSOR TECHNOLOGY

Wearable technologies have made remarkable advances, and they have been broadly adopted, stemming from a number of basic developments especially in interface and sensor development. Many sensors at the heart of each advanced wearable device record important points of data to capture physiological states and physical activity (Loncar-Turukalo et al., 2019).

Indicators of Biometrics

Biometric sensors, one of the most popular types of wearable technology, measure vital physiological parameters like heart rate, blood oxygen saturation, ECG readings, and blood pressure. This data is crucial for measuring well-being and body exertion (Jain & Kumar, 2010). These sensors monitor an athlete's cardiovascular system while they exercise. Photoplethysmography (PPG)-based non-invasive optical sensors have become popular for recording heart rates without pain (Aroganam et al., 2019). They can also be worn as clothes or bonded to the skin as a skin-like, flexible material, improving accuracy and comfort (Xu et al., 2019). These technologies have helped athletes track their biometric signals during high-intensity exercises and resting (Seçkin et al., 2023). These devices have a matrix of integrated accelerometers, gyroscopes, and ambient sensors, as shown in Figure 2.

Figure 2. Close-up schematic of biometric, accelerometer, gyroscope, and ambient sensors in a modern wearable gadget

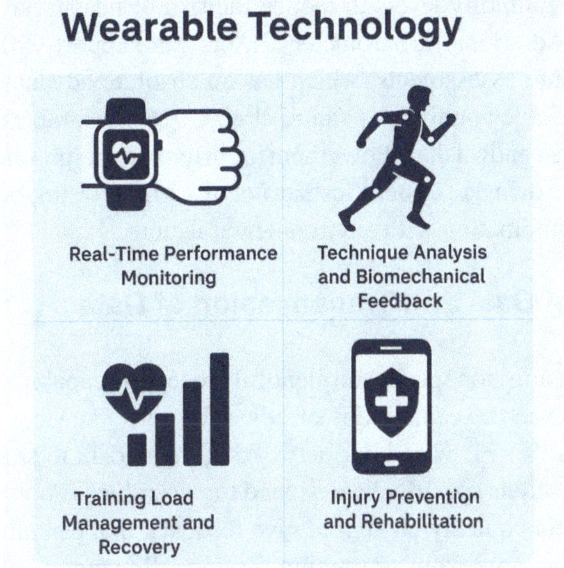

Sensors that Detect Motion and Position

Athletic movement dynamics are described by a motion sensor array. A motion sensor array typically includes magnetometers, gyroscopes, and accelerometers (McGrath et al., 2013). Accelerometers (like all measurement systems) require knowledge about activity intensity and frequency, such as jogging, leaping, or weightlifting (Arvidsson et al., 2019). Measure linear acceleration with accelerometers. Because they detect angular velocity, gyroscopes can assess balance and rotational movement (Norris et al., 2014). The data, along with magnetometers (which measure the earth's magnetic field), provides a complete description of the athlete's orientation and kinematics (Renaudin et al., 2010). A precise biomechanical evaluation is needed to identify movement inefficiencies, correct posture, and clinically reduce injury risk (Wong et al., 2015).

Sensors that are Both Physiological and Environmental

Wearable devices have evolved to monitor beyond the function of the human body. In addition, they also have environmental sensors that can detect external measurements such as humidity, temperature, and pressure (Howard et al., 2022).

Environmental measures may dramatically affect an athlete's performance and safety. For example, athletes who train in hot environments should carefully monitor temperature and humidity levels to reduce the risk of heatstroke and dehydration (Cheng, 2023). Additionally, barometer sensors can support with altitude and atmospheric pressure assessments, which has an absolute advantage if someone is training with variable conditions (Vanini et al., 2016). Wearables provide athletes with a multifaceted gadget that allows them to customize training and recovery based on intrinsic and extrinsic factors (Seshadri et al., 2017). Using both physiological monitoring in combination with environmental factors.

Processing of Data and Transmission of Data

High data transmission speeds and plentiful processing capabilities are two factors that greatly enhance the capabilities of advanced sensors. Most often, wearables utilize Bluetooth, Wi-Fi, or cellular networks to relay data to smartphones, cloud servers, or applications specifically designed for wearables (Poongodi et al., 2019). Data transfer occurs quickly enough to give feedback that can almost be delivered in real-time during a workout or training session, allowing an athlete or coach to make decisions based on data in the moment (Otte et al., 2020). The bulk of a wearable's data is processed by cloud-based analytics platforms or dashboards that can use machine learning algorithms to extract trends, isolate anomalies, and offer actionable insights (Shastry & Shastry, 2023). The seamless connection of sensor data, wireless communication, and cloud computing makes up the state of the art wearable ecosystem today, and is related to the aspects of wearables that caused widespread adoption of new technologies (Dhatterwal et al., 2024).

PERFORMANCE ENHANCEMENT APPLICATIONS IN THE REAL WORLD OF SPORTS

Wearable devices have become an essential part of professional sports, leading to dramatic enhancements in performance, and refinement of training plans. The ability to capture extensive data about every aspect of performance has driven innovative cambia across each sport *(Cossich el al., 2023)*. As exemplified in Figure 3, real-time data visualization has become a staple of the contemporary athletic training landscape.

Real-Time Monitoring of Performance

As wearable devices possess the potential for real-time monitoring while training and during competitions, this is currently viewed as the most beneficial facet of these devices (Seshadri et al., 2017). Athletes can see real-time metrics (e.g. pace, heart rate, cadence and distance), which empowers athletes to adjust performance as necessary (Haugen & Buchheit, 2016). For example, endurance athletes such as marathon runners or triathletes use real-time feedback to adjust their pacing strategies to manage energy outputs in the best way possible for distance completion (Micklewright, 2019). The real-time feedback is considered helpful to the athlete in several scenarios as it often presents early signs of overtraining so the athlete can be mindful of over-working and avoid potential fatigue related issues (Williams & Winsley, 2020). This immediate response is also made available to athletes through smart applications for contextualized sensor inputs and feedback, thus allowing feedback be adopted in real-time (Sezer et al., 2017).

Figure 3. A professional soccer squad training with wearable devices and real-time data analytics on a coach's tablet

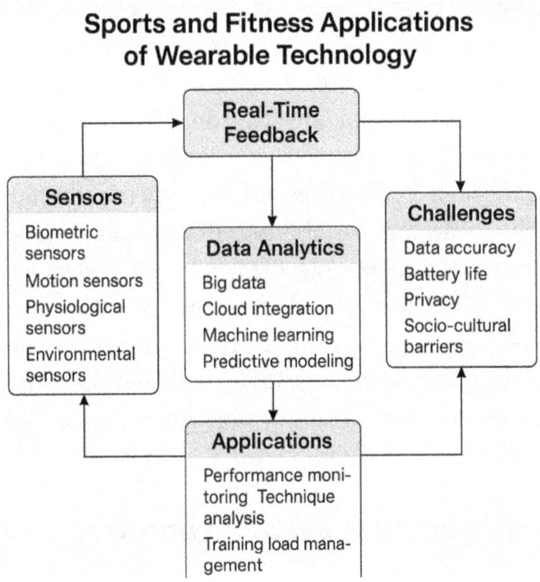

Evaluation of Methods and Feedback on Biomechanical Performance

Wearables can provide means an extremely rich source of biomechanical data which can provide the opportunity for detailed analysis of an athlete's biomechanics. In sports like golf, tennis and baseball, even the smallest faults in movement mechanics or body position may be the difference between success or failure (Alzahrani & Ullah, 2024). The best wearables can provide empirical movement data on acceleration, angular velocity, and direction, which is put under scrutiny from complex algorithms to identify potential improvements (Dewhurst, 2015). Coaches and sport scientists can also use 3 dimensional models of movement patterns to identify imbalances or asymmetries (Rana & Mittal, 2020). Unfortunately, the data derived from these processes is used not only for performance improvements, but also as a preventative measure to help reduce the risk of injury from repetitive strain or poor biomechanics (Hewett & Bates, 2017).

Instruction in the Management of Loads and Recovery

Maintaining a careful balance between rigorous exercise and rest is crucial to optimizing training loads. Wearable devices can protect an athlete from the pitfalls of overtraining. Wearable devices are an essential tool in quantifying the amount of training load an athlete is experiencing (Halson, 2014). In addition to quantifying the training load, wearable devices provide a more complete picture of an athlete's workload based on constant feedback on numerous metrics including, but not limited to, distance, intensity, heart rate variability (HRV), and muscular fatigue (Seshadri et al., 2019). Because of the continuous feedback provided from a wearable device, coaches can adjust training intensity and rest times with objective measures rather than subjective observations (Saw et al., 2016). Many elite sporting programs utilize trends in HRV and other stress markers in making decisions about changes in training loads. This process maximizes performance capabilities while potentially avoiding risk (Schneider et al., 2018).

Rehabilitation and Injury Prevention and Treatment

In terms of sports, one of the main uses of wearable technology is in injury prevention. Sensors that collect biomechanical data can detect deviations that may be associated with a higher likelihood of injury (Aroganam et al., 2019). With slight changes in gait, posture, or muscle engagement symmetry, there is potential to implement early intervention (Drijkoningen et al., 2015). There even comes the opportunities to implement the intervention. Furthermore, wearable technology has assisted injury rehabilitation. Patients who are recovering from accidents or surgery can assess their recovery, and they can confirm their rehabilitation activities are appropriately congruent and it avoids impacts to the injured areas of the body (Patel et al., 2012). Many athletes in high-performance sport now have a complete injury prevention and rehabilitation process (Coles, 2018). The process is achieved through continuous monitoring, early-warning analytics, and a feedback loop between the athlete and coach.

APPLICATIONS WITH REGARD TO PHYSICAL FITNESS FOR THE GENERAL PUBLIC

While it is accurate that professional athletes were the first to utilize wearable technology, the benefits extend to both normative and non-normative populations interested in fitness or general health and fitness. With the democratization of these devices, they have been infused into daily life, producing a better life for people across the globe (Aroganam et al., 2019).

Figure 4. A varied collection of people utilizing wearable gadgets during a community outdoor workout, demonstrating fitness technology democratization

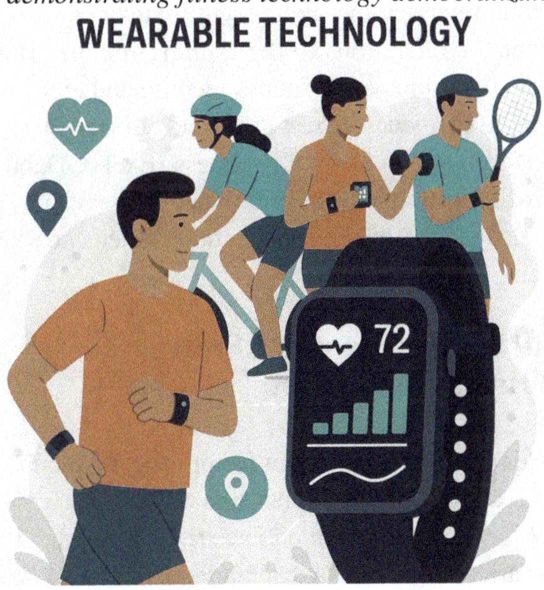

Providing Motivation for Active Lifestyles

Wearable technology is significantly influencing public health with the use of technology to encourage people to exercise regularly. Gamified challenges and reward systems act as motivators for fitness trackers and smart watches recipients to help them achieve daily step expectations and engage in a range of activities (Gao & Lee, 2019). Seeing progress toward individual fitness goals in real time brings a sense of accomplishment and propels an individual to continue using the technology and exercise regularly (Ramalho & Petrica, 2023). This instantaneous feedback loop is particularly important when individuals are transitioning from sedentary behaviors to more active routines that will benefit their overall health (Hamilton et al., 2008). The democratization of fitness technology has allowed for a larger uptake within the general population, as identified by the range of users in Figure 4.

Instructional and Coaching Programs Tailored to the Individual

Fitness apps which utilize data gathered from wearables have fundamentally changed the concept of personal coaching. Rather than following a generic template, contemporary fitness apps leverage a user's previous activity data to create tailored training plans for them (Kranz et al., 2013). The user's progress, levels of fatigue, and even variations in their day-to-day lifestyle are taken in to account when these systems make adaptable alterations to the training routine (Buttussi & Chittaro, 2008). These types of systems offer fitness recommendations in terms of the intensity of exercise, the frequency of rest days, and adjustments in dietary intake to ensure optimal performance and overall wellness (Exel & Dabnichki, 2024). This personalized approach to training not only enables users to achieve their fitness goals quicker, but also makes the experience more varied and sustainable over time (Buttussi & Chittaro, 2008).

Monitoring and Management of Chronic Diseases and Health Conditions

Wearable devices are also an important factor in health monitoring in addition to monitoring physical activity. Using fitness wearables as early warning systems for chronic diseases is increasingly commonplace (Mercer et al., 2016). Fitness wearables can assess a variety of physiological markers, including heart rate variability, sleep quality, stress levels, and more. Medical professionals and other professionals are now capturing wearable data as a part of their patient monitoring practice. By applying the patterns from daily activity logs, programs will search for possible issue like cardiovascular disease or bouts of poor sleep (Peake et al., 2018). The inclusion of technology to regular health monitoring has opened a whole new field of preventative medicine by enabling early interventions and ongoing health management (Oldenburg et al., 2015).

Community Involvement and Social Interaction

Another crucial element supporting the proliferation of wearables is the social dimension that it incorporates. There are many fitness platforms that allow users to not only share their achievements, compete in global challenges, but track their progress against peers (Lee et al., 2016). Such engagement in a group has been said to create community and challenge with mechanisms that support continued participation in healthy behaviours (Wallerstein et al., 2015). While social components developed in the platform may deliver participation benefits to individuals, they also foster support networks that contribute to the larger public health (Balatsoukas et al., 2015).

APPLICATIONS OF AI, DATA ANALYTICS, AND CLOUD INTEGRATION

Wearable technology is now a reality, thanks to data analytics, cloud computing, and artificial intelligence (AI) working together. Without algorithms to sort this data, these numbers would be meaningless and only go to waste (Putra et al., 2024). Current devices not only generate lots of data, but the sensors, too, have value.

The Use of Cloud Computing and Big Data

Wearables create real-time data streams and massive data sets that may be stored and analysed on the cloud (Greco et al., 2019). Cloud computing stores vast datasets, simplifying historical trend analysis and real-time feedback (Liu et al., 2018). These systems collect data from sports and fitness wearables and health monitoring devices to provide a complete health and performance picture (da Silva, 2024). Cloud-based platforms allow data sharing with coaches, health care providers, and friends and family, boosting wearable technology's value (Machorro-Cano et al., 2023).

Learning by Machines and Predictive Analytics Technology

When analysing the complex datasets collected from wearables, machine learning algorithms have become essential (Kubota et al., 2016). These prediction algorithms can detect anomalies, evaluate performance levels, and provide early warnings to injury potential (Claudino et al., 2019). With heart rate variability also being measured, machine learning algorithms can predict the onset of overtraining or identify the fatigue patterns developing (Leleu and Ward, 2024). A similar example is using pattern recognition algorithms to compare the present measures of performance to historical data, which inform individualized training modifications and injury avoidance recommendations (Van Eetvelde et al., 2021).

Feedback Systems That Are Real-Times

Real-time feedback systems for sports and fitness enthusiasts use data analytics and cloud computing (Xiao et al., 2023). Real-time feedback systems improve decision-making, whether they suggest a decreased workout intensity based on stress markers or a marathon speed shift (Smits et al., 2014). Due to artificial intelligence algorithms, users will soon have access to their performance data, which will be more accurate and relevant (Aminizadeh et al., 2024). Mobile apps and wearables enable this.

Digital Ecosystems that are Integrable

In brief, the health and fitness industry is in a digital transition that is seeing increases in interoperability across the many different digital systems. (Hermes et al. 2020). A holistic view of the user's health, fitness, and wellness can now be obtained by integrating wearable devices with nutrition apps, sleep trackers, and mental wellness programs (Del-Valle-Soto et al., 2024). Connected health systems can create comprehensive digital health platforms that can serve as a one-stop-shop for personalized health and wellness. These platforms are sustained by continuous data flow and cloud-based analytics (Land, 2018).

WEARABLE TECHNOLOGY HAS A NUMBER OF CHALLENGES AND LIMITATIONS

Wearable technology, although it has the potential to alter society, is confronted with a number of obstacles and constraints that can impede its efficiency and prevent it from gaining widespread adoption (Devi et al., 2023).

The Accuracy and Dependability of the Data

Among the many challenges faced by the industry of wearable technology, the validation and reliability of data retrieved from devices is one of the biggest challenges (Ginsburg et al., 2024). Variations in the data retrieved from a device can be attributed to variations in sensor quality, user mobility, and environmental disturbances (Calabrese et al., 2013). Despite the fact that developers and manufacturers are investing more money on more durable sensor technology and better calibration algorithms, they continue to search for consistency with their accuracy (Kanoun & Trankler, 2004).

Battery Capacity and Capacity for Use

Moreover, there is an important concern with the intrinsically linked nature of sensor performance and battery life (Huang et al., 2022). Many high-performance wearables require their batteries to be charged often to stay constant and usable, which is a challenging and disruptive time for the monitoring process. Some emergent and innovative strategies are in development, such as utilizing low-power sensors, harvesting energy with materials, and improving software on battery management (Chen, 2012). We include these strategies in our attempts to address the concerns laid out.

Right to Privacy, Data Ownership, and Data Security

With the large volume of personal data being collected, stored and shared between each other, it is no surprise that significant privacy and security issues would be raised (Arias et al., 2015). Users will typically be reluctant to adopt devices if they do not know how that data will be used, stored or shared when those devices are used. Additionally, while conversations about data ownership, particularly whether the data should be owned by the user or the device provider, continue to influence the legislative and ethical frameworks associated with this emerging industry (Hayes & Venkatraman, 2016).

Inequalities in Sociocultural Acceptance and Obstacles

One last thing to note is that typically technology breakthroughs do not deliver a promise of widespread adoption on their own. There are many socio-cultural barriers that drive the different responses to wearables from different cohorts (Bråtå et al., 2008). Barriers such as scepticism, complexity, and surveillance. Just as important as educating users about technology is educating fellow consumers about the value and risks of using wearable technology in their own lives (Aroganam et al., 2019).

Investigations of Real-World Applications and Case Studies

There are numerous real-world examples that clearly show us how wearable technology is changing the landscape of athletics and fitness as a whole. These case studies clearly illustrate the role of wearables in real world situations based on how they are used in actual practice.

Teams that Compete Professionally

Many professional sporting organizations have harnessed the use of wearable technology in both training and match-day protocols to enhance their performances. Taking professional soccer teams as an example, players are fitted with wearable sensors which can generate a large volume of physiological data and kinematic data across training and match situations (Aroganam et al., 2019). By analyzing the data they are able to tailor training programmes to players' individual needs and can optimize strategies by understanding players' fatigue and recovery during the match (Morais, 2023). In American football and rugby both sensor and internal sensors capture collision forces, loading and patterns of movement. This can provide immediate injury risk protection and optimize player engagement (Choudhury et al., 2024). The uses of wearable technology in professional sport have generated opportunities for enhanced individual performance, but the analyses have also enriched the strategy of teams and our understanding of injuries management and risk, demonstrating the diverse scope of even wearable technology in this paradigm (Aroganam et al., 2019). Here, Table 1, summarizes the diverse applications of wearable technology across professional sports, individual athletics, and health initiatives, demonstrating the versatility and impact of these devices in real-world scenarios

Table 1. Real-world applications of wearable technology across different sectors

Application Area	Technology Used	Key Benefits
Professional Soccer	Wearable sensors	Tailored training programs, strategy optimization
American Football/ Rugby	Collision sensors	Injury risk protection, player engagement optimization
Endurance Sports	GPS & speed sensors	Pacing optimization, recovery strategies
Corporate Wellness	Fitness trackers	Reduced healthcare costs, improved employee morale
Public Health	Aggregate wearable data	Optimized public spaces, population health monitoring

Competitors Who Compete on Their Own and Endurance Sports

Wearable devices are often utilized by endurance athletes for the quantification of training loads and pacing, and recovery components of training (Seshadri et al., 2019). Examples of endurance athletes include triathletes, bikers, and marathoners. Wearable devices permit endurance athletes to obtain highly specific data to facilitate the alteration of pacing strategies to optimize energy expenditure in the competition context, as well as timely and efficacious recovery strategies (Qi et al., 2024). Cyclists quantify whatever data they can obtain, for example bike and gear choice, using GPS data and speed sensors to make real-time and data-based decisions for modifying bi and gear choices based on current road conditions, and competing against their own success record (e.g., hill-climb distance, time, which suggests that sustainability plans could be relevant in both training and competition contexts) (Seshadri et al., 2021). It may be reasonable to suggest that the two examples presented illustrate how wearable technology limits injuries, limits training cycles, and sustain athletes' development by releasing performance and wellness outcomes *(Hu, 2024;* Seshadri et al., 2021).

Wellness Programs for Businesses and Community Health Initiatives

In addition to top or elite sport, other uses for wearable technology have emerged, notably as part of corporate wellness programs and public health initiatives. A number of companies have provided wearable fitness trackers to their employees as part of corporate wellness programs to help their employees with reducing stress, increasing their activity level, and improving their health overall. Employee measurable outcomes from these initiatives have resulted not only in the reduction of healthcare costs but also improved employee morale and performance (Doherty et al., 2025). Urban planners and public health department officials are also using aggregate data produced by wearables to develop city environments for people to live in. For example, they are finding ways to optimize public places for physical activity and track health trends for the entire population (Kohl et al., 2012). These types of initiatives show promise for the potential of wearable technology to assist many in improving their health beyond the individual user and under a larger population health strategy.

EMERGING TECHNOLOGIES AND EMERGING TRENDS IN THE FUTURE

At a time when wearables are rapidly evolving, there are a number of trends and innovations that will help to further elevate the level of wearables and their role in sports/fitness-related activities. The evolving nature of sensor technologies, paired with the expanding world of data analytics, as well as the rapid growth of immersive digital experiences will play a major role in shaping the future of wearables.

The Incorporation of Both Virtual Reality and Augmented Reality

The emergence of augmented reality (AR) and virtual reality (VR) systems is one of the most exciting and novel advancements in wearable technology (Kim et al., 2021). It is conceivable that in the future, wearables like smart glasses or contact lenses may be able to overlay real-time performance measures, or tactical information, directly into an athlete's field of view thus improving their feedback experience while performing (Geisen & Klatt, 2022). Like with virtual reality environments, virtual reality environments embedded with wearable data can give athletes the ability to recreate authentic game scenarios, providing a safe place to practice repetitive techniques and induce stress (Frangoudes et al.,, 2022). In terms of sports performance, these types of immersive technologies represent a new chapter in training methodology and are going to provide a multitude of sensory experiences (Cossich et al., 2023).

Recent Developments in Nanotechnology and Biosensors

Innovative biosensors that utilize the emerging applications of nanotechnology are about to change how health is monitored in sports and general fitness (Hosain, 2024) along with augmented and virtual reality. This new generation of sensors exists to provide an understanding of finer physiological responses at the cellular level. For example, with a sensor using nanotechnology, it may soon be able to determine possible dehydration, muscle overload, increased strain or tiredness; even before established metrics detect signal (Lee et al., 2017). Also, with the advancement of printed electronics and flexible circuits, this will give way to wearables that can be easily integrated into clothing or microelectronics that may take shape as skin patches for improved personal comfort and accuracy of data (Xu et al., 2019). The advances in augmenting reality with wearables will open up immersive training experiences with projected applications shown in Figure 5.

Figure 5. An athlete uses augmented reality smart glasses to track performance indicators while training, representing the future of immersive wearable technology

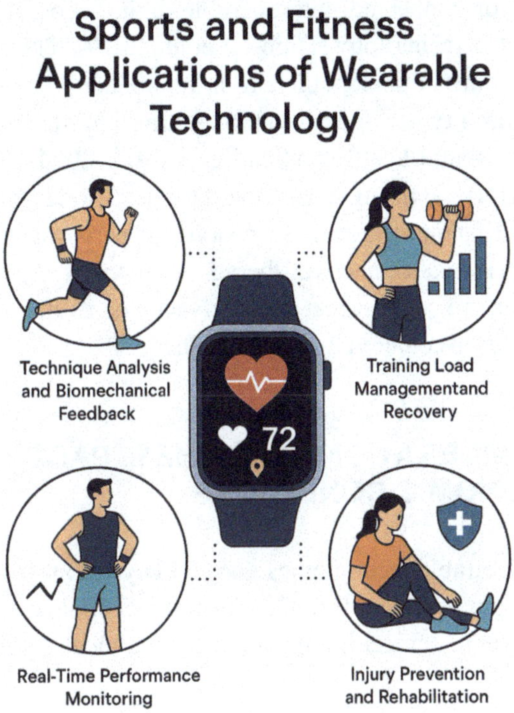

The Combination of Artificial Intelligence and Deterministic Modeling

In the coming years, the use of artificial intelligence in wearables will only increase (Gaffar & Gearhart, 2024). AI models are expected to support multiple sensor inputs to predict performance variability, predict injury risk and optimize the training program itself (Mateus et al., 2024). As deep learning and big data analytics find their way into the original AI models, this content will get even more personalized. The original user not only will see their physical parameters, but also the user's behavior and psychology (Boppiniti, 2020). Predictive modeling will change the landscape of wearables from reactive monitoring to proactive performance management (Ekundayo, 2024).

Interoperability and Digital Ecosystems that are Improved Together

The emergence of connected, interoperable digital ecosystems is a promising next step for wearable technology (Dinh-Le et al., 2019). The full development of a holistic understanding of personal health is versioned to be possible through the enhanced interoperability of fitness wearables, smart home devices, nutrition and sleep trackers, and mental health applications (Tariq, 2024). The ecosystem will ultimately take the form of a single dashboard for users to view all of their wellness information, supporting more informed decisions and personal treatments, ultimately leading to better performance and well-being (Spanakis et al., 2024). In health and fitness management, we suggest that these advancements may come in the way of a more integrative, user-centered approach (Zahab, 2024).

ETHICAL CONSIDERATIONS AND THE IMPACT ON SOCIOECONOMIC CONDITIONS

The impact of wearable technologies goes far beyond athletic performance and personal health; it is a full-blown trend. Wearable technologies will have massive impacts on the current socioeconomic conditions, ethical conditions, and current data practices.

Economic Advantages and the Development of Industry

The sector of wearables has been rapidly developing in recent years which has stirred economic activity across a range of industries including manufacture of tech devices, design of mobile applications, and also sports scientist consultancy (Gugliuzza & Drioli, 2013). Wearable devices have driven not only cross-sector collaboration between the technology sector and health care sector, but have also resulted in new entrepreneurial endeavors and revenue streams in the sports and recreation sector (Behr, 2018). In effect, the economic growth from wearables demonstrates not only technological success but consumer demand for greater personalization and connectivity in their fitness journeys (Sankar, 2024).

Data Ownership and Ethical Implications of the Situation

Ethical issues of privacy, data security, and data stewardship have emerged as paramount due to the staggering amounts of personal data produced by wearable technologies (Sivakumar et al., 2024). To ensure that the benefits of wearable technology do not undermine an individual's right to personal privacy, we need to develop clear and transparent data use policies, obtain proper user consent, and support robust cybersecurity protections. The challenge is to protect social rights and freedoms while also moving innovation forward with data-driven insights (Mone & Shakhlo, 2023). As policymakers and business leaders wrestle with the accountability issue of data ownership, this may very well be at the leading edge of workplace wearables.

Access and Fairness in Distribution

Wearable technology has enormous potential, but it can also compound existing inequalities. State-of-the-art devices and technologies are heavily dependent on wealth; the upper class or professional athletes have access to wearable technology (Hayes & Venkatraman, 2016). Yet, the digital divide can be partially addressed with programs, initiatives, and possibly, government incentives or subsidies, public health programs, corporate wellness programs (Roztocki et al., 2019). Ensuring fair access for the sake of social equity is important, but fair access is also important in order to maximize overall public health impact of wearable technology (Qi et al., 2024).

Acceptance in Society and Shifts in Cultural Norms

Issues with wearables' social and cultural acceptance exist in addition to issues related to technology and the economy (Lee et al., 2016). Users raised some questions around the benefits of continuous monitoring under the conditions that there are still issues related to surveillance and unwanted use of data (Sivakumar et al., 2024). Any strides that can be made in addressing issues of social and cultural acceptance certainly require some high level of education, transparency with manufacturers, and clear demonstration of value (e.g., performance enhancement or health benefits) when it comes to wearables (Nikitas et al., 2020). Once aware of which aspects of living in a data-driven world people can accept or reject, perhaps the broader use of wearable technology will evolve as society may see an increase in being comfortable in and about a world of possibility (Kalantari, 2017).

METHODS OF EXCELLENCE FOR THE IMPLEMENTATION OF WEARABLE TECHNOLOGY IN THE FIELD OF SPORTS AND FITNESS

In order to maximize the benefit of wearable technology, athletes, coaches, and everyday users must adopt best practices that ensure trustworthy data interpretation and a continuous improvement process.

Establishing Points of Reference and Keeping Track of Trends

Baseline measures in controlled environments are an essential step to take before incorporating wearables into a training plan (Cardinale & Varley, 2017). The sole purpose of baseline measures is to provide a reference point for assessing future improvements or initial signs of fatigue and overtraining *(Raslan, 2025)*. It is also crucial that continual monitoring and periodic reassessment of baseline values is carried out in order to adjust successfully to changes in physical characteristics and performance over time (Ajmal & Lodhi, 2015).

Approaches to Training that are Integrative

Wearable technology should not take the place of traditional approaches to coaching, nutrition, and mental support; they should coexist (Spender et al., 2019). A comprehensive training model wants to develop a complete picture of an athlete's performance state, by layering multiple variables together - physiological data, subjective measures of mood and energy (Wardiningsih, 2009). By combining sensor information with expert feedback, coaching models can be streamlined regarding training practices and become more perceptive to the nuances of an athlete's performance state, optimizing performance whilst minimizing risk of harm (Røsten et al., 2025).

Continuous Learning and Adjustment are Essential

The rapid rate of innovation in wearable technology means users need to continually educate themselves to stay aware of the new capacities and trends (Ometov et al., 2021). To support the utilization of emerging technology affordances, coaches, athletes, and fitness enthusiasts should be encouraged to engage in informal professional development. It does not matter whether the professional development need takes the form of workshops, engaging research, or co-developing tools with technology developers. Moreover, we can provide valuable contributions regularly through beta testing devices, or sharing experiences related to wearable technologies in user communities. This helps to tangibly impact the direction of future iterations of wearable devices so that technology innovation mirrors the everyday context of the users (Lu & O'Reilly, 2024).

CONCLUSION

Wearable technologies have changed the landscape for sports and fitness. Wearable technologies have come a long way from their simple beginnings as activity trackers. They have developed into sophisticated, data enabled, systems that allow athletes and health-conscious individuals looking to improve their performance, reduce their likelihood of injury, or improve their overall health. Progress in sensor technologies, real-time data analytic capabilities, and artificial intelligence have created possibilities for a new era of proactive training and individualized coaching. Looking ahead, wearable technologies will thrive with the advancements in augmented reality, nanotechnology-based biosensors, and collaborative digital ecosystems, allowing wearable technologies to remain agile and a part of contemporary sports and fitness performance. In addition, the potential for wearable technologies to achieve real improvement in public health and sport delivery are electrical! This enormity of potential must also be seen along with a number of issues related to data reliability, battery life, privacy, and access. The future of sports and fitness is going to be more individualized, inclusive, and revolutionary than it has ever been before. This is going to be accomplished by adopting a holistic strategy that incorporates technological innovation with traditional training methods and educated policy-making.

REFERENCES

Ajmal, M., & Lodhi, S. A. (2015). Organisational consciousness: A new paradigm for sustainable change management. *International Journal of Strategic Change Management*, *6*(3-4), 254–267. DOI: 10.1504/IJSCM.2015.075901

Alsubaei, F. S., Alshdadi, A. A., & Rizwanullah, M. (2025). Smart indoor monitoring for disabled individuals using an ensemble of deep learning models in an IoT environment. *Scientific Reports*, *15*(1), 1–16. DOI: 10.1038/s41598-025-00374-y

Alzahrani, A., & Ullah, A. (2024). Advanced biomechanical analytics: Wearable technologies for precision health monitoring in sports performance. *Digital Health*, *10*, 20552076241256745. DOI: 10.1177/20552076241256745

Aminizadeh, S., Heidari, A., Dehghan, M., Toumaj, S., Rezaei, M., Navimipour, N. J., & Unal, M. (2024). Opportunities and challenges of artificial intelligence and distributed systems to improve the quality of healthcare service. *Artificial Intelligence in Medicine*, *149*, 102779. DOI: 10.1016/j.artmed.2024.102779

Arias, O., Wurm, J., Hoang, K., & Jin, Y. (2015). Privacy and security in the Internet of Things and wearable devices. *IEEE Transactions on Multi-Scale Computing Systems*, *1*(2), 99–109. DOI: 10.1109/TMSCS.2015.2498605

Aroganam, G., Manivannan, N., & Harrison, D. (2019). Review on wearable technology sensors used in consumer sport applications. *Sensors (Basel)*, *19*(9), 1983. DOI: 10.3390/s19091983

Arvidsson, D., Fridolfsson, J., & Börjesson, M. (2019). Measurement of physical activity in clinical practice using accelerometers. *Journal of Internal Medicine*, *286*(2), 137–153. DOI: 10.1111/joim.12908

Balatsoukas, P., Kennedy, C. M., Buchan, I., Powell, J., & Ainsworth, J. (2015). The role of social network technologies in online health promotion: A narrative review of theoretical and empirical factors influencing intervention effectiveness. *Journal of Medical Internet Research*, *17*(6), e141. DOI: 10.2196/jmir.3662

. Behr, O. (2018). Fashion 4.0–digital innovation in the fashion industry. Journal of technology and innovation management, 2(1), 1-9.

Boppiniti, S. T. (2020). Big Data Meets Machine Learning: Strategies for Efficient Data Processing and Analysis in Large Datasets. *International Journal of Creative Research In Computer Technology and Design*, *2*(2).

Bråtå, H. O., Hagen, S. E., Hauge, A., Kotro, T., Orrenmaa, M., Power, D., & Repo, P. (2008). Users role in innovation processes in the sports equipment industry. *Industry and Innovation*, *15*(2), 179–197.

Buttussi, F., & Chittaro, L. (2008). MOPET: A context-aware and user-adaptive wearable system for fitness training. *Artificial Intelligence in Medicine*, *42*(2), 153–163. DOI: 10.1016/j.artmed.2007.11.004

Calabrese, F., Diao, M., Di Lorenzo, G., Ferreira, J.Jr, & Ratti, C. (2013). Understanding individual mobility patterns from urban sensing data: A mobile phone trace example. *Transportation Research Part C, Emerging Technologies*, *26*, 301–313. DOI: 10.1016/j.trc.2012.09.009

Cardinale, M., & Varley, M. C. (2017). Wearable training-monitoring technology: Applications, challenges, and opportunities. *International Journal of Sports Physiology and Performance*, *12*(s2), S2–S55. DOI: 10.1123/ijspp.2016-0423

Carminati, M., & Scandurra, G. (2021). Impact and trends in embedding field programmable gate arrays and microcontrollers in scientific instrumentation. *The Review of Scientific Instruments*, *92*(9), 091501. DOI: 10.1063/5.0050999

Chen, Y. K. (2012, January). Challenges and opportunities of internet of things. In 17th Asia and South Pacific design automation conference (pp. 383-388). IEEE. DOI: 10.1109/ASPDAC.2012.6164978

. Cheng, Y. (2023). Impact assessment of weather conditions on running performance and injury probability: An integrated analysis. Social Medicine and Health Management, 4(5).

Choudhury, R. D., Assumi, K. H., Nath, D., Kalai, S., & Nath, S. C. (2024). Progress of protective gear in preventing sports injuries. *Brazilian Journal of Development*, *10*(3), e67830–e67830. DOI: 10.34117/bjdv10n3-012

Claudino, J. G., Capanema, D. D. O., de Souza, T. V., Serrão, J. C., Machado Pereira, A. C., & Nassis, G. P. (2019). Current approaches to the use of artificial intelligence for injury risk assessment and performance prediction in team sports: A systematic review. *Sports Medicine - Open*, *5*(1), 1–12. DOI: 10.1186/s40798-019-0202-3

Coles, P. A. (2018). An injury prevention pyramid for elite sports teams. *British Journal of Sports Medicine*, *52*(15), 1008–1010. DOI: 10.1136/bjsports-2016-096697

Cossich, V. R., Carlgren, D., Holash, R. J., & Katz, L. (2023). Technological breakthroughs in sport: Current practice and future potential of artificial intelligence, virtual reality, augmented reality, and modern data visualization in performance analysis. *Applied Sciences (Basel, Switzerland)*, *13*(23), 12965. DOI: 10.3390/app132312965 da Silva, L. (2024). Wearable technology in sports monitoring performance and health metrics. [Journal of Sport Psychology]. *Revista de Psicología del Deporte*, *33*(2), 250–258.

Del-Valle-Soto, C., López-Pimentel, J. C., Vázquez-Castillo, J., Nolazco-Flores, J. A., Velázquez, R., Varela-Aldás, J., & Visconti, P. (2024). A comprehensive review of behavior change techniques in wearables and IoT: Implications for health and well-being. *Sensors (Basel)*, *24*(8), 2429. DOI: 10.3390/s24082429

Devi, D. H., Duraisamy, K., Armghan, A., Alsharari, M., Aliqab, K., Sorathiya, V., Das, S., & Rashid, N. (2023). 5G technology in healthcare and wearable devices: A review. *Sensors (Basel)*, *23*(5), 2519. DOI: 10.3390/s23052519

Dewhurst, P. (2015). *The science of the perfect swing*. Oxford University Press.

Dhatterwal, J. S., Kaswan, K. S., & Johri, P. (2024). IoT Sensor Communication Using Wireless Technology. In Wireless Communication Technologies (pp. 66-81). CRC Press. DOI: 10.1201/9781003389231-4

Dinh-Le, C., Chuang, R., Chokshi, S., & Mann, D. (2019). Wearable health technology and electronic health record integration: Scoping review and future directions. *JMIR mHealth and uHealth*, *7*(9), e12861. DOI: 10.2196/12861

Doherty, C., Baldwin, M., Lambe, R., Burke, D., & Altini, M. (2025). Readiness, recovery, and strain: An evaluation of composite health scores in consumer wearables. *Translational Exercise Biomedicine*, *2*(2), 128–144. DOI: 10.1515/teb-2025-0001

Dovgan, N. (2023). The Pivotal role of technology in enhancing Athletic Performance: Insights and future directions. Available at *SSRN* 4602857. DOI: 10.2139/ssrn.4602857

Drijkoningen, D., Caeyenberghs, K., Vander Linden, C., Van Herpe, K., Duysens, J., & Swinnen, S. P. (2015). Associations between muscle strength asymmetry and impairments in gait and posture in young brain-injured patients. *Journal of Neurotrauma*, *32*(17), 1324–1332. DOI: 10.1089/neu.2014.3787

Ekundayo, F. (2024). Real-time monitoring and predictive modelling in oncology and cardiology using wearable data and AI. *International Research Journal of Modernization in Engineering Technology and Science*, *6*(12).

Exel, J., & Dabnichki, P. (2024). Precision sports science: What is next for data analytics for athlete performance and well-being optimization? *Applied Sciences (Basel, Switzerland)*, *14*(8), 3361. DOI: 10.3390/app14083361

Fernández-Caramés, T. M., & Fraga-Lamas, P. (2018). Towards the Internet of smart clothing: A review on IoT wearables and garments for creating intelligent connected e-textiles. *Electronics (Basel)*, *7*(12), 405. DOI: 10.3390/electronics7120405

Frangoudes, F., Matsangidou, M., Schiza, E. C., Neokleous, K., & Pattichis, C. S. (2022). Assessing human motion during exercise using machine learning: A literature review. *IEEE Access : Practical Innovations, Open Solutions*, *10*, 86874–86903. DOI: 10.1109/ACCESS.2022.3198935

Fuller, D., Colwell, E., Low, J., Orychock, K., Tobin, M. A., Simango, B., Buote, R., Van Heerden, D., Luan, H., Cullen, K., Slade, L., & Taylor, N. G. (2020). Reliability and validity of commercially available wearable devices for measuring steps, energy expenditure, and heart rate: Systematic review. *JMIR mHealth and uHealth*, *8*(9), e18694. DOI: 10.2196/18694

Gaffar, S., & Gearhart, A. (2024). Artificial intelligence and wearable technology. In *Intelligence-based cardiology and cardiac surgery* (pp. 351–356). Academic Press. DOI: 10.1016/B978-0-323-90534-3.00056-1

Gagen, E. A. (2006). Measuring the soul: Psychological technologies and the production of physical health in Progressive Era America. *Environment and Planning. D, Society & Space*, *24*(6), 827–849. DOI: 10.1068/d373t

Gao, Z., & Lee, J. E. (2019). Emerging technology in promoting physical activity and health: Challenges and opportunities. *Journal of Clinical Medicine*, *8*(11), 1830. DOI: 10.3390/jcm8111830

Geisen, M., & Klatt, S. (2022). Real-time feedback using extended reality: A current overview and further integration into sports. *International Journal of Sports Science & Coaching*, *17*(5), 1178–1194. DOI: 10.1177/17479541211051006

Ginsburg, G. S., Picard, R. W., & Friend, S. H. (2024). Key issues as wearable digital health technologies enter clinical care. *The New England Journal of Medicine*, *390*(12), 1118–1127. DOI: 10.1056/NEJMra2307160

Greco, L., Ritrovato, P., & Xhafa, F. (2019). An edge-stream computing infrastructure for real-time analysis of wearable sensors data. *Future Generation Computer Systems*, *93*, 515–528. DOI: 10.1016/j.future.2018.10.058

Gugliuzza, A., & Drioli, E. (2013). A review on membrane engineering for innovation in wearable fabrics and protective textiles. *Journal of Membrane Science*, *446*, 350–375. DOI: 10.1016/j.memsci.2013.07.014

Halson, S. L. (2014). Monitoring training load to understand fatigue in athletes. *Sports Medicine (Auckland, N.Z.)*, *44*(S2, Suppl 2), 139–147. DOI: 10.1007/s40279-014-0253-z

Hamilton, M. T., Healy, G. N., Dunstan, D. W., Zderic, T. W., & Owen, N. (2008). Too little exercise and too much sitting: Inactivity physiology and the need for new recommendations on sedentary behavior. *Current Cardiovascular Risk Reports*, *2*(4), 292–298. DOI: 10.1007/s12170-008-0054-8

Haugen, T., & Buchheit, M. (2016). Sprint running performance monitoring: Methodological and practical considerations. *Sports Medicine (Auckland, N.Z.)*, *46*(5), 641–656. DOI: 10.1007/s40279-015-0446-0

Hayes, S. G., & Venkatraman, P. (Eds.). (2016). *Materials and technology for sportswear and performance apparel* (p. 293). CRC Press.

Heintzman, N. D. (2016). A digital ecosystem of diabetes data and technology: Services, systems, and tools enabled by wearables, sensors, and apps. *Journal of Diabetes Science and Technology*, *10*(1), 35–41. DOI: 10.1177/1932296815622453

Hermes, S., Riasanow, T., Clemons, E. K., Böhm, M., & Krcmar, H. (2020). The digital transformation of the healthcare industry: Exploring the rise of emerging platform ecosystems and their influence on the role of patients. *Business Research*, *13*(3), 1033–1069. DOI: 10.1007/s40685-020-00125-x

Hewett, T. E., & Bates, N. A. (2017). Preventive biomechanics: A paradigm shift with a translational approach to injury prevention. *The American Journal of Sports Medicine*, *45*(11), 2654–2664. DOI: 10.1177/0363546516686080

Hosain, M. N. (2024). IoT-enabled biosensors for real-time monitoring and early detection of chronic diseases. *Frontiers in Public Health*, •••, 12.

Howard, J., Murashov, V., Cauda, E., & Snawder, J. (2022). Advanced sensor technologies and the future of work. *American Journal of Industrial Medicine*, *65*(1), 3–11. DOI: 10.1002/ajim.23300

Hu, X., Ma, Z., Zhao, F., & Guo, S. (2024). Recent advances in self-powered wearable flexible sensors for human gait analysis. *Nanomaterials (Basel, Switzerland)*, *14*(14), 1173. DOI: 10.3390/nano14141173

Huang, J., Boles, S. T., & Tarascon, J. M. (2022). Sensing as the key to battery lifetime and sustainability. *Nature Sustainability*, *5*(3), 194–204. DOI: 10.1038/s41893-022-00859-y

. Jain, A. K., & Kumar, A. (2010). Biometrics of next generation: An overview. *Second generation biometrics, 12*(1), 2-3.

Kalantari, M. (2017). Consumers' adoption of wearable technologies: Literature review, synthesis, and future research agenda. *International Journal of Technology Marketing, 12*(3), 274–307. DOI: 10.1504/IJTMKT.2017.089665

Kanoun, O., & Trankler, H. R. (2004). Sensor technology advances and future trends. *IEEE Transactions on Instrumentation and Measurement, 53*(6), 1497–1501. DOI: 10.1109/TIM.2004.834613

Kim, H., Kwon, Y. T., Lim, H. R., Kim, J. H., Kim, Y. S., & Yeo, W. H. (2021). Recent advances in wearable sensors and integrated functional devices for virtual and augmented reality applications. *Advanced Functional Materials, 31*(39), 2005692. DOI: 10.1002/adfm.202005692

Kohl, H. W.III, Craig, C. L., Lambert, E. V., Inoue, S., Alkandari, J. R., Leetongin, G., & Kahlmeier, S. (2012). The pandemic of physical inactivity: Global action for public health. *Lancet, 380*(9838), 294–305. DOI: 10.1016/S0140-6736(12)60898-8

Kranz, M., Möller, A., Hammerla, N., Diewald, S., Plötz, T., Olivier, P., & Roalter, L. (2013). The mobile fitness coach: Towards individualized skill assessment using personalized mobile devices. *Pervasive and Mobile Computing, 9*(2), 203–215. DOI: 10.1016/j.pmcj.2012.06.002

Kubota, K. J., Chen, J. A., & Little, M. A. (2016). Machine learning for large-scale wearable sensor data in Parkinson's disease: Concepts, promises, pitfalls, and futures. *Movement Disorders, 31*(9), 1314–1326. DOI: 10.1002/mds.26693

Land, T. (2018). Powering up with technology to deliver transformative care and lower costs. *Frontiers of Health Services Management, 34*(3), 1–2. DOI: 10.1097/HAP.0000000000000029

Lee, E. C., Fragala, M. S., Kavouras, S. A., Queen, R. M., Pryor, J. L., & Casa, D. J. (2017). Biomarkers in sports and exercise: Tracking health, performance, and recovery in athletes. *Journal of Strength and Conditioning Research, 31*(10), 2920–2937. DOI: 10.1519/JSC.0000000000002122

Lee, J., Kim, D., Ryoo, H. Y., & Shin, B. S. (2016). Sustainable wearables: Wearable technology for enhancing the quality of human life. *Sustainability (Basel), 8*(5), 466. DOI: 10.3390/su8050466

Leleu, A., & Ward, T. E. (2024). Comparing automated and non-automated machine learning for binary classification of running fatigue using wearable sensors. In Advances in Digital Health and Medical Bioengineering (Vol. 111, pp. 401–408) DOI: 10.1007/978-3-031-62523-7_44

Liu, M., Xue, Z., Xu, X., Zhong, C., & Chen, J. (2018). Host-based intrusion detection system with system calls: Review and future trends. *ACM Computing Surveys, 51*(5), 1–36. DOI: 10.1145/3214304

Loncar-Turukalo, T., Zdravevski, E., da Silva, J. M., Chouvarda, I., & Trajkovik, V. (2019). Literature on wearable technology for connected health: Scoping review of research trends, advances, and barriers. *Journal of Medical Internet Research, 21*(9), e14017. DOI: 10.2196/14017

Lu, X., & O'Reilly, M. J. (2024). Applying User-Centered Design Principles to the Development of Wearable Technology in Fashion Design. *Studies in Art and Architecture, 3*(2), 10–16. DOI: 10.56397/SAA.2024.06.03

Machorro-Cano, I., Olmedo-Aguirre, J. O., Alor-Hernández, G., Rodríguez-Mazahua, L., Sánchez-Morales, L. N., & Pérez-Castro, N. (2023, December). Cloud-Based Platforms for Health Monitoring: A Review. In Informatics (Vol. 11, No. 1, p. 2). MDPI. DOI: 10.3390/informatics11010002

Mateus, N., Abade, E., Coutinho, D., Gómez, M. Á., Peñas, C. L., & Sampaio, J. (2024). Empowering the Sports Scientist with Artificial Intelligence in Training, Performance, and Health Management. *Sensors (Basel), 25*(1), 139. DOI: 10.3390/s25010139

. McGrath, M. J., Scanaill, C. N., McGrath, M. J., & Scanaill, C. N. (2013). Sensing and sensor fundamentals. Sensor technologies: Healthcare, wellness, and environmental applications, 15-50.

Mercer, K., Giangregorio, L., Schneider, E., Chilana, P., Li, M., & Grindrod, K. (2016). Acceptance of commercially available wearable activity trackers among adults aged over 50 and with chronic illness: A mixed-methods evaluation. *JMIR mHealth and uHealth, 4*(1), e4225. DOI: 10.2196/mhealth.4225

Micklewright, D. (2019). Decision-making, pacing, and performance in endurance sport. In *Endurance Performance in Sport* (pp. 47–69). Routledge. DOI: 10.4324/9781315167312-4

Mone, V., & Shakhlo, F. (2023). Health data on the go: Navigating privacy concerns with wearable technologies. *Legal Information Management, 23*(3), 179–188. DOI: 10.1017/S1472669623000427

Morais, J. E. (2023). Advances in Wearable Devices for Sports. *Applied Sciences (Basel, Switzerland), 13*(24), 13288. DOI: 10.3390/app132413288

Nash, C., Martindale, R., Collins, D., & Martindale, A. (2012). Parameterising expertise in coaching: Past, present and future. *Journal of Sports Sciences, 30*(10), 985–994. DOI: 10.1080/02640414.2012.682079

Nikitas, A., Michalakopoulou, K., Njoya, E. T., & Karampatzakis, D. (2020). Artificial intelligence, transport and the smart city: Definitions and dimensions of a new mobility era. *Sustainability (Basel), 12*(7), 2789. DOI: 10.3390/su12072789

Norris, M., Anderson, R., & Kenny, I. C. (2014). Method analysis of accelerometers and gyroscopes in running gait: A systematic review. *Proceedings of the Institution of Mechanical Engineers. Part P, Journal of Sports Engineering and Technology, 228*(1), 3–15. DOI: 10.1177/1754337113502472

Oldenburg, B., Taylor, C. B., O'Neil, A., Cocker, F., & Cameron, L. D. (2015). Using new technologies to improve the prevention and management of chronic conditions in populations. *Annual Review of Public Health, 36*(1), 483–505. DOI: 10.1146/annurev-publhealth-031914-122848

Oliveira, F., Costa, D. G., Assis, F., & Silva, I. (2024). Internet of Intelligent Things: A convergence of embedded systems, edge computing and machine learning. *Internet of Things : Engineering Cyber Physical Human Systems, 26*, 101153. DOI: 10.1016/j.iot.2024.101153

Ometov, A., Shubina, V., Klus, L., Skibińska, J., Saafi, S., Pascacio, P., Flueratoru, L., Gaibor, D. Q., Chukhno, N., Chukhno, O., Ali, A., Channa, A., Svertoka, E., Qaim, W. B., Casanova-Marqués, R., Holcer, S., Torres-Sospedra, J., Casteleyn, S., Ruggeri, G., & Lohan, E. S. (2021). A survey on wearable technology: History, state-of-the-art and current challenges. *Computer Networks, 193*, 108074. DOI: 10.1016/j.comnet.2021.108074

Otte, F. W., Davids, K., Millar, S. K., & Klatt, S. (2020). When and how to provide feedback and instructions to athletes?—How sport psychology and pedagogy insights can improve coaching interventions to enhance self-regulation in training. *Frontiers in Psychology, 11*, 1444. DOI: 10.3389/fpsyg.2020.01444

Patel, S., Park, H., Bonato, P., Chan, L., & Rodgers, M. (2012). A review of wearable sensors and systems with application in rehabilitation. *Journal of Neuroengineering and Rehabilitation, 9*(1), 1–17. DOI: 10.1186/1743-0003-9-21

Peake, J. M., Kerr, G., & Sullivan, J. P. (2018). A critical review of consumer wearables, mobile applications, and equipment for providing biofeedback, monitoring stress, and sleep in physically active populations. *Frontiers in Physiology*, *9*, 743. DOI: 10.3389/fphys.2018.00743

Poongodi, T., Krishnamurthi, R., Indrakumari, R., Suresh, P., & Balusamy, B. (2019). Wearable devices and IoT. In *A handbook of Internet of Things in biomedical and cyber physical system* (pp. 245–273). Springer International Publishing.

Putra, K. T., Arrayyan, A. Z., Hayati, N., Damarjati, C., Bakar, A., & Chen, H. C. (2024). A review on the application of internet of medical things in wearable personal health monitoring: A cloud-edge artificial intelligence approach. *IEEE Access : Practical Innovations, Open Solutions*, *12*, 21437–21452. DOI: 10.1109/ACCESS.2024.3358827

Qi, Y., Sajadi, S. M., Baghaei, S., Rezaei, R., & Li, W. (2024). Digital technologies in sports: Opportunities, challenges, and strategies for safeguarding athlete wellbeing and competitive integrity in the digital era. *Technology in Society*, *77*, 102496. DOI: 10.1016/j.techsoc.2024.102496

Ramalho, A., & Petrica, J. (2023). Knowledge in motion: A comprehensive review of evidence-based human kinetics. *International Journal of Environmental Research and Public Health*, *20*(11), 6020. DOI: 10.3390/ijerph20116020

Rana, M., & Mittal, V. (2020). Wearable sensors for real-time kinematics analysis in sports: A review. *IEEE Sensors Journal*, *21*(2), 1187–1207. DOI: 10.1109/JSEN.2020.3019016

Renaudin, V., Afzal, M. H., & Lachapelle, G. (2010, May). New method for magnetometers based orientation estimation. In *IEEE/ION Position, Location and Navigation Symposium* (pp. 348-356). IEEE. DOI: 10.1109/PLANS.2010.5507301

Røsten, S., Skirbekk, S., Tjønndal, A., Canini, D., Wågan, F., & Grut Vorpvik, J. (2025). Game changer: the technoscientific revolution in sports: by Rayvon Fouché, Johns Hopkins University Press, 2017.

Roztocki, N., Soja, P., & Weistroffer, H. R. (2019). The role of information and communication technologies in socioeconomic development: Towards a multidimensional framework. *Information Technology for Development*, *25*(2), 171–183. DOI: 10.1080/02681102.2019.1596654

Sankar, J. G. (2024). AI-driven marketing success stories: A case note of industry pioneers. In *AI-Driven Marketing Research and Data Analytics* (pp. 48–66). IGI Global. DOI: 10.4018/979-8-3693-2165-2.ch003

Saw, A. E., Main, L. C., & Gastin, P. B. (2016). Monitoring the athlete training response: subjective self-reported measures trump commonly used objective measures: a systematic review. *British Journal of Sports Medicine, 50*(5), 281–291. DOI: 10.1136/bjsports-2015-094758

Schneider, C., Hanakam, F., Wiewelhove, T., Döweling, A., Kellmann, M., Meyer, T., Pfeiffer, M., & Ferrauti, A. (2018). Heart rate monitoring in team sports—A conceptual framework for contextualizing heart rate measures for training and recovery prescription. *Frontiers in Physiology, 9*, 639. DOI: 10.3389/fphys.2018.00639

Seçkin, A. Ç., Ateş, B., & Seçkin, M. (2023). Review on Wearable Technology in sports: Concepts, Challenges and opportunities. *Applied Sciences (Basel, Switzerland), 13*(18), 10399. DOI: 10.3390/app131810399

Seshadri, D. R., Drummond, C., Craker, J., Rowbottom, J. R., & Voos, J. E. (2017). Wearable devices for sports: New integrated technologies allow coaches, physicians, and trainers to better understand the physical demands of athletes in real time. *IEEE Pulse, 8*(1), 38–43. DOI: 10.1109/MPUL.2016.2627240

Seshadri, D. R., Li, R. T., Voos, J. E., Rowbottom, J. R., Alfes, C. M., Zorman, C. A., & Drummond, C. K. (2019). Wearable sensors for monitoring the internal and external workload of the athlete. *NPJ Digital Medicine, 2*(1), 71. DOI: 10.1038/s41746-019-0149-2

Seshadri, D. R., Thom, M. L., Harlow, E. R., Gabbett, T. J., Geletka, B. J., Hsu, J. J., Drummond, C. K., Phelan, D. M., & Voos, J. E. (2021). Wearable technology and analytics as a complementary toolkit to optimize workload and to reduce injury burden. *Frontiers in Sports and Active Living, 2*, 630576. DOI: 10.3389/fspor.2020.630576

Sezer, O. B., Dogdu, E., & Ozbayoglu, A. M. (2017). Context-aware computing, learning, and big data in internet of things: A survey. *IEEE Internet of Things Journal, 5*(1), 1–27. DOI: 10.1109/JIOT.2017.2773600

. Shastry, K. A., & Shastry, A. (2023). An integrated deep learning and natural language processing approach for continuous remote monitoring in digital health. Decision analytics journal, 8, 100301.

Sivakumar, C. L. V., Mone, V., & Abdumukhtor, R. (2024). Addressing privacy concerns with wearable health monitoring technology. *Wiley Interdisciplinary Reviews. Data Mining and Knowledge Discovery, 14*(3), e1535. DOI: 10.1002/widm.1535

Smits, B. L., Pepping, G. J., & Hettinga, F. J. (2014). Pacing and decision making in sport and exercise: The roles of perception and action in the regulation of exercise intensity. *Sports Medicine (Auckland, N.Z.)*, *44*(6), 763–775. DOI: 10.1007/s40279-014-0163-0

Spanakis, M., Fragkiadaki, P., Renieri, E., Vakonaki, E., Fragkiadoulaki, I., Alegakis, A., Kiriakakis, M., Panagiotou, N., Ntoumou, E., Gratsias, I., Zoubaneas, E., Morozova, G. D., Ovchinnikova, M. A., Tsitsimpikou, C., Tsarouhas, K., Drakoulis, N., Skalny, A. V., & Tsatsakis, A. (2024). Advancing athletic assessment by integrating conventional methods with cutting-edge biomedical technologies for comprehensive performance, wellness, and longevity insights. *Frontiers in Sports and Active Living*, *5*, 1327792. DOI: 10.3389/fspor.2023.1327792

Spender, A., Bullen, C., Altmann-Richer, L., Cripps, J., Duffy, R., Falkous, C., Farrell, M., Horn, T., Wigzell, J., & Yeap, W. (2019). Wearables and the internet of things: Considerations for the life and health insurance industry. *British Actuarial Journal*, *24*, e22. DOI: 10.1017/S1357321719000072

Tariq, M. U. (2024). Advanced wearable medical devices and their role in transformative remote health monitoring. In *Transformative approaches to patient literacy and healthcare innovation* (pp. 308–326). IGI Global. DOI: 10.4018/979-8-3693-3661-8.ch015

Van Eetvelde, H., Mendonça, L. D., Ley, C., Seil, R., & Tischer, T. (2021). Machine learning methods in sport injury prediction and prevention: A systematic review. *Journal of Experimental Orthopaedics*, *8*(1), 1–15. DOI: 10.1186/s40634-021-00346-x

Vanini, S., Faraci, F., Ferrari, A., & Giordano, S. (2016). Using barometric pressure data to recognize vertical displacement activities on smartphones. *Computer Communications*, *87*, 37–48. DOI: 10.1016/j.comcom.2016.02.011

. Wallerstein, N., Minkler, M., Carter-Edwards, L., Avila, M., & Sanchez, V. (2015). Improving health through community engagement, community organization, and community building. Health behavior: theory, research and practice, 5.

Wardiningsih, W. (2009). Study of comfort properties of natural and synthetic knitted fabrics in different blend ratios for winter active sportswear (Doctoral dissertation, RMIT University).

Williams, C. A., & Winsley, R. (2020). Non-Functional Overreaching and Overtraining in Young Swimmers. In *High Performance Youth Swimming* (pp. 215–227). Routledge. DOI: 10.4324/9780429465598-18

Wong, C., Zhang, Z. Q., Lo, B., & Yang, G. Z. (2015). Wearable sensing for solid biomechanics: A review. *IEEE Sensors Journal*, *15*(5), 2747–2760. DOI: 10.1109/JSEN.2015.2393883

Xiao, L., Cao, Y., Gai, Y., Liu, J., Zhong, P., & Moghimi, M. M. (2023). Review on the application of cloud computing in the sports industry. *Journal of Cloud Computing (Heidelberg, Germany)*, *12*(1), 152. DOI: 10.1186/s13677-023-00531-6

Xu, K., Lu, Y., & Takei, K. (2019). Multifunctional skin-inspired flexible sensor systems for wearable electronics. *Advanced Materials Technologies*, *4*(3), 1800628. DOI: 10.1002/admt.201800628

. Zahab, M. (2024). The Potential of Smart Wearables on Health Management:-User-Centric Qualitative Research on Human-Technology Interaction based on the Theory of Planned Behavior.

Zainab, H., Khan, A. H., Khan, R., & Hussain, H. K. (2024). Integration of AI and wearable devices for continuous cardiac health monitoring. *International Journal of Multidisciplinary Sciences and Arts*, *3*(4), 123–139. DOI: 10.47709/ijmdsa.v3i4.4956

Compilation of References

Abdoli, S. (2022). Application of fuzzy-logic for design assessment of complex engineering systems in the early design stages. *Journal of Engineering Design*, *33*(3), 234–258. DOI: 10.1080/09544828.2022.2031918

Abernethy, A., Adams, L., Barrett, M., Bechtel, C., Brennan, P., Butte, A., Faulkner, J., Fontaine, E., Friedhoff, S., Halamka, J., Howell, M., Johnson, K., Long, P., McGraw, D., Miller, R., Lee, P., Perlin, J., Rucker, D., Sandy, L., & Valdes, K. (2022). The promise of digital health: Then, now, and the future. *NAM Perspectives*, *6*(22). Advance online publication. DOI: 10.31478/202206e PMID: 36177208

Abilevitch, L., Mizrahi, L., Cohen, G., Kenig, S., & Amir, E. (2023). Polyaniline for smart textile applications. In *IntechOpen eBooks*. DOI: 10.5772/intechopen.1001939

Adamu, M. F., Tesfaye, T., Berhanu, B., & Simegnaw, A. (2025). Electronic textiles. *Textile Progress*, *57*(1), 1–71. DOI: 10.1080/00405167.2025.2468595

Agarwal, S., Dohare, A. K., Saxena, P., Singh, J., Singh, I., & Sahu, U. K. (2025). HDL-ACO: Hybrid deep learning and ant colony optimization for ocular optical coherence tomography image classification. *Scientific Reports*, *15*(1), 5888. DOI: 10.1038/s41598-025-89961-7 PMID: 39966596

Ahmed, S. S. (2015). *A review of thermo regulated fabric using phase change material.* Textile Today.

Ahn, J., Yun, S., Kwon, J.-W., & Kim, W.-T. (2024). Literacy deep reinforcement learning-based federated digital twin scheduling for the software-defined factory. *Electronics (Basel)*, *13*(22), 4452. DOI: 10.3390/electronics13224452

Ahsan, M., Khan, S., & Ali, U. (2022). Explainable AI models for healthcare diagnostics. *International Journal of Advanced Sciences and Computing*, *1*(1), 8–15.

Ajmal, M., & Lodhi, S. A. (2015). Organisational consciousness: A new paradigm for sustainable change management. *International Journal of Strategic Change Management, 6*(3-4), 254–267. DOI: 10.1504/IJSCM.2015.075901

Alam, M. R., Islam, T., Repon, M. R., & Hoque, M. E. (2022). Carbonbased polymer nanocomposites for electronic textiles (etextiles). In *Elsevier eBooks* (pp. 443–482). https://doi.org/DOI: 10.1016/b9780128244920.000209

Alamer, F. A., Althagafy, K., Alsalmi, O., Aldeih, A., Alotaiby, H., Althebaiti, M., Alghamdi, H., Alotibi, N., Saeedi, A., Zabarmawi, Y., Hawsawi, M., & Alnefaie, M. A. (2022). Review on PEDOT:PSSBased Conductive Fabric. *ACS Omega, 7*(40), 35371–35386. DOI: 10.1021/acsomega.2c01834 PMID: 36249401

Alamer, F. A., & Alzahrani, A. (2025). Advancements and innovations in textile engineering: An InDepth analysis of SWCNTs and PANI integration in smart textile technologies. *Composites. Part A, Applied Science and Manufacturing, 108872*, 108872. Advance online publication. DOI: 10.1016/j.compositesa.2025.108872

Alanen, A. M., R€ Ais€ Anen, A. M., Benson, L. C., & Pasanen, K. (2021). *The use of inertial measurement units for analyzing change of direction movement in sports: A scoping review*. DOI: 10.1177/17479541211003064

Alghamdi, W., Mayakannan, S., Sivasankar, G. A., & Reddy, C. V. K. (2023). Turbulence modeling through deep learning: An in-depth study of Wasserstein GANs. In 2023 4th International Conference on Smart Electronics and Communication (ICOSEC) (pp. 793–797). IEEE. DOI: 10.1109/ICOSEC58147.2023.10275878

Ali, M., Giri, S., Liu, S., & Yang, Q. (2025). Digital twin-enabled real-time control in robotic additive manufacturing via soft actor-critic reinforcement learning. *arXiv preprint* arXiv:2501.18016. https://arxiv.org/abs/2501.18016

Ali, I., Islam, M. R., Yin, J., Eichhorn, S. J., Chen, J., Karim, N., & Afroj, S. (2024). Advances in smart photovoltaic textiles. *ACS Nano, 18*(5), 3871–3915. DOI: 10.1021/acsnano.3c10033 PMID: 38261716

Ali, J., Sharma, P., Kumar, V., Sharma, M., & Singh, R. (2024). Mitigating distributed denial of service (DDoS) attacks in cloud networks using neural networks. In *Proceedings of the 2024 Second International Conference on Advanced Computing & Communication Technologies (ICACCTech)* (pp. 161–166). IEEE. DOI: 10.1109/ICACCTech65084.2024.00036

Ali, J., Sharma, P., Kumar, V., Sharma, M., & Singh, R. (2024). Real-time anomaly detection and threat mitigation in IoT networks using convolutional neural networks (CNNs) for enhanced security. In *Proceedings of the 2024 Second International Conference on Advanced Computing & Communication Technologies (ICACCTech)* (pp. 619–624). IEEE. DOI: 10.1109/ICACCTech65084.2024.00104

Ali, S. M., Noghanian, S., Khan, Z. U., Alzahrani, S., Alharbi, S., Alhartomi, M., & Alsulami, R. (2025). Wearable and flexible sensor devices: Recent advances in designs, fabrication methods, and applications. *Sensors (Basel)*, *25*(5), 1377. Advance online publication. DOI: 10.3390/s25051377 PMID: 40096147

Aliyana, A. K., Yang, D., Tangsirinaruenart, O., & Stylios, G. K. (2024). A garment-integrated textile stitch-based strain sensor device, IoT-enabled for enhanced wearable sportswear applications. *Results in Engineering*, *23*, 102794. Advance online publication. DOI: 10.1016/j.rineng.2024.102794

Allison, L., Hoxie, S., & Andrew, T. L. (2017). Towards seamlessly integrated textile electronics: Methods to coat fabrics and fibers with conducting polymers for electronic applications. *Chemical Communications*, *53*(53), 7182–7193. DOI: 10.1039/C7CC02592K PMID: 28555696

Almeida, L., & Ramos, D. (2017). Health and safety concerns of textiles with nanomaterials. *IOP Conference Series. Materials Science and Engineering*, *254*, 102002. DOI: 10.1088/1757-899X/254/10/102002

Alowais, S. A., Alghamdi, S. S., Alsuhebany, N., Alqahtani, T., Alshaya, A. I., Almohareb, S. N., Aldairem, A., Alrashed, M., Bin Saleh, K., Badreldin, H. A., Al Yami, M. S., Al Harbi, S., & Albekairy, A. M. (2023). Revolutionising healthcare: The role of artificial intelligence in clinical practice. *BMC Medical Education*, *23*(1), 689. Advance online publication. DOI: 10.1186/s12909-023-04698-z PMID: 37740191

Alphin, K. L., Sisson, O. M., Hudgins, B. L., Noonan, C. D., & Bunn, J. A. (2020). Accuracy assessment of a GPS device for maximum sprint speed. *International Journal of Exercise Science*, *13*(4), 273.

Alsubaei, F. S., Alshdadi, A. A., & Rizwanullah, M. (2025). Smart indoor monitoring for disabled individuals using an ensemble of deep learning models in an IoT environment. *Scientific Reports*, *15*(1), 1–16. DOI: 10.1038/s41598-025-00374-y

Altun, F., Bayar, A., Hamzat, A. K., Asmatulu, R., Ali, Z., & Asmatulu, E. (2025). AI-Driven Innovations in 3D Printing: Optimization, Automation, and Intelligent Control. *Journal of Manufacturing and Materials Processing*, *9*(10), 329. DOI: 10.3390/jmmp9100329

Alzahrani, A., & Ullah, A. (2024). Advanced biomechanical analytics: Wearable technologies for precision health monitoring in sports performance. *Digital Health*, *10*, 20552076241256745. DOI: 10.1177/20552076241256745

Aminizadeh, S., Heidari, A., Dehghan, M., Toumaj, S., Rezaei, M., Navimipour, N. J., & Unal, M. (2024). Opportunities and challenges of artificial intelligence and distributed systems to improve the quality of healthcare service. *Artificial Intelligence in Medicine*, *149*, 102779. DOI: 10.1016/j.artmed.2024.102779

Amjadi, M., Kyung, K. U., Park, I., & Sitti, M. (2016). Stretchable, skin-mountable, and wearable strain sensors and their potential applications: A review. *Advanced Functional Materials*, *26*(11), 1678–1698. DOI: 10.1002/adfm.201504755

Anandhakrishnan, T., & Jaisakthi, S. M. (2022). Deep convolutional neural networks for image-based tomato leaf disease detection. *Sustainable Chemistry and Pharmacy*, *30*, 100793. DOI: 10.1016/j.scp.2022.100793

Angelucci, A., Cavicchioli, M., Cintorrino, I. A., Lauricella, G., Rossi, C., Strati, S., & Aliverti, A. (2021). Smart textiles and sensorized garments for physiological monitoring: A review of available solutions and techniques. *Sensors (Basel)*, *21*(3), 814. DOI: 10.3390/s21030814 PMID: 33530403

Anik, H. R., Tushar, S. I., Mahmud, S., Khadem, A. H., Sen, P., & Akter, M. (2024). Into the revolution of NanoFusion: Merging high performance and aesthetics by nanomaterials in textile finishes. *Advanced Materials Interfaces*, *12*(1), 2400368. Advance online publication. DOI: 10.1002/admi.202400368

Antony Jose, S., Tonner, A., Feliciano, M., Roy, T., Shackleford, A., & Menezes, P. L. (2025). Smart Manufacturing for High-Performance Materials: Advances, Challenges, and Future Directions. *Materials (Basel)*, *18*(10), 2255. DOI: 10.3390/ma18102255 PMID: 40428991

Apilo, O., Lasanen, M., Heikkilä, M., Kivimäki, A., & Viitala, R. (2019). Towards industry 4.0: NB-IoT and LTE-M for wearable IoT in sports. *Sensors (Basel)*, *19*(18), 3921. DOI: 10.3390/s19183921 PMID: 31514439

Apoorva, S., Nguyen, N.-T., & Sreejith, K. R. (2024). Recent developments and future perspectives of microfluidics and smart technologies in wearable devices. *Lab on a Chip*, *24*(7), 1833–1866. DOI: 10.1039/D4LC00089G PMID: 38476112

Apte, S., Karami, H., Vallat, C., Gremeaux, V., & Aminian, K. (2023). In-field assessment of change-of-direction ability with a single wearable sensor. *Scientific Reports*, *13*(1), 4518. DOI: 10.1038/s41598-023-30773-y

Ariansyah, D., Isnan, M., Rahutomo, R., & Pardamean, B. (2023). Digital twin (DT) smart city for air quality management. *Procedia Computer Science, 227*, 524–533. DOI: 10.1016/j.procs.2023.10.554

Arias, O., Wurm, J., Hoang, K., & Jin, Y. (2015). Privacy and security in the Internet of Things and wearable devices. *IEEE Transactions on Multi-Scale Computing Systems, 1*(2), 99–109. DOI: 10.1109/TMSCS.2015.2498605

Armanini, A., Colombo, A., Conci, N., Daldoss, M., Fontanelli, D., & Palopoli, L. (2012). Wireless sensor networks and video analysis for scalable people tracking. *2012 5th International Symposium on Communications, Control and Signal Processing*, 1–4.

Aroganam, G., Manivannan, N., & Harrison, D. (2019). Review on wearable technology sensors used in consumer sport applications. *Sensors (Basel), 19*(9), 1983. DOI: 10.3390/s19091983

Arulampalam, M. S., Maskell, S., Gordon, N., & Clapp, T. (2002). A tutorial on particle filters for online nonlinear/non-Gaussian Bayesian tracking. *IEEE Transactions on Signal Processing, 50*(2), 174–188. DOI: 10.1109/78.978374

Arvidsson, D., Fridolfsson, J., & Börjesson, M. (2019). Measurement of physical activity in clinical practice using accelerometers. *Journal of Internal Medicine, 286*(2), 137–153. DOI: 10.1111/joim.12908

Arya, C., Yamsani, N., Kumar, M., Singh, P., & Bhagat, V. K. (2023). A concise review of MRI feature extraction and classification with kernel functions. In 2023 3rd International Conference on Innovative Sustainable Computational Technologies (CISCT) (pp. 1–5). IEEE. DOI: 10.1109/CISCT57197.2023.10351471

Assenza, G., Fioravanti, C., Guarino, S., & Petrassi, V. (2020). New perspectives on wearable devices and electronic health record systems. In *2020 IEEE International Workshop on Metrology for Industry 4.0 & IoT* (pp. 740–745). IEEE. DOI: 10.1109/MetroInd4.0IoT48571.2020.9138170

Assuad, C. S. A., Leirmo, T., & Martinsen, K. (2022). Proposed framework for flexible de- and remanufacturing systems using cyber-physical systems, additive manufacturing, and digital twins. *Procedia CIRP, 112*, 226–231. DOI: 10.1016/j.procir.2022.09.076

Aughey, R. J. (2011). Applications of GPS technologies to field sports. *International Journal of Sports Physiology and Performance, 6*(3), 295–310. DOI: 10.1123/ijspp.6.3.295

Aungst, , T., Franzese, C., & Kim, Y. (2021). Digital health implications for clinical pharmacists services: A primer on the current landscape and future concerns. *Journal of the American College of Clinical Pharmacy: JAACP*, *4*(4), 514–524. DOI: 10.1002/jac5.1382

Avilés, R., Diego, B. S., Pino-Ortega, J., & Castellano, J. (2023). Assessment of a New Change of Direction Detection Algorithm Based on Inertial Data. *Sensors (Basel)*, *23*(6), 3095. https://www.proquest.com/scholarly-journals/assessment-new-change-direction-detection/docview/2791700160/se-2. DOI: 10.3390/s23063095

Axisa, F., Schmitt, P. M., Gehin, C., Delhomme, G., McAdams, E., & Dittmar, A. (2005). Flexible technologies and smart clothing for citizen medicine, home healthcare, and disease prevention. *IEEE Transactions on Information Technology in Biomedicine*, *9*(3), 325–336. DOI: 10.1109/TITB.2005.854505 PMID: 16167686

Azeem, M., Shahid, M., Masin, I., & Petru, M. (2024). Design and development of textilebased wearable sensors for realtime biomedical monitoring; a review. *Journal of the Textile Institute*, •••, 1–16. DOI: 10.1080/00405000.2024.2318500

Babiuch, M., Foltynek, P., & Smutny, P. (2019). Using the ESP32 microcontroller for data processing. *Proceedings of the 2019 20th International Carpathian Control Conference (ICCC)* (pp. 1–6). DOI: 10.1109/CarpathianCC.2019.8765944

Bafna, M., Thakur, H. K., Singh, J., Reddy, P. S., Ekvitayavetchanukul, P., & Rana, M. (2025). Predictive Modeling of Solar Power Generation Using Deep Learning for Real-Time Optimization in Photovoltaic System. In 2025 International Conference on Intelligent Control, Computing and Communications (IC3) (pp. 104-109). IEEE. DOI: 10.1109/IC363308.2025.10956811

Bah, A., & Davud, M. (2022). Analysis of breast cancer classification with machine learning based algorithms. In 2022 2nd International Conference on Computing and Machine Intelligence (ICMI). IEEE. DOI: 10.1109/ICMI55296.2022.9873696

Balatsoukas, P., Kennedy, C. M., Buchan, I., Powell, J., & Ainsworth, J. (2015). The role of social network technologies in online health promotion: A narrative review of theoretical and empirical factors influencing intervention effectiveness. *Journal of Medical Internet Research*, *17*(6), e141. DOI: 10.2196/jmir.3662

Balloch, A. S., Meghji, M., Newton, R. U., Hart, N. H., Weber, J. A., Ahmad, I., & Habibi, D. (2020). Assessment of a Novel Algorithm to Determine Change-of-Direction Angles While Running Using Inertial Sensors. *Journal of Strength and Conditioning Research*, *34*(1), 134–144. DOI: 10.1519/JSC.0000000000003064

Bangaru, S. S., Wang, C., & Aghazadeh, F. (2020). Data Quality and Reliability Assessment of Wearable EMG and IMU Sensor for Construction Activity Recognition. *Sensors 2020, Vol. 20, Page 5264, 20*(18), 5264. DOI: 10.3390/s20185264

Banos, O., Galvez, J. M., Damas, M., Pomares, H., & Rojas, I. (2014). Window Size Impact in Human Activity Recognition. *Sensors 2014, Vol. 14, Pages 6474-6499, 14*(4), 6474–6499. DOI: 10.3390/s140406474

Batwara, A., Kayande, R. A., & Kediya, S. (2025). Critical Drivers of AI Integration in Industrial Processes for Enhancing Smart Sustainable Manufacturing: Decision-Making Framework. Process Integration and Optimization for Sustainability, 1-23.

Bayindir, M., Abouraddy, A. F., Arnold, J., Joannopoulos, J. D., & Fink, Y. (2006). Thermal-sensing fiber devices by multimaterial codrawing. *Advanced Materials, 18*(7), 845–849. DOI: 10.1002/adma.200502106

Baz, A., Logeshwaran, J., Natarajan, Y., & Patel, S. (2024). Deep fuzzy nets approach for energy efficiency optimization in smart grids. *Applied Soft Computing, 161*, 111724. DOI: 10.1016/j.asoc.2024.111724

Beeby, S. P., Torah, R. N., Wagih, M., Isaia, B., Black, S., Saunders, J., & Yang, K. (2024). Heterogeneous E-Textiles: Materials, manufacturing and sustainability. *Advanced Materials Technologies*. Advance online publication. DOI: 10.1002/admt.202400844

Beke, A., & Kumbasar, T. (2023). More Than Accuracy: A Composite Learning Framework for Interval Type-2 Fuzzy Logic Systems. *IEEE Transactions on Fuzzy Systems, 31*(3), 734–744. DOI: 10.1109/TFUZZ.2022.3188920

Benson, L. C., Clermont, C. A., Bošnjak, E., & Ferber, R. (2018). The use of wearable devices for walking and running gait analysis outside of the lab: A systematic review. *Gait & Posture, 63*, 124–138. DOI: 10.1016/j.gaitpost.2018.04.047

Bhattacharjee, S., Macintyre, C. R., Wen, X., Bahl, P., Kumar, U., Chughtai, A. A., & Joshi, R. (2020). Nanoparticles incorporated graphenebased durable cotton fabrics. *Carbon, 166*, 148–163. DOI: 10.1016/j.carbon.2020.05.029

Bhatt, C., Kumar, I., Vijayakumar, V., Singh, K. U., & Kumar, A. (2021). The state of the art of deep learning models in medical science and their challenges. *Multimedia Systems, 27*(4), 599–613. DOI: 10.1007/s00530-020-00694-1

Bian, S., Liu, M., Zhou, B., & Lukowicz, P. (2022). The State-of-the-Art Sensing Techniques in Human Activity Recognition: A Survey. *Sensors 2022, Vol. 22, Page 4596, 22*(12), 4596. DOI: 10.3390/s22124596

Bijalwan, J. G., Ravi, V., Bijalwan, A., Alahmadi, T. J., & Diwakar, M. (2024). Navigating the future of secure and efficient intelligent transportation systems using AI and blockchain. The Open Transportation Journal, 18(1).

Bijalwan, J. G., Ravi, V., Bijalwan, A., Alahmadi, T. J., & Diwakar, M. (2024). Navigating the future of secure and efficient intelligent transportation systems using AI and blockchain. *The Open Transportation Journal, 18*(1). *(No DOI provided)*

Bishop, D. J., & Girard, O. (2013). Determinants of team-sport performance: Implications for altitude training by team-sport athletes. *British Journal of Sports Medicine, 47*(Suppl 1), i17–i21. DOI: 10.1136/bjsports-2013-092950

Bluetooth, S. I. G. (2019). *Bluetooth 5: Feature enhancements* (White paper).

Bol, D., Schramme, M., Moreau, L., Xu, P., Dekimpe, R., Saeidi, R., Haine, T., Frenkel, C., & Flandre, D. (2021). SleepRunner: A 28-nm FDSOI ULP cortex-M0 MCU with ULL SRAM and UFBR PVT compensation for 2.6–3.6-μW/DMIPS 40–80-MHz active mode and 131-nW/kB fully retentive deep-sleep mode. *IEEE Journal of Solid-State Circuits, 56*(7), 2256–2269. DOI: 10.1109/JSSC.2021.3056219

Bonilla, V., Campoverde, B., & Yoo, S. G. (2023). A systematic literature review of LoRaWAN: Sensors and applications. *Sensors (Basel), 23*(20), 8440. DOI: 10.3390/s23208440 PMID: 37896533

Boppiniti, S. T. (2020). Big Data Meets Machine Learning: Strategies for Efficient Data Processing and Analysis in Large Datasets. *International Journal of Creative Research In Computer Technology and Design*, 2(2).

Bråtå, H. O., Hagen, S. E., Hauge, A., Kotro, T., Orrenmaa, M., Power, D., & Repo, P. (2008). Users role in innovation processes in the sports equipment industry. *Industry and Innovation, 15*(2), 179–197.

Bressane, A., Da Silva Garcia, A., De Castro, M., Xerfan, S., Ruas, G., & Negri, R. (2024). Fuzzy Machine Learning Applications in Environmental Engineering: Does the Ability to Deal with Uncertainty Really Matter? *Sustainability (Basel), 16*(11), 4525. Advance online publication. DOI: 10.3390/su16114525

Buchheit, M., & Simpson, B. M. (2017). Player-tracking technology: Half-full or half-empty glass? *International Journal of Sports Physiology and Performance, 12*(s2), S2–S35. DOI: 10.1123/ijspp.2016-0499

Buechley, L., & Eisenberg, M. (2008). The LilyPad Arduino: Toward wearable engineering for everyone. *IEEE Pervasive Computing*, 7(2), 12–15. DOI: 10.1109/MPRV.2008.38

Bulić, P., Kojek, G., & Biasizzo, A. (2019). Data transmission efficiency in Bluetooth Low Energy versions. *Sensors (Basel), 19*(17), 3746. DOI: 10.3390/s19173746 PMID: 31470669

Buttussi, F., & Chittaro, L. (2008). MOPET: A context-aware and user-adaptive wearable system for fitness training. *Artificial Intelligence in Medicine, 42*(2), 153–163. DOI: 10.1016/j.artmed.2007.11.004

Calabrese, F., Diao, M., Di Lorenzo, G., Ferreira, J.Jr, & Ratti, C. (2013). Understanding individual mobility patterns from urban sensing data: A mobile phone trace example. *Transportation Research Part C, Emerging Technologies, 26*, 301–313. DOI: 10.1016/j.trc.2012.09.009

Camomilla, V., Bergamini, E., Fantozzi, S., & Vannozzi, G. (2018). Trends supporting the in-field use of wearable inertial sensors for sport performance evaluation: A systematic review. *Sensors (Basel), 18*(3), 873. DOI: 10.3390/s18030873

Canny, J. (1986). A Computational Approach to Edge Detection. In *IEEE TRANSACTIONS ON PATTERN ANALYSIS AND MACHINE INTELLIGENCE* (Issue 6). DOI: 10.1109/TPAMI.1986.4767851

Cardinale, M., & Varley, M. C. (2017). Wearable training-monitoring technology: Applications, challenges, and opportunities. *International Journal of Sports Physiology and Performance, 12*(s2), S2–S55. DOI: 10.1123/ijspp.2016-0423

Carminati, M., & Scandurra, G. (2021). Impact and trends in embedding field programmable gate arrays and microcontrollers in scientific instrumentation. *The Review of Scientific Instruments, 92*(9), 091501. DOI: 10.1063/5.0050999

Carvalho, L. I., & Sofia, R. C. (2020). A Review on Scaling Mobile Sensing Platforms for Human Activity Recognition: Challenges and Recommendations for Future Research. *IoT 2020, Vol. 1, Pages 451-473, 1*(2), 451–473. DOI: 10.3390/iot1020025

Castano, L. M., & Flatau, A. B. (2014). Smart fabric sensors and e-textile technologies: A review. *Smart Materials and Structures, 23*(5), 053001. DOI: 10.1088/0964-1726/23/5/053001

Chambers, R., Gabbett, T. J., Cole, M. H., & Beard, A. (2015). The Use of Wearable Microsensors to Quantify Sport-Specific Movements. *Sports Medicine (Auckland, N.Z.), 45*(7), 1065–1081. DOI: 10.1007/s40279-015-0332-9

Chaudhary, S., Verma, S., Sharma, V., Jain, S., & Kumar, P. (2024). A multi-model approach to hydrological event analysis and forecasting. In *2024 International Conference on Emerging Technologies and Innovation for Sustainability (EmergIN)* (pp. 485–490). IEEE. DOI: 10.1109/EmergIN63207.2024.10961633

Chen, H., Xu, J., Tao, Z., & Gao, W. (2020). Recent advances in self-powered wearable electronics. arXiv preprint arXiv:2009.05918.

Chen, Y. K. (2012, January). Challenges and opportunities of internet of things. In 17th Asia and South Pacific design automation conference (pp. 383-388). IEEE. DOI: 10.1109/ASPDAC.2012.6164978

Cheng, X., Shen, D., Zheng, K., Wu, Z., Shi, L., & Hu, X. (2025). A wearable strain sensor for medical rehabilitation based on piezoresistive knitting textile. *Sensors and Actuators. A, Physical*, *387*, 116379. DOI: 10.1016/j.sna.2025.116379

Chen, J., Huang, Y., Zhang, N., Zou, H., Liu, R., Tao, C., Fan, X., & Wang, Z. L. (2016). Micro-cable structured textile for simultaneously harvesting solar and mechanical energy. *Nature Energy*, *1*(10), 16138. DOI: 10.1038/nenergy.2016.138

Chen, M., Ma, Y., Li, Y., Wu, D., Zhang, Y., & Youn, C. H. (2024). Smart manufacturing in the context of Industry 4.0: A review. *Engineering*, *10*(1), 1–12. DOI: 10.1016/j.eng.2019.11.011 PMID: 32288964

Chen, P., Wang, D., & Gan, Z. (2023). Flexible and small textile antenna for UWB wireless body area network. *Micromachines*, *14*(4), 718. DOI: 10.3390/mi14040718 PMID: 37420951

Chen, X., Liu, A., Wei, Z., Ukkonen, L., & Virkki, J. (2017). Experimental study on strain reliability of embroidered passive UHF RFID textile tag antennas and interconnections. *Journal of Engineering*, *8493405*, 1–7. Advance online publication. DOI: 10.1155/2017/8493405

Chen, Z., Ming, X., Zhou, T., & Chang, Y. (2020). Sustainable supplier selection for smart supply chain considering internal and external uncertainty: An integrated rough-fuzzy approach. *Applied Soft Computing*, *87*, 106004. DOI: 10.1016/j.asoc.2019.106004

Cherenack, K., & van Pieterson, L. (2012). Smart textiles: Challenges and opportunities. *Journal of Applied Physics*, *112*(9), 091301. DOI: 10.1063/1.4742728

Choi, H. W., Shin, D., Yang, J., Lee, S., Figueiredo, C., Sinopoli, S., Ullrich, K., Jovančić, P., Marrani, A., Momentè, R., Gomes, J., Branquinho, R., Emanuele, U., Lee, H., Bang, S. Y., Jung, S., Han, S. D., Zhan, S., Harden-Chaters, W., & Kim, J. M. (2022). Smart textile lighting/display system with multifunctional fibre devices for large scale smart home and IoT applications. *Nature Communications*, *13*(1), 814. Advance online publication. DOI: 10.1038/s41467-022-28459-6 PMID: 35145096

Chollet, F. (2021). *Deep learning with Python* (2nd ed.). Manning Publications.

Choudhury, R. D., Assumi, K. H., Nath, D., Kalai, S., & Nath, S. C. (2024). Progress of protective gear in preventing sports injuries. *Brazilian Journal of Development, 10*(3), e67830–e67830. DOI: 10.34117/bjdv10n3-012

Chowdhury, S., Mishra, S., Sharma, N., Reddy, P. S., Sharma, Y. K., & Tyagi, A. (2024). Machine Learning Models for Enhancing Time Series Demand Forecasting in Business Intelligence. In *2024 International Conference on Emerging Technologies and Innovation for Sustainability (EmergIN)* (pp. 78-84). IEEE. DOI: 10.1109/EmergIN63207.2024.10961109

Clarke, R., Aspe, R. R., & Hughes, J. (n.d.). *Change of Direction Technical Models: Biomechanical Principles*. Retrieved 21 February 2024, from https://www.researchgate.net/publication/328561167

Claudino, J. G., Capanema, D. D. O., de Souza, T. V., Serrão, J. C., Machado Pereira, A. C., & Nassis, G. P. (2019). Current approaches to the use of artificial intelligence for injury risk assessment and performance prediction in team sports: A systematic review. *Sports Medicine - Open, 5*(1), 1–12. DOI: 10.1186/s40798-019-0202-3

Cleary, F., Srisa-An, W., Henshall, D. C., & Balasubramaniam, S. (2023). Emerging AI technologies inspiring the next generation of ETextiles. *IEEE Access : Practical Innovations, Open Solutions, 11*, 56494–56508. DOI: 10.1109/ACCESS.2023.3282184

Coles, P. A. (2018). An injury prevention pyramid for elite sports teams. *British Journal of Sports Medicine, 52*(15), 1008–1010. DOI: 10.1136/bjsports-2016-096697

Colyer, S. L., Evans, M., Cosker, D. P., & Salo, A. I. T. (2018). A Review of the Evolution of Vision-Based Motion Analysis and the Integration of Advanced Computer Vision Methods Towards Developing a Markerless System. *Sports Medicine - Open, 4*(1), 1–15. DOI: 10.1186/s40798-018-0139-y

Coosemans, J., Hermans, B., & Puers, R. (2006). Integrating wireless ECG monitoring in textiles. *Sensors and Actuators. A, Physical, 130–131*, 48–53. DOI: 10.1016/j.sna.2005.10.052

Cossich, V. R., Carlgren, D., Holash, R. J., & Katz, L. (2023). Technological breakthroughs in sport: Current practice and future potential of artificial intelligence, virtual reality, augmented reality, and modern data visualization in performance analysis. *Applied Sciences (Basel, Switzerland), 13*(23), 12965. DOI: 10.3390/app132312965

Cui, G., & Wang, C. (2025). Applications and development trends of textile materials in sports: A review. *Alexandria Engineering Journal, 126*, 491–506. DOI: 10.1016/j.aej.2025.04.094

Cummins, C., Orr, R., O'Connor, H., & West, C. (2013). Global positioning systems (GPS) and microtechnology sensors in team sports: A systematic review. *Sports Medicine (Auckland, N.Z.)*, *43*(10), 1025–1042. DOI: 10.1007/s40279-013-0069-2

da Silva, L. (2024). Wearable technology in sports monitoring performance and health metrics. [Journal of Sport Psychology]. *Revista de Psicología del Deporte*, *33*(2), 250–258.

Dalton-Brown, S. (2020). The ethics of medical AI and the physician-patient relationship. *Cambridge Quarterly of Healthcare Ethics*, *29*(1), 115–121. DOI: 10.1017/S0963180119000847 PMID: 31858938

Das, A. K. (2016). On partially included intuitionistic fuzzy rough relations. *Afrika Matematika*, *27*(5-6), 993–1001. DOI: 10.1007/s13370-016-0395-2

Das, A. K. (2018). Weighted fuzzy soft multiset and decision-making. *International Journal of Machine Learning and Cybernetics*, *9*(5), 787–794. DOI: 10.1007/s13042-016-0607-y

Das, A. K., Das, R., Das, S., Debnath, B. K., Granados, C., Shil, B., & Das, R. (2025c). A comprehensive study of neutrosophic SuperHyper BCI-semigroups and their algebraic significance. *Transactions on Fuzzy Sets and Systems*, *4*(2), 80–101. DOI: 10.71602/tfss.2025.1198050

Das, A. K., Das, R., Datta, R., & Granados, C. (2025e). An Innovative Approach to Group Decision-Making Based on Weighted Hypersoft Expert System. In Smarandache, F., Jdid, M., & Leyva-Vazquez, M. (Eds.), *Neutrosophic and Plithogenic Inventory Models for Applied Mathematics* (pp. 483–518). IGI Global Scientific Publishing., DOI: 10.4018/979-8-3693-3204-7.ch019

Das, A. K., & Granados, C. (2021). An advanced approach to fuzzy soft group decision-making using weighted average ratings. *SN Computer Science*, *2*(6), 471. DOI: 10.1007/s42979-021-00873-5

Das, A. K., & Granados, C. (2022a). A new fuzzy parameterized intuitionistic fuzzy soft multiset theory and group decision-making. *Journal of Current Science and Technology*, *12*(3), 547–567. DOI: 10.14456/jcst.2022.42

Das, A. K., & Granados, C. (2022b). FP-intuitionistic multi fuzzy N-soft set and its induced FP-Hesitant N soft set in decision-making. *Decision Making: Applications in Management and Engineering*, *5*(1), 67–89. DOI: 10.31181/dmame181221045d

Das, A. K., & Granados, C. (2023a). IFP-intuitionistic multi fuzzy N-soft set and its induced IFP-hesitant N-soft set in decision-making. *Journal of Ambient Intelligence and Humanized Computing*, *14*(8), 10143–10152. DOI: 10.1007/s12652-021-03677-w

Das, A. K., & Granados, C. (2023b). Preference intuitionistic fuzzy rough relation and its theoretical approach. *Journal of the Indian Mathematical Society*, *90*(3-4), 199–212.

Das, A. K., Granados, C., & Bhattacharya, J. (2022). Some new operations on fuzzy soft sets and their applications in decision-making. *Songklanakarin Journal of Science and Technology*, *44*(2), 440–449. DOI: 10.14456/sjst-psu.2022.61

Das, A. K., Gupta, N., & Granados, C. (2023). Weighted hesitant bipolar-valued fuzzy soft set in decision-making. *Songklanakarin Journal of Science and Technology*, *45*(6), 681–690. https://sjst.psu.ac.th/journal/45-6/10.pdf

Das, A. K., Gupta, N., Mahmood, T., Tripathy, B. C., Das, R., & Das, S. (2025a). An efficient water quality evaluation model using weighted hesitant fuzzy soft sets for water pollution rating. In Kumar, A., Kumar, P., Rathee, S., & Kumar, B. (Eds.), *Mechatronics: Concepts, Tools, Applications, and New Trends* (1st ed., p. 17). CRC Press., DOI: 10.1201/9781003494478-10

Das, A. K., Gupta, N., Mahmood, T., Tripathy, B. C., Das, R., & Das, S. (2025b). An innovative fuzzy multi-criteria decision making model for analyzing anthropogenic influences on urban river water quality. *Iran Journal of Computer Science*, *8*(1), 103–124. DOI: 10.1007/s42044-024-00211-x

Das, A. K., Mahmood, T., Das, R., & Das, S. (2025d). Optimizing Inventory Levels in Retail: Fuzzy Logic-Based Decision Support Systems for Adaptive Inventory Management. In Jana, S., Basu Mallik, B., Sarkar, A., & Jana, C. (Eds.), *Applications of Fuzzy Logic in Decision Making and Management Science. Information Systems Engineering and Management* (Vol. 29). Springer., DOI: 10.1007/978-3-031-77719-6_8

Das, D., Chatterjee, B., Maity, S., & Sen, S. (2019). Enabling covert body area networks using electro-quasistatic human body communication. *Scientific Reports*, *9*(1), 4160. DOI: 10.1038/s41598-018-38303-x PMID: 30858385

Davenport, T., & Kalakota, R. (2019). The potential for artificial intelligence in healthcare. *Future Healthcare Journal*, *6*(2), 94–98. DOI: 10.7861/futurehosp.6-2-94 PMID: 31363513

De Fazio, R., Mastronardi, V., De Vittorio, M., & Visconti, P. (2023). Wearable sensors and smart devices to monitor rehabilitation parameters and sports performance: An overview. *Sensors (Basel), 23*(4), 1856. DOI: 10.3390/s23041856 PMID: 36850453

De Iuliis, M., Kammouh, O., & Cimellaro, G. (2022). Measuring and improving community resilience: A fuzzy logic approach. *International Journal of Disaster Risk Reduction, 78*, 103118. Advance online publication. DOI: 10.1016/j.ijdrr.2022.103118

De Rossi, D., Lorussi, F., Mazzoldi, A., Orsini, P., & Scilingo, E. P. (2003). Active dressware: wearable kinesthetic systems. In *Sensors and Sensing in Biology and Engineering* (pp. 379–392). Springer Vienna. DOI: 10.1007/978-3-7091-6025-1_26

De, A., & Mitra, D. (2021). Bandwidth-enhanced ultra-wideband wearable textile antenna using denim substrate for various WBAN and IoT applications. *Radio Science, 56*(5), e2021RS007315. DOI: DOI: 10.1029/2021RS007315

Del-Valle-Soto, C., López-Pimentel, J. C., Vázquez-Castillo, J., Nolazco-Flores, J. A., Velázquez, R., Varela-Aldás, J., & Visconti, P. (2024). A comprehensive review of behavior change techniques in wearables and IoT: Implications for health and well-being. *Sensors (Basel), 24*(8), 2429. DOI: 10.3390/s24082429

Delves, R. I. M., Aughey, R. J., Ball, K., & Duthie, G. M. (2021). The Quantification of Acceleration Events in Elite Team Sport: a Systematic Review. *Sports Medicine - Open 2021 7:1, 7*(1), 1–35. DOI: 10.1186/s40798-021-00332-8

Demrozi, F., Pravadelli, G., Bihorac, A., & Rashidi, P. (2020). Human Activity Recognition Using Inertial, Physiological and Environmental Sensors: A Comprehensive Survey. *IEEE Access : Practical Innovations, Open Solutions, 8*, 210816–210836. DOI: 10.1109/ACCESS.2020.3037715

Devi, D. H., Duraisamy, K., Armghan, A., Alsharari, M., Aliqab, K., Sorathiya, V., Das, S., & Rashid, N. (2023). 5G technology in healthcare and wearable devices: A review. *Sensors (Basel), 23*(5), 2519. DOI: 10.3390/s23052519

Dewhurst, P. (2015). *The science of the perfect swing*. Oxford University Press.

Dhatterwal, J. S., Kaswan, K. S., & Johri, P. (2024). IoT Sensor Communication Using Wireless Technology. In Wireless Communication Technologies (pp. 66-81). CRC Press. DOI: 10.1201/9781003389231-4

Di Giammarco, M., Martinelli, F., Mercaldo, F., & Santone, A. (2022). High grade brain cancer segmentation by means of deep learning. *Procedia Computer Science, 207*, 1633–1640. DOI: 10.1016/j.procs.2022.09.220

Diab, M. S., & Rodriguez-Villegas, E. (2022). Embedded machine learning using microcontrollers in wearable and ambulatory systems for health and care applications: A review. *IEEE Access : Practical Innovations, Open Solutions*, *10*, 98450–98474. DOI: 10.1109/ACCESS.2022.3206782

Dinh-Le, C., Chuang, R., Chokshi, S., & Mann, D. (2019). Wearable health technology and electronic health record integration: Scoping review and future directions. *JMIR mHealth and uHealth*, *7*(9), e12861. DOI: 10.2196/12861

Doherty, C., Baldwin, M., Lambe, R., Burke, D., & Altini, M. (2025). Readiness, recovery, and strain: An evaluation of composite health scores in consumer wearables. *Translational Exercise Biomedicine*, *2*(2), 128–144. DOI: 10.1515/teb-2025-0001

Dolez, P. I. (2021). Energy harvesting Materials and structures for smart textile Applications: Recent progress and Path forward. *Sensors (Basel)*, *21*(18), 6297. DOI: 10.3390/s21186297 PMID: 34577509

Dolez, P. I., Decaens, J., Buns, T., Lachapelle, D., & Vermeersch, O. (2020, April). Applications of smart textiles in occupational health and safety. []. IOP Publishing.]. *IOP Conference Series. Materials Science and Engineering*, *827*(1), 012014. DOI: 10.1088/1757-899X/827/1/012014

Dong, H., Falis, M., Whiteley, W., Alex, B., Matterson, J., Ji, S., Chen, J., & Wu, H. (2022). Automated clinical coding: What, why, and where we are? *NPJ Digital Medicine*, *5*(1), 159. Advance online publication. DOI: 10.1038/s41746-022-00705-7 PMID: 36273236

Dos, T., Thomas, C., & Jones, P. A. (2021). The effect of angle on change of direction biomechanics: Comparison and inter-task relationships. *Journal of Sports Sciences*, *39*(22), 2618–2631. DOI: 10.1080/02640414.2021.1948258

Dovgan, N. (2023). The Pivotal role of technology in enhancing Athletic Performance: Insights and future directions. Available at *SSRN* 4602857. DOI: 10.2139/ssrn.4602857

Drijkoningen, D., Caeyenberghs, K., Vander Linden, C., Van Herpe, K., Duysens, J., & Swinnen, S. P. (2015). Associations between muscle strength asymmetry and impairments in gait and posture in young brain-injured patients. *Journal of Neurotrauma*, *32*(17), 1324–1332. DOI: 10.1089/neu.2014.3787

Dulal, M., Afroj, S., Ahn, J., Cho, Y., Carr, C., Kim, I., & Karim, N. (2022). Toward sustainable wearable electronic textiles. *ACS Nano*, *16*(12), 19755–19788. DOI: 10.1021/acsnano.2c07723 PMID: 36449447

Dunne, L. E., Brady, S., Smyth, B., & Diamond, D. (2005). Initial development and testing of a novel foam-based pressure sensor for wearable sensing. *Journal of Neuroengineering and Rehabilitation*, *2*(1), 4. DOI: 10.1186/1743-0003-2-4 PMID: 15740623

Dunn, J., Runge, R., & Snyder, M. (2018). Wearables and the medical revolution. *Personalized Medicine*, *15*(5), 429–448. DOI: 10.2217/pme-2018-0044 PMID: 30259801

Ehrmann, G., & Ehrmann, A. (2021). Electronic textiles. *Encyclopedia*, *1*(1), 115–130. DOI: 10.3390/encyclopedia1010013

Ekundayo, F. (2024). Real-time monitoring and predictive modelling in oncology and cardiology using wearable data and AI. *International Research Journal of Modernization in Engineering Technology and Science*, *6*(12).

Eom, W., Shin, H., Ambade, R. B., Lee, S. H., Lee, K. H., Kang, D. J., & Han, T. H. (2020). Largescale wetspinning of highly electroconductiveMXene fibers. *Nature Communications*, *11*(1), 2825. Advance online publication. DOI: 10.1038/s41467-020-16671-1 PMID: 32499504

Erdoğdu, A., Dayı, F., Yildiz, F., Yanik, A., & Ganji, F. (2025). Combining Fuzzy Logic and Genetic Algorithms to Optimize Cost, Time and Quality in Modern Agriculture. *Sustainability (Basel)*, *17*(7), 2829. Advance online publication. DOI: 10.3390/su17072829

Eshuis, R., Firat, M., & Kaymak, U. (2021). Modeling uncertainty in declarative artifact-centric process models using fuzzy logic. *Information Sciences*, *579*, 845–862. DOI: 10.1016/j.ins.2021.07.075

Esteva, A., Robicquet, A., Ramsundar, B., Kuleshov, V., DePristo, M., Chou, K., Cui, C., Corrado, G., Thrun, S., & Dean, J. (2019). A guide to deep learning in healthcare. *Nature Medicine*, *25*(1), 24–29. DOI: 10.1038/s41591-018-0316-z PMID: 30617335

ETSI. (2020). *EN 303 645 V2.1.1 (2020-06): Cyber Security for Consumer Internet of Things—Baseline requirements.*

European Commission. (2013). DEPHOTEX: Development of Photovoltaic Textiles. Cordis EU Projects. https://cordis.europa.eu/project/id/214459

Exel, J., & Dabnichki, P. (2024). Precision sports science: What is next for data analytics for athlete performance and well-being optimization? *Applied Sciences (Basel, Switzerland)*, *14*(8), 3361. DOI: 10.3390/app14083361

Fan, C., & Gao, F. (2021). Enhanced Human Activity Recognition Using Wearable Sensors via a Hybrid Feature Selection Method. *Sensors 2021, Vol. 21, Page 6434, 21*(19), 6434. DOI: 10.3390/s21196434

Farina, D., Merletti, R., & Enoka, R. M. (2004). The extraction of neural strategies from the surface EMG. *Journal of Applied Physiology, 96*(4), 1486–1495. DOI: 10.1152/japplphysiol.01070.2003 PMID: 15016793

Faruk, O., Ahmed, A., Khadem, A. H., Jia, L., & Sun, L. (2025). *GrapheneFunctionalized textile composites for wearable joule heating applications*. Advanced Nanocomposites., DOI: 10.1016/j.adna.2025.03.001

Fayed, K., Ragab, M. A. F., & Rodríguez, E. (2024). The Use of Wearable Technologies and Their Impact on Athletic Performance. *Studies in Systems. Decision and Control, 550*, 225–238. DOI: 10.1007/978-3-031-66218-8_17

Fernández-Caramés, T. M., & Fraga-Lamas, P. (2018). Towards the Internet of smart clothing: A review on IoT wearables and garments for creating intelligent connected e-textiles. *Electronics (Basel), 7*(12), 405. DOI: 10.3390/electronics7120405

Fleury, A., Noury, N., & Vuillerme, N. (2007). A fast algorithm to track changes of direction of a person using magnetometers. *Annual International Conference of the IEEE Engineering in Medicine and Biology - Proceedings*, 2311–2314. DOI: 10.1109/IEMBS.2007.4352788

Frangoudes, F., Matsangidou, M., Schiza, E. C., Neokleous, K., & Pattichis, C. S. (2022). Assessing human motion during exercise using machine learning: A literature review. *IEEE Access : Practical Innovations, Open Solutions, 10*, 86874–86903. DOI: 10.1109/ACCESS.2022.3198935

Fuller, D., Colwell, E., Low, J., Orychock, K., Tobin, M. A., Simango, B., Buote, R., Van Heerden, D., Luan, H., Cullen, K., Slade, L., & Taylor, N. G. (2020). Reliability and validity of commercially available wearable devices for measuring steps, energy expenditure, and heart rate: Systematic review. *JMIR mHealth and uHealth, 8*(9), e18694. DOI: 10.2196/18694

Gaffar, S., & Gearhart, A. (2024). Artificial intelligence and wearable technology. In *Intelligence-based cardiology and cardiac surgery* (pp. 351–356). Academic Press. DOI: 10.1016/B978-0-323-90534-3.00056-1

Gagen, E. A. (2006). Measuring the soul: Psychological technologies and the production of physical health in Progressive Era America. *Environment and Planning. D, Society & Space, 24*(6), 827–849. DOI: 10.1068/d373t

Gangadharan, S. M. P., Dharani, M., Thapliyal, N., Yamsani, N., & Singh, P. (2023). Comparative analysis of deep learning-based brain tumor prediction models using MRI scan. In 2023 3rd International Conference on Innovative Sustainable Computational Technologies (CISCT) (pp. 1–6). IEEE. DOI: 10.1109/CISCT57197.2023.10351227

Gao, R., Wang, L., & Xu, X. (2024). Digital twin-driven smart manufacturing: A review. *Journal of Manufacturing Systems*, *62*, 1–13. DOI: 10.1016/j.jmsy.2024.01.001

Gao, Z., & Lee, J. E. (2019). Emerging technology in promoting physical activity and health: Challenges and opportunities. *Journal of Clinical Medicine*, *8*(11), 1830. DOI: 10.3390/jcm8111830

Garg, N., Kukreja, S., Saini, M., Singh, I., & Mall, S. (2023). IoT-enhanced agricultural water management system utilizing conventional machine learning methods. In 2023 International Conference on Modeling, Simulation & Intelligent Computing (MoSICom) (pp. 554–559). IEEE. DOI: 10.1109/MoSICom59118.2023.10458818

Garg, N., Saini, M., Singh, I., & Mall, S. (2023). Designing smart grid and irrigation management systems for information sharing for enhanced efficiency and sustainability. In 2023 International Conference on Modeling, Simulation & Intelligent Computing (MoSICom) (pp. 532–537). IEEE. DOI: 10.1109/MoSICom59118.2023.10458827

Garg, K. K., Keshari, N., Biswas, S., Majumder, J., & Gangopadhyay, S. (2026). Predicting Consumer Buying Behavior Using Hybrid Machine Learning Models: A Multi-dimensional Analysis of Online Retail Data. In Swaroop, A., Virdee, B., Correia, S. D., & Polkowski, Z. (Eds.), *Proceedings of Data Analytics and Management. ICDAM 2025. Lecture Notes in Networks and Systems* (Vol. 1615). Springer., DOI: 10.1007/978-3-032-04222-4_13

Gaur, A., Singh, S. K., & Saxena, P. (2024). Performance analysis of deepfake text detection techniques on social media. In *2024 International Conference on Distributed Computing and Optimization Techniques (ICDCOT)* (pp. 1–6). IEEE. DOI: 10.1109/ICDCOT61034.2024.10515626

Geisen, M., & Klatt, S. (2022). Real-time feedback using extended reality: A current overview and further integration into sports. *International Journal of Sports Science & Coaching*, *17*(5), 1178–1194. DOI: 10.1177/17479541211051006

Ginsburg, G. S., Picard, R. W., & Friend, S. H. (2024). Key issues as wearable digital health technologies enter clinical care. *The New England Journal of Medicine*, *390*(12), 1118–1127. DOI: 10.1056/NEJMra2307160

Godara, A., Mezzo, L., Luizi, F., Warrier, A., Lomov, S., Van Vuure, A., Gorbatikh, L., Moldenaers, P., & Verpoest, I. (2009). Influence of carbon nanotube reinforcement on the processing and the mechanical behaviour of carbon fiber/epoxy composites. *Carbon*, *47*(12), 2914–2923. DOI: 10.1016/j.carbon.2009.06.039

Gong, Y., Liu, G., Xue, Y., Li, R., & Meng, L. (2023). A survey on dataset quality in machine learning. *Information and Software Technology*, *162*, 107268. DOI: 10.1016/j.infsof.2023.107268

Grancarić, A. M., Jerković, I., Koncar, V., Cochrane, C., Kelly, F. M., Soulat, D., & Legrand, X. (2017). Conductive polymers for smart textile applications. *Journal of Industrial Textiles*, *48*(3), 612–642. DOI: 10.1177/1528083717699368

Gray, A. J., Jenkins, D., Andrews, M. H., Taaffe, D. R., & Glover, M. L. (2010). Validity and reliability of GPS for measuring distance travelled in field-based team sports. *Journal of Sports Sciences*, *28*(12), 1319–1325. DOI: 10.1080/02640414.2010.504783

Greco, L., Ritrovato, P., & Xhafa, F. (2019). An edge-stream computing infrastructure for real-time analysis of wearable sensors data. *Future Generation Computer Systems*, *93*, 515–528. DOI: 10.1016/j.future.2018.10.058

Gugliuzza, A., & Drioli, E. (2013). A review on membrane engineering for innovation in wearable fabrics and protective textiles. *Journal of Membrane Science*, *446*, 350–375. DOI: 10.1016/j.memsci.2013.07.014

Gumaei, A., Al-Rakhami, M., AlSalman, H., Mizanur Rahman, S. M., & Alamri, A. (2020). DL-HAR: Deep Learning-Based Human Activity Recognition Framework for Edge Computing. *Computers, Materials & Continua*, *65*(2), 1033–1057. DOI: 10.32604/cmc.2020.011740

Gupta, M., Singh, M., Sharma, A., Sukhija, N., Aggarwal, P. K., & Jain, P. (2023). Unification of machine learning and blockchain technology in healthcare industry. In *Innovations in healthcare informatics: From interoperability to data analysis* (pp. 185–206). Institution of Engineering and Technology. DOI: 10.1049/PBHE041E_ch6

Gupta, P., Sharma, P., Kumar, V., Sharma, M., & Singh, R. (2025). Type 2 Diabetes Detection Using a Bio-inspired Machine Learning Approach. In Kumar, R., Verma, A. K., Verma, O. P., & Rajpurohit, J. (Eds.), *Soft Computing: Theories and Applications. SoCTA 2024. Lecture Notes in Networks and Systems* (Vol. 1343). Springer.

Gupta, S. (2021). Deep learning based human activity recognition (HAR) using wearable sensor data. *International Journal of Information Management Data Insights*, *1*(2), 100046. DOI: 10.1016/j.jjimei.2021.100046

Haake, S. J. (2009). The impact of technology on sporting performance in Olympic sports. *Journal of Sports Sciences*, *27*(13), 1421–1431. DOI: 10.1080/02640410903062019

Haghi, M., Thurow, K., & Stoll, R. (2017). Wearable devices in medical internet of things: Scientific research and commercially available devices. *Healthcare Informatics Research*, *23*(1), 4–15. DOI: 10.4258/hir.2017.23.1.4 PMID: 28261526

Hajiaghajani, A., Afandizadeh Zargari, A. H., Dautta, M., Jimenez, A., Kurdahi, F., & Tseng, P. (2021). Textile-integrated metamaterials for near-field multibody area networks. *Nature Electronics*, *4*(11), 808–817. DOI: 10.1038/s41928-021-00663-0

Halson, S. L. (2014). Monitoring training load to understand fatigue in athletes. *Sports Medicine (Auckland, N.Z.)*, *44*(S2, Suppl 2), 139–147. DOI: 10.1007/s40279-014-0253-z

Hamid, A., Rafiq, D., Nahvi, S. A., & Bazaz, M. A. (2023). Deep learning assisted surrogate modeling of large-scale power grids. Sustainable Energy. *Grids and Networks*, *34*, 101031. DOI: 10.1016/j.segan.2023.101031

Hamilton, M. T., Healy, G. N., Dunstan, D. W., Zderic, T. W., & Owen, N. (2008). Too little exercise and too much sitting: Inactivity physiology and the need for new recommendations on sedentary behavior. *Current Cardiovascular Risk Reports*, *2*(4), 292–298. DOI: 10.1007/s12170-008-0054-8

Harding, J. W., Mackintosh, C. G., Hahn, A. G., & James, D. A. (2008). Classification of aerial acrobatics in elite half-pipe snowboarding using body mounted inertial sensors. *The Engineering of Sport*, *7*(2), 447–456. DOI: 10.1007/978-2-287-09413-2_55

Harding, J. W., Small, J. W., & James, D. A. (2007). Feature extraction of performance variables in elite half-pipe snowboarding using body mounted inertial sensors. *BioMEMS and Nanotechnology III*, *6799*, 332–343.

Haris, M., Saad, S., Ammad, S., & Rasheed, K. (2024). AI in fabrication and construction. In *AI in Material Science* (pp. 169–192). CRC Press. DOI: 10.1201/9781003438489-8

Hartzog, W. (2018). *PrivacyOs Blueprint: The Battle to Control the Design of New Technologies*. Harvard University Press.

Hassabo, A. G., & Mohamed, A. L. (2019, April 12). Enhancement of thermoregulating textile materials using phase change material (PCM). Evolution in Polymer Technology Journal.

Haugen, T., & Buchheit, M. (2016). Sprint running performance monitoring: Methodological and practical considerations. *Sports Medicine (Auckland, N.Z.)*, *46*(5), 641–656. DOI: 10.1007/s40279-015-0446-0

Hayes, S. G., & Venkatraman, P. (Eds.). (2016). *Materials and technology for sportswear and performance apparel* (p. 293). CRC Press.

He, G., Zhu, C., Shi, Y., Yu, Y., Wu, Y., Soutis, C., Cao, L., & Liu, X. (2024). Development of a MusselInspired conductive graphene coated cotton yarn for wearable sensors. *iScience*, *28*(2), 111711. DOI: 10.1016/j.isci.2024.111711 PMID: 39898041

Heikenfeld, J., Jajack, A., Rogers, J., Gutruf, P., Tian, L., Pan, T., Li, R., Khine, M., Kim, J., Wang, J., & Kim, J. (2018). Wearable sensors: Modalities, challenges, and prospects. *Lab on a Chip*, *18*(2), 217–248. DOI: 10.1039/C7LC00914C PMID: 29182185

Heintzman, N. D. (2016). A digital ecosystem of diabetes data and technology: Services, systems, and tools enabled by wearables, sensors, and apps. *Journal of Diabetes Science and Technology*, *10*(1), 35–41. DOI: 10.1177/1932296815622453

Hermes, S., Riasanow, T., Clemons, E. K., Böhm, M., & Krcmar, H. (2020). The digital transformation of the healthcare industry: Exploring the rise of emerging platform ecosystems and their influence on the role of patients. *Business Research*, *13*(3), 1033–1069. DOI: 10.1007/s40685-020-00125-x

Hewett, T. E., & Bates, N. A. (2017). Preventive biomechanics: A paradigm shift with a translational approach to injury prevention. *The American Journal of Sports Medicine*, *45*(11), 2654–2664. DOI: 10.1177/0363546516686080

He, Z., Wu, Y., & Zhang, X. (2023). Artificial intelligence in wearable healthcare: From data analytics to personalized monitoring. *IEEE Reviews in Biomedical Engineering*, *16*, 120–138.

Hilmi, A. (2025, October 21). Wearables for defense: the next frontier in military technology. *Sensio*. https://www.sensio-ai.in/post/wearables-for-defense-the-next-frontier-in-military-technology

Holley, K., & Becker, S. (2021). *AI-first healthcare: AI applications in the business and clinical management of health* (1st ed.). O'Reilly Media.

Horrocks, A. R., & Anand, S. C. (2000). *Handbook of technical textiles*. Woodhead Publishing.

Hosain, M. N. (2024). IoT-enabled biosensors for real-time monitoring and early detection of chronic diseases. *Frontiers in Public Health*, •••, 12.

Hossain, M. M., & Bradford, P. D. (2020). Durability of smart electronic textiles. In Elsevier eBooks (pp. 27–53). DOI: 10.1016/B978-0-12-820777-2.00003-0

Hossain, M. M., Lubna, M. M., & Bradford, P. D. (2023). Multifunctional and washable Carbon NanotubeWrapped textile yarns for wearable ETextiles. *ACS Applied Materials & Interfaces*, *15*(2), 3365–3376. DOI: 10.1021/acsami.2c19826 PMID: 36622361

Hou, C. (2020). A study on IMU-based human activity recognition using deep learning and traditional machine learning. *2020 5th International Conference on Computer and Communication Systems (ICCCS)*, 225–234.

Howard, J., Murashov, V., Cauda, E., & Snawder, J. (2022). Advanced sensor technologies and the future of work. *American Journal of Industrial Medicine*, *65*(1), 3–11. DOI: 10.1002/ajim.23300

Hsu, P., Liu, X., Liu, C., Xie, X., Lee, H. R., Welch, A. J., Zhao, T., & Cui, Y. (2014). Personal thermal management by Metallic NanowireCoated Textile. *Nano Letters*, *15*(1), 365–371. DOI: 10.1021/nl5036572 PMID: 25434959

Huang, J., Boles, S. T., & Tarascon, J. M. (2022). Sensing as the key to battery lifetime and sustainability. *Nature Sustainability*, *5*(3), 194–204. DOI: 10.1038/s41893-022-00859-y

Huang, Z., Shen, Y., Li, J., Fey, M., & Brecher, C. (2021). A survey on AI-driven digital twins in industry 4.0: Smart manufacturing and advanced robotics. *Sensors (Basel)*, *21*(19), 6340. DOI: 10.3390/s21196340 PMID: 34640660

Hulsen, T. (2023). Explainable artificial intelligence (XAI): Concepts and challenges in healthcare. *AI*, *4*(3), 652–666. DOI: 10.3390/ai4030034

Hu, Q., Shen, X., Qian, X., Huang, G., & Yuan, M. (2022). The personal protective equipment (PPE) based on individual combat: A systematic review and trend analysis. *Defence Technology*, *28*, 195–221. DOI: 10.1016/j.dt.2022.12.007

Hussain, I., Amna, R., Kalidasan, B., Rani, G. M., Bandi, H., Venkateswarlu, S., Khan, S. A., Mohapatra, D., & Zhang, K. (2025). MXEnes for various applications: Recent trends and future aspects. *SmartMat*, *6*(2), e70012. Advance online publication. DOI: 10.1002/smm2.70012

Hu, X., Fan, Q., Wang, S., Chen, Y., Wang, D., Chen, K., Ge, F., Zhou, W., & Liang, K. (2024). TwoDimensional MXENES: Innovative materials for efficient thermal management and safety solutions. *Research*, *7*, 0542. Advance online publication. DOI: 10.34133/research.0542 PMID: 39703779

Hu, X., Ma, Z., Zhao, F., & Guo, S. (2024). Recent advances in self-powered wearable flexible sensors for human gait analysis. *Nanomaterials (Basel, Switzerland), 14*(14), 1173. DOI: 10.3390/nano14141173

Imamguluyev, R., Hashim, S., & Hajiyev, I. (2024). Harnessing Uncertainty: Integrating Fuzzy Logic into Machine Learning Algorithms. 2024 4th International Conference on Pervasive Computing and Social Networking (ICPCSN), 510-514. DOI: 10.1109/ICPCSN62568.2024.00086

Islam, M. M., Poly, T. N., Yang, H. C., & Li, Y. C. (2021). Deep into laboratory: An artificial intelligence approach to recommend laboratory tests. *Diagnostics (Basel), 11*(6), 990. Advance online publication. DOI: 10.3390/diagnostics11060990 PMID: 34072571

Ismail, S. I., Nunome, H., Lysdal, F. G., Kersting, G., & Tamura, Y. (2022). *Futsal playing surface characteristics significantly affect perceived traction and change of direction performance among experienced futsal players*. DOI: 10.1080/14763141.2022.2143415

Jain, P., Sharma, S., Arora, S., Shokeen, V., Aggarwal, P. K., & Singh, M. (2024). Edge computing-based design for IoT security. In Network Optimization in Intelligent Internet of Things Applications: Principles and Challenges (pp. 298–309). CRC Press. DOI: 10.1201/9781003405535-22

Jaiswal, P., Kaushik, A., Lawless, F., Malaquias, T., & McCaffery, F. (2025). *Preliminary Investigation on Machine Learning and Deep Learning Models for Change of Direction Classification in Running*. 180–191. DOI: 10.1007/978-3-031-77731-8_17

Jalali, E., Erasmus, E., Schutte-Smith, M., & Visser, H. G. (2024). Fixation of nanoparticles on fabric: Applications in general health management. *Materials Today. Communications, 41*, 110577. DOI: 10.1016/j.mtcomm.2024.110577

James, D. A., & Petrone, N. (2016). *Sensors and Wearable Technologies in Sport*. DOI: 10.1007/978-981-10-0992-1

Janiesch, C., Zschech, P., & Heinrich, K. (2021). Machine learning and deep learning. *Electronic Markets, 31*(3), 685–695. DOI: 10.1007/s12525-021-00475-2

Jee, H., Kwon, O. Y., Park, Y. H., Kim, Y. W., & Lee, J. H. (2019). A wearable smart shirt for monitoring respiration, ECG, and activity. *IEEE Sensors Journal, 19*(24), 11734–11744. DOI: 10.1109/JSEN.2019.2935203

Jiang, Z. ;, Hao, Y. ;, Jin, N. ;, Li, Y. A., Baran, J., Leszczak, J., Jiang, Z., Hao, Y., Jin, N., & Li, Y. (2022). A Systematic Review of the Relationship between Workload and Injury Risk of Professional Male Soccer Players. *International Journal of Environmental Research and Public Health 2022, Vol. 19, Page 13237, 19*(20), 13237. DOI: 10.3390/ijerph192013237

Jiang, W., & Yin, Z. (2015, October). Human activity recognition using wearable sensors by deep convolutional neural networks. In *Proceedings of the 23rd ACM international conference on Multimedia* (pp. 1307-1310). DOI: 10.1145/2733373.2806333

Jiang, Y. (2020). Combination of wearable sensors and internet of things and its application in sports rehabilitation. *Computer Communications, 150*, 167–176. DOI: 10.1016/j.comcom.2019.11.021

Jin, C., Chen, X., & Duan, Y. (2022). Smart textiles for personalized thermal management: Current advances and future challenges. arXiv preprint arXiv:2201.04453.

Jones, P. A., & Dos'Santos, T. (2023). *Multidirectional speed in sport: research to application*. Taylor & Francis. DOI: 10.4324/9781003267881

K, N., & Rout, C. S. (2021). Conducting polymers: a comprehensive review on recent advances in synthesis, properties and applications. *RSC Advances, 11*(10), 5659–5697. .DOI: 10.1039/D0RA07800J

Kalantari, M. (2017). Consumers' adoption of wearable technologies: Literature review, synthesis, and future research agenda. *International Journal of Technology Marketing, 12*(3), 274–307. DOI: 10.1504/IJTMKT.2017.089665

Kamal, B., Ullah, S., & Khan, S. (2016). A NOVEL EXPERIMENTAL PROCEDURE FOR DIGITIZING FLEXIBLE WEARABLE ANTENNAS FOR BODY AREA NETWORKS (BANS). *Jurnal Teknologi (Sciences & Engineering), 78*(4-3).

Kanoun, O., & Trankler, H. R. (2004). Sensor technology advances and future trends. *IEEE Transactions on Instrumentation and Measurement, 53*(6), 1497–1501. DOI: 10.1109/TIM.2004.834613

Kartsakli, E., Lalos, A. S., Antonopoulos, A., Tennina, S., Di Renzo, M., Alonso, L., & Verikoukis, C. (2014). A survey on M2M systems for mHealth: A wireless communications perspective. *Sensors (Basel), 14*(10), 18009–18052. DOI: 10.3390/s141018009 PMID: 25264958

Keaney, E. M., & Reid, M. (2018). Quantifying hitting activity in tennis with racket sensors: New dawn or false dawn? *Sports Biomechanics*.

Keawploy, N., Venkatkarthick, R., Wangyao, P., & Qin, J. (2020). Screen printed textile electrodes using graphene and carbon nanotubes with silver for flexible supercapacitor applications. *Journal of Metals Materials and Minerals*, *30*(4), 39–44. DOI: 10.55713/jmmm.v30i4.892

Khalifeh, A., Mazunga, F., Nechibvute, A., & Nyambo, B. M. (2022). Microcontroller unit-based wireless sensor network nodes: A review. *Sensors (Basel)*, *22*(22), 8937. Advance online publication. DOI: 10.3390/s22228937 PMID: 36433537

Khan, M., Refati, M. F. a. D., Arup, M. M. R., Islam, M. A., &Mobarak, M. H. (2025). Conductive Polymer-Based electronics in additive manufacturing: materials, processing, and applications. *Advances in Polymer Technology, 2025*(1)..DOI: 10.1155/adv/4234491

Khan, S., Hussain, A., & Lee, S. (2024). Privacy and security challenges in IoT-enabled wearable systems: Toward regulatory compliance in healthcare. *ACM Computing Surveys*, *56*(1), 1–32. DOI: 10.1145/3640311

Khan, Y., Han, D., Pierre, A., & Arias, A. C. (2020). Towards conformable hybrid electronics for wearable health applications. *Advanced Functional Materials*, *30*(47), 1909475. DOI: 10.1002/adfm.201909475

Khan, Y., Ostfeld, A. E., Lochner, C. M., Pierre, A., & Arias, A. C. (2020). Monitoring of vital signs with flexible and wearable medical devices. *Advanced Materials*, *32*(15), 1905279. DOI: 10.1002/adma.201905279 PMID: 31742812

Khatun, M. A., Yousuf, M. A., Ahmed, S., Uddin, M. Z., Alyami, S. A., Al-Ashhab, S., Akhdar, H. F., Khan, A., Azad, A., & Moni, M. A. (2022). Deep CNN-LSTM With Self-Attention Model for Human Activity Recognition Using Wearable Sensor. *IEEE Journal of Translational Engineering in Health and Medicine*, *10*, 1–16. Advance online publication. DOI: 10.1109/JTEHM.2022.3177710

Khdoudi, A., Masrour, T., El Hassani, I., & El Mazgualdi, C. (2024). A deep-reinforcement-learning-based digital twin for manufacturing process optimization. *Systems*, *12*(2), 38. DOI: 10.3390/systems12020038

Kim, H., Kwon, Y. T., Lim, H. R., Kim, J. H., Kim, Y. S., & Yeo, W. H. (2021). Recent advances in wearable sensors and integrated functional devices for virtual and augmented reality applications. *Advanced Functional Materials*, *31*(39), 2005692. DOI: 10.1002/adfm.202005692

Kim, H.-S., Kumar, S., & Culler, D. E. (2019). Thread/OpenThread: A compromise in low-power wireless multihop network architecture for the Internet of Things. *IEEE Communications Magazine*, *57*(7), 55–61. DOI: 10.1109/MCOM.2019.1800788

Kini, A. S., Gopal Reddy, A. N., Kaur, M., Satheesh, S., Singh, J., Martinetz, T., & Alshazly, H. (2022). Ensemble deep learning and internet of things-based automated COVID-19 diagnosis framework. *Contrast Media & Molecular Imaging, 2022*(1), 7377502. DOI: 10.1155/2022/7377502 PMID: 35280708

Kivikangas, M. (2022). *Estimating three-dimensional motion using IMU sensors.*

Kohl, H. W.III, Craig, C. L., Lambert, E. V., Inoue, S., Alkandari, J. R., Leetongin, G., & Kahlmeier, S. (2012). The pandemic of physical inactivity: Global action for public health. *Lancet, 380*(9838), 294–305. DOI: 10.1016/S0140-6736(12)60898-8

Koncar, V. (Ed.). (2016). *Smart textiles and their applications.* Woodhead Publishing. DOI: 10.1016/B978-0-08-100574-3.00001-1

Kranz, M., Möller, A., Hammerla, N., Diewald, S., Plötz, T., Olivier, P., & Roalter, L. (2013). The mobile fitness coach: Towards individualized skill assessment using personalized mobile devices. *Pervasive and Mobile Computing, 9*(2), 203–215. DOI: 10.1016/j.pmcj.2012.06.002

Kubota, K. J., Chen, J. A., & Little, M. A. (2016). Machine learning for large-scale wearable sensor data in Parkinson's disease: Concepts, promises, pitfalls, and futures. *Movement Disorders, 31*(9), 1314–1326. DOI: 10.1002/mds.26693

Kumar, R., & Sharma, A. (2021). Power management in embedded systems for wearable devices: An overview. *Journal of Low Power Electronics and Applications, 11*(1), 1. Advance online publication. DOI: 10.3390/jlpea11010001

Kumar, V., Dandotiya, M., Ekvitayavetchanukul, P., Rana, M., & Singh, B. (2026). Deep Learning-Powered Diagnostic Model for Early Detection and Prognosis of Bone Diseases Using Radiographic Imaging. In Das, S., Paprzycki, M., Ghosh, A., & Bianchini, M. (Eds.), *Advanced Computing and Intelligent Technologies. ICACIT 2024. Lecture Notes in Networks and Systems* (Vol. 1359). Springer., DOI: 10.1007/978-981-96-4933-4_5

Kumar, V., Wahi, C., Sagar, B. B., & Manjul, M. (2024). Ensemble learning based intrusion detection for wireless sensor network environment. *International Journal of Performability Engineering, 20*(9), 541–550. DOI: 10.23940/ijpe.24.09.p2.541551

Kuncar, A., Sysel, M., & Urbanek, T. (2011). *Calibration of Triaxial Accelerometer and Triaxial Magnetometer for Tilt Compensated Electronic Compass.*

Kuncar, A., Sysel, M., & Urbanek, T. (2016). Calibration of triaxial accelerometer and triaxial magnetometer for tilt compensated electronic compass. *Automation Control Theory Perspectives in Intelligent Systems: Proceedings of the 5th Computer Science On-Line Conference 2016 (CSOC2016),* Vol 3 5, 45–52.

Kutty, A., Kucukvar, M., Onat, N., Ayvaz, B., & Abdella, G. (2023). Measuring sustainability, resilience and livability performance of European smart cities: A novel fuzzy expert-based multi-criteria decision support model. *Cities (London, England)*, *137*, 104293. Advance online publication. DOI: 10.1016/j.cities.2023.104293

Land, T. (2018). Powering up with technology to deliver transformative care and lower costs. *Frontiers of Health Services Management*, *34*(3), 1–2. DOI: 10.1097/HAP.0000000000000029

Langås, E. F., Zafar, M. H., & Sanfilippo, F. (2025). Exploring the synergy of human-robot teaming, digital twins, and machine learning in Industry 5.0: A step towards sustainable manufacturing. *Journal of Intelligent Manufacturing*. Advance online publication. DOI: 10.1007/s10845-025-02580-x

Leclercq, C., Witt, H., Hindricks, G., Katra, R. P., Albert, D., Belliger, A., Cowie, M. R., Deneke, T., Friedman, P., Haschemi, M., Lobban, T., Lordereau, I., McConnell, M. V., Rapallini, L., Samset, E., Turakhia, M. P., Singh, J. P., Svennberg, E., Wadhwa, M., & Weidinger, F. (2022). Wearables, telemedicine, and artificial intelligence in arrhythmias and heart failure: Proceedings of the European Society of Cardiology Cardiovascular Round Table. *Europace : European Pacing, Arrhythmias, and Cardiac Electrophysiology : Journal of the Working Groups on Cardiac Pacing, Arrhythmias, and Cardiac Cellular Electrophysiology of the European Society of Cardiology*, *24*(9), 1372–1383. DOI: 10.1093/europace/euac052 PMID: 35640917

Lee, E. C., Fragala, M. S., Kavouras, S. A., Queen, R. M., Pryor, J. L., & Casa, D. J. (2017). Biomarkers in sports and exercise: Tracking health, performance, and recovery in athletes. *Journal of Strength and Conditioning Research*, *31*(10), 2920–2937. DOI: 10.1519/JSC.0000000000002122

Lee, J., Kim, D., Ryoo, H. Y., & Shin, B. S. (2016). Sustainable wearables: Wearable technology for enhancing the quality of human life. *Sustainability (Basel)*, *8*(5), 466. DOI: 10.3390/su8050466

Lee, S., An, S., & Kim, H. (2020). Wash durability of conductive fabrics for wearable electronics. *Textile Research Journal*, *90*(7–8), 857–870. DOI: 10.1177/0040517519881570

Lee, S., & Park, S. (2024). Optimizing washing conditions for smart fabrics: A comprehensive study. *RSC Advances*, *14*(54), 40098–40116. DOI: 10.1039/D4RA07365G PMID: 39717812

Le, K., Narayana, H., Servati, A., Bahi, A., Soltanian, S., Servati, P., & Ko, F. (2022). Electronic textiles for electrocardiogram monitoring: A review on the structure–property and performance evaluation from fiber to fabric. *Textile Research Journal*, *93*(3–4), 878–910. DOI: 10.1177/00405175221108208

Leleu, A., & Ward, T. E. (2024). Comparing automated and non-automated machine learning for binary classification of running fatigue using wearable sensors. In Advances in Digital Health and Medical Bioengineering (Vol. 111, pp. 401–408) DOI: 10.1007/978-3-031-62523-7_44

Lewy, H. (2015). Wearable technologies—Future challenges for implementation in healthcare services. *Healthcare Technology Letters*, *2*(1), 2–5. DOI: 10.1049/htl.2014.0104 PMID: 26609396

Liang, Z., Deng, Y., Shi, Z., Liao, X., Zong, H., Ren, L., Li, X., Zeng, X., Hu, P., Ke, W., Wu, B., Wang, K., Qian, J., Xu, W., & Chen, F. (2025). AI-driven design of powder-based nanomaterials for smart textiles: From data intelligence to system integration. *Advanced Powder Materials*, *5*(1), 100356. DOI: 10.1016/j.apmate.2025.100356

Liao, X., Zhang, Z., Liang, Q., Liao, Q., & Zhang, Y. (2015). Extremely elastic wearable carbon nanotube fiber strain sensor for monitoring of human motion. *ACS Nano*, *9*(6), 5929–5936. DOI: 10.1021/acsnano.5b00599 PMID: 26038807

Li, H., Song, B., Yang, Z., & Wang, Y. (2019). Wearable sensors integration: Design strategies and approaches. *IEEE Sensors Journal*, *19*(20), 9345–9355. DOI: 10.1109/JSEN.2019.2929173

Li, J., Chen, Y., & Tao, X. (2022). Recent progress in smart textiles for personalized healthcare and environmental monitoring. *Advanced Materials Technologies*, *7*(12), 2200643.

Lin, R., Kim, H.-J., Achavananthadith, S., Kurt, S. A., Tan, S. C. C., Yao, H., & Lee, J. K. W. (2020). Wireless battery-free body sensor networks using near-field-enabled clothing. *Nature Communications*, *11*(1), 444. DOI: 10.1038/s41467-020-14311-2 PMID: 31974376

Liu, H., Shi, Y., Pan, Y., Wang, Z., & Wang, B. (2025). Sensory interactive fibers and textiles. *NPJ Flexible Electronics*, *9*(1), 23. Advance online publication. DOI: 10.1038/s41528-025-00398-4

Liu, M., Xue, Z., Xu, X., Zhong, C., & Chen, J. (2018). Host-based intrusion detection system with system calls: Review and future trends. *ACM Computing Surveys*, *51*(5), 1–36. DOI: 10.1145/3214304

Liu, X., Wang, H., Li, Z., & Qin, L. (2021). Deep learning in ECG diagnosis: A review. *Knowledge-Based Systems*, *227*, 107187. DOI: 10.1016/j.knosys.2021.107187

Liu, Y., Zhang, L., Yang, Y., & Wang, X. (2024). Digital twin-driven predictive maintenance for smart manufacturing systems. *Journal of Manufacturing Processes*, *80*, 1–10. DOI: 10.1016/j.jmapro.2024.08.012

Li, X., Xu, B., & Gao, Y. (2020). Encapsulation techniques for durable wearable biosensors. *Advanced Healthcare Materials*, *9*(15), 2000605. DOI: 10.1002/adhm.202000605 PMID: 32529775

Lodhi, S. K., Gill, A. Y., & Hussain, I. (2024). AI-powered innovations in contemporary manufacturing procedures: An extensive analysis. *International Journal of Multidisciplinary Sciences and Arts*, *3*(4), 15–25. DOI: 10.47709/ijmdsa.v3i4.4616

Loncar-Turukalo, T., Zdravevski, E., da Silva, J. M., Chouvarda, I., & Trajkovik, V. (2019). Literature on wearable technology for connected health: Scoping review of research trends, advances, and barriers. *Journal of Medical Internet Research*, *21*(9), e14017. DOI: 10.2196/14017

López-Nava, I. H., Arnrich, B., Munoz-Melendez, A., & Güneysu, A. (2017). Variability analysis of therapeutic movements using wearable inertial sensors. *Journal of Medical Systems*, *41*(1), 7. DOI: 10.1007/s10916-016-0645-8 PMID: 27848176

Losada-Benitez, J. A., Nuñez-Sánchez, F. J., & Barbero-Álvarez, J. C. (2023). Quantifying technical load and physical activity in professional soccer players during pre-season matches with IMU technology. *Frontiers in Physiology*, *14*, 1274171. DOI: 10.3389/fphys.2023.1274171

Luptáková, I. D., Kubovčík, M., & Pospíchal, J. (2022). Wearable Sensor-Based Human Activity Recognition with Transformer Model. *Sensors 2022, Vol. 22, Page 1911, 22*(5), 1911. DOI: 10.3390/s22051911

Lu, X., & O'Reilly, M. J. (2024). Applying User-Centered Design Principles to the Development of Wearable Technology in Fashion Design. *Studies in Art and Architecture*, *3*(2), 10–16. DOI: 10.56397/SAA.2024.06.03

Machorro-Cano, I., Olmedo-Aguirre, J. O., Alor-Hernández, G., Rodríguez-Mazahua, L., Sánchez-Morales, L. N., & Pérez-Castro, N. (2023, December). Cloud-Based Platforms for Health Monitoring: A Review. In Informatics (Vol. 11, No. 1, p. 2). MDPI. DOI: 10.3390/informatics11010002

Maji, P. K., BISwas, R., and Roy, A. R. (2001). Fuzzy soft sets. *Journal of Fuzzy Mathematics*, *9*(3), 589–602.

Mamun, M. A. A., & Yuce, M. R. (2020). Recent progress in nanomaterial enabled chemical sensors for wearable environmental monitoring applications. *Advanced Functional Materials*, *30*(51), 2005703. DOI: 10.1002/adfm.202005703

Mandal, A. (2025). Design of a multipurpose communicable handglove based on navigation tools and microcontroller unit. *2025 4th International Conference on Range Technology (ICORT)* (pp. 1–5). DOI: 10.1109/ICORT64008.2025.11115501

Mann, S. (1997). Wearable computing: A first step toward personal imaging. *Computer*, *30*(2), 25–32. DOI: 10.1109/2.566147

Mann, S., Nolan, J., & Wellman, B. (2003). Sousveillance: Inventing and using wearable computing devices for data collection in surveillance environments. *Surveillance & Society*, *1*(3), 331–355. DOI: 10.24908/ss.v1i3.3344

Marterer, V., Radouchová, M., Soukup, R., Hipp, S., & Blecha, T. (2024). Wearable textile antennas: Investigation on material variants, fabrication methods, design and application. *Fashion and Textiles*, *11*(1), 9. Advance online publication. DOI: 10.1186/s40691-023-00369-1

Martínez-Mireles, J. R., Rodríguez-Flores, J., García-Márquez, M. A., Austria-Cornejo, A., & Figueroa-Diaz, R. A. (2025). AI in Smart Manufacturing. In Machine and Deep Learning Solutions for Achieving the Sustainable Development Goals (pp. 463-494). IGI Global Scientific Publishing. DOI: 10.4018/979-8-3693-8161-8.ch023

Martín-Moya, R., & González-Fernández, F. T. (2022). Test for the improvement and evaluation of change of direction in team sports: A systematic review. *Journal of Physical Education and Sport*, *22*(7), 1716–1722. DOI: 10.7752/JPES.2022.07215

Mata, A., Moffat, D., Almeida, S., Radeta, M., Jay, W., Mortimer, N., Awty-Carroll, K., Thomas, O. R., Brotas, V., & Groom, S. (2024). Drone imagery and deep learning for mapping the density of wild Pacific oysters to manage their expansion into protected areas. *Ecological Informatics*, *82*, 102708. DOI: 10.1016/j.ecoinf.2024.102708

Mateus, N., Abade, E., Coutinho, D., Gómez, M. Á., Peñas, C. L., & Sampaio, J. (2024). Empowering the Sports Scientist with Artificial Intelligence in Training, Performance, and Health Management. *Sensors (Basel)*, *25*(1), 139. DOI: 10.3390/s25010139

Mattila, H. (Ed.). (2006). *Intelligent textiles and clothing*. Woodhead Publishing. https://www.amazon.in/Intelligent-Textiles-Clothing-Woodhead-Publishing/dp/1845690052

Mattila, H. (2006). *Intelligent textiles and clothing*. Woodhead Publishing.

Mattmann, C., Amft, O., Harms, H., Tröster, G., & Clemens, F. (2008). Recognizing upper body postures using textile strain sensors. In *2008 12th IEEE International Symposium on Wearable Computers* (pp. 29–36). IEEE. DOI: 10.1109/ISWC.2008.4911581

Mattmann, C., Clemens, F., & Tröster, G. (2008). Sensor for measuring strain in textile. *Sensors (Basel), 8*(6), 3719–3732. DOI: 10.3390/s8063719 PMID: 27879904

Meghji, M., Balloch, A., Habibi, D., Ahmad, I., Hart, N., Newton, R., Weber, J., & Waqar, A. (2019). An Algorithm for the Automatic Detection and Quantification of Athletes' Change of Direction Incidents Using IMU Sensor Data. *IEEE Sensors Journal, 19*(12), 4518–4527. DOI: 10.1109/JSEN.2019.2898449

Mendel, J. (2000). Uncertainty, fuzzy logic, and signal processing. *Signal Processing, 80*(6), 913–933. DOI: 10.1016/S0165-1684(00)00011-6

Meng, T., Huang, J., Chen, Y., Li, H., Li, X., Zhang, B., Hao, J., Chew, C., Yang, D., & Zhong, Z. (2023). New Framework for Fuzzy Logic Reasoning: A Robust Control Theoretic Approach. *International Journal of Fuzzy Systems, 26*(2), 463–481. DOI: 10.1007/s40815-023-01606-x

Mercer, K., Giangregorio, L., Schneider, E., Chilana, P., Li, M., & Grindrod, K. (2016). Acceptance of commercially available wearable activity trackers among adults aged over 50 and with chronic illness: A mixed-methods evaluation. *JMIR mHealth and uHealth, 4*(1), e4225. DOI: 10.2196/mhealth.4225

Micklewright, D. (2019). Decision-making, pacing, and performance in endurance sport. In *Endurance Performance in Sport* (pp. 47–69). Routledge. DOI: 10.4324/9781315167312-4

Migliaccio, G. M., Padulo, J., & Russo, L. (2024). The Impact of Wearable Technologies on Marginal Gains in Sports Performance: An Integrative Overview on Advances in Sports, Exercise, and Health. *Applied Sciences 2024, Vol. 14, Page 6649, 14*(15), 6649. DOI: 10.3390/app14156649

Mikołajewska, E., Mikołajewski, D., Mikołajczyk, T., & Paczkowski, T. (2025). Generative AI in AI-based digital twins for fault diagnosis for predictive maintenance in Industry 4.0/5.0. *Applied Sciences (Basel, Switzerland), 15*(6), 3166. DOI: 10.3390/app15063166

Min, Q., Lu, Y., Liu, Z., Su, C., & Wang, B. (2019). Machine learning-based digital twin framework for production optimization in petrochemical industry. *International Journal of Information Management, 49*, 502–519. DOI: 10.1016/j.ijinfomgt.2019.05.020

Mirza, M., Anjum, R., Min, H., Alkahtani, B., & Anjum, M. (2025). The Linear Diophantine Fuzzy Z-Numbers Set: Development and Application to Decision Making in Textile Engineering Using the Bonferroni Mean Operator. *IEEE Access : Practical Innovations, Open Solutions, 13*, 70512–70532. DOI: 10.1109/ACCESS.2025.3561136

Mishra, R. K., Sarkar, J., Verma, K., Chianella, I., Goel, S., & Nezhad, H. Y. (2024). Exploring transformative and multifunctional potential of MXenes in 2D materials for nextgeneration technology. *Open Ceramics, 18*, 100596. DOI: 10.1016/j.oceram.2024.100596

Mohamed, M., Almalaq, A., Abdullah, H., Alnowibet, K., Alrasheedi, A., & Zaindin, M. (2021). A Distributed Stochastic Energy Management Framework Based-Fuzzy-PDMM for Smart Grids Considering Wind Park and Energy Storage Systems. *IEEE Access : Practical Innovations, Open Solutions, 9*, 46674–46685. DOI: 10.1109/ACCESS.2021.3067501

Molodtsov, D. (1999). Soft set theory: First results. *Computers & Mathematics with Applications (Oxford, England), 37*(4–5), 19–31. DOI: 10.1016/S0898-1221(99)00056-5

Mone, V., & Shakhlo, F. (2023). Health data on the go: Navigating privacy concerns with wearable technologies. *Legal Information Management, 23*(3), 179–188. DOI: 10.1017/S1472669623000427

Morais, J. E. (2023). Advances in Wearable Devices for Sports. *Applied Sciences (Basel, Switzerland), 13*(24), 13288. DOI: 10.3390/app132413288

Moraru, L., Moldovan, C., & Munteanu, R. (2020). Washable wearable UHF RFID tags: Reliability assessment. *Sensors (Basel), 20*(12), 3451. DOI: 10.3390/s20123451 PMID: 32570940

Moreno, J., Sanchez, M., Mendoza, O., Díaz, A., Castillo, O., Melin, P., & Castro, J. (2020). Design of an interval Type-2 fuzzy model with justifiable uncertainty. *Information Sciences, 513*, 206–221. DOI: 10.1016/j.ins.2019.10.042

Morral-Yepes, M., Moras, G., Bishop, C., & Gonzalo-Skok, O. (2022). Assessing the reliability and validity of agility testing in team sports: A systematic review. *Journal of Strength and Conditioning Research, 36*(7), 2035–2049. DOI: 10.1519/JSC.0000000000003753

Mukherjee, A., & Das, A. K. (2015b). Interval valued intuitionistic fuzzy soft multi set theoretic approach to decision making problems, 2015 International Conference on Computer, Communication and Control (IC4), Indore, India, 2015, pp. 1-5, DOI: 10.1109/IC4.2015.7375640

Mukherjee, A., & Das, A. K. (2016a). Application of fuzzy soft multi sets in decision-making problems. In: Nagar, A., Mohapatra, D., Chaki, N. (eds) Proceedings of 3rd International Conference on Advanced Computing, Networking and Informatics. Smart Innovation, Systems and Technologies, vol. 43. Springer, New Delhi. DOI: 10.1007/978-81-322-2538-6_3

Mukherjee, A., & Das, A. K. (2023). Essentials of fuzzy soft multisets: Theory and applications, Springer Nature Singapore Pte Ltd., 1-151. ISBN: 978-981-19-2759-1, DOI: 10.1007/978-981-19-2760-7

Mukherjee, A., & Das, A. K. (2015a). Application of interval valued intuitionistic fuzzy soft set in investment decision making, *2015 Fifth International Conference on Advances in Computing and Communications (ICACC)*, Kochi, India, 2015, pp. 61-64, DOI: 10.1109/ICACC.2015.37

Mukherjee, A., & Das, A. K. (2015c). Relations on intuitionistic fuzzy soft multi sets. In Kim, K. (Ed.), *Information Science and Applications. Lecture Notes in Electrical Engineering* (Vol. 339). Springer., DOI: 10.1007/978-3-662-46578-3_71

Mukherjee, A., & Das, A. K. (2016b). Topological structure formed by intuitionistic fuzzy rough relations. *Journal of the Indian Mathematical Society*, *83*(1-2), 135–144.

Mukherjee, A., & Das, A. K. (2022). Einstein operations on fuzzy soft multisets and decision making. *Boletim da Sociedade Paranaense de Matematica*, *40*, 1–10. DOI: 10.5269/bspm.32546

Muruganandam, S., Salameh, A. A., Pozin, M. A. A., Manikanthan, S. V., & Padmapriya, T. (2023). Sensors and machine learning and AI operation-constrained process control method for sensor-aided industrial internet of things and smart factories. *Measurement. Sensors*, *25*, 100668. DOI: 10.1016/j.measen.2023.100668

Nahavandi, D., Alizadehsani, R., Khosravi, A., & Acharya, U. R. (2022). Application of artificial intelligence in wearable devices: Opportunities and challenges. *Computer Methods and Programs in Biomedicine*, *213*, 106541. DOI: 10.1016/j.cmpb.2021.106541

Najjar, M.-C. (2023). Legal and Ethical Issues Arising from the Application of Data Analytics and Artificial Intelligence to Traditional Sports. *Albany Law Journal of Science & Technology*, *33*, •••. https://heinonline.org/HOL/Page?handle=hein.journals/albnyst33&id=57&div=&collection=

Nash, C., Martindale, R., Collins, D., & Martindale, A. (2012). Parameterising expertise in coaching: Past, present and future. *Journal of Sports Sciences*, *30*(10), 985–994. DOI: 10.1080/02640414.2012.682079

NATO Smart Textiles Lab Report. (2021). Electrochromic camouflage uniforms in forest environments.

Nayak, S., Panigrahi, C. R., Pati, B., Nanda, S., & Hsieh, M. Y. (2022). Comparative Analysis of HAR Datasets Using Classification Algorithms. *Computer Science and Information Systems*, *19*(1), 47–63. DOI: 10.2298/CSIS201221043N

Nidhi, S., P., Kumar, V., Sharma, M., & Singh, R. (2024). A comprehensive evaluation of radial basis function networks for feature selection and classification. In Proceedings of the 2024 Second International Conference on Advanced Computing & Communication Technologies (ICACCTech) (pp. 706–710). IEEE. DOI: 10.1109/ICACCTech65084.2024.00118

Nikitas, A., Michalakopoulou, K., Njoya, E. T., & Karampatzakis, D. (2020). Artificial intelligence, transport and the smart city: Definitions and dimensions of a new mobility era. *Sustainability (Basel)*, *12*(7), 2789. DOI: 10.3390/su12072789

NIST. (2020). *NISTIR 8259: Foundational cybersecurity activities for IoT device manufacturers.* DOI: DOI: 10.6028/NIST.IR.8259

Norris, M., Anderson, R., & Kenny, I. C. (2014). Method analysis of accelerometers and gyroscopes in running gait: A systematic review. *Proceedings of the Institution of Mechanical Engineers. Part P, Journal of Sports Engineering and Technology*, *228*(1), 3–15. DOI: 10.1177/1754337113502472

Novakovsky, G., Dexter, N., Libbrecht, M. W., Wasserman, W. W., & Mostafavi, S. (2022). Obtaining genetics insights from deep learning via explainable artificial intelligence. *Nature Reviews Genetics 2022 24:2*, *24*(2), 125–137. DOI: 10.1038/s41576-022-00532-2

Oatley, G., Choudhury, T., & Buckman, P. (2021). Smart textiles for improved quality of life and cognitive assessment. *Sensors (Basel)*, *21*(23), 8008. DOI: 10.3390/s21238008 PMID: 34884010

Ojstršek, A., Jug, L., & Plohl, O. (2022). A review of electro conductive textiles utilizing the DipCoating technique: Their functionality, durability and sustainability. *Polymers, 14*(21), 4713. DOI: 10.3390/polym14214713 PMID: 36365707

Ojuroye, O., Torah, R., Beeby, S., & Wilde, A. (2017). Smart textiles for smart home control and enriching future wireless sensor network data. In *Sensors for Everyday Life* (pp. 159–183). Healthcare Settings. DOI: 10.1007/978-3-319-47319-2_9

Oldenburg, B., Taylor, C. B., O'Neil, A., Cocker, F., & Cameron, L. D. (2015). Using new technologies to improve the prevention and management of chronic conditions in populations. *Annual Review of Public Health, 36*(1), 483–505. DOI: 10.1146/annurev-publhealth-031914-122848

Oliveira, F., Costa, D. G., Assis, F., & Silva, I. (2024). Internet of Intelligent Things: A convergence of embedded systems, edge computing and machine learning. *Internet of Things : Engineering Cyber Physical Human Systems, 26*, 101153. DOI: 10.1016/j.iot.2024.101153

Ometov, A., Shubina, V., Klus, L., Skibińska, J., Saafi, S., Pascacio, P., Flueratoru, L., Gaibor, D. Q., Chukhno, N., Chukhno, O., Ali, A., Channa, A., Svertoka, E., Qaim, W. B., Casanova-Marqués, R., Holcer, S., Torres-Sospedra, J., Casteleyn, S., Ruggeri, G., & Lohan, E. S. (2021). A survey on wearable technology: History, state-of-the-art and current challenges. *Computer Networks, 193*, 108074. DOI: 10.1016/j.comnet.2021.108074

Orasugh, J. T., Botlhoko, O. J., Temane, L. T., & Ray, S. S. (2024). Progress in polymer nonwoven textile materials in electromagnetic interference shielding applications. *Functional Composite Materials, 5*(1), 5. Advance online publication. DOI: 10.1186/s42252-024-00054-6

Otte, F. W., Davids, K., Millar, S. K., & Klatt, S. (2020). When and how to provide feedback and instructions to athletes?—How sport psychology and pedagogy insights can improve coaching interventions to enhance self-regulation in training. *Frontiers in Psychology, 11*, 1444. DOI: 10.3389/fpsyg.2020.01444

Outlast Technologies. (2020). Phase change materials in smart textiles. NASA Spinoff Report. https://spinoff.nasa.gov/Spinoff2020/ps_1.html

Ozyagcilar, T. (2012). Implementing a tilt-compensated ecompass using accelerometer and magnetometer sensors. *Freescale Semiconductor, AN, 4248.*

Pal, P., Landivar-Bowles, J., Landivar-Scott, J., Duffield, N., Nowka, K., Jung, J., Chang, A., Lee, K., Zhao, L., & Bhandari, M. (2025). Unmanned aerial system and machine learning driven digital-twin framework for in-season cotton growth forecasting. *Computers and Electronics in Agriculture, 228*, 109589. DOI: 10.1016/j.compag.2024.109589

Panigutti, C., Perotti, A., & Pedreschi, D. (2020). Doctor XAI An ontology-based approach to black-box sequential data classification explanations. *FAT* 2020 - Proceedings of the 2020 Conference on Fairness, Accountability, and Transparency*, 629–639. DOI: 10.1145/3351095.3372855

Papi, E., Belsi, A., McGregor, A. H., & McCarthy, C. (2019). Smart socks for gait analysis: Reliability and clinical applications. *Gait & Posture, 72*, 55–60. DOI: 10.1016/j.gaitpost.2019.06.008

Park, S., & Jayaraman, S. (2003). Enhancing the quality of life through wearable technology. *IEEE Engineering in Medicine and Biology Magazine, 22*(3), 41–48. DOI: 10.1109/MEMB.2003.1213625 PMID: 12845818

Park, S.-Y., Lee, C., Jeong, S., Lee, J., Kim, D., Jang, Y., Seol, W., Kim, H., & Ahn, S.-H. (2024). Digital twin and deep reinforcement learning-driven robotic automation system for confined workspaces: A nozzle dam replacement case study in nuclear power plants. *International Journal of Precision Engineering and Manufacturing–Green Technology, 11*(3), 939–962. DOI: 10.1007/s40684-023-00593-6

Patel, S., Dholariya, K., Bhatt, N. V., & Thyagarajan, K. (2025, April). Data Analytics and Data Science in Diagnosing and Managing Sleep Disorders Using Wearable Devices. In *2025 Seventh International Conference on Computational Intelligence andCommunication Technologies (CCICT)* (pp. 594-600). IEEE. DOI: 10.1109/CCICT65753.2025.00095

Patel, S., Park, H., Bonato, P., Chan, L., & Rodgers, M. (2012). A review of wearable sensors and systems with application in rehabilitation. *Journal of Neuroengineering and Rehabilitation, 9*(1), 1–17. DOI: 10.1186/1743-0003-9-21

Paterson, T. E., Hagis, N., Boufidis, D., Wang, Q., Moore, S. R., Da Silva, A. C., Mitchell, R. L., Alix, J. J. P., & Minev, I. R. (2022). Monitoring of hand function enabled by low complexity sensors printed on textile. *Flexible and Printed Electronics, 7*(3), 035003. DOI: 10.1088/2058-8585/ac7dd1

Patil, D., Rane, N. L., Desai, P., & Rane, J. (2024). Machine learning and deep learning: Methods, techniques, applications, challenges, and future research opportunities. In *Trustworthy Artificial Intelligence in Industry and Society*. Deep Science Publishing., DOI: 10.70593/978-81-981367-4-9_2

Patro, K. K., Allam, J. P., Neelapu, B. C., Tadeusiewicz, R., Acharya, U. R., Hammad, M., Yildirim, O., & Pławiak, P. (2023). Application of Kronecker convolutions in deep learning technique for automated detection of kidney stones with coronal CT images. *Information Sciences*, *640*, 119005. DOI: 10.1016/j.ins.2023.119005

Peake, J. M., Kerr, G., & Sullivan, J. P. (2018). A critical review of consumer wearables, mobile applications, and equipment for providing biofeedback, monitoring stress, and sleep in physically active populations. *Frontiers in Physiology*, *9*, 743. DOI: 10.3389/fphys.2018.00743

Pereira, C., Pereira, A. M., Freire, C., Pinto, T. V., Costa, R. S., & Teixeira, J. S. (2020). Nanoengineered textiles: from advanced functional nanomaterials to groundbreaking highperformance clothing. In *Elsevier eBooks* (pp. 611–714). https://doi .org/DOI: 10.1016/b9780128167878.000211

Polglaze, T., Dawson, B., & Peeling, P. (2016, May). Gold Standard or Fool's Gold? The Efficacy of Displacement Variables as Indicators of Energy Expenditure in Team Sports. *Sports Medicine (Auckland, N.Z.)*, *46*(5), 657–670. Advance online publication. DOI: 10.1007/s40279-015-0449-x

Poongodi, T., Krishnamurthi, R., Indrakumari, R., Suresh, P., & Balusamy, B. (2019). Wearable devices and IoT. In *A handbook of Internet of Things in biomedical and cyber physical system* (pp. 245–273). Springer International Publishing.

Pu, J., Ma, K., Luo, Y., Tang, S., Liu, T., Liu, J., Leung, M., Yang, J., Hui, R., Xiong, Y., & Tao, X. (2023). Textile electronics for wearable applications. *International Journal of Extreme Manufacturing*, *5*(4), 042007. DOI: 10.1088/2631-7990/ace66a

Putra, K. T., Arrayyan, A. Z., Hayati, N., Damarjati, C., Bakar, A., & Chen, H. C. (2024). A review on the application of internet of medical things in wearable personal health monitoring: A cloud-edge artificial intelligence approach. *IEEE Access : Practical Innovations, Open Solutions*, *12*, 21437–21452. DOI: 10.1109/ACCESS.2024.3358827

Qi, Y., Sajadi, S. M., Baghaei, S., Rezaei, R., & Li, W. (2024). Digital technologies in sports: Opportunities, challenges, and strategies for safeguarding athlete wellbeing and competitive integrity in the digital era. *Technology in Society*, *77*, 102496. DOI: 10.1016/j.techsoc.2024.102496

Rahman, F., Ahmad, A., & Yadav, S. S. (2020). Low-power microcontroller-based system design for health monitoring applications. *International Journal of Embedded Systems and Applications*, *10*(3), 13–21.

Ramalho, A., & Petrica, J. (2023). Knowledge in motion: A comprehensive review of evidence-based human kinetics. *International Journal of Environmental Research and Public Health, 20*(11), 6020. DOI: 10.3390/ijerph20116020

Ramanujam, E., Perumal, T., & Padmavathi, S. (2021). Human activity recognition with smartphone and wearable sensors using deep learning techniques: A review. *IEEE Sensors Journal, 21*(12), 13029–13040. DOI: 10.1109/JSEN.2021.3069927

Rana, M., & Mittal, V. (2020). Wearable sensors for real-time kinematics analysis in sports: A review. *IEEE Sensors Journal, 21*(2), 1187–1207. DOI: 10.1109/JSEN.2020.3019016

Rather, H. A., & Dar, M. A. (2024). Smart Textiles with Integrated Biosensors for Real-time Health Monitoring. *Biosensors and Nanotheranostics, 3*(1), 1–9.

Reilly, B., Morgan, O., Czanner, G., & Robinson, M. A. (2021a). Automated Classification of Changes of Direction in Soccer Using Inertial Measurement Units. *Sensors 2021, Vol. 21, Page 4625, 21*(14), 4625. DOI: 10.3390/s21144625

Renaudin, V., Afzal, M. H., & Lachapelle, G. (2010, May). New method for magnetometers based orientation estimation. In *IEEE/ION Position, Location and Navigation Symposium* (pp. 348-356). IEEE. DOI: 10.1109/PLANS.2010.5507301

Rodgers, M. M., Pai, V. M., & Conroy, R. S. (2015). Recent advances in wearable sensors for health monitoring. *IEEE Sensors Journal, 15*(6), 3119–3126. DOI: 10.1109/JSEN.2014.2357257

Røsten, S., Skirbekk, S., Tjønndal, A., Canini, D., Wågan, F., & Grut Vorpvik, J. (2025). Game changer: the technoscientific revolution in sports: by Rayvon Fouché, Johns Hopkins University Press, 2017.

Roztocki, N., Soja, P., & Weistroffer, H. R. (2019). The role of information and communication technologies in socioeconomic development: Towards a multi-dimensional framework. *Information Technology for Development, 25*(2), 171–183. DOI: 10.1080/02681102.2019.1596654

Rubežienė, V., Varnaitė Žuravliova, S., Sankauskaitė, A., Pupeikė, J., Ragulis, P., & Abraitienė, A. (2023). The Impact of Structural Variations and Coating Techniques on the Microwave Properties of Woven Fabrics Coated with PEDOT: PSS Composition. *Polymers, 15*(21), 4224. DOI: 10.3390/polym15214224 PMID: 37959904

Ryalat, M., Franco, E., Elmoaqet, H., Almtireen, N., & Al-Refai, G. (2024). The integration of advanced mechatronic systems into industry 4.0 for smart manufacturing. *Sustainability (Basel), 16*(19), 8504. DOI: 10.3390/su16198504

Sadi, M. S., & Kumpikaitė, E. (2022). Advances in the robustness of wearable electronic textiles: Strategies, stability, washability and perspective. *Nanomaterials (Basel, Switzerland)*, *12*(12), 2039. DOI: 10.3390/nano12122039 PMID: 35745378

Sahu, D. P., Sharma, P., Kumar, V., Sharma, M., & Singh, R. (2024). Exploring person-job fit influence on innovation behavior through machine learning approach in public sector banking institutions. In Proceedings of the 2024 International Conference on IoT, Communication and Automation Technology (ICICAT) (pp. 844–848). IEEE. DOI: 10.1109/ICICAT62666.2024.10923296

Sahu, D. P., Murkute, S., Yadav, U., Agarwal, M., & Kumar, P. (2024). Deep learning based bug detection in Solidity smart contracts. In *Recent Trends in Image Processing and Pattern Recognition* (Vol. 2026). Springer., DOI: 10.1007/978-3-031-53082-1_9

Sajid, M., Singh, J., Haidri, R. A., Prasad, M., Varadarajan, V., Kotecha, K., & Garg, D. (2021). A novel algorithm for capacitated vehicle routing problem for smart cities. *Symmetry*, *13*(10), 1923. DOI: 10.3390/sym13101923

Saleem, H., & Zaidi, S. (2020). Sustainable use of nanomaterials in textiles and their environmental impact. *Materials (Basel)*, *13*(22), 5134. DOI: 10.3390/ma13225134 PMID: 33203051

Sanei, S., & Chambers, J. A. (2013). *EEG signal processing*. John Wiley & Sons.

Sankar, J. G. (2024). AI-driven marketing success stories: A case note of industry pioneers. In *AI-Driven Marketing Research and Data Analytics* (pp. 48–66). IGI Global. DOI: 10.4018/979-8-3693-2165-2.ch003

Sankar, K. N. A., & Mohanta, K. (2017). Preparation of highly conductive yarns by an optimized impregnation process. *Journal of Electronic Materials*, *47*(3), 1970–1978. DOI: 10.1007/s11664-017-5998-3

Saraswat, M., Sharma, P., Kumar, V., Sharma, M., & Singh, R. (2024). Optimizing Distribution Grid Operations with Deep Reinforcement Learning for a Novel Approach to Managing Renewable Energy Integration. In 2024 4th International Conference on Innovative Sustainable Computational Technologies (CISCT) (pp. 1-6). IEEE.

Saw, A. E., Main, L. C., & Gastin, P. B. (2016). Monitoring the athlete training response: subjective self-reported measures trump commonly used objective measures: a systematic review. *British Journal of Sports Medicine*, *50*(5), 281–291. DOI: 10.1136/bjsports-2015-094758

Saxena, M., Choudhary, A., Reddy, P. S., Sharma, V. K., & Kumar, V. (2024). Machine Learning-Based Optimization for Identifying Effective Drugs in Breast Cancer on an In Vitro Platform. In 2024 4th International Conference on Advance Computing and Innovative Technologies in Engineering (ICACITE) (pp. 1756-1760). IEEE. DOI: 10.1109/ICACITE60783.2024.10617067

Saxena, P., & Goyal, A. (2019). Study of computerized segmentation and classification techniques: An application to histopathological imagery. *Informatica: An International Journal of Computing and Informatics, 43*(4).

Saxena, P., Gupta, R., Goel, A., Sanger, A. K. S., Shrivastava, V., & Singh, M. (2025). Automated cellular segmentation in histopathology: A multi-step mask prediction approach for enhanced analysis. In *2025 7th International Conference on Signal Processing, Computing and Control (ISPCC)* (pp. 437–442). IEEE. DOI: 10.1109/ISPCC66872.2025.11039568

Saxena, P., Jain, P., Aggarwal, P. K., Singh, M., Goel, S., & Batra, M. (2024). Communication Requirements and Performance Metrics for Electric Vehicle Charging. In Optimized Energy Management Strategies for Electric Vehicles (1st ed., pp. 35–60). DOI: 10.4018/979-8-3693-6844-2.ch002

Saxena, P., Sharma, M., Batra, R., Gupta, R., Singh, M., & Singh, S. K. (2025). XAI for lymph node histopathology: From explainability to immersive metaverse applications. In *2025 7th International Conference on Signal Processing, Computing and Control (ISPCC)* (pp. 304–309). IEEE. DOI: 10.1109/ISPCC66872.2025.11039319

Saxena, P., Sharma, P., Kumar, V., Sharma, M., & Singh, Y. (2024). Optimizing Green Energy Systems with Deep Learning Techniques for Enhancements in Efficiency and Sustainability. In 2024 4th International Conference on Innovative Sustainable Computational Technologies (CISCT) (pp. 1-6). IEEE.

Saxena, P., & Goyal, A. (2023). Computer-assisted grading of follicular lymphoma: A classification based on SVM, machine learning, and transfer learning approaches. *Imaging Science Journal, 70*(1), 30–45. DOI: 10.1080/13682199.2022.2162663

Saxena, P., Goyal, A., Bivi, M. A., Singh, S. K., & Rashid, M. (2023). Segmentation of nucleus and cytoplasm from H&E-stained follicular lymphoma. *Electronics (Basel), 12*(3), 651. DOI: 10.3390/electronics12030651

Saxena, P., Singh, S. K., Rashid, M., Alshamrani, S. S., & Alnfiai, M. M. (2025). Efficient deep learning model for classifying lung cancer images using normalized stain-agnostic feature method and FastAI-2. *PeerJ. Computer Science, 11*, e2903. DOI: 10.7717/peerj-cs.2903 PMID: 40567713

Saxena, P., Sinha, A., & Singh, S. K. (2024). Computer-assisted interpretation, in-depth exploration and single cell type annotation of RNA sequence data using k-means clustering algorithm. *Computer Methods in Biomechanics and Biomedical Engineering*, •••, 1–11. DOI: 10.1080/10255842.2023.2300685 PMID: 38235728

Saxena, P., Sinha, A., & Singh, S. K. (2024). Review of computer-assisted diagnosis model to classify follicular lymphoma histology. *Cell Biochemistry and Function*, *42*(5), e4088. DOI: 10.1002/cbf.4088 PMID: 38973163

Schiavone, P. D., Rossi, D., Di Mauro, A., Gurkaynak, F. K., Saxe, T., Wang, M., Yap, K. C., & Benini, L. (2021). Arnold: An eFPGA-augmented RISC-V SoC for flexible and low-power IoT end nodes. *IEEE Transactions on Very Large Scale Integration (VLSI). IEEE Transactions on Very Large Scale Integration (VLSI) Systems*, *29*(4), 677–690. DOI: 10.1109/TVLSI.2021.3058162

Schneider, C., Hanakam, F., Wiewelhove, T., Döweling, A., Kellmann, M., Meyer, T., Pfeiffer, M., & Ferrauti, A. (2018). Heart rate monitoring in team sports—A conceptual framework for contextualizing heart rate measures for training and recovery prescription. *Frontiers in Physiology*, *9*, 639. DOI: 10.3389/fphys.2018.00639

Seçkin, A. Ç., Ateş, B., & Seçkin, M. (2023). Review on Wearable Technology in sports: Concepts, Challenges and opportunities. *Applied Sciences (Basel, Switzerland)*, *13*(18), 10399. DOI: 10.3390/app131810399

Seshadri, D. R., Li, R. T., Voos, J. E., Rowbottom, J. R., Alfes, C. M., Zorman, C. A., & Drummond, C. K. (2019). Wearable sensors for monitoring the internal and external workload of the athlete. *Npj Digital Medicine 2019 2:1*, *2*(1), 1–18. DOI: 10.1038/s41746-019-0149-2

Seshadri, D. R., Drummond, C., Craker, J., Rowbottom, J. R., & Voos, J. E. (2017). Wearable devices for sports: New integrated technologies allow coaches, physicians, and trainers to better understand the physical demands of athletes in real time. *IEEE Pulse*, *8*(1), 38–43. DOI: 10.1109/MPUL.2016.2627240

Seshadri, D. R., Thom, M. L., Harlow, E. R., Gabbett, T. J., Geletka, B. J., Hsu, J. J., Drummond, C. K., Phelan, D. M., & Voos, J. E. (2021). Wearable technology and analytics as a complementary toolkit to optimize workload and to reduce injury burden. *Frontiers in Sports and Active Living*, *2*, 630576. DOI: 10.3389/fspor.2020.630576

Seyam, A. F. M., & Hamouda, T. (2013). Smart textiles: Evaluation of optical fibres as embedded sensors for structure health monitoring of fibre reinforced composites. *Journal of the Textile Institute*, *104*(8), 892–899. DOI: 10.1080/00405000.2013.765087

Sezer, O. B., Dogdu, E., & Ozbayoglu, A. M. (2017). Context-aware computing, learning, and big data in internet of things: A survey. *IEEE Internet of Things Journal, 5*(1), 1–27. DOI: 10.1109/JIOT.2017.2773600

Shah, A. (2024). The future of manufacturing with AI and data analytics. In *The Future of Manufacturing with AI and Data Analytics* (pp. 1–20). Wiley., DOI: 10.1002/9781394303601.ch23

Shahidi, S., & Moazzenchi, B. (2018). Carbon nanotube and its applications in textile industry – A review. *Journal of the Textile Institute, 109*(12), 1653–1666. DOI: 10.1080/00405000.2018.1437114

Shahid, M. A., Rahman, M. M., Hossain, M. T., Hossain, I., Sheikh, M. S., Rahman, M. S., Uddin, N., Donne, S. W., & Hoque, M. I. U. (2025). Advances in conductive PolymerBased flexible electronics for multifunctional applications. *Journal of Composites Science, 9*(1), 42. DOI: 10.3390/jcs9010042

Shaik, S., Ahad, A., Murugavalli, N., Kavitha, V., Suryanarayana, S., & Raj, M. (2024). *ODESMAN: Optimizing Decision-Making in Complex Environments: Integrating Neutrosophic and Fuzzy Logic for Advanced System Modeling*. International Journal of Neutrosophic Science., DOI: 10.54216/IJNS.240226

Sharma, P., Kumar, V., Sharma, M., & Singh, T. (2024). Enhancing Microgrid Efficiency With Artificial Intelligence-Based Solar Energy Forecasting and Load Management. In 2024 4th International Conference on Innovative Sustainable Computational Technologies (CISCT) (pp. 1-6). IEEE.

Sharma, A., Singh, M., Gupta, M., Sukhija, N., & Aggarwal, P. K. (2022). IoT and blockchain technology in 5G smart healthcare. In *Blockchain Applications for Healthcare Informatics: beyond 5G* (pp. 137–161). Elsevier., DOI: 10.1016/B978-0-323-90615-9.00004-9

Sharma, R., & Gupta, H. (2024). Leveraging cognitive digital twins in Industry 5.0 for achieving Sustainable Development Goal 9: An exploration of inclusive and sustainable industrialization strategies. *Journal of Cleaner Production, 448*, 141364. DOI: 10.1016/j.jclepro.2024.141364

Sharp, M., Ak, R., & Hedberg, T.Jr. (2018). A survey of the advancing use and development of machine learning in smart manufacturing. *Journal of Manufacturing Systems, 48*, 170–179. DOI: 10.1016/j.jmsy.2018.02.004 PMID: 31092965

Shateri-Khalilabad, M., & Yazdanshenas, M. E. (2013). Fabricating electroconductive cotton textiles using graphene. *Carbohydrate Polymers, 96*(1), 190–195. DOI: 10.1016/j.carbpol.2013.03.052 PMID: 23688469

Shcherbina, A. D., Mattsson, M. L., & McCarthy, J. M. (2021). Effect of wearable technology on patient readmission: A meta-analysis. *Journal of Medical Internet Research, 23*(2). https://www.jmir.org/2021/2/e20697

Shi, B., Chen, X., Yue, Z., Yin, S., Weng, Q., Zhang, X., Wang, J., & Wen, W. (2019). Wearable ankle robots in post-stroke rehabilitation of gait: A systematic review. *Frontiers in Neurorobotics, 13*, 63. DOI: 10.3389/fnbot.2019.00063 PMID: 31456681

Shi, J., Liu, S., Zhang, L., Yang, B., Shu, L., Yang, Y., Ren, M., Wang, Y., Chen, J., Chen, W., Chai, Y., & Tao, X. (2020). Smart textile-integrated microelectronic systems for wearable applications. *Advanced Materials, 32*(5), 1901958. DOI: 10.1002/adma.201901958 PMID: 31273850

Shi, Q., Sun, J., Hou, C., Li, Y., Zhang, Q., & Wang, H. (2019). Advanced functional fiber and smart textile. *Advanced Fiber Materials, 1*(1), 3–31. DOI: 10.1007/s42765-019-0002-z

Shivahare, B. D., Ravi, V., Chandan, R. R., Alahmadi, T. J., Singh, P., & Diwakar, M. (2024). Delving into machine learning's influence on disease diagnosis and prediction. *The Open Public Health Journal, 17*(1), e18749445297804. DOI: 10.2174/0118749445297804240401061128

Siddiqi, S. J., Saleh, S., Jan, M. A., & Tariq, M. (2025). Securing the metaverse: Web 3.0 for decentralized digital twin-enhanced vehicle–road safety. *Future Generation Computer Systems, 164*, 107555. DOI: 10.1016/j.future.2024.107555

Siddle, J., Weaver, K., Greig, M., Harper, D., & Brogden, C. M. (2024). A low-volume Nordic hamstring curl programme improves change of direction ability, despite no architectural, strength or speed adaptations in elite youth soccer players. *Research in Sports Medicine, 32*(1), 49–60. DOI: 10.1080/15438627.2022.2079984

Siekkinen, M., Hiienkari, M., Nurminen, J. K., & Nieminen, J. (2012). How low energy is Bluetooth Low Energy? Comparative measurements with ZigBee/802.15.4. *2012 IEEE WCNC Workshops*, 232–237. DOI: DOI: 10.1109/WCNCW.2012.6215496

Silvestre De Oliveira, M., Fidelis Peixer, G., Forcellini, F. A., Riso Barbosa Jr, J., & Lozano Cadena, J. A. (2023). Model-based trade-off curves to support the set-based concurrent engineering of highly innovative projects. *Concurrent Engineering, Research and Applications, 31*(3-4), 89–100. DOI: 10.1177/1063293X231209680

Singh, B. P., Dandotiya, M., Ekvitayavetchanukul, P., Rana, M., & Singh, B. (2025). Predicting blood glucose levels in type 1 diabetes using deep learning and regression techniques. In 2025 International Conference on Intelligent Control, Computing and Communications (IC3) (pp. 86–91). IEEE. DOI: 10.1109/IC363308.2025.10956664

Singh, M., & Malik, A. (2024). Multi-Hop Routing Protocol in SDN-Based Wireless Sensor Network. In Software-Defined Network Frameworks (1st ed., pp. 121–141). CRC Press. DOI: 10.1201/9781040018323-8

Singh, M., Sharma, M., Gupta, I., Sharma, C., Kaur, G., & Saxena, P. (2025). Energy Aware Routing Protocols Using Modified Ant Colony Optimization for High-Mobility MANETs in Disaster Recovery Scenarios. *2025 7th International Conference on Signal Processing, Computing and Control (ISPCC)*, 443–448. DOI: 10.1109/ISPCC66872.2025.11039378

Singh, M., Sukhija, N., Sharma, A., Gupta, M., & Aggarwal, P. K. (2021). Security and Privacy Requirements for IoMT-Based Smart Healthcare System: Challenges, Solutions, and Future Scope. In Big Data Analysis for Green Computing: Concepts and Applications. DOI: 10.1201/9781003032328-2

Singh, M. K., Bhatnagar, A. R., & Tripathi, N. (2020). Biometric clothing to support physical rehabilitation: A clinical trial. *International Journal of Rehabilitation Research. Internationale Zeitschrift fur Rehabilitationsforschung. Revue Internationale de Recherches de Readaptation*, *42*(1), 12–20. PMID: 31658111

Singh, M., Gupta, M., Sharma, A., Jain, P., & Aggarwal, P. K. (2023). Role of deep learning in healthcare industry: Limitations, challenges and future scope. In *Deep learning for healthcare services* (pp. 1–22). Bentham Science., DOI: 10.2174/9789815080230123020003

Singh, M., & Malik, A. (2025). Energy-Efficient Cluster Head Selection in Heterogeneous WSNs Using Aquila Optimizer and Fuzzy Logic (AO-FL). *International Journal of Communication Systems*, *38*(16), 1–22. DOI: 10.1002/dac.70272

Singh, M., Sharma, R., Singh, R., Malik, A., & Aggarwal, P. K. (2024). Advancements in renewable energy harvesting for EV charging infrastructure. In *Optimized Energy Management Strategies for Electric Vehicles* (pp. 119–152). IGI Global., DOI: 10.4018/979-8-3693-6844-2.ch005

Singh, N. P., Hasan, D. S., Shelke, N. A., Namdev, A., & Girdhar, N. (2024). An integrated machine learning and IoT based approach for enhanced healthcare efficiency and personalized treatment. In Abraham, A., Bajaj, A., Hanne, T., & Siarry, P. (Eds.), *Intelligent Systems Design and Applications* (Vol. 1046). Springer., DOI: 10.1007/978-3-031-64813-7_16

Singh, R. K., Pappinisseri Puluckul, P., Berkvens, R., & Weyn, M. (2020). Energy consumption analysis of LPWAN technologies and lifetime estimation for IoT application. *Sensors (Basel)*, *20*(17), 4794. DOI: 10.3390/s20174794 PMID: 32854350

Sivakumar, C. L. V., Mone, V., & Abdumukhtor, R. (2024). Addressing privacy concerns with wearable health monitoring technology. *Wiley Interdisciplinary Reviews. Data Mining and Knowledge Discovery*, *14*(3), e1535. DOI: 10.1002/widm.1535

Smarandache, F. (1998). *A unifying field in logics: Neutrosophy: Neutrosophic probability, set, and logic*. American Research Press.

Smarandache, F. (2005). Neutrosophic set: A generalization of the intuitionistic fuzzy sets. *International Journal of Pure and Applied Mathematics*, *24*, 287–297.

Smith, A. A., Li, R., & Tse, Z. T. H. (2023). Reshaping healthcare with wearable biosensors. *Scientific Reports*, *13*(1), 4998. DOI: 10.1038/s41598-022-26951-z PMID: 36973262

Smits, B. L., Pepping, G. J., & Hettinga, F. J. (2014). Pacing and decision making in sport and exercise: The roles of perception and action in the regulation of exercise intensity. *Sports Medicine (Auckland, N.Z.)*, *44*(6), 763–775. DOI: 10.1007/s40279-014-0163-0

Sousa Lima, W., Souto, E., El-Khatib, K., Jalali, R., & Gama, J. (2019). Human activity recognition using inertial sensors in a smartphone: An overview. *Sensors (Basel)*, *19*(14), 3213. DOI: 10.3390/s19143213

Spanakis, M., Fragkiadaki, P., Renieri, E., Vakonaki, E., Fragkiadoulaki, I., Alegakis, A., Kiriakakis, M., Panagiotou, N., Ntoumou, E., Gratsias, I., Zoubaneas, E., Morozova, G. D., Ovchinnikova, M. A., Tsitsimpikou, C., Tsarouhas, K., Drakoulis, N., Skalny, A. V., & Tsatsakis, A. (2024). Advancing athletic assessment by integrating conventional methods with cutting-edge biomedical technologies for comprehensive performance, wellness, and longevity insights. *Frontiers in Sports and Active Living*, *5*, 1327792. DOI: 10.3389/fspor.2023.1327792

Spender, A., Bullen, C., Altmann-Richer, L., Cripps, J., Duffy, R., Falkous, C., Farrell, M., Horn, T., Wigzell, J., & Yeap, W. (2019). Wearables and the internet of things: Considerations for the life and health insurance industry. *British Actuarial Journal*, *24*, e22. DOI: 10.1017/S1357321719000072

Spiteri, T., Newton, R. U., Binetti, M., Hart, N. H., Sheppard, J. M., & Nimphius, S. (2015). Mechanical determinants of faster change of direction and agility performance in female basketball athletes. *Journal of Strength and Conditioning Research*, *29*(8), 2205–2214. DOI: 10.1519/JSC.0000000000000876

Spiteri, T., Newton, R. U., & Nimphius, S. (2015). Neuromuscular strategies contributing to faster multidirectional agility performance. *Journal of Electromyography and Kinesiology, 25*(4), 629–636. DOI: 10.1016/j.jelekin.2015.04.009

Spörk, M., Boano, C. A., & Römer, K. (2019). Performance and trade-offs of the new PHY modes of BLE 5. *Proceedings of the 1st Workshop on PERSIST-IoT*, 7–12. DOI: DOI: 10.1145/3331052.3332471

Spörk, M., Classen, J., Boano, C. A., Hollick, M., & Römer, K. (2020). Improving the reliability of Bluetooth Low Energy connections in the presence of interference. In *Proceedings of the 17th International Conference on Embedded Wireless Systems and Networks* (pp. 144–155). Junction Publishing. (No DOI)

Stavrakis, A. K., Simić, M., & Stojanović, G. M. (2021). Electrical characterization of conductive threads for textile electronics. *Electronics (Basel), 10*(8), 967. DOI: 10.3390/electronics10080967

Stoppa, M., & Chiolerio, A. (2014). Wearable electronics and smart textiles: A critical review. *sensors, 14*(7), 11957-11992.

Stoppa, M.. (2022). Electromyography-based monitoring using smart textile garments for sports training. *Journal of Sports Engineering and Technology, 236*(3), 234–246.

Stoppa, M., & Chiolerio, A. (2014). Wearable electronics and smart textiles: A critical review. *Sensors (Basel), 14*(7), 11957–11992. DOI: 10.3390/s140711957 PMID: 25004153

Stoppa, M., & Chiolerio, A. (2023). Sustainable smart textiles: Advances in eco-friendly materials and manufacturing for wearable applications. *Journal of Cleaner Production, 412*, 137256.

Sugiyama, T., Maeo, S., Kurihara, T., Kanehisa, H., & Isaka, T. (2021). Change of Direction Speed Tests in Basketball Players: A Brief Review of Test Varieties and Recent Trends. *Frontiers in Sports and Active Living, 3*, 645350. DOI: 10.3389/fspor.2021.645350

Sultanpure, K. A., Bagade, J., Bangare, S. L., Bangare, M. L., Bamane, K. D., & Patankar, A. J. (2024). Internet of things and deep learning-based digital twins for diagnosis of brain tumor by analyzing MRI images. *Measurement. Sensors, 33*, 101220. DOI: 10.1016/j.measen.2024.101220

Sun, Y., Van, M., McIlvanna, S., Nhat, N. M., Olayemi, K., Close, J., & McLoone, S. (2024). Digital twin-driven reinforcement learning for obstacle avoidance in robot manipulators: A self-improving online training framework. *arXiv preprint* arXiv:2403.13090. https://arxiv.org/abs/2403.13090

Suo, X., Tang, W., & Li, Z. (2024). Motion Capture Technology in Sports Scenarios: A Survey. *Sensors 2024, Vol. 24, Page 2947, 24*(9), 2947. DOI: 10.3390/s24092947

Su, Y., Li, Y., & Cao, Z. (2023). Gait-based privacy protection for smart wearable devices. *IEEE Internet of Things Journal, 11*(2), 3497–3509. DOI: 10.1109/JIOT.2023.3296650

Syduzzaman, M., Hassan, A., Anik, H. R., Akter, M., & Islam, M. R. (2023). Nanotechnology for High-Performance Textiles: A promising frontier for innovation. *ChemNanoMat : Chemistry of Nanomaterials for Energy, Biology and More*, *9*(9), e202300205. Advance online publication. DOI: 10.1002/cnma.202300205

Takabayashi, K., Tanaka, S., & Sakakibara, K. (2022). Influence of on-body propagation on UWB wearable device communications. *Electronics (Basel), 11*(21), 3491. DOI: 10.3390/electronics11213491

Tao, X. (2005). *Wearable electronics and photonics*. Woodhead Publishing., DOI: 10.1533/9781845690868

Tariq, M. U. (2024). Advanced wearable medical devices and their role in transformative remote health monitoring. In *Transformative approaches to patient literacy and healthcare innovation* (pp. 308–326). IGI Global. DOI: 10.4018/979-8-3693-3661-8.ch015

Tat, T., Chen, G., Zhao, X., Zhou, Y., Xu, J., & Chen, J. (2022). Smart textiles for healthcare and sustainability. *ACS Nano*, *16*(9), 13301–13313. DOI: 10.1021/acsnano.2c06287 PMID: 35969207

Thrun, S. (2002). Probabilistic robotics. *Communications of the ACM*, *45*(3), 52–57. DOI: 10.1145/504729.504754

Tseghai, G. B., Mengistie, D. A., Malengier, B., Fante, K. A., & Van Langenhove, L. (2020). PEDOT:PSS Based Conductive Textiles and their Applications. *Sensors (Basel), 20*(7), 1881. DOI: 10.3390/s20071881 PMID: 32231114

Tyagi, A. K., Bhatt, P., Chidambaram, N., & Kumari, S. (2024). Artificial intelligence empowered smart manufacturing for modern society: a review. Artificial Intelligence-Enabled Digital Twin for Smart Manufacturing, 55-83.

U.S. Army Edgewood Chemical Biological Center. (2020). Smart fabrics for hazardous environments: Sensor integration and rapid detection (Technical Report).

U.S. Army Natick Soldier Research, Development and Engineering Center. (2021). *SMART e-textile program: Reducing heat stress in military personnel.* U.S. Department of Defense.

U.S. DARPA. (2019). Smart wound monitoring bandage initiative: Budget overview. *Defense Research Funding Reports*, *12*, 45–48.

Ugwuanyi, J. C., Sigweni, B., & Abu-Mahfouz, A. M. (2021). Performance analysis of LoRaWAN and NB-IoT for smart city applications. *Electronics (Basel)*, *10*(18), 2224. DOI: 10.3390/electronics10182224

Umek, A., & Kos, A. (2016). The role of high performance computing and communication for real-time biofeedback in sport. *Mathematical Problems in Engineering*, *2016*, 2016. DOI: 10.1155/2016/4829452

Van Eetvelde, H., Mendonça, L. D., Ley, C., Seil, R., & Tischer, T. (2021). Machine learning methods in sport injury prediction and prevention: A systematic review. *Journal of Experimental Orthopaedics*, *8*(1), 1–15. DOI: 10.1186/s40634-021-00346-x

Vanini, S., Faraci, F., Ferrari, A., & Giordano, S. (2016). Using barometric pressure data to recognize vertical displacement activities on smartphones. *Computer Communications*, *87*, 37–48. DOI: 10.1016/j.comcom.2016.02.011

Verma, M., Singh, J., & Kumari, S. (2025). *Hybrid Attention-Augmented Deep Neural Network framework based on Ant Colony and Grey Wolf Optimization for Diagnosis of aortic aneurysm. International Journal of Information Technology.* Springer., DOI: 10.1007/s41870-025-02812-6

Vilas-Boas, J. L., Rodrigues, J. J. P. C., & Alberti, A. M. (2022). Convergence of distributed ledger technologies with digital twins, IoT, and AI for fresh food logistics: Challenges and opportunities. *Journal of Industrial Information Integration*, *31*, 100393. DOI: 10.1016/j.jii.2022.100393

Virkki, J., Björninen, T., Kellomäki, T., Merilampi, S., Ukkonen, L., & Sydänheimo, L. (2014). Reliability of washable, wearable, screen-printed UHF RFID tags. *Microelectronics and Reliability*, *54*(4), 840–846. DOI: 10.1016/j.microrel.2013.12.011

Wahyono, H., Khumaidi, A., & Setyawati, K. (2024). *Application of Fuzzy Logic in PEOPLES Framework for Community Resilience Measurement in Flood Disaster. Jurnal Ilmiah Merpati*. Menara Penelitian Akademika Teknologi Informasi., DOI: 10.24843/JIM.2024.v12.i02.p05

Wang, B., Zhou, H., Yang, G., Li, X., & Yang, H. (2022). Human digital twin (HDT)-driven human–cyber–physical systems: Key technologies and applications. [English Edition]. *Chinese Journal of Mechanical Engineering*, *35*(1), 11. Advance online publication. DOI: 10.1186/s10033-022-00680-w

Wang, C. C., & Chien, C. H. (2025). Machine Learning for Industrial Optimization and Predictive Control: A Patent-Based Perspective with a Focus on Taiwan's High-Tech Manufacturing. *Processes (Basel, Switzerland), 13*(7), 2256. DOI: 10.3390/pr13072256

Wang, H., Zhang, Y., Liang, X., & Zhang, Y. (2021). Smart fibers and textiles for personal health management. *ACS Nano, 15*(8), 12497–12508. DOI: 10.1021/acsnano.1c06230 PMID: 34398600

Wang, H., Zhang, Y., & Yu, D. (2022). Embroidery-based e-textiles for wearable sensing applications: A review. *Textile Research Journal, 92*(13–14), 2062–2078.

Wang, J., Xu, L., Memon, M., & Hassan, M. M. (2021). Smart wearables for the Internet of Things: Current applications and challenges. *IEEE Internet of Things Journal, 8*(10), 7789–7810. DOI: 10.1109/JIOT.2021.3052804

Wang, L., Zhao, F., & Zhang, G. (2022). Analysis on the Impact of Large-Scale Sports Events on Regional Economy Based on SWOT-PEST Model. *Journal of Mathematics, 2022*(1), 7769128. Advance online publication. DOI: 10.1155/2022/7769128

Wang, M., Chen, C., Wu, H., Zhang, J., Zhou, S., & Wang, G. (2024, October). Will Smart Ring be Next Wave of Wearables? In *2024 IEEE Biomedical Circuits and Systems Conference (BioCAS)* (pp. 1-5). IEEE. DOI: 10.1109/BioCAS61083.2024.10798374

Wang, X., Zhang, Y., & Li, H. (2024). Integration of digital twin and AI for smart manufacturing: A review. *Journal of Manufacturing Science and Engineering, 146*(1), 1–13. DOI: 10.1115/1.4051234

Wang, Z. L., & Chen, J. (2016). Wearable triboelectric nanogenerators for motion energy harvesting and self-powered sensors. *Advanced Materials, 28*(22), 4523–4535.

Waqar, A., Ahmad, I., Habibi, D., Hart, N., & Phung, Q. V. (2021). Enhancing athlete tracking using data fusion in wearable technologies. *IEEE Transactions on Instrumentation and Measurement, 70*, 1–13. DOI: 10.1109/TIM.2021.3069520

Wardiningsih, W. (2009). Study of comfort properties of natural and synthetic knitted fabrics in different blend ratios for winter active sportswear (Doctoral dissertation, RMIT University).

Weng, J. K., Virk, R., Kaiser, K., Hoffman, K. E., Goodman, C. R., Mitchell, M., Shaitelman, S., Schlembach, P., Reed, V., Wu, C.-F., Xiao, L., Smith, G. L., & Smith, B. D. (2024). Automated, real-time integration of biometric data from wearable devices with electronic medical records: A feasibility study. *JCO Clinical Cancer Informatics, 8*(8), e2400040. DOI: 10.1200/CCI.24.00040 PMID: 39348612

Williams, C. A., & Winsley, R. (2020). Non-Functional Overreaching and Overtraining in Young Swimmers. In *High Performance Youth Swimming* (pp. 215–227). Routledge. DOI: 10.4324/9780429465598-18

Wong, C., Zhang, Z. Q., Lo, B., & Yang, G. Z. (2015). Wearable sensing for solid biomechanics: A review. *IEEE Sensors Journal, 15*(5), 2747–2760. DOI: 10.1109/JSEN.2015.2393883

Xiang, H., Li, Y., Liao, Q., Xia, L., Wu, X., Zhou, H., Li, C., & Fan, X. (2024). Recent advances in smart fabric-type wearable electronics toward comfortable wearing. *Energies, 17*(11), 2627. DOI: 10.3390/en17112627

Xiao, L., Cao, Y., Gai, Y., Liu, J., Zhong, P., & Moghimi, M. M. (2023). Review on the application of cloud computing in the sports industry. *Journal of Cloud Computing (Heidelberg, Germany), 12*(1), 152. DOI: 10.1186/s13677-023-00531-6

Xu, H., Omitaomu, F., Sabri, S., Zlatanova, S., Li, X., & Song, Y. (2024). Leveraging generative AI for urban digital twins: A scoping review on the autonomous generation of urban data, scenarios, designs, and 3D city models for smart city advancement. *arXiv preprint* arXiv:2405.19464. https://arxiv.org/abs/2405.19464

Xu, K., Lu, Y., & Takei, K. (2019). Multifunctional skin-inspired flexible sensor systems for wearable electronics. *Advanced Materials Technologies, 4*(3), 1800628. DOI: 10.1002/admt.201800628

Yamamoto, K.. (2019). Sweat biosensors for wearable fitness monitoring. *Biosensors & Bioelectronics, 124*, 89–97.

Yang, D., Maity, S., & Sen, S. (2022). Physically secure wearable–wearable through-body interhuman body communication. *Frontiers in Electronics, 2*, 807051. DOI: 10.3389/felec.2021.807051

Yang, L., Amin, O., & Shihada, B. (2024a). Intelligent Wearable Systems: Opportunities and Challenges in Health and Sports. *ACM Computing Surveys, 56*(7), 1–42. DOI: 10.1145/3648469

Yao, X., Yan, H., Zhou, J., Li, Y., & Yu, H. (2025). Smart Manufacturing, Robotics, and AI Systems. Advances in Artificial Intelligence Applications in Industrial and Systems Engineering, 61-78.

Yao, S., Yang, J., Poblete, F. R., Hu, X., & Zhu, Y. (2019). Multifunctional electronic textiles using Silver nanowire composites. *ACS Applied Materials & Interfaces, 11*(34), 31028–31037. DOI: 10.1021/acsami.9b07520 PMID: 31373192

Yazici, İ., Shayea, I., & Din, J. (2023). A survey of applications of artificial intelligence and machine learning in future mobile networks-enabled systems. *Engineering Science and Technology, an International Journal*, *44*, 101455. Advance online publication. DOI: 10.1016/j.jestch.2023.101455

Yin, J., Li, J., Reddy, V. S., Ji, D., Ramakrishna, S., & Xu, L. (2023). Flexible Textile-Based sweat sensors for wearable applications. *Biosensors (Basel)*, *13*(1), 127. DOI: 10.3390/bios13010127 PMID: 36671962

Yin, R., Wang, D., Zhao, S., Lou, Z., & Shen, G. (2021). Wearable sensors-enabled human–machine interaction systems: From design to application. *Advanced Functional Materials*, *31*(11), 2008936. DOI: 10.1002/adfm.202008936

Younes, B. (2023). Smart Etextiles: A review of their aspects and applications. *Journal of Industrial Textiles*, *53*, 15280837231215493. Advance online publication. DOI: 10.1177/15280837231215493

Young, W., Dos'Santos, T., Harper, D., Jefferys, I., & Talpey, S. (2022). Agility in Invasion Sports: Position Stand of the IUSCA. *International Journal of Strength and Conditioning*, *2*(1). Advance online publication. DOI: 10.47206/ijsc.v2i1.126

Zadeh, L. (1983). The role of fuzzy logic in the management of uncertainty in expert systems. *Fuzzy Sets and Systems*, *11*(1-3), 197–198. DOI: 10.1016/S0165-0114(83)80081-5

Zadeh, L. A. (1965). Fuzzy sets. *Information and Control*, *8*(3), 338–353. DOI: 10.1016/S0019-9958(65)90241-X

Zainab, H., Khan, A. H., Khan, R., & Hussain, H. K. (2024). Integration of AI and wearable devices for continuous cardiac health monitoring. *International Journal of Multidisciplinary Sciences and Arts*, *3*(4), 123–139. DOI: 10.47709/ijmdsa.v3i4.4956

Zaman, S. U., Tao, X., Cochrane, C., & Koncar, V. (2022). Smart E-Textile Systems: A review for healthcare applications. *Electronics (Basel)*, *11*(1), 99. DOI: 10.3390/electronics11010099

Zeng, W., Shu, L., Li, Q., Chen, S., Wang, F., & Tao, X. M. (2014). Fiber-based wearable electronics: A review of materials, fabrication, devices, and applications. *Advanced Materials*, *26*(31), 5310–5336. DOI: 10.1002/adma.201400633 PMID: 24943999

Zhai, H., Liu, J., Liu, Z., & Li, Y. (2025). Functional graphene fiber materials for advanced wearable applications. *Advanced Fiber Materials*, *7*(2), 443–468. Advance online publication. DOI: 10.1007/s42765-025-00512-1

Zhang, M., & Sawchuk, A. A. (2012). A Feature Selection-Based Framework for Human Activity Recognition Using Wearable Multimodal Sensors. *BODYNETS 2011 - 6th International ICST Conference on Body Area Networks*, 92–98. DOI: 10.4108/icst.bodynets.2011.247018

Zhang, H., Shen, Q., Zheng, P., Wang, H., Zou, R., Zhang, Z., Pan, Y., Zhi, J.-Y., & Xiang, Z. R. (2024). Harvesting inertial energy and powering wearable devices: A review. *Small Methods*, *8*(1), 2300771. DOI: 10.1002/smtd.202300771 PMID: 37853661

Zhang, H., Xiao, H., & Long, J. (2025). Preparation, structure, property and application of MXene in fabricating functional and intelligent textiles: A comprehensive review. *Composites. Part B, Engineering*, *112461*, 112461. Advance online publication. DOI: 10.1016/j.compositesb.2025.112461

Zhang, S., Li, Y., Zhang, S., Shahabi, F., Xia, S., Deng, Y., & Alshurafa, N. (2022). Deep learning in human activity recognition with wearable sensors: A review on advances. *Sensors (Basel)*, *22*(4), 1476. DOI: 10.3390/s22041476

Zhang, Y., Li, B., & Wang, J. (2024). A review on digital twin technology in smart manufacturing. *Robotics and Computer-integrated Manufacturing*, *78*, 102456. DOI: 10.1016/j.rcim.2024.102456

Zhang, Z., Chen, P., Guo, K., Wang, C., & Zhang, X. (2015). A colour-tunable, weavable fibre-shaped polymer light-emitting electrochemical cell. *Nature Photonics*, *9*(4), 233–238. DOI: 10.1038/nphoton.2015.37

Zhao, N., Tian, Y., Zhang, H., & Herrera-Viedma, E. (2024). Fuzzy-based adaptive event-triggered control for nonlinear cyber-physical systems against deception attacks via a single parameter learning method. *Information Sciences*, *657*, 119948. DOI: 10.1016/j.ins.2023.119948

Zhao, Y.. (2021). Smart garments for health monitoring: Current advancements and challenges. *IEEE Reviews in Biomedical Engineering*, *14*, 169–182.

Zhao, Y., Guo, X., & Zhou, M. (2021). Data security and privacy in smart textiles: A survey. *IEEE Internet of Things Journal*, *8*(14), 11181–11195. DOI: 10.1109/JIOT.2021.3065584

Zheng, H., Xie, W., Ryzhov, I. O., & Choy, K. (2025). Digital twin calibration with model-based reinforcement learning. *arXiv preprint* arXiv:2501.02205. https://arxiv.org/abs/2501.02205

Zheng, Y., Zhang, Q., Jin, W., Jing, Y., Chen, X., Han, X., Bao, Q., Liu, Y., Wang, X., Wang, S., Qiu, Y., Di, C., & Zhang, K. (2019). Carbon nanotube yarn based thermoelectric textiles for harvesting thermal energy and powering electronics. *Journal of Materials Chemistry. A, Materials for Energy and Sustainability*, *8*(6), 2984–2994. DOI: 10.1039/C9TA12494B

Zhou, M., Gao, Q., & Wang, S. (2020). Energy harvesting for wearable and stretchable electronics: From flexible supercapacitors to self-powered sensors. *Nano Energy*, *72*, 104676. DOI: 10.1016/j.nanoen.2020.104676

Zhou, Y., Chen, L., & Mathur, A. (2020). Privacy and security in wearable health monitoring systems: A review. *ACM Computing Surveys*, *53*(6), 126. Advance online publication. DOI: 10.1145/3417984

Zjavka, L. (2022). Power quality 24-hour prediction using differential, deep and statistics machine learning based on weather data in an off-grid. *Journal of the Franklin Institute*, •••, 1–25. DOI: 10.1016/j.jfranklin.2022.06.048

About the Contributors

Pranshu Saxena is an Assistant Professor at Bennett University, Greater Noida, with a Ph.D. in Computer Science and Engineering from Punjab Technical University, Jalandhar. With 11+ years of academic and research experience, his expertise includes medical image processing, automated image segmentation, texture analysis, and pattern recognition. He has published high-impact SCI and SCOPUS-indexed research papers in reputed journals and authored books in computer-assisted medical image analysis. Recognized for his contributions to histopathological image analysis and AI-driven medical diagnostics, he continues to make significant strides in research, teaching, and academic leadership.

Mandeep Singh is an accomplished academician and Assistant Professor in the School of Computer Science, Engineering, and Technology (SCSET) at Bennett University, Greater Noida, India. He holds a Master's degree with Distinction in Information Security from Ambedkar Institute of Advanced Communication Technologies and Research (AIACTR) (now NSUT East Campus). With over nine years of teaching experience, he is currently pursuing a Ph.D. in Computer Science & Engineering from Dr. B.R. Ambedkar National Institute of Technology (NIT) Jalandhar, where he completed his coursework with Distinction (8.00 CGPA). Mr. Singh's expertise spans Computer Networks, Mobile Ad Hoc Networks (MANETs), Machine Learning, and Deep Learning. His research contributions include book chapters, international journals, and conference proceedings, alongside a notable track record of patent publications. Additionally, he actively reviews papers for international conferences and journals and has held special sessions at prestigious events. His dedication to excellence has earned him awards and recognition from the Institute of Industrial Development (IID) and MSME, Government of India for contributions to research and entrepreneurship.

Sanjay Kumar Singh is currently working as an assistant professor in university school of automation and Robotics, Guru Gobind Singh Indraprastha University,

East Delhi campus, Delhi. He has more than 12 years of experience in teaching and received integrated postgraduate (B. Tech + M. Tech) degree from ABV-Indian Institute of Information Technology & Management, Gwalior and Ph. D. degree from IKG-Punjab Technical University. He has published more than 50 papers in reputed journals and international conferences. His area of research in the field of machine learning, deep learning, soft computing and intelligent systems.

Mamoon Rashid is currently working as an Assistant Professor at the School of Information Communication and Technology, Bahrain Polytechnic, Bahrain. He received his Ph.D. from the Department of Computer Engineering, Punjabi University, Punjab, India in the field of Medical Imaging Informatics. He is carrying 10+ years of teaching experience at university level as an Assistant Professor. He has published 150+ papers indexed in SCI/SCIE journals and conferences of international repute with JCR Impact Factor of 300+. His area of interest includes Data Analytics, Computational Medical Imaging- (fMRI, MRI, 3D-Amplified MRI, CT Scan Imaging) and Computer Vision.

***

Manish Baboo Agarwal has completed M. Tech in Materials from IIT Bombay and Ph.D. in Physics (Nanotechnology) from IIT Dhanbad. Publication in Scopus, Sci journals, published many book chapters in Scopus journals and reputed journals, having national and international Patents also publisher of the book on Nanomaterials.

Mansi Agarwal is currently working as an Assistant Professor at Sanskar College of Management and Professional Studies, Ghaziabad, India. She holds an MBA with dual specialization in Human Resource Management and Finance. With a strong academic foundation and a passion for innovation in business practices, her research interests lie at the intersection of technology, modern management practices and trends, and global industry trends. She is actively engaged in guiding MBA students in research projects and academic development. Her current research focuses on AI-Driven Adaptation and Management Strategies for Advanced smart Fabrication in Global Manufacturing Enterprises, exploring how artificial intelligence can enhance operational efficiency and strategic decision-making. As a dedicated academic and emerging researcher, she strives to contribute to the evolving landscape of smart manufacturing and its implications for future-ready enterprises.

Seema Agarwal has completed Ph.D. in Mathematics from Dr. B.R. Ambedkar University Agra. publications in renowned journals including book chapter, research papers, patents (National and International)

Shahjad Ali has completed M. Tech in Environmental Sciences and Ph.D. in Environmental Sciences from IIT Dhanbad. Has many papers, book chapters published in SCi, Scopus journals including Springer Nature (Q1).

Monika Bansal is currently working as an Assistant Professor in the Computer Science & Engineering Department, School of Engineering & Technology at Vivekananda Institute of Professional Studies – Technical Campus (VIPS-TC). Dr. Bansal received her Ph.D. degree in Computer Science from Shiv Nadar University, and her M.Tech. degree in Information Technology from Guru Gobind Singh Indraprastha University in 2022 and 2012, respectively. She has more than 14 years of teaching and 6 years of research experience. She has written and published many papers in reputed national and international conferences such as the CODS-COMAD Conference (a third-tier conference in computer science rankings), ANTS, and IKE. In addition, she has published papers in several SCI/SCIE and Scopus-indexed journals with high impact factors (i.e., greater than 5), such as Information Processing & Management, Data Mining & Knowledge Discovery, and Neurocomputing. She has reviewed many conference and journal papers during her career. She has also convened and organized several conferences, seminars, and co-curricular/extracurricular events for the benefit of students. Her research interests include machine learning and data science. Being a passionate teacher, she believes that teaching is not merely restricted to helping students understand the underlying concepts of a course, but also involves developing critical thinking and evaluating alternative approaches to problem-solving. She always puts in consistent efforts toward the overall development of her students.

Vaishnavi Bhamburkar is Master of design student from Maharashtra, India, currently pursuing a Master of Design (M.Des.) in User Experience design at the National Institute of Fashion Technology (NIFT), India, for the 2024–2026 batch. Prior to her postgraduate studies, she completed her Bachelor's degree at Vishwakarma university Pune in Bachelor of design [product design] Vaishnavi maintains a linkdin account she has 1 year experience as user experience design and worked on several projects. Her academic background and creative pursuits indicate a strong foundation in both technical and aesthetic aspects of design

Mehnaz Fathima C is working as an Assistant Professor in the Department of Electronics & Communication Engineering, Sahyadri College of Engineering & Management. The areas of interest are VLSI & Embedded System, Signal Processing.

Ritam Dutta has 16+ years of Teaching and Research Experience (05 SCI, 55 Scopus, 01 Patent Grant), 10+ years of Administrative Experience and 03+ year of Post Ph.D. Experience, served in several reputed institutions across India in the capacity of Assistant Prof., Associate Prof. and Professor. Dr. Dutta is a Member of Professional Bodies: IE(India) (M'19), IEEE (M'25), ISTE (M'19). Dr. Dutta has also acted as Invited Speaker, Guest Editor, Reviewer, Session Chair of many Scopus indexed Conferences and Journals at National as well as international level. Earlier Dr. Dutta has earned his Ph.D (Engg.) from Sikkim Manipal Institute of Technology, SMU, Sikkim and M.Tech from SRM Institute of Science and Technology, Chennai, currently pursuing Post Doc. in Nanodevice Modelling for Healthcare applications from Handong Global University, Pohang, South Korea.

Babitha Hemanth is an accomplished Electrical and Electronics Engineer with a Master's degree in Industrial Automation and Robotics. She combines extensive teaching and industry experience, bringing both practical knowledge and academic depth to her role as an Assistant Professor at Sahyadri College of Engineering and Management. Babitha is also a published author, recognized for her contributions to the field of engineering and automation.

Nishi Jain is currently serving as an Assistant Professor in the Department of Computer Science and Engineering. She is currently pursuing her Ph.D. in the Department of Computer Science and Engineering, specializing in " Agriculture Crop Yield Prediction using Machine Learning," from GGSIPU, Delhi. She completed her Bachelor's degree in Software Engineering from Delhi Technological University, in 2016. She has qualified UGC NET in 2019. She has published SCI journal papers and patents. Her research interests include Machine Learning, Design and Analysis of Algorithm and Data Structure.

Swati Jain is currently working as an Assistant Professor at Sanskar College of Management and Professional Studies, Ghaziabad, India. She holds an MBA with dual specialization in Human Resource Management and Finance. Her teaching and research interests include Human Resource Strategic applications in management, organizational strategies, and financial decision-making. She is actively engaged in guiding MBA students in research projects and academic development. Her current research focuses on AI-driven adaptation and management strategies for advanced smart fabrication in global manufacturing enterprises. She has contributed to scholarly publications and continues to explore the integration of emerging technologies with modern management practices.

Aatif Jamshed is an accomplished academician and researcher serving as an Associate Professor in the Department of Computer Science at KIET Group of

Institutions, Ghaziabad, affiliated with Dr. APJ Abdul Kalam Technical University (AKTU), Lucknow. With over 16 years of teaching and industrial experience, he has contributed significantly to the fields of Data Mining, Deep Learning, and Machine Learning. Dr. Jamshed has held faculty positions at several esteemed institutions, including ABES Engineering College, Galgotias University, and Vidya Institute of Engineering & Technology. He is also the founder and Director of FusionTech Innovations LLP, a startup focused on technological innovations. His academic journey culminated in a Ph.D. in Computer Science and Engineering from Veer Madho Singh Bhandari Uttarakhand Technical University, Dehradun. Dr. Jamshed has authored over 40 research papers published in reputable international journals and conferences. Dr. Jamshed has authored and co-authored several books, with four of them available on Amazon. He has also published multiple utility and design patents, showcasing his commitment to innovation and practical applications of his research. He has been awarded a funded project under the TIDE 2.0 scheme, a government initiative aimed at promoting technology innovation and entrepreneurship. In recognition of his research excellence, he received the "Research Award 2022" from ABES Engineering College. ORCID: 0000-0003-3152-6147

Manu Mehrotra has completed his Ph.D. in Chemistry from Jodhpur University, India. Has many National and International Patents. Research publications in Sci, Scopus journals of research papers, book chapters etc.

Arya Nambiar is a pre-final year student pursuing a Bachelor of Fashion Technology (BFTech) at the National Institute of Fashion Technology (NIFT), Kannur. She has a keen interest in the intersection of technology and fashion, particularly in areas such as inventory management, automation, and sustainable production. Arya is a detail-oriented and motivated learner, consistently exploring innovative solutions to enhance efficiency in the apparel and retail industries.

Abhishek Nayak holds a Master's degree in Electronics and Communication, with a strong academic and professional foundation in both digital and analog systems. His expertise spans across modern technologies, with a particular focus on machine learning and its wide-ranging applications in real-world problem-solving. Passionate about bridging innovation with practicality, Abhishek has also dedicated his research and projects to the development and utilization of drones, exploring their potential in diverse domains such as surveillance, environmental monitoring, and automation. His work reflects a unique blend of technical depth and forward-looking vision, making significant contributions to the integration of emerging technologies in engineering solutions. With a commitment to continuous learning and innovation, Abhishek strives to advance knowledge in his fields of expertise while inspiring others to explore the evolving intersections of electronics, communication, and intelligent systems.

Ayona Padhi is a pre-final year student pursuing a Bachelor of Fashion Technology (B.F.Tech) with a specialization in Fashion Management. She is currently engaged in her Graduation Project, which allows her to apply academic knowledge to real-world industry practices. Her work focuses on exploring innovative approaches within the fashion business domain, emphasizing strategic thinking, market analysis, and sustainable practices

Rupayan Roy is currently serving as an Assistant Professor at the National Institute of Fashion Technology (NIFT), Kannur. He holds a Ph.D. and Master's degree from IIT Delhi, and has a distinguished academic background in textiles and fashion technology. Dr. Roy has authored 50+ research articles and book chapters, and has participated in over 10 national and international conferences. His research interests include smart textiles, nonwoven composites, ergonomic workstations, sustainability in fashion, and IoT applications in apparel production. He is also involved in curriculum development and innovation-led projects, with a strong focus on industry-oriented teaching.

Hemant Sahu is currently working as an Assistant Professor in the Department of Computer Science & IT at Janardan Rai Nagar Rajasthan Vidyapeeth (Deemed to be University), Udaipur. He also serves as the Coordinator of the University's IQAC Cell. With over 17 years of teaching experience, he has published numerous research papers in SCOPUS and UGC-indexed journals and has presented his work at various national and international conferences. His research interests include network security, multimedia, and machine learning.

Kapil Kumar Sharma is a dedicated academician and researcher in the field of Computer Science. He holds a Master of Computer Applications (MCA) degree and is currently pursuing his Ph.D., with a strong focus on emerging technologies such as the Internet of Things (IoT) and Machine Learning. His research work primarily explores innovative approaches to data-driven systems, energy management, and sustainable computing solutions. Mr. Sharma has published various research papers and book chapters in reputed national and international journals and edited books, contributing valuable insights to the academic and research community. In recognition of his expertise, he also serves as a reviewer for several research papers, evaluating scholarly work and supporting the advancement of scientific knowledge. With a deep interest in technology, teaching, and research, Kapil Kumar Sharma continues to inspire students and fellow researchers through his commitment to excellence and innovation in computer science education and research. His professional journey reflects his dedication to continuous learning and contribution to the growth of the academic and scientific community.

Meenakshi Sharma received the B.E. degree from Dr. Babasaheb Ambedkar Marathwada University, Aurangabad, Maharashtra, in 1996, and the M.Tech. degree in communication systems and the Ph.D. degree in microwave engineering. She is currently a Professor and the Head of the Department of Electronics and Communication Engineering, Inderprastha Engineering College, Ghaziabad. Besides this, she is the Dean Student, a Welfare, and the Coordinator of Internal Quality Assurance Cell, IPEC. She has more than 26 years of teaching and one year of industrial experience to her credit. She is the Founder Faculty of IPEC and has been working with IPEC since past 23 years in various capacities. She has published a good number of research articles with journals of repute, including SCI publications. She has authored a book titled "Electronics Engineering Made Easy" with Cengage Learning (formerly Thomson Press) under technological university series. She has one granted patent and six published patents to her credit. She has been a resource person for various workshops and faculty development programs. She has also convened numerous national level conferences and seminars.

Priyanka Verma is a dedicated and dynamic academician with over 20 years of experience in higher education, research, and academic leadership. Currently a Professor at Poornima University, Jaipur, she has made significant contributions to the domains of computer science, artificial intelligence, web mining, and green and sustainable technologies. Dr. Verma's academic background is diverse and interdisciplinary, having completed a B.Sc. (Hons) in Botany from Delhi University, followed by M.Sc. (IT), MCA, and a Ph.D. in Computer Science. This unique blend of biological sciences and advanced computing enables her to apply technological solutions to sustainability challenges with a strong understanding of environmental systems. Dr. Verma's research spans areas such as data mining, AI-based water resource optimization, web usage analytics, green IoT, and sentiment analysis using chatbots. Her work reflects a strong focus on environmental sustainability through computational intelligence. She has authored more than 15 peer-reviewed research papers, many of which are indexed in Scopus and Web of Science, and published in high-impact journals.

Index

Symbols

A

B

C

D

E

F

G

H

I

L

M

N

P

Q

R

S

T

U

W

www.ingramcontent.com/pod-product-compliance
Lightning Source LLC
LaVergne TN
LVHW061529290926
849162LV00004B/727